A
COMPLETE GCSE
MATHEMATICS

HIGHER COURSE

Second Edition

By the same author

A Complete GCSE Mathematics — General Course
A Complete GCSE Mathematics — Basic Course

A Complete O-level Mathematics
New Comprehensive Mathematics for O-level
CSE Mathematics 1 The Core Course
CSE Mathematics 2 The Special Topics
A Concise CSE Mathematics

Arithmetic for Commerce
Skills in Numeracy
A First Course in Statistics

Revision Practice in Algebra
Revision Practice in Arithmetic
Revision Practice in Geometry and Trigonometry
Revision Practice in Multiple-Choice Maths Questions
Revision Practice in Short-Answer Questions in O-level Maths
Revision Practice in Statistics

with S. Llewellyn   Mathematics The Basic Skills

with G.W. Taylor   BTEC First Mathematics for Technicians
                     BTEC National NII Mathematics for Technicians
                     BTEC National NIII Mathematics for Technicians

A
COMPLETE GCSE
MATHEMATICS

HIGHER COURSE

A. Greer

Formerly Senior Lecturer,
Gloucestershire College of Arts and Technology

SECOND EDITION

Stanley Thornes (Publishers) Ltd

First published in 1987 by
Stanley Thornes (Publishers) Ltd
Old Station Drive
Leckhampton
CHELTENHAM GL53 0DN

Reprinted 1988 (twice)
Second Edition 1989

British Library Cataloguing in Publication Data

Greer, A. (Alex)
    A complete GCSE mathematics: a higher
    course—2nd edition
    I. Title
    510

    ISBN 0-7487-0074-9 (paperback)
    ISBN 0-7487-0101-X (casebound)

Typeset by Tech-Set, Gateshead, Tyne & Wear in 10/12 Century.
Printed and bound in Great Britain at The Bath Press, Avon.

Contents

Preface

This book completes a series of three books intended for students taking the General Certificate of Secondary Education (GCSE) in Mathematics. It attempts to cover all the topics prescribed in the syllabuses of the five regional examining groups. It is suitable for those students who expect to gain a grade A, B or C, the target grade being B.

Because this is a revision book, the sections on Arithmetic, Algebra, Geometry, Graphical Work and Statistics have been dealt with separately. The emphasis is on a simple approach and a teacher and the class may work, if desired, through the book, chapter by chapter. I am conscious that many students, after leaving their school, will take courses in Colleges of Further Education in order to improve their grades. Thus a large number of examples and exercises have been included which supplement and amplify the text, and many practical examples of mathematics in the real world have been used.

At the end of most chapters there is a 'miscellaneous exercise', often divided into two parts, Section A and Section B. This is because all the examining groups have devised overlapping examination papers. The lower level students will take papers 1 and 2, the general level students will take papers 2 and 3 and the higher level students will take papers 3 and 4. In the miscellaneous exercises, Section B in the basic book has exercises similar in standard to Section A in the general book, while Section B in the general book has exercises similar in standard to Section A in the higher book.

Also at the end of many chapters are sets of multi-choice questions and mental tests. It is hoped that these, together with the miscellaneous exercises, will give the students confidence in answering examination papers.

The author would like to thank Mrs Anne Smith for her invaluable work in checking the answers to the exercises and also the publishers who have worked extremely hard to ensure that this book is published in time for it to be used for the first GCSE examinations.

A. Greer   Gloucester 1987

Coursework

We are aware of the demands made upon GCSE candidates and it is not suggested that they should attempt all the tasks which are offered in the chapter on coursework. The investigations and other studies are provided as a guide for those students who are not sure what is expected of them in this part of the examination.

The readers may like to choose a few of the suggestions for practice. We hope the topics will provide some inspiration when candidates attempt their own particular coursework projects.

I was pleased to accept the author's invitation to provide a section dealing with coursework and it is a source of personal satisfaction to be associated with this very successful series of books.

C.H. Hopkins   Gloucester 1

Acknowledgements

The author and publishers are grateful to the following who provided material for inclusion:

Cheltenham and Gloucester Omnibus Co Ltd. for the timetable and route map (1980), page 97.

National Welsh Omnibus Services Ltd for the timetable and map, page 102.

1 Operations in Arithmetic

Types of Number

Counting numbers or natural numbers are the numbers $1, 2, 3, 4, 5$, etc.

Whole numbers are the numbers $0, 1, 2, 3, 4, 5$, etc.

Positive numbers are numbers whose value is greater than zero. They either have no sign attached to them or a $+$ sign. Some examples of positive numbers are $5, +9, 8, +21, +293$ and 564.

Negative numbers have a value which is less than zero. They are always written with a minus sign in front of them. Some examples of negative numbers are $-4, -15, -168$ and -783.

Integers are whole numbers but they include negative numbers. Some examples of integers are $-28, -15, 0, 2, +9, 24, +371, -892$.

Directed numbers are numbers which have either a positive or a negative sign attached to them.

Directed Numbers in Practical Situations

Directed numbers are used:

(1) To show Celsius temperatures above and below freezing point (Fig. 1.1).

(2) On graphs (Fig. 1.2).

(3) In sport. Golf scores are often stated above and below par for the course.

Thus $+4$ means a score of 4 over par and -3 would indicate a score of 3 below par.

(4) To represent time. For example -12 seconds to blast-off and 5 seconds into free flight.

(5) In business calculations. A profit can be regarded as positive money whilst a loss is then regarded as negative money.

Fig. 1.1

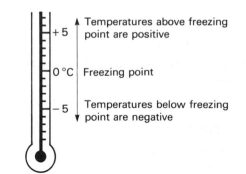

Fig. 1.2

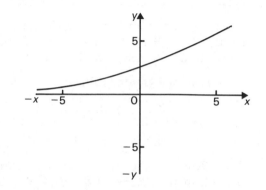

The Number Line

Directed numbers may be represented on a number line (Fig. 1.3). Numbers to the right of zero are positive whilst those to the left of zero are negative numbers.

Fig. 1.3

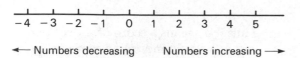

← Numbers decreasing   Numbers increasing →

Note that negative numbers become lower in value the further we move to the left of zero. For instance -8 is lower than -3 and -56 is lower in value than -27.

Positive numbers, however, become greater in value the further we move to the right of zero. For instance 12 has a greater value than 7 and $+89$ has a greater value than $+34$.

Example 1

Rearrange in order of size, smallest first, the numbers

$$-3, 5, -7, -2, +8, -4 \text{ and } +3$$

The order is:

$$-7, -4, -3, -2, +3, 5 \text{ and } +8$$

Exercise 1.1

1. Draw a number line covering the range from -10 to $+10$. On it show the numbers $-6, 3, +7, -8, -2, 0, 5$ and 4.

2. Which has the greater value:
   (a) -6 or $+3$?
   (b) -57 or -63?
   (c) $+72$ or 48?

3. Write down all the whole numbers from the following: $7, -4, +3, -6, 2, 0, -9, 5$ and -2.

4. Rearrange in order of size, smallest first, the numbers $-3, -5, -1, -9$ and -4.

5. Rearrange in order of size, largest first, the numbers $3, -8, -15, +25, 18, -6, -2$ and 12.

6. From the following set of numbers write down the positive integers: $+5, -3, -1, 7, +11$ and -15.

7. Plants in a greenhouse will die if the temperature falls below $-2\,°C$. Which of the following temperatures mean death for the plants: $-1, 3, -5$ and $-9\,°C$?

Rounding Off

Calculators save a lot of time and effort when used for various arithmetic operations like addition, subtraction, multiplication and division.

Calculators which are in good order do not make mistakes but human beings do. Therefore, we need to be able to estimate the size of an answer to give a check on the one produced by the calculator.

One way of doing this is to round off the numbers involved. Generally it is good enough to round off

> a number between 10 and 100 to the nearest ten;
>
> a number between 100 and 1000 to the nearest hundred;
>
> a number between 1000 and 10 000 to the nearest thousand;

and so on.

Example 2

(a) Round off the number 67 to the nearest number of tens.

> 67 is roughly 7 tens, so 67 is rounded up to 70.

(b) Round off 362 to the nearest number of hundreds.

> 362 is roughly 4 hundreds, so 362 is rounded up to 400.

(c) Round off 8317 to the nearest number of thousands.

> 8317 is roughly 8 thousands, so 8317 is rounded down to 8000.

(d) Round off 75 to the nearest number of tens.

> 75 lies exactly half way between 70 and 80. In such cases we always round up, so that 75 is rounded up to 80.

Exercise 1.2

Round off the following numbers to the nearest number of tens:

1. 42   2. 63   3. 85
4. 92   5. 76

Round off the following numbers to the nearest number of hundreds:

6. 698   7. 703   8. 886
9. 350   10. 885

Round off the following numbers to the nearest number of thousands:

11. 2631   12. 5378   13. 8179
14. 8500   15. 7500

Some Definitions

The result obtained by adding numbers is called the **sum** or **total**. Thus the sum of 4, 3 and 8 is $4 + 3 + 8 = 15$. The order in which numbers are added is not important so that

$$4 + 3 + 8 = 8 + 3 + 4$$
$$= 3 + 8 + 4$$
$$= 15$$

The **difference** between two numbers is the larger number minus the smaller number. Thus the difference between 10 and 15 is $15 - 10 = 5$. The order in which we subtract is very important. $7 - 3$ is not the same as $3 - 7$.

The result obtained by multiplying numbers is called the **product**. Thus the product of 8 and 7 is $8 \times 7 = 56$. The order in which we multiply is not important so that

$$3 \times 4 \times 6 = 4 \times 3 \times 6$$
$$= 6 \times 3 \times 4$$
$$= 72$$

The result obtained by division is called the **quotient**. The quotient of $8 \div 4$ is 2. The order in which we divide is very important so that $12 \div 3$ is not the same as $3 \div 12$.

Exercise 1.3

This exercise should be done mentally.

1. Find the sum of 3, 6 and 7.

2. Add 6, 8 and 4.

3. Find the total of 6, 7 and 4.

4. Calculate the sum of 2, 5, 8 and 15.

5. Find the difference between 7 and 19.

6. What is 21 minus 5?

7. Find the value of $24 - 18$.

8. Find the product of 4 and 9.

9. Work out $7 \times 3 \times 4$.

10. Calculate 6 times 8.

11. What is the product of 4, 5 and 7?

12. Work out $5 \times 6 \times 8$.

Adding Integers

To add integers we make use of the number line.

Example 3

Use a number line to find the value of $3 + 4$.

In Fig. 1.4 we measure 3 units to the right of 0 to represent $+3$. Then, from this point we measure 4 units to the right to represent $+4$. We see from the diagram that

$$3 + 4 = 7$$

Fig. 1.4

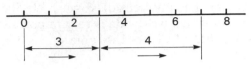

Example 4

Use a number line to find the value of $-3 + (-2)$.

Measure 3 units to the left of 0 to represent -3 (Fig. 1.5). Then from this point measure 2 units to the left to represent -2. We see from the diagram that

$$-3 + (-2) = -5$$

Fig. 1.5

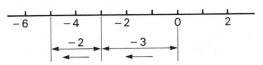

Note that $-3 + (-2)$ is usually written $-3 - 2$ and hence

$$-3 - 2 = -5$$

We see that the rules for the addition of integers having the same sign are as follows:

(1) If all the numbers to be added are positive then add them together and place a plus sign (or no sign) in front of the sum.

(2) If all the numbers to be added are negative then add the numbers together and place a negative sign in front of the sum.

Example 5

(a) $3 + 5 + 6 + 2 = 16$

(b) $-4 - 6 - 9 - 3 = -22$

Example 6

Use a number line to find the value of $4 - 6$.

Looking at Fig. 1.6 we see that

$$4 - 6 = -2$$

When the signs of the two numbers are different find the difference between them and place the sign of the larger number in front of the difference.

Fig. 1.6

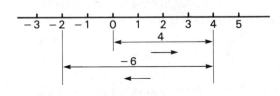

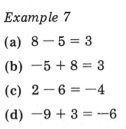

Example 7

(a) $8 - 5 = 3$

(b) $-5 + 8 = 3$

(c) $2 - 6 = -4$

(d) $-9 + 3 = -6$

If there are more than two numbers first add the positive numbers together. Next add the negative numbers together. The set of numbers is thereby reduced to two numbers, one positive and the other negative. We can then proceed as for two numbers.

Example 8

$$-5 + 8 - 7 + 3 + 6 - 9 + 4 + 1 - 2$$
$$= (8 + 3 + 6 + 4 + 1)$$
$$\quad - (5 + 7 + 9 + 2)$$
$$= 22 - 23$$
$$= -1$$

Exercise 1.4

This exercise should be done mentally.

1. $8 + 5$
2. $-6 - 3$
3. $-14 - 18$
4. $-7 - 6 - 3$
5. $8 - 12$
6. $9 - 17$
7. $8 - 7 - 15$
8. $-6 + 10$
9. $-3 + 5 + 7 - 4 - 2$
10. $6 + 4 - 3 - 5 - 7 + 2$

Subtracting integers

Example 9

Using a number line find the value of $-(+5) - 4$.

To represent $+5$ we measure 5 units to the right of 0. Therefore to represent $-(+5)$ we must reverse direction and measure 5 units to the left of 0 as shown in Fig. 1.7. Hence $-(+5)$ is the same as -5.

Therefore

$$-(+5) - 4 = -5 - 4$$
$$= -9$$

To subtract a directed number change its sign and add the resulting number.

Fig. 1.7

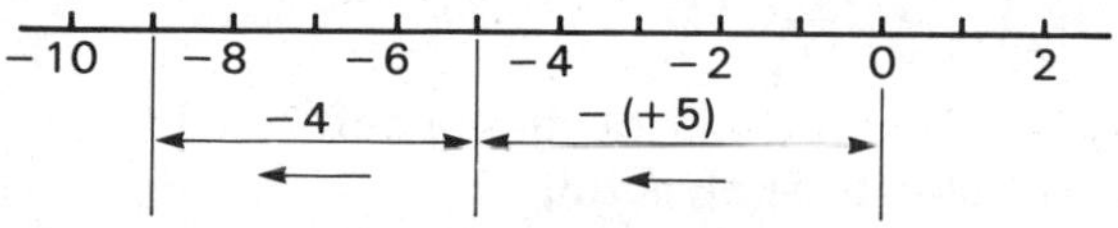

Example 10

(a) $5 - (-3) = 5 + 3 = 8$
(b) $-7 - (+4) = -7 - 4 = -11$
(c) $8 - (+5) = 8 - 5 = 3$

Exercise 1.5

This exercise should be done mentally.

1. $7 - (+6)$
2. $-5 - (-6)$
3. $5 - (-3)$
4. $-3 - (-4)$
5. $-10 - (-4)$
6. $8 - (+6)$
7. $-3 - (-7)$
8. $-10 - (-5)$

Multiplication of Integers

$$4 + 4 + 4 = 12$$
$$3 \times 4 = 12$$

Hence two positive numbers multiplied together give a positive product.

$$-4 - 4 - 4 = -12$$
$$3 \times (-4) = -12$$

Hence a positive number multiplied by a negative number gives a negative product.

Suppose that we wish to find the value of $(-3) \times (-4)$.

$$-3 \times (4 - 4) = 3 \times 0$$
$$= 0$$
$$-3 \times 4 + (-3) \times (-4) = 0$$
$$-12 + (-3) \times (-4) = 0$$
$$\therefore \ (-3) \times (-4) = 12$$

Hence a negative number multiplied by a negative number gives a positive product.

Therefore the rules for multiplication are:

(1) Numbers with like signs give a positive product.

(2) Numbers with unlike signs give a negative product.

Example 11

(a) $2 \times 3 = 6$

(b) $(-2) \times 3 = -6$

(c) $3 \times (-2) = -6$

(d) $(-2) \times (-3) = 6$

Division of Integers

The rules for division must be similar to those for multiplication. Since

$$3 \times (-4) = -12$$

$$-12 \div 3 = -4$$

and $$-12 \div (-4) = 3$$

Therefore the rules are:

(1) Numbers with like signs, when divided, give a positive quotient.

(2) Numbers with unlike signs give a negative quotient.

Exercise 1.6

This exercise should be done mentally.

Find the value of the following:

1. $7 \times (-6)$ 2. $(-6) \times 7$

3. $(-2) \times (-3)$ 4. $(-5) \times 3$

5. $(-2) \times (-5) \times (-6)$

6. $4 \times (-3) \times (-2)$ 7. $2 \times (-5) \times 4$

8. $(-3) \times (-4) \times 5$ 9. $8 \div (-2)$

10. $(-9) \div 3$ 11. $(-8) \div (-2)$

12. $(-4) \times 6 \times 5$ 13. $18 \div (-6)$

14. $(-24) \div (-12)$ 15. $22 \div (-11)$

The Sequence of Arithmetical Operations

Numbers are often combined in a series of arithmetical operations. When this happens a definite sequence must be observed.

(1) Brackets are used if there is any danger of ambiguity. The contents of any brackets must be evaluated before performing any other operation.

Thus

$$2 \times (7 + 4) = 2 \times 11$$

$$= 22$$

$$15 - (2 - 7) = 15 - (-5)$$

$$= 15 + 5$$

$$= 20$$

We get the same answer and sometimes make the work easier if we expand the bracket as shown below.

$$5 \times (3 + 6) = 5 \times 3 + 5 \times 6$$

$$= 15 + 30$$

$$= 45$$

Sometimes the multiplication sign is omitted.

$$5(8 - 4) \text{ means } 5 \times (8 - 4)$$

$$= 5 \times 8 + 5 \times (-4)$$

$$= 20$$

(2) Multiply and/or divide before adding and/or subtracting.

Example 12

Find the value of $11 - 12 \div 4 + 3 \times (6 - 2)$.

$$11 - 12 \div 4 + 3 \times (6 - 2)$$

$$= 11 - 12 \div 4 + 3 \times 4$$

(by working out the contents of the bracket)

$$= 11 - 3 + 12$$

(by multiplying and dividing)

$$= 20$$

(by adding and subtracting)

Sometimes several brackets are used in an expression. The contents of each should be worked out before performing any other operation.

Example 13

Work out the value of
$[15 \times (-3) \times 2] \div [(-5) \times (-6)]$

$$[15 \times (-3) \times 2] \div [(-5) \times (-6)]$$

$$= (-90) \div 30$$

$$= -3$$

Exercise 1.7

Work out the value of each of the following:

1. $7 + 4 \times 3$
2. $2 \times 6 - 3$
3. $5 \times 4 - 3 \times 6 + 5$
4. $8 \times 5 - 15 - 5 + 7$
5. $3 + 6 \times (3 + 2)$
6. $9 - 4 \times (3 - 2)$
7. $10 - 12 \div 6 + 3(8 - 3)$
8. $15 \div (4 + 1) - 9 \times 3 + 7(4 + 3)$
9. $35 \div (20 - 25)$
10. $(-24) \div (-4) - 24 \div 6$
11. $3(2 - 5) + 4$
12. $25 \div (-5) - 7$
13. $3 - (-12) \times 4 - 15 \div (-3)$
14. $36 \div (-4) + 36 \div (-6)$
15. $4(2 - 8) + (-12) \div (-3)$
16. $[(-3) \times (-4) \times (-2)] \div [3 \times 4]$
17. $[4 \times (-6) \times (-8)] \div [(1 - 3) \times (-2) \times (-4)]$
18. $[5 \times (-3) \times 6] \div [(2 - 12) \times (5 - 8)]$
19. $[6 \times (-4) \times 5(2 - 12)] \div [(-8) \times 3 \times (-2)]$
20. $[(-9) \times (-5) \times 2(4 - 6)] \div [3 \times 5(7 - 4) \times (-2)]$

Some Important Facts

(1) When zero is added to any number the sum is the number.

$$873 + 0 = 873 \quad \text{and} \quad 0 + 74 = 74$$

(2) When a number is multiplied by 1 the product is the number.

$$9483 \times 1 = 9483 \quad \text{and}$$

$$1 \times 534 = 534$$

(3) When a number is multiplied by zero the product is zero.

$$178 \times 0 = 0 \quad \text{and} \quad 0 \times 2398 = 0$$

(4) It is impossible to divide a number by zero. Thus, for instance, $8 \div 0$ is meaningless.

(5) When zero is divided by any number the result is zero.

$$0 \div 79 = 0 \quad \text{and} \quad 0 \div 7 = 0$$

Exercise 1.8

This exercise should be done mentally.

Work out the value of each of the following:

1. $768 + 0$
2. 256×1
3. 0×50
4. $0 \times 4 \times 5$
5. $3 \times (-6) \times (-1)$
6. $(-7) \times 0 \times (-3)$
7. $16 - 8 \times 0 - 3$
8. $12 - 24 - (-6) + 7(5 - 5)$
9. $0 - 32$
10. $(7 - 7) \times (-3)$
11. $7 \times (3 - 9) + 0 \div 7$
12. $4(2 - 8) + (-12) - (-3) + 0 \times (-5)$

Sequences

A set of numbers connected by some definite law is called a sequence of numbers. Each of the numbers in the sequence is called a term in the sequence.

Example 14

(a) Write down the next two terms of the sequence $4, 8, 12, 16, \ldots$

> Each term in the sequence is formed by adding 4 to the previous term.
>
> Therefore
>
> $$5\text{th term} = 16 + 4$$
> $$= 20$$
> $$6\text{th term} = 20 + 4$$
> $$= 24$$

(b) Find the next two terms of the sequence $5, 2, -1, -4, \ldots$

> Each term in the sequence is formed by subtracting 3 from the previous term.

Therefore

$$5\text{th term} = -4 - 3$$
$$= -7$$
$$6\text{th term} = -7 - 3$$
$$= -10$$

(c) In the sequence $1, 3, 9, ?, 81, ?$ write down the terms denoted by the question marks.

> Each term is formed by multiplying the previous term by 3.
>
> Therefore
>
> $$4\text{th term} = 9 \times 3$$
> $$= 27$$
> $$6\text{th term} = 81 \times 3$$
> $$= 243$$

(d) Find the next two terms of the sequence $1, 2, 3, 5, 8, \ldots$

> Each term is formed by adding the two previous terms.
>
> Therefore
>
> $$6\text{th term} = 5 + 8$$
> $$= 13$$
> $$7\text{th term} = 8 + 13$$
> $$= 21$$

(e) Write down the next two terms of the sequence $2, 7, 17, 37, \ldots$

> Each term is formed by multiplying the previous term by 2 and adding 3 to the product.
>
> Therefore
>
> $$5\text{th term} = 37 \times 2 + 3$$
> $$= 77$$
> $$6\text{th term} = 77 \times 2 + 3$$
> $$= 157$$

Exercise 1.9

Write down the next two terms for each of the following sequences:

1. $2, 10, 50, \ldots$ 2. $1, 5, 9, 13, \ldots$

3. $3, 9, 15, 21, \ldots$ 4. $176, 88, 44, \ldots$

5. $1, 3, 6, 10, \ldots$ 6. $4, -1, -6, \ldots$

7. $1, 1, 2, 3, 5, 8, 13, \ldots$

8. $3, 8, 18, 38, \ldots$ 9. $5, 9, 21, 57, \ldots$

10. $-2, 6, -18, 54, \ldots$

11. $2, 0, -2, -4, \ldots$

12. $-5, -3, -1, 1, \ldots$

Write down the terms denoted by the question marks in the following sequences:

13. $0, 2, 4, ?, ?, 10, 12, 14, \ldots$

14. $3, 5, 8, 13, ?, 34, 55, ?$

15. $2, 4, 10, 28, ?, ?, 730$

16. $5, 11, 29, 83, ?, 731, ?$

17. $243, 81, 27, ?, 3, ?$

Miscellaneous Exercise 1

Section A

1. Write down the next three terms of the following sequences:

    (a) $21, 17, 13, \ldots$

    (b) $40, 32, 24, \ldots$

2. Subtract one hundred thousand from one million.

3. $53 + 29 = 37 + ?$ Write down the missing number.

4. A car travels 9 kilometres on 1 litre of petrol. How many litres will be needed for a journey of 324 kilometres?

5. Write down the next two numbers of the sequence $-9, 9, -9, \ldots$

6. Find the value of:

    (a) $(-5) \times (-8)$ (b) $-5 - 8$

    (c) $8 \div (-2)$ (d) $30 \times (-10)$

7. Write down in order of size, smallest first, $-3, 0, -5$ and 3.

8. A wall is 342 bricks long. It has 23 courses of bricks. How many bricks have been used in its construction?

9. Which is the larger and by how much: $(5 + 2) \times (7 \quad 4)$ or $5 + 2(7 - 4)$?

10. Write down the terms denoted by the question marks in the sequence: $?, ?, 12, 24, 48, ?$

11. Work out the answers to the following:

    (a) $(7 - 3) \times (7 + 4)$

    (b) $7 - 3(7 + 4)$

12. In each of the following insert either a plus sign or a minus sign to make the statement correct:

    (a) $27 + 5 - 12 = 27 + (5 \ldots 12)$

    (b) $13 - 4 + 9 = 13 \ldots (4 - 9)$

Section B

1. Find the values of the following:

    (a) $11 - 12 \div 6 + 3(6 - 2)$

    (b) $5 - 2(3 - 8) + 8 \div (2 - 6) + 7$

2. Work out the value of:

    $[(2 - 20) \div (-3)] \div [(4 - 16) \div (-6)]$

3. Write down the next two terms of the following sequences:

    (a) $2, -2, 2, -2, \ldots$

    (b) $-5, -3, -1, \ldots$

4. (a) Round off 1745 to the nearest thousand.

    (b) Round off 855 to the nearest hundred.

5. (a) Find the sum of 136, 549 and 876.

(b) What is the difference between 87 and 152?

(c) Find the product of $-3, -6$ and 7.

6. Find the value of:
$3(4-8) - 4(2-5)$.

7. Work out:
$[3(2-5) + (-16)] \div (-5) + 7 \times (3-3)$.

8. Find the next two terms of the sequence:
$1, -2, -11, -38 \ldots$

Multi-Choice Questions 1

1. The next number in the sequence 1, 3, 6, 10 is

A 11   B 15   C 16   D 20

2. $3 + 7 \times 4$ is equal to

A 25   B 31   C 40   D 84

3. The missing number in the sequence 55, 45, 36, ?, 21, 15, 10 is

A 27   B 28   C 29   D 30

4. The next two numbers in the sequence 5, 4, 6, 5, 7, 6, 8, ... are

A 9 and 7   B 7 and 6

C 9 and 6   D 7 and 9

5. $(-2) \times (-3)$ is equal to

A -6 B -5 C $+5$ D $+6$

6. The value of $[(-1) - (-1)] - 1$ is

A -2 B -1 C 0   D 2

7. The product of 4, -6 and -2 is

A 48   B 18   C 11   D -11

8. $7 \times 6 - 12 \div 3 + 1$ is equal to

A 40   B 39   C 21   D -44

9. The sum of $-3, -5$ and -7 is

A -15 B -21 C -35 D -105

Mental Test 1

Try to answer the following questions without writing anything down except the answer.

1. Find the sum of 4, 6, 5 and 8.

2. Find the sum of $-2, -4, 6$ and 7.

3. Find the difference between 5 and -7.

4. Find the product of -2 and -5.

5. Divide 72 by -8.

6. Find the product of $-2, 3$ and -5.

7. Work out $15 - 3(5-2)$.

8. Find the value of $3 \times 4 - 15 \div 3 + 2$.

9. Work out $-6 - 7 - 8$.

10. Subtract 8 from -5.

Factors and Multiples

Odd and Even Numbers

Counting in twos, starting with zero the sequence is $0, 2, 4, 6, 8, 10, 12, 14, \ldots$ Continuing the sequence, we discover that each number in it ends in either $0, 2, 4, 6$ or 8. Such numbers are called **even numbers**. Thus $930, 282, 654, 8736$ and 98 are all even numbers.

If we now start at 1 and count in twos the sequence is $1, 3, 5, 7, 9, 11, 13, 15, 17, 19, \ldots$ Carrying on the sequence, we discover that each number in it ends in $1, 3, 5, 7$ or 9. Such numbers are called **odd numbers**. Thus $91, 103, 825, 96\,327$ and 59 are all odd numbers.

Note that any number which when divided by 2 has no remainder is an even number; if it leaves a remainder of 1 it is an odd number.

Powers of Numbers

The quantity $5 \times 5 \times 5$ is usually written 5^3. 5^3 is called the third power of 5. The number 3, which indicates the number of fives to be multiplied together is called the index (plural: indices).

$$2^6 = 2 \times 2 \times 2 \times 2 \times 2 \times 2$$
$$= 64$$
$$3^4 = 3 \times 3 \times 3 \times 3$$
$$= 81$$

Squares of Numbers

When a number is multiplied by itself it is called the square of the number. Thus the square of a number is the number raised to the power of 2.

$$\text{Square of } 3 = 3^2$$
$$= 3 \times 3$$
$$= 9$$
$$\text{Square of } 15 = 15^2$$
$$= 15 \times 15$$
$$= 225$$

Cubes of Numbers

The cube of a number is the number raised to the power of 3.

$$\text{Cube of } 7 = 7^3$$
$$= 7 \times 7 \times 7$$
$$= 343$$
$$\text{Cube of } 85 = 85^3$$
$$= 85 \times 85 \times 85$$
$$= 614\,125$$

Square Roots of Numbers

The square root of a number is the number whose square equals the given number.

Since $5^2 = 25$, the square root of 25 is 5. The sign $\sqrt{}$ is used to denote a positive square root and we write $\sqrt{25} = 5$.

Similarly, since $7^2 = 49$, $\sqrt{49} = 7$.

However, since, for instance, $(-3)^2 = 9$ and $(+3)^2 = 9$ we write $\sqrt{9} = \pm 3$.

Similarly $\sqrt{81} = \pm 9$.

If only a positive square root is needed then we write, for instance, $\sqrt{36} = 6$.

Square Root of a Product

The square root of a product is the product of the square roots of the individual numbers.

Example 1

(a) Find the square root of 4×9.
$$\sqrt{4 \times 9} = \sqrt{4} \times \sqrt{9}$$
$$= 2 \times 3$$
$$= 6$$

(b) Find the square root of $16 \times 36 \times 81$.
$$\sqrt{16 \times 36 \times 81} = \sqrt{16} \times \sqrt{36} \times \sqrt{81}$$
$$= 4 \times 6 \times 9$$
$$= 216$$

Cube Roots of Numbers

The cube root of a number is the number whose cube equals the given number.

Since $4^3 = 64$, the cube root of 64 is 4. The sign $\sqrt[3]{}$ is used to denote a cube root and we write $\sqrt[3]{64} = 4$.

Similarly, since $12^3 = 1728$, $\sqrt[3]{1728} = 12$.

Questions 1 to 20 should be done mentally. State whether the following numbers are even or odd:

1. 95   2. 936

3. 852   4. 5729

5. 8887   6. 63 752

7. 456 310   8. 9367

9. From the set of numbers 17, 36, 59, 98, 121, 136 and 259 write down

   (a) all the odd numbers

   (b) all the even numbers.

Find values of the following:

10. The square of 6

11. The cube of 5

12. The square of -7

13. The cube of -5

14. (a) 8^2 (b) 2^3 (c) 3^4 (d) 2^5

15. (a) $\sqrt{16}$ (b) $\sqrt{144}$ (c) $\sqrt{169}$

16. (a) $\sqrt[3]{8}$ (b) $\sqrt[3]{27}$ (c) $\sqrt[3]{216}$

17. The square root of 9×36

18. $\sqrt{64 \times 64}$

19. $\sqrt{9 \times 25 \times 36}$

20. $\sqrt{4 \times 9 \times 25}$

21. Find the sum of 2^6 and 6^2.

22. Find the product of 2^3 and 6^2.

23. Find the difference between 2^5 and 5^2.

24. Find the value of $(-5)^2 + (-3)^3$.

25. Find the difference between $(-7)^2$ and 5^3.

Factors

A number is a factor of another number if it divides into that number without leaving a remainder.

Since 63 is divisible by 7, 7 is a factor of 63. 63 has other factors, namely 1, 3, 9, 21 and 63, since each of these numbers divides into 63 without leaving a remainder.

Exercise 2.2

This exercise should be done mentally.

1. 6 is a factor of some of the following numbers. Write down these numbers.

   (a) 12   (b) 31   (c) 42

   (d) 66   (e) 86

2. Is 5 a factor of

   (a) 8   (b) 12   (c) 25

   (d) 32   (e) 50   (f) 157

   (g) 295   (h) 700?

3. Is 9 a factor of

   (a) 30   (b) 45   (c) 72

   (d) 128   (e) 135?

4. Which of the following numbers are factors of 42:

   2, 3, 4, 5, 6, 7, 8 and 9?

5. Which of the following numbers are factors of 72:

   2, 3, 4, 5, 6, 7, 8, 9, 11, 12, 13, 14?

Prime Numbers

Every number has itself and 1 as factors. If it has no other factors it is said to be a prime number. It is useful to learn all of the prime numbers up to 100. They are:

   2, 3, 5, 7, 11, 13, 17, 19, 23, 29,
   31, 37, 41, 43, 47, 53, 59, 61, 67,
   71, 73, 79, 83, 89, 97

Notice that, with the exception of 2, all the prime numbers are odd numbers. Also note that although 1 is a factor of all other numbers it is not regarded as being a prime number.

Prime Factors

A factor which is a prime number is called a prime factor. In the statement $63 = 3 \times 21$, 3 is a prime factor of 63 but 21 is not a prime factor of 63 since it equals 3×7.

Example 2

Find all the prime factors of 54.

$$54 = 2 \times 3 \times 3 \times 3$$

Since $3 \times 3 \times 3 = 3^3$ we may write

$$54 = 2 \times 3^3$$

2 and 3^3 are the prime factors of 54.

The prime factors of any number can be found by successive division as shown in Example 3. Note that we start by dividing by the smallest of the prime factors and end by dividing by the largest of them.

Example 3

Find the prime factors of 420.

2	420
2	210
3	105
5	35
7	7
	1

Hence

$$420 = 2 \times 2 \times 3 \times 5 \times 7$$
$$= 2^2 \times 3 \times 5 \times 7$$

Example 4

Find all the prime factors of 60.

The factors are:

If we pair off from each end as shown above and multiply them together we get

$$1 \times 60 = 60 \qquad 2 \times 30 = 60$$
$$3 \times 20 = 60 \qquad 4 \times 15 = 60$$
$$5 \times 12 = 60 \qquad 6 \times 10 = 60$$

We see that each pair gives a product of 60. We can use this method of pairing to check that all the factors of a number have been found.

Negative Factors

In finding the prime factors of a number we should also include its negative factors.

Consider the number 6. Since $6 = 2 \times 3$, 2 and 3 are the prime factors of 6.

However $6 = (-2) \times (-3)$ and both -2 and -3 are also factors of 6.

Hence all the prime factors of 6 are 2, -2, 3 and -3. For brevity we write:

the prime factors of 6 are ± 2 and ± 3

Example 5

Write down all the prime factors of 36.

$$36 = 2 \times 2 \times 3 \times 3$$
$$= 2^2 \times 3^2$$
$$= (-2) \times (-2) \times (-3) \times (-3)$$
$$= (-2)^2 \times (-3)^2$$
$$= (-2) \times 2 \times (-3) \times 3$$

Multiples

Numbers which are divisible by a second number are said to be multiples of this second number.

Thus 20, 28, 30 and 44 are all multiples of 2 and 35, 50, 70, 95 and 205 are all multiples of 5.

Example 6

Write down all the multiples of 7 between 10 and 50.

The multiples are 14, 21, 28, 35, 42 and 49 because 7 will divide exactly into all of these numbers.

Lowest Common Multiple (LCM)

The multiples of 3 are 3, 6, 9, 12, 15, 18, 21, 24, 27, 30, 33, 36, 39, 42, etc.

The multiples of 4 are 4, 8, 12, 16, 20, 24, 28, 32, 36, 40, 44, etc.

We see that 12, 24 and 36 are multiples which are common to both 3 and 4. The lowest of these common multiples is 12 and we say that the LCM of 3 and 4 is 12. Note that 12 is the lowest number into which both 3 and 4 will divide.

Example 7

Find the LCM of 4, 10 and 12.

$$4 = 2^2$$
$$10 = 2 \times 5$$
$$12 = 2^2 \times 3$$

The LCM is the product of the highest power of each prime factor. Hence

$$LCM = 2^2 \times 3 \times 5$$
$$= 60$$

We see that 60 is the lowest number into which 4, 10 and 12 will divide exactly.

Highest Common Factor (HCF)

The factors of 30 are 1, 2, 3, 5, 6, 10, 15 and 30.

The factors of 40 are 1, 2, 4, 5, 8, 10, 20 and 40.

We see that 1, 2, 5 and 10 are factors which are common to both 30 and 40. The highest of these common factors is 10. We say that the HCF of 30 and 40 is 10. Note that 10 is the highest number which will divide exactly into 30 and 40.

Example 8

Find the HCF of 42, 98 and 112.

$$42 = 2 \times 3 \times 7$$
$$98 = 2 \times 7^2$$
$$112 = 2^4 \times 7$$

We see that only 2 and 7 are factors of all three numbers. Hence

$$\text{HCF of } 42, 98 \text{ and } 112 = 2 \times 7$$
$$= 14$$

Exercise 2.3

1. Write down all the factors of
   (a) 15   (b) 64   (c) 48
   (d) −24   (e) −98

2. Write down the next two prime numbers larger than 19.

3. From the set of numbers 24, 27, 33, 45, 49 and 61 write down:
   (a) an even number
   (b) an odd number
   (c) a prime number
   (d) a multiple of 7
   (e) a factor of 135.

4. Consider the numbers 11, 21, 31, 77 and 112.
   (a) Two of these numbers are prime. What is the number which lies exactly half-way between them?
   (b) Three of these numbers have a common factor. What is it?

5. Express each of these numbers as a product of its prime factors:
   (a) 24   (b) 72   (c) 45

6. Find the LCM of the following numbers:
   (a) 8 and 12   (b) 3, 4 and 5
   (c) 2, 6 and 12   (d) 3, 6 and 8
   (e) 2, 8 and 10   (f) 20 and 25
   (g) 20 and 32   (h) 10, 15 and 40
   (i) 12, 42, 60 and 70
   (j) 18, 30, 42 and 48

7. Write down all the factors of 20 and 24. Hence find the common factors and write down the HCF of 20 and 24.

8. Write down all the multiples of 6 and 10 less than 61. Find the common multiples and hence write down the LCM of 6 and 10.

9. Find the HCF of:
   (a) 8 and 12   (b) 24 and 36
   (c) 10, 15 and 30   (d) 26, 39 and 52
   (e) 12, 18, 30 and 42
   (f) 28, 42, 84, 98 and 112

10. Find all the prime factors of:
    (a) 180   (b) 7560   (c) 1575

11. Find all the factors of −84.

12. Write down all the factors of 120.

Miscellaneous Exercise 2

Section A

1. Given the numbers 49, 55, 60, 63, 81 and 122 write down:

   (a) the multiples of 5

   (b) the multiples of 3

   (c) the even numbers.

2. Express 360 as the product of prime numbers.

3. Find the LCM of the following sets of numbers:

   (a) 10 and 15   (b) 12 and 15

4. Write down:

   (a) all the positive whole numbers which are factors of 42

   (b) all the multiples of 5 between 11 and 29.

5. Three prime numbers are 11, 13 and 37.

   (a) Calculate their sum. Is the sum a prime number?

   (b) Calculate their product. Is the product a prime number?

6. (a) Find the sum of 2^5 and 5^2.

   (b) Calculate the product of 3^2 and 2^3.

7. List the following numbers:

   (a) the prime factors of 200

   (b) the multiples of 6 less than 55

   (c) the next three prime numbers after 25.

Section B

1. Consider the numbers 2, 3, 4, 5, 6, 12, 18 and 24.

   (a) Which of them are factors of 12?

   (b) Which of them are multiples of 4?

2. Express as a product of prime factors:

   (a) 56   (b) 132

3. Find the LCM of 2700 and 4500.

4. Find the values of:

   (a) $2^3 + 3^2$ (b) $5^3 - 6^2$

   (c) $8^2 \times 5^3$ (d) $4^3 \div 2^2$

5. Find all the factors, both positive and negative of the following numbers:

   (a) -98 (b) 48   (c) -252

6. Complete this list of factors:

   $1, \ldots, \ldots, 4, \ldots, \ldots, \ldots, 18, \ldots,$
   $36, \ldots, 108$

7. Complete this list of multiples:

   $6, \ldots, \ldots, 24, \ldots, \ldots, \ldots, \ldots, 54$

8. Work out the HCF of 10, 25 and 60.

Multi-Choice Questions 2

1. Which of the following is a prime number?

   A 15   B 27   C 39   D 41

2. 2, 3 and 5 are the factors of

   A 6   B 10   C 15   D 30

3. 54 is a multiple of

   A 3   B 4   C 5   D 7

4. Consider the numbers 11, 21, 31, 77 and 112. Three of these numbers have a common factor. It is

   A 2   B 7   C 11   D 14

5. The LCM of 2, 4, 5 and 6 is

   A 240   B 60   C 6   D 5

6. The HCF of 20, 30 and 60 is

   A 2   B 10   C 20   D 60

7. The difference between 3^4 and 4^3 is

   A 0   B 1   C 7   D 17

8. The number 400, when expressed as a product of its prime factors, is

   A $2^4 \times 5^2$ B 2×5

   C 3×2 D $2 \times 5 \times 10$

9. What is the value of $4^2 + 7^2$?

    A 11   B 14   C 28   D 65

10. What is the square root of 16×64?

    A 32   B 40   C 128   D 256

Mental Test 2

Try to answer the following questions without writing anything down except the answer.

1. Calculate the square of 9.

2. What is the cube of -5?

3. Calculate the value of 2^4.

4. Determine the cube root of 64.

5. Find the square root of 36×36.

6. Find the difference between 2^3 and 3^2.

7. What is 243 expressed as a power of 3?

8. Write down the factors (positive and negative) of 48.

9. Write down all the multiples of 7 up to 41.

10. Write down the prime factors of 18.

11. What is the LCM of 6 and 8?

12. What is the HCF of 6 and 15?

Fractions

Introduction

In a fraction the top number is called the **numerator** and the bottom number is called the **denominator**. In the fraction $\frac{5}{8}$, 5 is the numerator and 8 is the denominator.

The denominator shows the number of equal parts into which the whole has been divided whilst the numerator tells how many of these equal parts are taken. In the fraction $\frac{9}{11}$, the whole has been divided into 11 equal parts and 9 of these equal parts have been taken.

Types of Fraction

In a **proper fraction** the top number (the numerator) is less than the bottom number (the denominator). Thus $\frac{3}{7}$ and $\frac{24}{37}$ are both proper fractions.

In an **improper fraction** the top number is larger than the bottom number. Thus $\frac{15}{2}$ and $\frac{29}{7}$ are both improper fractions.

A **mixed number** is the sum of a whole number and a proper fraction. Thus $5 + \frac{7}{8}$ (usually written $5\frac{7}{8}$) is a mixed number.

A mixed number can be converted into an improper fraction and vice versa.

Example 1

(a) Convert $8\frac{2}{3}$ into an improper fraction.

$$8\frac{2}{3} = \frac{(8 \times 3) + 2}{3}$$

$$= \frac{26}{3}$$

(b) Express $\frac{27}{4}$ as a mixed number.

$$\frac{27}{4} = 6\frac{3}{4} \quad \text{(since } 27 \div 4 = 6 \text{ remainder 3)}$$

Equivalent Fractions

Two fractions are **equivalent** if they have the same value.

The value of a fraction remains the same if both its numerator and its denominator are multiplied or divided by the same number provided that the number is not zero.

$$\frac{5}{8} \text{ is equivalent to } \frac{5 \times 3}{8 \times 3} = \frac{15}{24}$$

$$\frac{12}{18} \text{ is equivalent to } \frac{12 \div 6}{18 \div 6} = \frac{2}{3}$$

Lowest Terms

A fraction is said to be in its **lowest terms** when it is impossible to find a number which will divide exactly into both its numerator and denominator.

The fractions $\frac{5}{7}$ and $\frac{11}{19}$ are both in their lowest terms but the fraction $\frac{6}{10}$ is not in its lowest terms because it can be reduced to $\frac{3}{5}$ by dividing top and bottom numbers by 2.

Example 2

Reduce $\frac{21}{35}$ to its lowest terms.

$$\frac{21}{35} \text{ is equivalent to } \frac{21 \div 7}{35 \div 7} = \frac{3}{5}$$

Exercise 3.1

Express each of the following as mixed numbers:

1. $\frac{9}{2}$ 2. $\frac{11}{3}$ 3. $\frac{23}{5}$

4. $\frac{19}{7}$ 5. $\frac{39}{8}$

Express each of the following as improper fractions:

6. $3\frac{1}{4}$ 7. $2\frac{3}{5}$ 8. $5\frac{7}{8}$

9. $3\frac{7}{20}$ 10. $5\frac{1}{8}$

Reduce each of the following fractions to its lowest terms:

11. $\frac{6}{15}$ 12. $\frac{15}{20}$ 13. $\frac{6}{18}$

14. $\frac{18}{30}$ 15. $\frac{25}{60}$

Comparing the Size of Fractions

When the values of two or more fractions are to be compared, express each of the fractions with the same denominator. This common denominator should be the LCM of the denominators of the fractions to be compared. It is sometimes called the **lowest common denominator**.

Example 3

Arrange the fractions $\frac{5}{6}, \frac{8}{9}, \frac{7}{8}$ and $\frac{11}{12}$ in order of size beginning with the smallest.

The LCM of the denominators 6, 8, 9 and 12 is 72, i.e. the lowest common denominator is 72.

$\frac{5}{6}$ is equivalent to $\dfrac{5 \times 12}{6 \times 12} = \dfrac{60}{72}$

$\frac{8}{9}$ is equivalent to $\dfrac{8 \times 8}{9 \times 8} = \dfrac{64}{72}$

$\frac{7}{8}$ is equivalent to $\dfrac{7 \times 9}{8 \times 9} = \dfrac{63}{72}$

$\frac{11}{12}$ is equivalent to $\dfrac{11 \times 6}{12 \times 6} = \dfrac{66}{72}$

Because all the fractions have been expressed with the same denominator all that we have to do is to compare the numerators. Therefore the order of size is

$\frac{60}{72}, \frac{63}{72}, \frac{64}{72}$ and $\frac{66}{72}$ or $\frac{5}{6}, \frac{7}{8}, \frac{8}{9}$ and $\frac{11}{12}$

Adding and Subtracting Fractions

Two fractions which have the same denominator can be added together by adding their numerators. Thus

$$\frac{3}{11} + \frac{5}{11} = \frac{3 + 5}{11}$$

$$= \frac{8}{11}$$

When two fractions have different denominators they cannot be added together directly. However, if we express the fractions with the same denominator they can be added.

Example 4

Add $\frac{2}{5}$ and $\frac{3}{7}$.

The lowest common denominator of 5 and 7 is 35.

$\frac{2}{5}$ is equivalent to $\dfrac{2 \times 7}{5 \times 7} = \dfrac{14}{35}$

$\frac{3}{7}$ is equivalent to $\dfrac{3 \times 5}{7 \times 5} = \dfrac{15}{35}$

$$\frac{2}{5} + \frac{3}{7} = \frac{14}{35} + \frac{15}{35}$$

$$= \frac{14 + 15}{35}$$

$$= \frac{29}{35}$$

If desired, the work may be set out as follows:

$$\tfrac{2}{5} + \tfrac{3}{7} = \frac{(2 \times 7) + (3 \times 5)}{35}$$

$$= \frac{14 + 15}{35}$$

$$= \tfrac{29}{35}$$

When mixed numbers are to be added together, the whole numbers and the fractions are added separately.

Example 5

Add $4\tfrac{2}{3}$ and $2\tfrac{3}{5}$

$$4\tfrac{2}{3} + 2\tfrac{3}{5} = 6 + \tfrac{2}{3} + \tfrac{3}{5}$$

$$= 6 + \tfrac{10}{15} + \tfrac{9}{15}$$

$$= 6 + \tfrac{19}{15}$$

$$= 6 + 1\tfrac{4}{15}$$

$$= 6 + 1 + \tfrac{4}{15}$$

$$= 7\tfrac{4}{15}$$

Two fractions which have the same denominator can be subtracted by finding the difference of their numerators.

$$\tfrac{7}{9} - \tfrac{2}{9} = \frac{7 - 2}{9}$$

$$= \tfrac{5}{9}$$

When the fractions to be subtracted do not have the same denominator a method similar to that for addition is used.

Example 6

Subtract $\tfrac{3}{4}$ from $\tfrac{5}{6}$.

The lowest common denominator is 12.

$\tfrac{3}{4}$ is equivalent to $\dfrac{3 \times 3}{12} = \tfrac{9}{12}$

$\tfrac{5}{6}$ is equivalent to $\dfrac{5 \times 2}{12} = \tfrac{10}{12}$

$$\tfrac{5}{6} - \tfrac{3}{4} = \tfrac{10}{12} - \tfrac{9}{12}$$

$$= \frac{10 - 9}{12}$$

$$= \tfrac{1}{12}$$

If desired the work may be set out as follows:

$$\tfrac{5}{6} - \tfrac{3}{4} = \frac{(5 \times 2) - (3 \times 3)}{12}$$

$$= \frac{10 - 9}{12}$$

$$= \tfrac{1}{12}$$

When mixed numbers are involved first subtract the whole numbers and then deal with the fractional parts.

Example 7

Subtract $4\tfrac{1}{3}$ from $6\tfrac{3}{4}$.

$$6\tfrac{3}{4} - 4\tfrac{1}{3} = 2 + \tfrac{3}{4} - \tfrac{1}{3}$$

$$= 2 + \tfrac{9}{12} - \tfrac{4}{12}$$

$$= 2 + \tfrac{5}{12}$$

$$= 2\tfrac{5}{12}$$

Example 8

Take $3\tfrac{3}{5}$ from $7\tfrac{1}{2}$.

$$7\tfrac{1}{2} - 3\tfrac{3}{5} = 4 + \tfrac{1}{2} - \tfrac{3}{5}$$

$$= 4 + \frac{5 - 6}{10}$$

$$= 4 - \tfrac{1}{10}$$

$$= 3 + 1 - \tfrac{1}{10}$$

$$= 3 + \tfrac{9}{10}$$

$$= 3\tfrac{9}{10}$$

Exercise 3.2

Arrange the following fractions in order of size beginning with the smallest:

1. $\frac{2}{5}, \frac{3}{7}$ and $\frac{1}{3}$
2. $\frac{7}{10}, \frac{2}{3}$ and $\frac{3}{4}$
3. $\frac{3}{4}, \frac{11}{16}, \frac{7}{8}$ and $\frac{21}{32}$
4. $\frac{4}{5}, \frac{13}{20}, \frac{7}{10}$ and $\frac{3}{4}$

Add together:

5. $\frac{2}{3} + \frac{1}{5}$
6. $\frac{1}{4} + \frac{3}{8}$
7. $\frac{5}{8} + \frac{1}{9}$
8. $\frac{1}{2} + \frac{2}{3} + \frac{3}{4}$
9. $\frac{3}{8} + \frac{2}{5} + \frac{1}{3}$
10. $2\frac{5}{8} + 3\frac{1}{4}$
11. $3\frac{2}{3} + 4\frac{3}{5}$
12. $2\frac{3}{8} + 4\frac{2}{7} + 1\frac{3}{4}$
13. $5\frac{1}{2} + 4\frac{1}{3} + 3\frac{1}{5}$
14. $2\frac{5}{8} + 4\frac{1}{4} + \frac{9}{10} + 5\frac{1}{2}$

Subtract the following:

15. $\frac{2}{3} - \frac{1}{4}$
16. $\frac{5}{8} - \frac{1}{3}$
17. $\frac{3}{4} - \frac{2}{5}$
18. $3\frac{3}{4} - 2\frac{3}{8}$
19. $5\frac{7}{16} - 3\frac{1}{3}$
20. $2\frac{3}{8} - \frac{9}{16}$
21. $5\frac{1}{3} - 2\frac{3}{4}$
22. $7\frac{2}{5} - 6\frac{5}{8}$

Multiplication of Fractions

To multiply fractions first multiply their numerators and then multiply their denominators.

Example 9

Multiply $\frac{3}{8}$ by $\frac{5}{7}$.

$$\frac{3}{8} \times \frac{5}{7} = \frac{3 \times 5}{8 \times 7}$$
$$= \frac{15}{56}$$

If any factors are common to a numerator and a denominator they should be cancelled before multiplying.

Example 10

Find the value of $\frac{2}{3} \times \frac{5}{7} \times \frac{21}{32}$

$$\frac{\overset{1}{\cancel{2}}}{\cancel{3}_1} \times \frac{5}{\cancel{7}_1} \times \frac{\overset{3}{\cancel{21}}{}^1}{\cancel{32}_{16}} = \frac{1 \times 5 \times 1}{1 \times 1 \times 16}$$
$$= \frac{5}{16}$$

Mixed numbers must be converted into improper fractions before multiplying.

Example 11

Multiply $1\frac{3}{8}$ by $2\frac{1}{3}$.

$$1\frac{3}{8} \times 2\frac{1}{3} = \frac{11}{8} \times \frac{7}{3}$$
$$= \frac{11 \times 7}{8 \times 3}$$
$$= \frac{77}{24}$$
$$= 3\frac{5}{24}$$

In problems with fractions the word 'of' is frequently used. It should always be taken as meaning 'multiply'.

Example 12

Find $\frac{2}{3}$ of $\frac{4}{5}$.

$$\frac{2}{3} \text{ of } \frac{4}{5} = \frac{2}{3} \times \frac{4}{5}$$
$$= \frac{2 \times 4}{3 \times 5}$$
$$= \frac{8}{15}$$

Division of Fractions

To divide by a fraction, invert it and multiply.

Example 13

Divide $\frac{3}{5}$ by $\frac{7}{8}$.

$$\frac{3}{5} \div \frac{7}{8} = \frac{3}{5} \times \frac{8}{7}$$
$$= \frac{3 \times 8}{5 \times 7}$$
$$= \frac{24}{35}$$

Exercise 3.3

Multiply the following:

1. $\frac{5}{8} \times \frac{4}{7}$ 2. $\frac{1}{4} \times \frac{3}{5}$

3. $\frac{5}{9} \times 2\frac{2}{3}$ 4. $1\frac{3}{4} \times \frac{5}{8}$

5. $\frac{2}{5} \times \frac{10}{11} \times \frac{3}{4}$ 6. $2\frac{1}{4} \times \frac{8}{27} \times \frac{5}{7}$

7. $3\frac{2}{3} \times 3\frac{3}{5} \times 1\frac{1}{6}$ 8. $3\frac{1}{3} \times 4\frac{1}{2} \times 7\frac{1}{5}$

Find the values of:

9. $\frac{2}{3}$ of 27   10. $\frac{3}{5}$ of 120

11. $\frac{3}{4}$ of 16   12. $\frac{1}{2}$ of $\frac{2}{5}$

13. $\frac{2}{3}$ of $\frac{27}{64}$

Divide each of the following:

14. $\frac{5}{6} \div \frac{2}{3}$ 15. $\frac{7}{8} \div \frac{3}{4}$

16. $1\frac{1}{8} \div \frac{3}{4}$ 17. $2\frac{2}{5} \div \frac{3}{10}$

18. $\frac{3}{8} \div 2\frac{1}{4}$ 19. $1\frac{1}{2} \div 2\frac{1}{4}$

20. $1\frac{2}{3} \div 2\frac{2}{9}$

Operations with Fractions

The sequence of operations when dealing with fractions is the same as that used when dealing with whole numbers.

(1) First work out the contents of any brackets.

(2) Multiply and/or divide.

(3) Add and/or subtract.

Example 14

Simplify $(2\frac{1}{2} - 1\frac{1}{3}) \div 1\frac{5}{9}$.

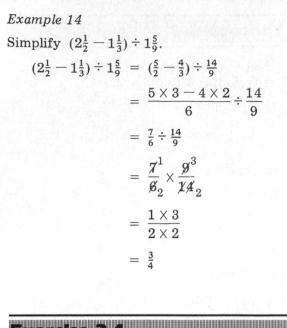

$$(2\tfrac{1}{2} - 1\tfrac{1}{3}) \div 1\tfrac{5}{9} = (\tfrac{5}{2} - \tfrac{4}{3}) \div \tfrac{14}{9}$$

$$= \frac{5 \times 3 - 4 \times 2}{6} \div \frac{14}{9}$$

$$= \tfrac{7}{6} \div \tfrac{14}{9}$$

$$= \frac{7^1}{6_2} \times \frac{9^3}{14_2}$$

$$= \frac{1 \times 3}{2 \times 2}$$

$$= \tfrac{3}{4}$$

Exercise 3.4

Work out:

1. $\frac{1}{4} \div (1\frac{1}{8} \times \frac{2}{5})$ 2. $1\frac{2}{3} \div (\frac{3}{5} \div \frac{9}{10})$

3. $(1\frac{7}{8} \times 2\frac{2}{5}) - 3\frac{2}{3}$ 4. $(2\frac{2}{3} + 1\frac{1}{5}) \div 5\frac{4}{5}$

5. $3\frac{2}{3} \div (\frac{2}{3} + \frac{4}{5})$ 6. $\frac{2}{5} \times (\frac{2}{3} - \frac{1}{4}) + \frac{1}{2}$

7. $2\frac{8}{9} \div (1\frac{2}{3} + \frac{1}{2})$ 8. $(2\frac{1}{2} - 1\frac{3}{8}) \times 1\frac{1}{3}$

Practical Applications of Fractions

The examples which follow are all of a practical nature. They depend upon fractions for their solution.

Example 15

(a) A girl spends $\frac{3}{4}$ of her pocket money and has 90 p left. How much did she have to start with?

The whole amount of her pocket money is represented by 1, so the amount left is represented by

$$1 - \tfrac{3}{4} = \tfrac{1}{4}$$

$\frac{1}{4}$ represents 90 p

1 represents $4 \times 90\,p = 360\,p$

The girl had £3.60 to start with.

(b) A group of school children went to a hamburger bar. $\frac{2}{5}$ of them bought hamburgers only, $\frac{1}{4}$ bought chips only, and the remainder bought drinks only.

(i) What fraction bought food?

(ii) What fraction bought drinks?

(i) Fraction who bought food

$$= \frac{2}{5} + \frac{1}{4}$$

$$= \frac{8 + 5}{20}$$

$$= \frac{13}{20}$$

(ii) The whole group of school children is a whole unit, i.e. 1. Therefore

Fraction who bought drinks

$$= 1 - \frac{13}{20}$$

$$= \frac{20 - 13}{20}$$

$$= \frac{7}{20}$$

Exercise 3.5

1. Calculate $\frac{3}{16}$ of £800.

2. Jane takes $5\frac{3}{4}$ minutes to iron a blouse. How many blouses can she iron in 23 minutes?

3. At a youth club $\frac{2}{5}$ of those present were playing darts and $\frac{1}{4}$ were playing other games.

   (a) What fraction were playing games?

   (b) What fraction were not playing games?

4. A watering can holds $12\frac{1}{2}$ litres. It is filled 11 times from a tank containing 400 litres. How much water is left in the tank?

5. A school has 600 pupils, of which $\frac{1}{5}$ are in the upper school, $\frac{1}{4}$ in the middle school and the remainder in the lower school. How many pupils are in the lower school?

6. A boy spends $\frac{5}{8}$ of his pocket money and has 60 p left. How much money did he have to start with?

7. During 'bob a job week' a boy scout decided to earn money by cleaning shoes. It takes him $2\frac{1}{2}$ minutes to clean one pair. At one house he was given 12 pairs to clean. How long did it take him to complete the task?

8. The profits of a business are £29 000. It is shared between two partners A and B. If A receives $\frac{9}{20}$ of the profits, how much money does B receive?

Powers of Fractions

To raise a fraction to a power we raise both the numerator and the denominator to that power. Thus

$$(\tfrac{2}{5})^3 = \frac{2^3}{5^3}$$

$$= \tfrac{8}{125}$$

Square Roots of Fractions

To find the square root of a fraction we find the square roots of the numerator and denominator separately. Thus

$$\sqrt{\frac{16}{81}} = \frac{\sqrt{16}}{\sqrt{81}}$$

$$= \frac{4}{9}$$

Exercise 3.6

Find values for each of the following:

1. $(\frac{1}{2})^4$ 2. $(\frac{3}{4})^2$

3. $(\frac{3}{7})^2$ 4. $(\frac{2}{3})^3$

5. $(1\frac{1}{3})^2$ 6. $(2\frac{2}{5})^2$

7. $(-\frac{3}{4})^3$ 8. $(-\frac{5}{6})^2$

9. $2 \times (\frac{4}{5})^2$ 10. $\frac{3}{4} \times (\frac{2}{3})^3$

Find the square roots of the following:

11. $\frac{4}{9}$ 12. $\frac{81}{100}$

13. $\frac{25}{49}$ 14. $\frac{100}{256}$

15. $\frac{1}{36}$ 16. $(25 + 144)$

17. $(169 - 25)$ 18. $(25 - 16)$

19. What is the value of $9 \times \sqrt{1\frac{7}{9}}$?

20. Work out the value of $2\frac{1}{2} \times \sqrt{\frac{16}{25}}$.

Miscellaneous Exercise 3

Section A

1. Consider the fractions $\frac{3}{5}, \frac{3}{4}, \frac{7}{10}$ and $\frac{13}{20}$.

   (a) Which fraction is the largest?

   (b) What is the difference between the largest and smallest fractions?

   (c) Calculate the product of $\frac{3}{5}$ and $\frac{3}{4}$.

   (d) Divide $\frac{3}{5}$ by $\frac{13}{20}$.

2. Give the answers to the following in their lowest terms:

   (a) $\frac{3}{5} \times \frac{10}{21}$ (b) $\frac{5}{8} \div 2\frac{3}{4}$

   (c) $\frac{7}{8} - \frac{2}{3}$ (d) $1\frac{1}{3} + 2\frac{3}{4}$

3. Find $\frac{3}{4}$ of $\frac{2}{3}$.

4. Find the next three terms in the sequence $2\frac{1}{4}, 2\frac{5}{8}, 3, \ldots$

5. Find $\frac{5}{8}$ of 32.

6. Write down the next fraction in the sequence $\frac{3}{4}, \frac{9}{16}, \frac{3}{8}, \ldots$

7. Express $\frac{1}{2} - \frac{3}{4} + \frac{5}{8} - \frac{7}{16} + \frac{19}{32}$ as a single fraction.

8. Express as a single fraction $(2\frac{1}{2} + 1\frac{1}{4}) \div 2\frac{1}{2}$.

9. A man left $\frac{3}{8}$ of his money to his wife and half the remainder to his son. If he left £8000, how much did his son receive?

10. Find $\frac{3}{4}$ of $7\frac{1}{3}$.

Section B

1. Work out $\dfrac{\frac{1}{4} + \frac{1}{3}}{\frac{1}{2} \times \frac{1}{3}}$.

2. Find $\frac{5}{9}$ of £1350.

3. $\frac{5}{8}$ of a fence has been built. If there is still 40 ft to be built, how long will the fence be?

4. Three people A, B and C share a sum of money. A takes one sixth of it, B takes one-fifth of the remainder and C takes what is left. If the amount of money to be divided is £1200, how much does each person receive?

5. Express as a single fraction in its lowest terms $(3\frac{1}{2} + 2\frac{1}{4}) \div 2\frac{1}{2}$.

6. Next year a woman will receive a rise amounting to one-eighth of her weekly wage. Her weekly wage will then be £180. What is her present weekly wage?

7. A man left three-eighths of his money to his wife and half the remainder to his son. The rest was divided equally between his five daughters. Find what fraction of the money each daughter received.

8. A man sells his car for £1620 and, as a result, loses one-tenth of the price he paid for it. What price did he pay for it?

9. A, B and C started a business and in the first year the profits were £11 500. A received one-fifth of this profit and B received seven-twentieths. How much money did C receive?

10. What is the value of $(2\frac{1}{2})^2 \times 5$?

Multi-Choice Questions 3

1. $(1\frac{2}{3})^2$ is equal to

   A $2\frac{7}{9}$ B $2\frac{4}{9}$ C $1\frac{2}{3}$ D $1\frac{4}{9}$

2. $\dfrac{10^2 - 5^2}{10^2 + 5^2}$ is equal to

   A $\frac{3}{5}$ B $\frac{2}{5}$ C $\frac{1}{3}$ D $\frac{1}{5}$

3. During an epidemic $\frac{2}{5}$ of the people in a certain African village died and 600 were left. How many people died?

   A 240   B 360   C 400   D 900

4. Expressed in its simplest form $(3\frac{1}{3} - 2\frac{1}{2}) \div \frac{5}{12}$ is equal to

   A $\frac{5}{12}$ B $\frac{5}{6}$ C 1   D 2

5. $(\frac{1}{3} - \frac{1}{6}) \div (\frac{1}{3} + \frac{1}{6})$ is equal to

   A $\frac{1}{6}$ B $\frac{1}{3}$ C $\frac{1}{2}$ D 2

6. $6\frac{1}{5} - 2\frac{3}{4} + 1\frac{1}{6}$ is equal to

   A $2\frac{17}{60}$ B $4\frac{17}{60}$ C $4\frac{37}{60}$ D $5\frac{11}{30}$

7. What is the value of $12 \times \sqrt{\frac{4}{16}}$?

   A 6   B 4   C 3   D $\frac{1}{2}$

8. The next number in the series $3, -1, \frac{1}{3}, -\frac{1}{9}, \ldots$ is

   A $\frac{1}{27}$ B $-\frac{1}{27}$ C $\frac{2}{9}$ D $-\frac{2}{9}$

Mental Test 3

Try to answer the following questions without writing anything down except the answer.

Work out the values of each of the following:

1. $\frac{3}{8} + \frac{1}{4}$ 2. $\frac{2}{5} + \frac{3}{20}$

3. $\frac{7}{10} - \frac{2}{5}$ 4. $\frac{1}{2} + \frac{1}{4} + \frac{1}{8}$

5. $\frac{1}{2} + \frac{3}{4} - \frac{5}{8}$ 6. $\frac{1}{2} \times \frac{1}{5}$

7. $\frac{2}{3} \times \frac{3}{7}$ 8. $\frac{1}{4} - 2$

9. $\frac{2}{5} \times \frac{15}{16}$ 10. $\frac{1}{4} - \frac{3}{4}$

11. What is the difference between $\frac{3}{4}$ and $\frac{5}{8}$?

12. Calculate $\frac{2}{3}$ of 12.

13. Work out $\frac{3}{5}$ of 20.

14. What is the value of $\sqrt{\frac{25}{64}}$?

15. Write down $3\frac{4}{5}$ as an improper fraction.

16. Write down $\frac{12}{32}$ as a fraction in its lowest terms.

17. Calculate the product of $\frac{2}{3}$ and $\frac{4}{5}$.

18. Find the sum of $\frac{3}{8}$ and $\frac{5}{16}$.

The Decimal System

Place Value

The number 66.66 is $60 + 6 + \frac{6}{10} + \frac{6}{100}$.

Reading from left to right each figure 6 is ten times the value of the following 6. The decimal point separates the whole numbers from the fractional parts.

We can write $\frac{3}{10}$ as .3

However when there are no whole numbers it is usual to write a zero in front of the decimal point so that .3 would be written 0.3.

If we want to write seven hundred and eight we write 708, the zero keeping the place for the missing tens.

In a similar way

$$\frac{3}{10} + \frac{7}{1000} = 0.307$$

The zero keeps the place for the missing hundredths.

Numbers such as 9.65, 83.274 and 0.05 are called decimal numbers.

Example 1

Write down the place value of the figure 9 in

(a) 32.89   (b) 0.904   (c) 79.3

    (a) The place value is $\frac{9}{100}$.

    (b) The place value is $\frac{9}{10}$.

    (c) The place value is 9 units.

1. Read off as decimal numbers:

    (a) $\frac{7}{10}$ (b) $\frac{3}{10} + \frac{8}{100}$

    (c) $\frac{8}{100} + \frac{7}{1000}$ (d) $\frac{1}{100} + \frac{7}{1000}$

    (e) $\frac{9}{100}$ (f) $8 + \frac{6}{100}$

    (g) $826 + \frac{3}{1000}$ (h) $24 + \frac{2}{100} + \frac{9}{1000}$

2. Write down, in fractional form, the place value of the figure 7 in each of the following decimal numbers:

    (a) 0.007   (b) 32.87

    (c) 3.758

3. Read off the following decimal numbers in fractional form with denominators 10, 100 or 1000:

    (a) 0.2   (b) 4.6

    (c) 3.58

4. Put the following numbers in order of size, smallest first:

    (a) 203, 2.03, 20.3 and 0.203

    (b) 0.263, 26.3, 0.002 63, 263 and 2630

    (c) 53.4, 0.534, 5.34 and 0.0534

Negative Decimals

As with integers, decimal numbers may be either positive or negative.

    5.368 is a positive decimal number
    −9.47 is a negative decimal number

Fig. 4.1 shows a number line for decimal numbers. The numbers −1.3, 2.4 and 5.8 have been shown.

Fig. 4.1

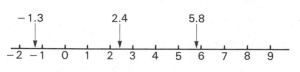

Adding and Subtracting Decimal Numbers

When adding or subtracting, the decimal points must be placed underneath each other.

Example 2

(a) Find the sum of 358.28, 0.732 and 15.2.

$$\begin{array}{r} 358.28 \\ 0.732 \\ +15.2 \\ \hline 374.212 \end{array}$$

(b) Calculate the difference between 27.295 and 18.24.

$$\begin{array}{r} 27.295 \\ -18.24 \\ \hline 9.055 \end{array}$$

Multiplication of Decimal Numbers

To multiply two decimal numbers:

(1) Multiply the two numbers disregarding the decimal points.

(2) Add the number of decimal places (the number of figures that follow the decimal point) in the first number to that in the second number.

(3) Place the decimal point in the product so that this has the same number of decimal places as the sum found in (2).

Example 3

(a) Find the product of 2.34 and 5.6.

$$\begin{array}{r} 2.34 \quad \text{two decimal places} \\ \times 5.6 \quad \text{one decimal place} \\ \hline 13.104 \quad \text{three decimal places} \end{array} \Big\} \; 2 + 1 = 3$$

(b) Multiply 0.3 by 0.005.

$$\begin{array}{r} 0.3 \quad \text{one decimal place} \\ \times 0.005 \quad \text{three decimal places} \\ \hline 0.0015 \quad \text{four decimal places} \end{array}$$

Division of Decimal Numbers

To divide one decimal number by another:

(1) Move the decimal point in the divisor until it is a whole number.

(2) Compensate for (1) by moving the decimal point in the dividend (the number being divided) by exactly the same number of places.

(3) Divide.

Example 4

Divide 1.428 by 0.35

$$1.428 \div 0.35 = 142.8 \div 35$$
$$= 4.08$$

A calculator will generally be used to add, subtract, multiply and divide decimal numbers. However, the rules given above are useful when mental arithmetic is required to produce the answers.

Example 5

Divide 0.000 49 by 0.007.

$$0.000\,49 \div 0.007 = 0.49 \div 7$$
$$= 0.07$$

Multiplying by Powers of 10

When a number is multiplied by 10^n all the figures are moved n places to the left.

Example 6

(a) $93.587 \times 10 = 93.587 \times 10^1$
$$= 935.87$$

(b) $93.587 \times 100 = 93.587 \times 10^2$
$$= 9358.7$$

(c) $93.587 \times 1000 = 93.587 \times 10^3$
$$= 93\,587$$

Dividing by Powers of 10

When a number is divided by 10^n all the figures are moved n places to the right.

Example 7

(a) $18.2 \div 10 = 18.2 \div 10^1$
$$= 1.82$$

(b) $18.2 \div 100 = 18.2 \div 10^2$
$$= 0.182$$

(c) $18.2 \div 1000 = 18.2 \div 10^3$
$$= 0.0182$$

Exercise 4.2

Using a calculator, work out:

1. $3.462 + 0.794 + 25.6$
2. $0.689 + 0.088 + 0.007 + 1.362$
3. $27.6 - 5.32$
4. $769.078 - 85.981$
5. 509.1×18
6. 58.23×0.46
7. $3.362 \times 0.075 \times 15.16$
8. $40.8 \div 3$
9. $33.12 \div 24$
10. $1.0116 \div 0.009$
11. Find the sum of 875.2, 27.93 and 8.385.
12. Calculate the total of 0.357, 18.6, 9.172 and 23.28.
13. What is the difference between 15.26 and 117.12?
14. Find the value of 0.394 minus 0.0063.
15. Find the product of 0.27, 3.58 and 11.7.
16. Multiply 27.26 by 0.039.
17. Divide 0.0034 by 2.5.

The following questions should be done mentally:

18. Multiply each of the following by (i) 10, (ii) 100, (iii) 1000:
    (a) 0.25 (b) 5.92
    (c) 38.173 (d) 481.2
19. Divide each of the following by (i) 10, (ii) 100, (iii) 1000:
    (a) 278 (b) 29.23
    (c) 638.42 (d) 0.057
20. Work out the value of each of the following:
    (a) $5.6 + 7.3$ (b) $0.36 + 1.47$
    (c) $0.005 + 0.073$ (d) $2.16 + 0.09$
    (e) $7.5 - 2.6$ (f) $0.78 - 0.03$
    (g) $0.082 - 0.076$ (h) 2.3×4
    (i) 0.62×3 (j) 0.03×0.2
    (k) 0.005×0.4 (l) 0.003×0.002
    (m) $15.5 \div 5$ (n) $0.006 \div 0.3$
    (o) $0.008 \div 0.02$

Decimal Places

The number of decimal places in a decimal number is the number of figures which follow the decimal point. Thus 18.36 has two decimal places and 0.0046 has four decimal places.

Example 8

Divide 15.187 by 3.57.

Using an eight-digit calculator,

$$15.187 \div 3.57 = 4.254\,061\,6$$

This is the limit of accuracy of the calculator but it may not be the correct answer. It is the answer correct to 7 decimal places because this is the number of figures following the decimal point.

For most purposes in arithmetic, numbers can be approximated by stating them to so many decimal places (d.p. for short).

To correct a number to so many decimal places, if the first figure to be discarded is 5 or more, the previous figure is increased by 1.

Example 9

(a) $93.7254 = 93.725$ correct to 3 d.p.

$ = 93.73$ correct to 2 d.p.

$ = 93.7$ correct to 1 d.p.

(b) $0.007\,362 = 0.0074$ correct to 4 d.p.

$ = 0.007$ correct to 3 d.p.

$ = 0.01$ correct to 2 d.p.

(c) $7.601 = 7.60$ correct to 2 d.p.

$ = 7.6$ correct to 1 d.p.

Note carefully how zeros must be kept to show the position of the decimal point or to indicate that it is one of the decimal places.

Significant Figures

A second way of approximating a number is to use significant figures. In the number 2179, the figure 2 is the most significant figure because it has the greatest value. The figure 1 is the next most significant figure whilst 9 is the least significant figure because it has the least value.

The rules regarding significant figures (s.f. for short) are as follows:

(1) If the first figure to be discarded is 5 or more, the previous figure is increased by 1.

$$7.192\,53 = 7.1925 \text{ correct to 5 s.f.}$$

$ = 7.193 \text{ correct to 4 s.f.}$

$ = 7.19 \text{ correct to 3 s.f.}$

$ = 7.2 \text{ correct to 2 s.f.}$

(2) Zeros must be kept to show the position of the decimal point or to indicate that zero is a significant figure.

$$35\,291 = 35\,290 \text{ correct to 4 s.f.}$$

$ = 35\,300 \text{ correct to 3 s.f.}$

$ = 35\,000 \text{ correct to 2 s.f.}$

$ = 40\,000 \text{ correct to 1 s.f.}$

$$0.0739 = 0.074 \text{ correct to 2 s.f.}$$

$ = 0.07 \text{ correct to 1 s.f.}$

$$18.403 = 18.40 \text{ correct to 4 s.f.}$$

$ = 18.4 \text{ correct to 3 s.f.}$

Necessary and Unnecessary Zeros

A possible source of confusion is deciding which zeros are necessary and which are not.

(1) Zeros are not needed after the last non-zero figure in a number unless it is a significant figure.

$$6.3000 = 6.3$$
$$4.70 = 4.7$$
$$0.004\,50 = 0.0045$$

The number half-way between 5.4 and 5.8 is 5.6.

The number half-way between 5.9 and 6.1 is 6.0. In this case the zero is a significant figure and it should be left.

(2) Zeros are needed to keep the place for any missing hundreds, tens, units, tenths, hundredths, etc. In the following numbers the zeros are needed:

70; 7.205; 0.004; 9.005; 5000

(3) One zero is usually written before the decimal point if the number possesses no whole numbers but this is not essential. Thus

.837 is usually written 0.837

.058 is usually written 0.058

(4) Sometimes zeros are written in front of integers. For instance the early pages of a document are often written 002, 036, etc. whilst James Bond is also known as 007!

Correcting Answers

The answer to a calculation should not contain more significant figures than the least number of significant figures used amongst the given numbers.

Example 10

Find the product of 1.384, 7.23 and 1.246.

The least number of significant figures amongst the given numbers is three (for the number 7.23). Hence the product should only be stated correct to 3 s.f.

$$1.384 \times 7.23 \times 1.246 = 12.5$$

correct to 3 s.f.

Exercise 4.3

Write down the following correct to the number of decimal places stated:

1. 19.372
    (a) 2 d.p.   (b) 1 d.p.

2. 0.007 519
    (a) 5 d.p.   (b) 3 d.p.   (c) 2 d.p.

3. 4.9703
    (a) 3 d.p.   (b) 2 d.p.

4. 153.2617
    (a) 3 d.p.   (b) 2 d.p.   (c) 1 d.p.

Use a calculator to work out:

5. $18.89 \div 14.2$ correct to 2 d.p.

6. $0.0396 \div 2.51$ correct to 3 d.p.

7. 7.217×3.26 correct to 2 d.p.

8. $(184.3 \times 0.000\,116) \div (11.49 \times 0.7362)$ correct to 4 d.p.

Write down the following correct to the number of significant figures stated:

9. 24.935
    (a) 4 s.f.   (b) 2 s.f.

10. 0.007 326
    (a) 3 s.f.   (b) 2 s.f.   (c) 1 s.f.

11. 35.604
    (a) 4 s.f.   (b) 3 s.f.

12. 35 681
    (a) 4 s.f.   (b) 3 s.f.   (c) 2 s.f.

13. 13 359 285
    (a) 4 s.f.   (b) 3 s.f.   (c) 2 s.f.

14. Write down the following numbers without unnecessary zeros:
    (a) 48.90   (b) 4.000
    (c) 0.5000   (d) 600.00
    (e) 0.007 30   (f) 108.070

Each number in the following is correct to the number of significant figures shown. Use a calculator to work out the answers to the correct number of significant figures.

15. 15.64×19.75

16. $14.6 \times 5.73 \times 2.68$

17. $13.96 \div 0.42$

18. $43.5 \times 0.87 \times 1.23$

19. $(15.76 \times 8.3) \div 9.725$

20. $(11.29 \times 3.2734) \div (77.23 \times 0.0068)$

Estimation

Estimation is used to make sure that the answer to a calculation is sensible particularly if it has been produced by a calculator. Although a calculator in good condition does not make mistakes, human beings do. In estimating an answer either rounding or significant figures may be used.

Whichever method is used, try to choose numbers which are easy to add, subtract and multiply. If division is needed try to select numbers which will cancel or divide out exactly.

Example 11

(a) Multiply 32.4 by 0.259.

For a rough estimate we will take
$$32 \times 0.25 = 32 \div 4$$
$$= 8$$

Accurate calculation:

$32.4 \times 0.259 = 8.39$ (correct to 3 s.f.)

The rough estimate shows that the answer is sensible.

(b) Find the total of 5.32, 0.925 and 17.81.

For a rough estimate we will take
$$5 + 1 + 18 = 24$$

Accurate answer:
$$5.32 + 0.925 + 17.81 = 24.055$$

The rough estimate again shows that the answer is sensible.

(c) Work out $(47.5 \times 36.52) \div (11.3 \times 2.75)$.

For a rough estimate we will take:

$$\frac{50 \times 36}{10 \times 3} = 60$$

Accurate calculation:

$$\frac{47.5 \times 36.52}{11.3 \times 2.75} = 55.8$$

The rough estimate shows that the answer is sensible.

Exercise 4.4

Do a rough estimate for each of the following and then, using a calculator, work out an answer correct to the required number of significant figures:

1. $18.25 + 39.3 + 429.8$

2. $76.815 - 57.23 - 9.63$

3. 22×0.57

4. 41.35×0.26

5. $0.732 \times 0.098 \times 2.17$

6. $92.17 \div 31.45$

7. $0.092 \div 0.035$

8. $(27.18 \times 29.19) \div 0.037$

9. $(1.456 \times 0.0125) \div 0.0532$

10. $(29.92 \times 31.32) \div (10.89 \times 2.95)$

Fraction-to-Decimal Conversion

The line separating the numerator and denominator of a fraction acts like a division sign. Hence

$\frac{1}{4}$ is equivalent to $1 \div 4$

$\frac{7}{8}$ is equivalent to $7 \div 8$

Example 12

(a) Convert $\frac{7}{8}$ into a decimal number

$$\frac{7}{8} = 7 \div 8$$
$$= 0.875$$

(b) Convert $2\frac{9}{16}$ into a decimal number.

With mixed numbers we need only convert the fractional part to a decimal number. Thus

$$2\frac{9}{16} = 2 + \frac{9}{16}$$
$$= 2 + (9 \div 16)$$
$$= 2 + 0.5625$$
$$= 2.5625$$

Recurring Decimals

Converting $\frac{3}{4}$ into a decimal number:

$$3 \div 4 = 0.75 \text{ exactly}$$

Converting $\frac{2}{3}$ into a decimal number:

Using a calculator,

$$2 \div 3 = 0.666\,666\,66$$

We see that we will continue to obtain sixes for evermore and we say that the 6 recurs.

Now converting $\frac{7}{11}$ to a decimal number

$$7 \div 11 = 0.636\,363\,63$$

We see that the 63 recurs.

Sometimes it is one figure which recurs and sometimes it is a group of figures. If one

figure or a group of figures recurs we are said to have a recurring decimal.

To save writing so many figures the dot notation is used. For example:

$$\frac{1}{3} = 1 \div 3$$
$$= 0.\dot{3} \text{ (meaning 0.333 333 ...)}$$
$$\frac{1}{6} = 1 \div 6$$
$$= 0.16 \text{ (meaning 0.166 666 ...)}$$

When two figures have dots over them the meaning is as shown below:

$$0.3\dot{1}\dot{8} = 0.318\,181 \ldots$$
$$0.4\dot{1}2\dot{7} = 0.412\,712\,712 \ldots$$

For all practical purposes we do not need recurring decimals. What we need is a decimal number stated to so many decimal places or significant figures.

Thus

$$\frac{2}{3} = 0.67 \text{ correct to 2 s.f.}$$
$$\frac{7}{11} = 0.636 \text{ correct to 3 d.p.}$$

Converting Decimals into Fractions

It will be recalled that decimals are fractions with denominators of 10, 100, 1000, etc. Thus

$$0.53 = \frac{53}{100}$$
$$0.625 = \frac{625}{1000}$$
$$= \frac{5}{8}$$

Example 13

Subtract $2\frac{5}{16}$ from 2.3214.

$$2\frac{5}{16} = 2.3125$$
$$\text{Difference} = 2.3214 - 2.3125$$
$$= 0.0089$$

Exercise 4.5

1. Using a calculator, convert the following fractions into decimal numbers:

   (a) $\frac{1}{4}$ (b) $\frac{7}{8}$ (c) $2\frac{19}{32}$ (d) $3\frac{15}{64}$

2. Write down the following recurring decimals correct to

   (i) 5 decimal places,

   (ii) 6 significant figures:

   (a) $0.\dot{5}$ (b) $0.\dot{1}\dot{7}$

   (c) $0.\dot{3}\dot{6}$ (d) $0.\dot{2}\dot{1}$

   (e) $0.\dot{4}2\dot{8}$ (f) $0.5\dot{6}\dot{3}$

   (g) $0.5\dot{6}7\dot{1}$ (h) $0.03\dot{2}$

3. Convert each of the following fractions into a recurring decimal:

   (a) $\frac{2}{9}$ (b) $\frac{5}{11}$ (c) $\frac{2}{15}$ (d) $\frac{9}{22}$

4. Convert the following decimal numbers into fractions in their lowest terms:

   (a) 0.3   (b) 0.65

   (c) 0.375   (d) 0.4375

   (e) 1.75   (f) 7.36

5. Work out $\frac{3}{16} - 0.17$.

6. What is 3.627 minus $5\frac{3}{8}$?

7. Find the sum of $2\frac{13}{16}$ and 1.782.

8. Subtract 0.395 from $\frac{43}{64}$.

Types of Number

We have already seen (in Chapter 1) that:

Counting numbers are the numbers 1, 2, 3, 4, ...

Whole numbers are the numbers 0, 1, 2, 3, 4, ...

Integers are whole numbers but they include negative numbers. Thus ..., −2, −1, 0, 1, 2, 3, are integers.

Directed numbers have a sign attached to them. Thus −75 is a negative directed number or a negative integer.

$\frac{2}{3}$ is a positive fraction whilst $-\frac{5}{6}$ is a negative fraction.

0.325 is a positive decimal number whilst −0.074 is a negative decimal number.

Fractional and decimal numbers, both positive and negative, as well as integers can be shown on a number line (Fig. 4.2).

Fig. 4.2

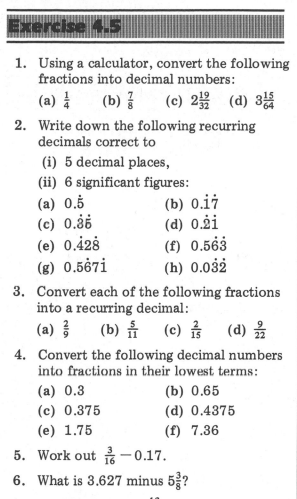

A **rational number** is any number which can be expressed as a fraction. Thus any terminating decimal number is a rational number. Recurring decimals and integers are also rational numbers because they can all be expressed as fractions.

Irrational numbers cannot be expressed as fractions. For instance, $\sqrt{2} = 1.414\,213\ldots$ and it never terminates. Hence $\sqrt{2}$ is an irrational number. Another example of an irrational number is $\pi = 3.141\,592\ldots$

Not all square roots are irrational, for instance $\sqrt{9} = 3$ and $\sqrt{2.25} = 1.5$.

Exercise 4.6

1. From the following numbers, write down those which are positive integers:

   $$5, -8, \tfrac{3}{8}, 1\tfrac{1}{4}, 2.74, 198$$

2. From the following set of numbers, write down those which are negative integers:

   $$8.5, -9, 7, -8, 11, -4$$

3. From the following set of numbers, write down those which are rational:

   $$1.57, \tfrac{1}{7}, -5.625, \sqrt{16}, \sqrt{15}, -3\tfrac{1}{2}$$

4. From the following numbers, write down those which are whole numbers:

   $$\tfrac{1}{4}, -2, 0, 6, 3.16, -0.4, 10$$

5. Arrange the following numbers in order of size, smallest first:

$$3, -4, \tfrac{5}{8}, -3\tfrac{3}{4}, 0, 7, -3.4, 2.5$$

Reciprocals

The reciprocal of a number is

$$\frac{1}{\text{number}} = 1 \div \text{number}.$$

Thus the reciprocal of 5 is $\tfrac{1}{5}$ and the reciprocal of 21.3 is $\dfrac{1}{21.3}$.

Most calculators have a reciprocal key but if yours has not, a reciprocal can be found by dividing 1 by the number.

Example 14

(a) The reciprocal of $0.362 = \dfrac{1}{0.362}$

$$= 2.762$$

(b) The reciprocal of $956.3 = \dfrac{1}{956.3}$

$$= 0.001\,05$$

Example 15

Use a calculator to find the value of

$$\frac{1}{\sqrt{7.517}} + \frac{1}{3.625^2}.$$

Input	Display
7.517	7.517
$\sqrt{}$	2.7417 …
1/x	0.3647 …
M+	0.3647 …
3.625	3.625
×	3.625
=	13.1406 …
1/x	0.0760 …
+	0.0760 …
MR	0.3647
=	0.4408

The above has been programmed for a Casio fx-102 scientific calculator but the method

will be similar for any other type. The answer is 0.4408 correct to 4 significant figures.

Using a calculator, find the reciprocals of the following numbers correct to 4 significant figures.

1. 8.19   2. 9.239
3. 89.2   4. 7142
5. 0.1537   6. 0.039 47
7. 0.001 56   8. 16 342

Using a calculator, find the values of each of the following correct to 4 significant figures.

9. $\dfrac{1}{15.28^2}$ 10. $\dfrac{1}{0.1372^2}$

11. $\dfrac{1}{\sqrt{18.73}}$ 12. $\dfrac{1}{\sqrt{0.017\,98}}$

13. $\dfrac{1}{30.15^2 + 8.29^2}$ 14. $\dfrac{1}{8.2} + \dfrac{1}{9.9}$

15. $\dfrac{1}{0.7325} + \dfrac{1}{0.9817}$

16. $\dfrac{1}{\sqrt{7.517}} + \dfrac{1}{8.209^2}$

17. $\dfrac{1}{71.36} + \dfrac{1}{\sqrt{863.5}} - \dfrac{1}{7.589^2}$

18. $\dfrac{1}{\sqrt{0.069\,43}} + \dfrac{1}{0.087\,64}$

Miscellaneous Exercise 4

Section A

1. From the following numbers, write down those which are

(a) counting numbers,

(b) integers,

(c) rational numbers:

$$-100, -58.3, 0, \tfrac{1}{2}, 3, 19, 14.3,$$
$$97.2, \sqrt{5}$$

2. An eight-digit calculator gives the value of $\sqrt{11}$ as 3.316 624 8. Write down the value of $\sqrt{11}$ correct to
   (a) 3 decimal places
   (b) 3 significant figures.

3. Express as fractions in their lowest terms
   (a) 0.88   (b) 0.026

4. (a) Express $3\tfrac{1}{2} - (1\tfrac{1}{5} \times \tfrac{1}{2})$ as a mixed number.
   (b) Convert this mixed number to a decimal.

5. (a) Write down $0.1\dot{7}$ correct to 4 d.p.
   (b) State $0.2\dot{1}\dot{5}$ correct to 5 s.f.

6. (a) Reduce $\tfrac{20}{24}$ to its lowest terms.
   (b) Convert this fraction to a decimal number correct to 3 s.f.

7. Write down the next two numbers in the sequence $5.2, -1.56, 0.468, \ldots$

8. Find the difference between $\tfrac{7}{16}$ and 0.47.

Section B

1. Which of the following are
   (a) counting numbers
   (b) integers
   (c) rational numbers?
   $$-97, -85.2, 0, \tfrac{3}{5}, 13, 72, 14.8, 105$$

2. A calculator fails to show the decimal point when doing the calculation: $12.3102 \div 0.042$. It gives the answer as 2931. Without using a calculator place the decimal point in the correct position.

3. Write down the number 0.070 408 correct to
   (a) 3 decimal places
   (b) 4 significant figures.

4. Write as recurring decimals:
   (a) $\tfrac{2}{9}$ (b) $\tfrac{8}{11}$ (c) $\tfrac{22}{7}$

5. Convert the following decimal numbers to fractions in their lowest terms:
   (a) 0.8   (b) 3.65
   (c) 0.375   (d) 0.6875

6. (a) Express $1\tfrac{2}{3} - (\tfrac{1}{2} + \tfrac{5}{9})$ as a proper fraction.
   (b) Convert this fraction to a recurring decimal using the dot notation to show the figures which recur.
   (c) State this recurring decimal number correct to 5 decimal places.

7. Rearrange the following numbers in order of size, starting with the largest:
   $$8.3, \quad -0.7, \quad 3\tfrac{15}{16}, \quad -7.2, \quad -\tfrac{5}{6}$$

Multi-Choice Questions 4

1. The number of people attending a football match is quoted as 27 000 correct to 2 significant figures. The greatest possible attendance shown by this figure is
   A 26 999   B 27 000
   C 27 499   D 27 599

2. Which one of the following is irrational?
   A $\sqrt{9}$ B $\sqrt{0.9}$
   C $\sqrt{0.09}$ D $\sqrt{0.0009}$

3. Express as a decimal number
   $$1\tfrac{1}{2} \times (\tfrac{2}{3} - \tfrac{1}{6})$$
   A 0.125   B 0.5
   C 0.75   D 1.5

4. What is 0.0063 correct to 2 decimal places?
   A 0.006   B 0.01
   C 0.06   D 0.10

5. Which of the following is the closest approximation to $\sqrt{0.91}$?

   A 0.95   B 0.45

   C 0.31   D 0.09

6. To which set of numbers does $\frac{1}{2}$ belong?

   A integers

   B irrational numbers

   C whole numbers

   D rational numbers

7. $(\frac{1}{3} - \frac{1}{6}) \div (\frac{1}{3} + \frac{1}{6})$ equals

   A $0.1\dot{6}$ B $0.\dot{3}$

   C 0.5   D 2

8. Correct to 2 significant figures, 3.0394 equals

   A 3.0   B 3.03

   C 3.039   D 3.4

Mental Test 4

Try to answer the following questions without writing anything down except the answer.

1. Find the sum of 1, 3 and 2.54.

2. Work out the product of 0.04 and 0.003.

3. Find the difference between 0.04 and 0.4.

4. Divide 10 by 0.02.

5. Find the value of $(-0.3)^2$.

6. Write down 0.075 538 correct to

   (a) 2 decimal places

   (b) 2 significant figures.

7. Calculate the value of $(0.5)^2 \times 100$.

8. Read off $\frac{3}{20}$ as a decimal number.

9. Write down the value of $(0.1)^3$.

10. Read off $2\frac{1}{5}$ as a decimal number.

11. Write down $0.\dot{6}$ as a fraction.

12. Write down, in fractional form, the place value of the digit 8 in the number 7.6832.

13. What is the reciprocal of 5?

14. Find the reciprocal of 20.

15. The reciprocal of a number is 0.25. What is the number?

Measurement

Measurement of Length

In the metric system the standard unit of length is the metre (abbreviation m). For some purposes the metre is too large a unit and it is therefore split up into smaller units as follows:

$$1 \text{ metre (m)} = 10 \text{ decimetres (dm)}$$
$$= 100 \text{ centimetres (cm)}$$
$$= 1000 \text{ millimetres (mm)}$$

In dealing with large distances the kilometre is used such that

$$1 \text{ kilometre (km)} = 1000 \text{ metres (m)}$$

Because the metric system is essentially a decimal system it is easy to convert from one unit to another by multiplying and dividing by 10, 100 or 1000.

Example 1

(a) Convert 28.35 m into millimetres.
$$28.35 \text{ m} = 28.35 \times 1000 \text{ mm}$$
$$= 28\,350 \text{ mm}$$

(b) Convert 879 cm into metres.
$$879 \text{ cm} = (879 \div 100) \text{ m}$$
$$= 8.79 \text{ m}$$

(c) Convert 734 mm into centimetres.
$$734 \text{ mm} = (734 \div 10) \text{ cm}$$
$$= 73.4 \text{ cm}$$

(d) Convert 87 600 mm into kilometres.
$$87\,600 \text{ mm} = (87\,600 \div 1000) \text{ m}$$
$$= 87.6 \text{ m}$$
$$= (87.6 \div 1000) \text{ km}$$
$$= 0.0876 \text{ km}$$

In the Imperial system lengths are measured in inches, feet, yards, and miles such that

$$12 \text{ inches (in)} = 1 \text{ foot (ft)}$$
$$3 \text{ feet (ft)} = 1 \text{ yard (yd)}$$
$$1760 \text{ yards (yd)} = 1 \text{ mile}$$

Example 2

(a) Change 468 inches into feet.
$$468 \text{ in} = (468 \div 12) \text{ ft}$$
$$= 39 \text{ ft}$$

(b) Change 1548 inches into yards.
$$1548 \text{ in} = (1548 \div 12) \text{ ft}$$
$$= 129 \text{ ft}$$
$$= (129 \div 3) \text{ yd}$$
$$= 43 \text{ yd}$$

(c) Change 22 352 yards into miles.
$$22\,352 \text{ yd} = 22\,352 \div 1760 \text{ miles}$$
$$= 12.7 \text{ miles}$$

(d) Change 32 yards into inches.

$$32\,yd = 32 \times 3\,ft$$
$$= 96\,ft$$
$$= 96 \times 12\,in$$
$$= 1152\,in$$

Measurement of Mass

In the metric system light objects are weighed using grams or milligrams. Heavier objects are weighed using kilograms or tonnes.

$$1\,gram\,(g) = 1000\,milligrams\,(mg)$$
$$1\,kilogram\,(kg) = 1000\,grams\,(g)$$
$$1\,tonne\,(t) = 1000\,kilograms\,(kg)$$

Example 3

(a) Change 900 g into kilograms.

$$900\,g = (900 \div 1000)\,kg$$
$$= 0.9\,kg$$

(b) Change 8 kg into grams.

$$8\,kg = (8 \times 1000)\,g$$
$$= 8000\,g$$

(c) Change 47 000 mg into kilograms.

$$47\,000\,mg = (47\,000 \div 1000)\,g$$
$$= 47\,g$$
$$= (47 \div 1000)\,kg$$
$$= 0.047\,kg$$

In the Imperial system very light objects are weighed in ounces. Heavier objects are measured in pounds whilst very heavy objects are weighed in hundredweight or tons such that

$$16\,ounces\,(oz) = 1\,pound\,(lb)$$
$$112\,pounds\,(lb) = 1\,hundredweight\,(cwt)$$
$$20\,hundredweight\,(cwt) = 1\,ton$$
$$1\,ton = 2240\,pounds\,(lb)$$

Example 4

(a) How many ounces are there in 7.3 lb?

$$7.3\,lb = 7.3 \times 16\,oz$$
$$= 116.8\,oz$$

(b) Change 308 224 oz into tons.

$$308\,224\,oz = 308\,224 \div 16\,lb$$
$$= 19\,264\,lb$$
$$= 19\,264 \div 2240\,tons$$
$$= 8.6\,tons$$

Measurement of Capacity

Fluids of various kinds are usually stored in tins, bottles and tanks. The amount of fluid that a container will hold, when full, is called its capacity.

In the metric system capacities are measured in millilitres, centilitres and litres such that:

$$1\,litre\,(\ell) = 100\,centilitres\,(c\ell)$$
$$= 1000\,millilitres\,(m\ell)$$

Example 5

(a) How many centilitres are there in 2.7 litres?

$$2.7\,litres = 2.7 \times 100\,centilitres$$
$$= 270\,centilitres$$

(b) Change 45 000 millilitres into litres.

$$45\,000\,millilitres = 45\,000 \div 1000\,litres$$
$$= 45\,litres$$

In the Imperial system capacities are measured in fluid ounces, pints and gallons such that:

$$20\,fluid\,ounces\,(fl\,oz) = 1\,pint\,(pt)$$
$$8\,pints\,(pt) = 1\,gallon\,(gal)$$

(N.B. A U.S. pint is less than an Imperial pint, and there are 16 U.S. fluid ounces in one U.S. pint.)

Example 6

(a) Change 7 gallons into pints.

$$7 \text{ gal} = 7 \times 8 \text{ pt}$$
$$= 56 \text{ pt}$$

(b) Change 8960 fluid ounces into gallons.

$$8960 \text{ fl oz} = 8960 \div 20 \text{ pt}$$
$$= 448 \text{ pt}$$
$$= 448 \div 8 \text{ gal}$$
$$= 56 \text{ gal}$$

Exercise 5.1

1. Convert to millimetres:
   (a) 4.8 cm   (b) 8.9 cm
   (c) 15.3 dm   (d) 8.3 m
   (e) 28.1 m   (f) 19.4 dm

2. Convert the following to centimetres:
   (a) 8 m   (b) 8.9 m
   (c) 3.25 dm   (d) 800 mm
   (e) 8964 mm   (f) 5.4 mm

3. Convert the following to kilometres:
   (a) 9000 m   (b) 58 m
   (c) 800 cm   (d) 47 000 mm
   (e) 520 mm

4. Change to feet:
   (a) 39 in   (b) 74.4 in
   (c) 9.2 yd   (d) 39.4 yd

5. Change to yards:
   (a) 158.4 in   (b) 8.7 ft
   (c) 1209 ft

6. Change to inches:
   (a) 15.8 ft   (b) 5.1 yd
   (c) 7.25 miles

7. Change to miles:
   (a) 11 264 yd   (b) 49 104 ft
   (c) 551 232 in

8. Change to kilograms:
   (a) 6980 g   (b) 11.2 t
   (c) 567 900 mg

9. Change to grams:
   (a) 0.75 kg   (b) 1830 mg
   (c) 6.3 t

10. Change to milligrams:
    (a) 4.9 g   (b) 0.009 g
    (c) 0.35 kg

11. Change to tonnes:
    (a) 8570 kg   (b) 45 kg
    (c) 197 000 g

12. Change to ounces:
    (a) 4.5 lb   (b) 0.26 lb

13. Change to pounds:
    (a) 52 oz   (b) 5.2 cwt
    (c) 7.3 tons

14. Change to tons:
    (a) 13 440 lb (b) 184 cwt

15. Change to hundredweight:
    (a) 436.8 lb   (b) 12.4 tons

16. Change to centilitres:
    (a) 4.3 litres (b) 800 millilitres

17. Change to millilitres:
    (a) 0.3 litres (b) 3.4 centilitres

18. Change to litres:
    (a) 550 centilitres (b) 90 millilitres

19. Change to pints:
    (a) 25 fl oz   (b) 6.25 gal

20. Change to fluid ounces:
    (a) 1.3 pt   (b) 0.45 pt
    (c) 0.3 gal

21. Change to gallons:
    (a) 5.2 pt   (b) 320 fl oz

Conversion of Metric and Imperial Units

Conversions from metric units to Imperial units and vice versa are often needed. Sometimes an accurate conversion is required but frequently an approximate conversion is good enough.

Metric Equivalents
1 inch = 2.54 centimetres = $2\frac{1}{2}$ cm approx. (i.e. 2 in ≈ 5 cm)
1 foot = 30.48 centimetres = 30 cm approx.
1 yard = 0.91 metres = 1 m approx.
1 kilometre = 0.621 miles = $\frac{5}{8}$ mile approx. (i.e. 5 miles ≈ 8 km)
1 kilogram = 2.205 pounds = $2\frac{1}{4}$ lb approx. (i.e. 4 kg ≈ 9 lb)
1 fluid ounce = 28.41 millilitres = 30 mℓ approx.
1 litre = 1.760 pints = $1\frac{3}{4}$ pt approx. (i.e. 4ℓ ≈ 7 pt)
1 gallon = 4.546 litres = $4\frac{1}{2}$ ℓ approx. (i.e. 2 gal ≈ 9 ℓ)

Example 7

(a) Given that 1 kg = 2.205 lb, convert 50 kg to pounds.

$$50\,kg = 50 \times 2.205\,lb$$

$$= 110.25\,lb$$

Note that 50 kg is 1 cwt approximately and potatoes, for instance, are often sold in 50 kg bags rather than 1 cwt sacks.

(b) Taking 8 kilometres to be 5 miles, convert 135 miles to kilometres.

$$135\,miles = (135 \div 5) \times 8\,km$$

$$= 216\,km$$

1. (a) Taking 1 litre = 1.760 pt, convert 5.23 litres into pints.

    (b) Taking 1 ft = 30.48 cm, convert 8.1 ft into centimetres.

2. 1 fl oz = 30 millilitres approximately. Use this information to convert 203 millilitres to the nearest fluid ounce.

3. Using 1 km = 0.621 mile, convert 32 km to the nearest mile.

4. A woman buys 2 kg of tomatoes. Taking 4 kg to be 9 lb, calculate the number of pounds of tomatoes she bought.

5. 1 km = $\frac{5}{8}$ mile approximately. Use this conversion to find the number of kilometres in 25 miles.

6. Using 1 gal = 4.55 litres, convert 12.3 litres into gallons. State the answer correct to 2 decimal places.

7. A consignment of steel weighing 5 tonnes is delivered to an engineering factory. Using 1 kg = 2.205 lb, work out the weight of steel, in tons, delivered to the factory.

The Arithmetic of Metric Quantities

Metric quantities are added, subtracted, multiplied and divided in the same way as for decimal numbers. However, it is important that all the quantities used are in the same units.

Example 8

(a) Add 15.2 m, 39.2 cm and 150.2 mm, stating the answer in metres.

$$15.2\,m + 39.2\,cm + 150.2\,mm$$

$$= 15.2\,m + 0.392\,m + 0.1502\,m$$

$$= 15.7422\,m$$

(b) Subtract 158 mm from 73.5 cm stating the difference in millimetres.

$$73.5 \, cm - 158 \, mm = 735 \, mm - 158 \, mm$$
$$= 577 \, mm$$

(c) 57 lengths of wood each 95 cm long are required by a builder. Assuming no waste in cutting the timber, calculate the total length of wood required. State the length in metres.

$$\text{Total length required} = 57 \times 95 \, cm$$
$$= 57 \times 0.95 \, m$$
$$= 54.15 \, m$$

(d) Frozen peas are packed in bags containing 450 g. How many full bags can be filled from 2 t of peas?

$$450 \, g = 0.45 \, kg \quad \text{and} \quad 2 \, t = 2000 \, kg$$
$$\text{Number of bags filled} = 2000 \div 0.45$$
$$= 4444$$

Exercise 5.3

1. Add 47 cm, 5.83 m and 15 mm, stating the sum in metres.

2. Add 93 km, 462 m and 5 cm, stating the answer in metres.

3. Add 792 g, 15 000 mg and 1.265 kg, giving the sum in kilograms.

4. Subtract 15.2 cm from 0.78 m, stating the difference in centimetres.

5. A length of ribbon is 2.5 m long. It has lengths of 25 cm, 863 mm and 0.36 m cut from it. What length of ribbon, in centimetres, remains?

6. 95 lengths of steel bar each 127 cm long are ordered by a toy manufacturer. Find, in metres, the total length of bar required.

7. 209 lengths of cloth each 135 cm long are cut from a bale containing 300 m. What length of cloth remains?

8. **(a)** How many lengths of string each 53 cm long can be cut from a ball containing 25 m?

   (b) What length of string, in millimetres, remains?

9. A spice is packed in jars containing 32 grams. How many jars can be filled from 15 kg of the spice?

10. A certain type of tablet has a weight of 8.2 mg. How many of these tablets weigh 41 grams?

Money

The British system of currency uses the pound sterling as the basic unit. The only sub-unit used is pence such that:

One pound (£1) = 100 pence (100 p)

A decimal point is used to separate the pounds from the pence and, for instance, £3.58 means three pounds and fifty-eight pence.

There are two ways of expressing amounts less than one pound. For example, 74 pence may be written 74p or £0.74. 5 pence may be written 5p or £0.05.

The addition, subtraction, multiplication and division of sums of money are performed in exactly the same way as for decimal numbers.

Example 9

(a) Add £8.94, £3.28 and 95p.

$$£8.34 + £3.28 + 95p = £8.94 + £3.28$$
$$+ £0.95$$
$$= £13.17$$

(b) A five-pound note is tendered for purchases of 84p. How much change should be obtained?

$$£5.00 - 84p = £5.00 - £0.84$$
$$= £4.16$$

(c) An article costs 82p. How much do 35 of these articles cost?

$$\text{Total cost} = 35 \times 82\text{p}$$
$$= 35 \times £0.82$$
$$= £28.70$$

(d) 37 similar articles cost £34.78. How much does each article cost?

$$\text{Cost of each article} = £34.78 \div 37$$
$$= £0.94 \quad \text{or} \quad 94\text{p}$$

Exercise 5.4

1. Add £2.15, £7.28 and £6.54.

2. Find the total of 27p, 82p, 97p and £2.36.

3. Add £15.36, £10.42, 75p, 86p and £73.25.

4. Subtract 87p from £3.26.

5. A lady goes shopping in a department store where she has a credit rating of £200. She spends £43.64, £59.76 and £87.49. How much more can she spend on credit in the store?

6. A man deposited £540 in his bank account but filled out cheques for £138.26, £57.49 and £78.56 before depositing a further £436. How much money is standing to his account?

7. An article costs £12.63. How much do 87 of these articles cost?

8. Bread rolls cost 7p each. 140 are needed for a dinner party. How much do the rolls cost?

9. An electricity bill for a quarter (13 weeks) costs £74.36. What is the average cost of electricity per week?

10. 41 similar articles cost £102.09. How much does each article cost?

Miscellaneous Exercise 5

Section A

1. Convert each of the following to metres:
   (a) 2.8 km (b) 2 km 56 m
   (c) 672 cm (d) 28 400 mm

2. How many 25 ml doses can be obtained from a bottle containing 1.7 litres of medicine?

3. How many
   (a) milligrams are there in 3 grams,
   (b) centimetres are there in 9 metres,
   (c) litres are there in 3500 millilitres?

4. How many lengths of tape, each 75 cm long, can be cut from a reel 10 m long? What is the length of the piece left over?

5. Express 45 cm as a fraction of 1 metre. Express the fraction in its lowest terms.

6. Write in order of size, smallest first, 80 g, 0.8 kg, 800 mg and 0.88 g.

7. A cup costs 48p and a saucer costs 31p. How much change should be obtained from a ten-pound note after purchasing 8 cups and 8 saucers?

8. A bank delivers £50 worth of two-pence pieces to a supermarket. Calculate the weight of these coins, in kilograms, if each coin weighs 6.92 grams.

Section B

1. 15.4 tonnes of a medicine is to be made into tablets. If each tablet weighs 7 mg, calculate the number of tablets that can be made. You may assume that there is no waste.

2. A pile of 500 sheets of paper is 4 cm thick. Calculate, in millimetres, the thickness of each piece of paper.

3. (a) Express 50 km in millimetres.

    (b) Express 4 mm in kilometres.

4. A signpost in France states 'Brest 45 km'. Taking 5 miles = 8 kilometres, find, to the nearest mile, the distance to Brest.

5. A firm decides to use metric measures for each of its products. At present a medicine is sold in bottles containing 5 fl oz. If 1 fl oz = 28.41 millilitres, work out to the nearest centilitre the amount to be stated on the bottle.

6. For a household, the bill per quarter (13 weeks) for gas is £125.83 whilst the electricity bill for the same period is £92.48. Calculate, to the nearest penny, the average cost per week of gas and electricity.

7. At a supermarket a woman bought 5 packs of butter at 59 p per pack, 4 lb of carrots at 12 p per pound, 3 packets of soap at 37 p per packet and 7 cartons of yoghurt at 17 p per carton. For these purchases she tendered a ten-pound note. How much change should she have received?

Multi-Choice Questions 5

1. Four packets have weights marked 2 kg, 500 g, 3.5 kg and 250 g. Their total weight, in kilograms, is

    A 5.25 B 6
    C 6.25 D 75.55

2. A metre rule is broken into four equal parts. The length of each part is

    A 10 cm B 20 cm
    C 25 cm D 50 cm

3. Before baking, a clay pot weighed 2 kg. After baking it weighed 1.8 kg. What was the loss of weight?

    A 0.2 g B 2 g
    C 20 g D 200 g

4. A table is 1.22 m wide. How many millimetres is this?

    A 12.2 B 122
    C 1220 D 12 200

5. A piece of ribbon 4 m long is cut into 16 pieces of equal length. What is the length of each piece?

    A 12.5 cm B 16 cm
    C 24 cm D 25 cm

6. If 1 km = 0.6 miles, then 60 miles is equivalent to

    A 36 km B 100 km
    C 360 km D 1000 km

7. What is the cost of 1 kg of tea if a packet weighing 125 g costs 21 p?

    A 84 p B £1.67
    C £1.68 D £1.71

8. How many packets of tea each containing 120 g can be made up from a tea chest containing 60 kg of tea?

    A 5 B 50
    C 500 D 5000

Mental Test 5

Try to answer the following questions without writing anything down except the answer.

1. Convert 8 m into centimetres.

2. Change 45 m into millimetres.

3. Change 45 cm into metres.

4. Convert 7 mm into metres.

5. Change 820 cm into millimetres.

6. Convert 19.7 km into metres.

7. Change 48 in into feet.

8. Change 144 in into yards.

9. How many feet are there in 1 mile?

10. Change 80 g into kilograms.

11. How many grams are equivalent to 0.73 kg?

12. Convert 18 700 mg into grams.

13. Convert 265 000 mg into kilograms.

14. How many ounces are there in 25 lb?

15. How many pounds are there in 5 cwt?

16. How many fluid ounces are there in 8 pt?

17. Change 120 fl oz into pints.

18. Change 320 fl oz into gallons.

19. If 1 in = 2.5 cm, convert 8 in into centimetres.

20. 1 fl oz = 30 millilitres approximately. Approximately how many centilitres are equivalent to 600 fl oz?

21. Add 400 cm and 6 m stating the sum in metres.

22. Subtract 50 mm from 8 cm, stating the difference in millimetres.

23. How many doses of 5 millilitres can be obtained from a medicine bottle holding half a litre?

24. Add £3.47 and 52p.

25. Subtract 87p from £4.00.

Ratio and Proportion

Ratio

Concrete can be made by mixing sand and cement in the ratio $3:1$ (three to one) and then adding water. As long as sand and cement are kept in these proportions and mixed with water they will make concrete.

We could use one bucketful of cement to three bucketfuls of sand or one bag of cement to three bags of sand. Since these ratios are both $3:1$, when mixed with water they will make concrete. The actual amounts of sand and cement used only affect the amount of concrete made.

Simplifying Ratios

The ratios $3:7$ and $9:4$ are in their simplest terms because there is no whole number which will divide exactly into both sides.

The ratio $8:6$ is not in its simplest terms because 2 will divide into both sides to give $4:3$ which is equivalent to $8:6$.

Example 1

Put the ratio $72:84$ in its simplest terms.

$72:84$ is equivalent to
$72 \div 12 : 84 \div 12$

Therefore $72:84$ is equivalent to $6:7$

This is the ratio $72:84$ in its simplest terms. Example 1 shows the similarity between ratios and fractions. Compare, for instance,

$$42:49 = 6:7 \text{ with } \tfrac{42}{49} = \tfrac{6}{7}$$

and $\qquad 20:5 = 4:1 \text{ with } \tfrac{20}{5} = \tfrac{4}{1}$

Note that when converting from the form $a:b$ to a fraction:

(1) The number on the left-hand side of the ratio becomes the numerator of the fraction.

(2) The two quantities must be stated in the same units.

(3) The fraction should be written in its lowest terms.

Example 2

(a) Express the ratio $40:216$ as a fraction in its lowest terms.

$$40:216 = \tfrac{40}{216}$$
$$= \tfrac{5}{27}$$

(b) Express $15\,\text{cm}:6\,\text{m}$ as a fraction in its lowest terms.

$$15\,\text{cm}:6\,\text{m} = 15\,\text{cm}:600\,\text{cm}$$
$$= \tfrac{15}{600}$$
$$= \tfrac{1}{40}$$

Exercise 6.1

Express each of the following ratios in its simplest terms:

1. $5:25$
2. $4:12$
3. $54:36$
4. $35:42$
5. $64:56$
6. $60:48$
7. $3:6:12$
8. $20:25:30:35$

Express each of the following ratios as fractions in their lowest terms:

9. $5:7$
10. $6:12$
11. $32:48$
12. $30p:£6$
13. $8\,cm:4\,m$
14. $6\,kg:500\,g$
15. $5\,kg:2\,t$

Fractional Ratios

To simplify a ratio such as $1\frac{1}{4}:\frac{1}{3}$ change the mixed number into an improper fraction and then express each fraction with the same denominator.

This common denominator should be the LCM of the original denominators.

$1\frac{1}{4}:\frac{1}{3}$ is equivalent to $\frac{5}{4}:\frac{1}{3}$

The LCM of 4 and 3 is 12. Expressing each fraction with a denominator of 12 gives

$\frac{5}{4}:\frac{1}{3}$ is equivalent to $\frac{15}{12}:\frac{4}{12}$

Now multiply each side of the ratio by 12.

$\frac{15}{12}:\frac{4}{12}$ is equivalent to $15:4$

Sometimes a ratio is expressed in decimal form.

Example 3

Simplify the ratio $7.4:1$.

$$7.4:1 = 74:10 \text{ (multiplying both sides by 10)}$$
$$= 37:5 \text{ (dividing each side by 2)}$$

Exercise 6.2

Simplify each of the following ratios:

1. $\frac{3}{4}:\frac{1}{3}$
2. $\frac{2}{5}:\frac{3}{10}$
3. $\frac{1}{2}:\frac{3}{8}$
4. $\frac{5}{8}:\frac{2}{3}$
5. $2\frac{3}{4}:1\frac{1}{2}$
6. $1\frac{1}{3}:3\frac{1}{4}$
7. $2\frac{2}{5}:5\frac{7}{8}$
8. $9.4:1$
9. $18.6:1$
10. $27.8:1$

Proportional Parts

The line AB (Fig. 6.1), whose length is 15 cm, has been divided into two parts in the ratio $2:3$. The line has been divided into its proportional parts and, as can be seen from the diagram, the line has been divided into a total of 5 parts. AC contains 2 of these parts and the length BC contains 3 of them. Each part is

$$15 \div 5 = 3 \text{ cm long}$$
$$AC = 2 \times 3 \text{ cm}$$
$$= 6 \text{ cm}$$
$$BC = 3 \times 3 \text{ cm}$$
$$= 9 \text{ cm}$$

Fig. 6.1

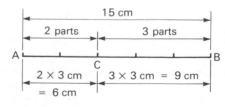

The problem of dividing the line AC into two proportional parts in the ratio $2:3$ could be solved like this:

$$\text{Total number of parts} = 2 + 3$$
$$= 5$$
$$\text{Length of each part} = 15\,\text{cm} \div 5$$
$$= 3\,\text{cm}$$
$$\text{Length of AC} = 2 \times 3\,\text{cm}$$
$$= 6\,\text{cm}$$
$$\text{Length of BC} = 3 \times 3\,\text{cm}$$
$$= 9\,\text{cm}$$

Example 4

Divide £240 in the ratio $3:4:5$.

$$\text{Total number of parts} = 3 + 4 + 5$$
$$= 12$$
$$\text{Amount of each part} = \pounds240 \div 12$$
$$= \pounds20$$
$$\text{Amount of first part} = 3 \times \pounds20$$
$$= \pounds60$$
$$\text{Amount of second part} = 4 \times \pounds20$$
$$= \pounds80$$
$$\text{Amount of third part} = 5 \times \pounds20$$
$$= \pounds100$$

Example 5

Two lengths are in the ratio $8:5$. If the first length is 120 metres, what is the second length?

The first length is represented by 8 parts. Therefore

$$\text{Length of each part} = 120 \div 8$$
$$= 15\,\text{m}$$

The second length is represented by 5 parts. Therefore

$$\text{Length of second part} = 5 \times 15\,\text{m}$$
$$= 75\,\text{m}$$

Alternatively, the problem may be solved by using fractions:

$$\text{Second length} = \tfrac{5}{8} \text{ of first length}$$
$$= \tfrac{5}{8} \times 120\,\text{m}$$
$$= 75\,\text{m}$$

Exercise 6.3

1. Divide £800 in the ratio $5:3$.

2. Divide 120 kg in the ratio $2:3:5$.

3. A line 1.68 m long is to be divided into three parts in the ratio $2:7:11$. Find, in millimetres, the length of each part.

4. A sum of money is divided into two parts in the ratio $5:7$. If the smaller amount is £200, find the larger amount.

5. A metal alloy consists of copper, zinc and tin in the ratio $2:3:5$. Find the amount of each constituent in 75 kg of the alloy.

6. A sum of money is divided into three parts in the ratio $2:4:5$. If the largest share is £40, calculate the total amount shared.

7. Four friends contribute sums of money to a charitable organisation in the ratio $2:4:5:7$. If the largest amount contributed is £4.20, find the total amount contributed by the four people.

8. An aircraft carries fuel in three tanks whose capacities are in the ratio $3:4:5$. The capacity of the smallest tank is 720 litres. Calculate:

    (a) the capacity of the larger tank

    (b) the total capacity of the three tanks.

Direct Proportion

Two quantities are said to be in direct proportion if they increase or decrease at the same rate. Thus the quantity of petrol used by a car and the distance travelled are in direct proportion.

Again, if we buy potatoes at 20p for 2 kg then we expect to pay 40p for 4 kg and 10p for 1 kg. If we double the quantity bought we double the cost; if we halve the quantity bought then we halve the cost. The quantity of potatoes and their cost are in direct proportion.

Example 6

If 7 pens cost 56p, find the cost of 6 pens.

> *Method 1* (the unitary method)
>
> 7 pens cost 56p
>
> 1 pen costs 56p ÷ 7 = 8p
>
> 6 pens cost 8p × 6 = 48p

> *Method 2* (the fractional method)
>
> 7 pens cost 56p
>
> 6 pens cost $\frac{6}{7}$ × 56p = 48p

Inverse Proportion

If an increase (or decrease) in one quantity produces a decrease (or increase) in a second quantity in the same ratio, the two quantities are in inverse proportion.

Example 7

Five men building a wall take 20 days to complete it. How long would it take 4 men to complete it?

Method 1 (the unitary method)

5 men take 20 days

   1 man takes 5 × 20 days = 100 days

(1 man takes longer, so multiply by 5.)

        4 men take 100 ÷ 4 = 25 days

(4 men take less time than 1 man, so divide by 4.)

Method 2 (fractional method)

The time ratio is 5 : 4 (4 men take longer than 5 men.)

   Time taken by 4 men = $\frac{5}{4}$ × 20 days

                       = 25 days

Exercise 6.4

1. A car does 8 km on one litre of petrol. How far will it go on 6 litres?

2. A train travels 200 km in 4 hours. If it travels at the same rate, how long will it take to complete a journey of 350 km?

3. A book weighs 800 g. What is the weight, in kilograms, of 20 similar books?

4. A farmer employs 12 men to harvest his potato crop. They take 9 days to complete the work. If the farmer had employed 8 men, how long would it have taken them?

5. 4 people can clean an office in 6 hours. How many people would be needed to clean the office in 4 hours?

6. 8 people take 5 hours to pick a row of raspberries. How long would it take 4 people to do the work?

7. If 15 kg of apples cost £3.00, calculate the cost of

   (a) 5 kg   (b) 40 kg

8. A bag contains sweets. When divided amongst 8 children each child receives 9 sweets. If the sweets were divided amongst 12 children, how many sweets would each receive?

Measures of Rate

If a car travels 8 km on 1 litre of petrol its fuel consumption is 8 km per litre (8 km/ℓ) of petrol. This is the rate at which it consumes petrol as it travels along.

The flow of water from a tap or a pipe is usually measured in gallons per minute (gal/min) or litres per minute (ℓ/min). This is the rate of flow of water.

When a car has a speed of 40 miles per hour (40 mile/h) its rate of travel is 40 miles in one hour.

Example 8

(a) A car has a fuel consumption of 30 mile/gal. How much fuel will be needed for a journey of 150 miles?

$$\text{Fuel required} = \frac{\text{Length of journey}}{\text{Fuel consumption}}$$

$$= \frac{150}{30} \text{ gal}$$

$$= 5 \text{ gal}$$

(b) The flow of water from a pipe is 20 gal/min. How long will it take to fill a tank with a capacity of 70 gallons?

$$\text{Time taken} = \frac{\text{Capacity of container}}{\text{Rate of flow}}$$

$$= \frac{70}{20} \text{ min}$$

$$= 3.5 \text{ min}$$

We see from Example 7 that if we know the fuel consumption we can work out the amount of fuel required for any length of journey. We can also work out the length of journey obtained with any given quantity of fuel.

Similarly, if we know the rate of flow of water we can work out the time taken to fill a container of known capacity. We can also calculate the volume of water delivered in a given time.

Exercise 6.5

1. A car has a fuel consumption of 25 mile/gal.

    (a) How far will it travel on 3 gallons of petrol?

    (b) Find the amount of fuel needed for a journey of 150 miles.

2. The flow of water from a tap is measured as 60 litres in 5 minutes.

    (a) Work out the rate of flow of water.

    (b) Find the volume of water that will flow in 7 minutes.

    (c) How long will it take to fill a tank with a capacity of 90 litres?

3. A car travels 60 miles on 2 gallons of petrol.

    (a) Calculate the fuel consumption.

    (b) How far would the car be expected to travel on 5 gallons of petrol?

    (c) How much fuel would be needed for a journey of 135 miles?

4. A train travels at an average speed of 50 km/h.

    (a) How far will the train travel in 3 hours if this speed is maintained?

    (b) How long will it take the train to travel 175 km?

5. The density of aluminium is 2700 kilograms per cubic metre.

    (a) Work out the weight, in tonnes, of 7 cubic metres.

    (b) A piece of aluminium weighs 120 kg. Calculate its volume, in cubic metres, correct to 2 significant figures.

6. A schoolboy found that he walked 100 m in 50 s.

    (a) Calculate his speed in metres per second.

    (b) If he continues to walk at this speed, how long will it take him to walk a further 350 m?

7. The tensile strength of polystyrene is 48 newtons per square millimetre.

    (a) Calculate the force, in newtons, that a bar of polystyrene will withstand if the cross-sectional area of the bar is 5 square millimetres.

    (b) A bar of polystyrene is to withstand a stress of 120 newtons per square millimetre. What cross-sectional area is required?

Rate of Exchange

Every country has its own monetary system. If there is to be trade and travel between any two countries there must be a rate at which the money of one country can be converted into money of the other country. This rate is called the rate of exchange.

Foreign Exchange Rates at June 1985

Country	Rate of exchange
Belgium	77.75 francs = £1
France	11.73 francs = £1
Germany	3.85 marks = £1
Greece	170 drachmas = £1
Italy	2450 lire = £1
Spain	220 pesetas = £1
United States	$1.27 = £1

Example 9

(a) If 220 Spanish pesetas = £1, find to the nearest penny, the value in sterling of 5000 pesetas.

$$5000 \text{ pesetas} = £\frac{5000}{220}$$

$$= £22.73$$

(b) A tourist changes traveller's cheques for £50 into French francs. How many francs does she get?

$$£50 = 50 \times 11.73 \text{ francs}$$

$$= 586.5 \text{ francs (587 francs to the nearest whole number)}$$

Exercise 6.6

Use a calculator and where necessary state the answer correct to 2 decimal places.

Using the exchange rates given above find:

1. The number of German marks equivalent to £15.

2. The number of Spanish pesetas equivalent to £25.

3. The number of United States dollars equivalent to £80.

4. The number of pounds sterling equivalent to 225 U.S. dollars.

5. The number of Belgian francs equivalent to £98.50.

6. The number of pounds sterling equivalent to 8960 Italian lire.

7. A transistor radio costs £52.60 in London. An American visitor wishes to buy the radio but wants to pay in dollars. How much, in dollars, will he pay?

8. A tourist changes traveller's cheques for £100 into Greek currency at 170 drachmas to the pound sterling. She spends 14 600 drachmas and changes the remainder back into British currency at £1 = 155 drachmas. How much will the tourist get for her drachmas?

Miscellaneous Exercise 6

Section A

1. In a school the ratio of the number of pupils to the number of teachers is 18:1. If the number of pupils is 540, how many teachers are there?

2. £979 is to be divided into three parts in the ratio 6:3:2. Calculate the amount of the smallest part.

3. A car travels 12 km on a litre of petrol. How many whole litres will be needed to make sure of completing a journey of 100 km?

4. A councillor gets a car allowance of 17.4p per mile. How much will he be paid for a journey of 20 miles?

5. A vending machine needs 20 litres of orangeade to fill 50 cups. How many litres are needed to fill 60 cups?

6. Change £150 into U.S. dollars when the exchange rate is $1.25 to £1.

7. If two men can paint a fence in 6 hours, how long will it take three men to paint it?

8. 2 dratt equal 1 assam and 1 dratt equals 4 yeda. How many yeda equal 1 assam?

9. Taking £1 as being equivalent to 12.06 French francs, express 5.50 francs as pence to the nearest whole number.

10. A car which travels 10 km on a litre of petrol requires 22.5 litres for a journey. A second car has a fuel consumption rate of 4 km/ℓ. How much petrol would the second car need to make the same journey?

11. The annual rent of a field is £93.50. The rent is shared by two farmers in the ratio 15:7. Find the difference between the amounts of their shares of the rent.

12. Twelve bottles of claret cost £33.96. How many bottles can be purchased for £76.41?

Section B

1. Express in their lowest terms the ratios
   (a) 15:25:50   (b) 8 cm:9 m
   (c) 17.8:1

2. Simplify the ratio $\frac{5}{8}:1\frac{5}{12}$.

3. Divide £24 in the ratio
   (a) 1:3:5   (b) 3:5:8:9

4. A rectangular photograph has sides in the ratio 7:5.
   (a) If the length of the photograph is 10.5 cm and its width is shorter, find its width.
   (b) The photograph is enlarged so that its width is 8.75 cm. Find the length of the enlarged photograph.

5. The amount of commission earned by a salesman is directly proportional to the amount of his sales. In a certain week he was paid £193.50 commission on sales of £3870. How much commission will he expect to earn on sales of £4960 during another week?

6. In a school the ratio of the numbers of pupils to the number of teachers is 18.2:1.
   (a) Find the smallest possible number of pupils in the school.
   (b) If the number of teachers and pupils together is 960, find the number of teachers.

7. A person on holiday in France changed
    £400 into francs when the exchange
    rate was 10.50 = £1. His hotel expen-
    ses were 250 francs per night for eight
    nights and his other expenses were
    1896 francs. On returning home he
    changed what francs he had left into
    sterling at a rate of 10.30 francs = £1.
    Calculate:

    (a) The number of francs received for
    the £400.

    (b) The total amount spent, in francs.

    (c) The amount in sterling obtained for
    the francs he had left.

8. In 1973, the cost of building a garage
    was divided between materials, labour
    and overheads in the ratio 10:13:2
    and the garage cost £240.

    (a) Calculate the respective costs of
    materials, labour and overheads.

    (b) By 1985, the cost of materials had
    risen by 15 times, the cost of labour by
    25 times and the cost of overheads by
    11 times. Work out the cost of building
    a similar garage in 1985.

Multi-Choice Questions 6

1. In Holland there are 100 cents in a
    florin. There are 5 florins to a pound
    sterling. How many pence is a 25-cent
    coin worth?

    A 1   B 2   C 3   D 5

2. The ratio of A's share to B's share in a
    business is 5:4. If the total profit is
    £360 then A's share is

    A £90   B £200
    C £225   D £250

3. If £120 is divided in the ratio 2:3 then
    the smaller share is

    A £40   B £48   C £72   D £80

4. The ratio of the shares of two partners
    A and B in the profit of a business is
    5:3. How much will A receive when B
    receives £480?

    A £800   B £3000
    C £288   D £216

5. A car has a petrol consumption of
    12 km/ℓ. How many complete litres
    of petrol will be needed for a journey
    of 200 km?

    A 17   B 18
    C 24   D 240

6. Change £150 into dollars when the
    exchange rate is $2.00 = £1.

    A $475   B $125
    C $300   D $600

7. If £90 is divided in the ratio 4:5:6
    then the largest share is

    A £22.50   B £33.75
    C £36   D £45

8. A mortar mixture is made of cement,
    sand and water mixed by weight in the
    ratio 1:3:5. What weight of sand is
    contained in 63 kg of the mixture?

    A 7 kg   B 21 kg
    C 35 kg   D 42 kg

9. Which one of the following ratios is
    not in the ratio 45:72?

    A 5:9   B 10:16
    C 25:40   D 65:104

10. If $2.80 = £1 and £1 = 9.92 francs,
    how many francs can be exchanged for
    $700?

    A 1692   B 1695
    C 2480   D 6944

Mental Test 6

Try to answer the following questions without writing anything down except the answer.

1. An alloy is made by mixing copper and zinc in the ratio 5:2. If 10 kg of zinc is used, how much copper is in the alloy?

2. Write the ratio 12:30 in its simplest terms.

3. Express the ratio 6:9 as a fraction in its lowest terms.

4. Write the ratio £4:20p as a fraction in its lowest terms.

5. Divide £400 in the ratio 7:3.

6. If 8 rubbers cost 72p, how much do 3 rubbers cost.

7. 2 men digging a ditch take 16 days to complete the job. How long would 8 men take?

8. A car has a fuel consumption of 30 miles per gallon. How many gallons of petrol are needed for a journey of 240 miles?

9. If the exchange rate is 4 German marks = £1, how many marks are equivalent to £20?

10. Write down in its simplest terms the ratio 7.3:1.

7 Percentages

Introduction

When comparing fractions it is often convenient to express them with a denominator of 100. Thus

$$\tfrac{1}{2} = \tfrac{50}{100} \quad \text{and} \quad \tfrac{2}{5} = \tfrac{40}{100}$$

Fractions with a denominator of 100 are called percentages.

$$\tfrac{1}{4} = \tfrac{25}{100}$$
$$= 25 \text{ per cent}$$
$$\tfrac{3}{10} = \tfrac{30}{100}$$
$$= 30 \text{ per cent}$$

The symbol % is usually used instead of the words per cent. Thus

$$\tfrac{3}{20} = \tfrac{15}{100}$$
$$= 15\%$$

Changing Fractions into Percentages

To convert a fraction into a percentage multiply it by 100.

Example 1

(a)
$$\tfrac{17}{20} = \tfrac{17}{20} \times 100\%$$
$$= 85\%$$

(b)
$$0.3 = 0.3 \times 100\%$$
$$= 30\%$$

Not all percentages are whole numbers.

For instance

$$\tfrac{3}{8} = \tfrac{3}{8} \times 100\%$$
$$= 37\tfrac{1}{2}\%$$

and $0.036 = 0.036 \times 100\% = 3.6\%$.

Banks, finance houses and building societies frequently state their interest rates as mixed numbers or decimal numbers.

For instance $7\tfrac{1}{4}\%$, $12\tfrac{1}{2}\%$ and 9.36%.

Changing Percentages into Fractions

To change a percentage into a fraction or a decimal divide it by 100.

Example 2

(a)
$$15\tfrac{1}{2}\% = 15\tfrac{1}{2} \div 100$$
$$= \tfrac{31}{200}$$

(b)
$$9.2\% = 9.2 \div 100$$
$$= 0.092$$

Exercise 7.1

The following table gives corresponding fractions, decimals and percentages. Copy the table and write in the figures which should be placed in each of the spaces filled by a question mark. Put fractions in their lowest terms.

	Fraction	Decimal	Percentage
1.	$\frac{1}{4}$	?	?
2.	$\frac{11}{20}$	?	?
3.	$\frac{7}{8}$	?	?
4.	$\frac{2}{3}$	?	?
5.	?	0.08	?
6.	?	0.192	?
7.	?	?	15
8.	?	?	27
9.	?	?	62.5
10.	?	?	$8\frac{1}{4}$
11.	?	?	$9\frac{7}{8}$
12.	?	?	$6\frac{2}{3}$

Percentage of a Quantity

To find the percentage of a quantity we first convert the percentage to a decimal number.

Example 3

(a) What is 10% of £60?

Since $10\% = 10 \div 100 = 0.1$

10% of £60 $= 0.1 \times £60 = £6$

(b) Calculate $8\frac{1}{4}\%$ of 65 metres.

$$8\frac{1}{4}\% = 8.25\%$$
$$= 8.25 \div 100$$
$$= 0.0825$$
$$8\frac{1}{4}\% \text{ of } 65\,\text{m} = 0.0825 \times 65\,\text{m}$$
$$= 5.3625\,\text{m}$$

(c) 22% of a certain length is 55 cm. What is the complete length?

The complete length is represented by 100%. Hence

$$\text{Complete length} = \frac{100}{22} \times 55\,\text{cm}$$
$$= 250\,\text{cm}$$

(d) What percentage is 37 of 264? State the answer correct to 5 significant figures.

$$\text{Percentage} = \frac{37}{264} \times 100\%$$
$$= 14.015\%$$

Exercise 7.2

1. Calculate:
   (a) 20% of 50
   (b) 30% of 80
   (c) 5% of 120
   (d) 12% of 20
   (e) 20.3% of 105
   (f) 3.7% of 68
   (g) $12\frac{1}{2}\%$ of 32
   (h) $6\frac{1}{4}\%$ of 48

2. What percentage is:
   (a) 25 of 200
   (b) 30 of 150
   (c) 24 of 150
   (d) 29 of 178
   (e) 15 of 33?

Where necessary state the percentage correct to 5 significant figures.

3. In a test a girl scores 36 marks out of 60.
   (a) What is her percentage mark?
   (b) The percentage needed to pass the test is 45%. What is the pass mark?

4. If 20% of a length is 23 cm, what is the complete length?

5. Given that 13.3 cm is 15% of a certain length, what is the complete length?

6. What is

    (a) 9% of £80

    (b) 12% of £110

    (c) 75% of £250?

7. 27% of a consignment of fruit is bad.
    If the consignment weighs 800 kg,
    how much fruit is good?

8. In a certain county the average number
    of children eating lunches at school is
    29 336, which represents 74% of the
    total number of children attending
    school. Calculate the total number of
    children attending school in the county.

Percentage Change

An increase of 5% in a number means that
if the number is represented by 100, the
increase is 5 and the new number is 105.
The ratio of the new number to the old
number is 105 : 100.

Example 4

An increase of 10% in salaries makes the
wage bill for a business £55 000 per week.

(a) What was the wage bill before the
increase?

(b) What was the amount of the increase?

    (a) If 100% represents the wage bill
    before the increase then 110% rep-
    resents the wage bill after the increase.

$$\text{Old wage bill} = \frac{100}{110} \times £55\,000$$

$$= £50\,000$$

    (b) Amount of

$$\text{increase} = 10\% \text{ of } £50\,000$$

$$= 0.1 \times £50\,000$$

$$= £5000$$

Example 5

When a sum of money is decreased by 20%
it becomes £40. What was the original sum?

If 100% represents the original sum
then the sum after the decrease of 20%
is represented by 80%. Therefore

$$\text{Original sum} = \frac{100}{80} \times £40$$

$$= £50$$

Exercise 7.3

1. The duty on an article is 20% of its
    value. If the price of the article after
    the duty has been paid is £960, find
    the price before duty.

2. When a sum of money is decreased by
    10% it becomes £18. What was the
    original sum?

3. A man sells his car for £850 thus losing
    15% of what he paid for it. How much
    did the car cost him?

4. During an epidemic 40% of the people
    in a town in Africa died and 1200 were
    left. How many people died?

5. At the end of a year the value of a
    machine has depreciated by 15% of its
    value at the beginning of the year. If its
    value at the end of the year was £1360,
    what was its value at the beginning of
    the year?

6. 25% of a consignment of fruit was bad.
    If 1500 kg was good, how much did the
    consignment weigh?

7. A man pays 20% of his salary in income
    tax. If his salary, after the tax had been
    paid, was £6400 per annum what was
    his salary before tax?

Miscellaneous Exercise 7

Section A

1. Find 5% of £260.

2. Express $\frac{135}{150}$ as a percentage.

3. After prices have been raised by 8% the new price of an article is £70.47. Calculate the original price.

4. Next year a man will receive a wage increase of 12% and his weekly wage will then be £161.28. What is his present weekly wage?

5. The number of people working for a company at the end of 1984 was 1210. This was an increase of 10% on the number working for the company at the beginning of 1984. How many people worked for the company at the beginning of 1984?

6. The entry fee for an examination was £5.00 in 1984 and it rose to £6.40 in 1985. Express the increase in the fee as a percentage of the fee in 1984.

7. 8% of a sum of money is equal to £9.60.

   (a) What is the sum of money?

   (b) Calculate 92% of this sum of money.

8. A girl scores 66 marks out of 120 in an examination.

   (a) What is her percentage mark?

   (b) If the percentage required to pass the examination is 45%, how many marks are required for a pass?

Section B

1. The value of a car, after each year's use, decreases by a fixed percentage of its value at the beginning of that year. If a car costs £5120 when new and its value after one year was £4160, by what percentage has the value decreased? Calculate the value of the car after a further year's use.

2. On 1st January, a man's shares were worth £500. By 1st February, their value had fallen by 20% but on 1st March their value was 120% of their value on 1st February. Calculate the value of the shares:

   (a) on 1st February

   (b) on 1st March?

3. In a sale, the price of a bed, after being reduced by 17%, is £149.40. Calculate the price before the reduction.

4. A woman sells her car for £2430 and, as a result, loses 10% of the price she paid for it. What price did she pay for it?

5. When a petrol tank on a car is 75% full, it contains 35 litres. How much does the tank hold when full?

6. The statistics published by an examining board show that the number of entries for English was 7962 in the year 1983, and 9519 in the following year.

   (a) Find the percentage increase correct to 1 decimal place.

   (b) If the increase in 1985 was 24% of the entry in 1984, find the entry in 1985.

7. A certain mixture of glycerine and water contains 35% by weight of glycerine. To every 100 grams of the mixture are added 25 grams of water. Find what percentage of the weight of the diluted mixture is water.

Multi-Choice Questions 7

1. One of the following is not equal to 35%. Which one?

   A $\frac{35}{100}$ B $\frac{7}{20}$ C $\frac{35}{10}$ D 0.35

2. $\frac{11}{25}$ is the same as

   A 4.4%   B 25%   C 44%   D 440%

3. 30% of a certain length is 600 mm. The complete length is

    A 2 mm   B 20 mm

    C 200 mm   D 2000 mm

4. A boy scored 70% in a test. If the maximum mark was 40, then the boy's mark was

    A 4   B 10   C 28   D 35

5. During a sale a shop reduced the price of everything by 10%. Find the sale price of an article originally priced at £43.

    A £34   B £38.70

    C £39.70   D £47.30

6. A special offer of 4p off the normal price of 50p for a packet of biscuits is made. What percentage reduction does this represent?

    A 2%   B 4%   C 8%   D 25%

7. $\frac{1}{4}$% of £480 is

    A £1.20   B £12

    C £19.20   D £120

8. A drum contains 18 litres of oil. If 6.75 litres are taken out, what percentage of oil is left?

    A 62.5%   B 61.1%

    C 37.5%   D 33.3%

Mental Test 7

Try to answer the following questions without writing anything down except the answer.

1. Convert 0.6 into a percentage.

2. Change 40% into a fraction in its lowest terms.

3. Convert 35% into a decimal number.

4. Calculate 10% of 980.

5. What percentage is 4 of 25?

6. Calculate 75% of 40.

7. 25% of a length is 10 cm. What is the complete length?

8. What is 0.25% of £2000?

9. A girl scores 80 marks out of 120 in a test. What was her percentage mark?

10. In an examination 60% of the maximum mark is required for a pass. If the maximum mark is 200, what is the pass mark?

11. In a sale a shopkeeper reduces all his articles by 20%. What is the sale price of an article originally costing £150?

12. A consignment of perishable goods weighs 300 kg. 30% of the consignment is unsaleable. What weight is saleable?

Wages and Salaries

Introduction

Everyone who works for an employer receives a wage or salary in return for his or her labour. The payment may be made in several different ways but wages are usually paid weekly and salaries monthly.

Payment by the Hour

Many people who work, for instance, in factories, in the transport industry, in construction and building, are paid so much for each hour that they work. Most employees work a basic week of so many hours and it is this basic week which fixes the hourly (or basic) rate of pay. The basic week and the basic rate of pay are often fixed by negotiation between the employer and the trade union which represents the workers.

Example 1

(a) A man works a basic week of 35 hours and his weekly wage is £175. Work out his hourly rate of pay.

$$\text{Hourly rate of pay} = \frac{\text{Weekly wage}}{\text{Basic week}}$$
$$= \frac{£175}{35}$$
$$= £5$$

(b) A woman works a basic week of 38 hours and her hourly rate of pay is £3. Calculate her weekly wage.

$$\text{Weekly wage} = £3 \times 38$$
$$= £114$$

Overtime

Hourly paid workers are usually paid extra for working more hours than the basic week requires. These extra hours of work are called overtime. Overtime is usually paid at time-and-a-quarter ($1\frac{1}{4}$ times the basic rate), time-and-a-half ($1\frac{1}{2}$ times the basic rate) or double time (twice the basic rate).

Example 2

A girl is paid a basic rate of £3 per hour. Calculate her hourly overtime rate of pay when this is paid at

(a) time-and-a-quarter,

(b) time-and-a-half and

(c) double time.

(a) $\text{Overtime rate} = 1\frac{1}{4} \times £3$
$$= £3.75 \text{ per hour}$$

(b) $\text{Overtime rate} = 1\frac{1}{2} \times £3$
$$= £4.50 \text{ per hour}$$

(c) $\text{Overtime rate} = 2 \times £3$
$$= £6 \text{ per hour}$$

Example 3

Peter Taylor works a 40-hour week for which he is paid £180. He works 5 hours' overtime at time-and-a-half. What is his total wage for the week?

$$\text{Basic rate of pay} = \frac{\text{Weekly wage}}{\text{Basic week}}$$

$$= \frac{£180}{40}$$

$$= £4.50$$

$$\text{Overtime rate} = 1\tfrac{1}{2} \times £4.50$$

$$= £6.75$$

$$\text{Payment for overtime} = 5 \times £6.75$$

$$= £33.75$$

$$\text{Total wage for the week} = £180$$
$$+ £33.75$$

$$= £213.75$$

Piecework

Some workers are paid a fixed amount for each article or piece of work that they make. Often if they make more than a certain number of articles they are paid a bonus.

Example 4

A pieceworker is paid 8p for every handle he fixes to an electric iron up to 200 per day. For each handle over 200 that he fixes he is paid a bonus of 2p. If he fixes 250 handles in a day, work out how much he earns.

$$\text{Money earned on first 200} = 200 \times 8\text{p}$$

$$= £16.00$$

$$\text{Money earned on next 50} = 50 \times (8 + 2)\text{p}$$

$$= 50 \times 10\text{p}$$

$$= £5$$

$$\text{Total amount earned} = £16 + £5$$

$$= £21$$

Commission

Shop assistants, salesmen and representatives are often paid commission on top of their basic wage. This is calculated as a small percentage of the value of the goods that they have sold.

Example 5

A salesman is paid a weekly wage of £120. In addition he is paid a commission of 2% on the goods which he sells. During a certain week he sold goods to the value of £4000. What are his earnings for the week?

$$\text{Commission} = 2\% \text{ of } £4000$$

$$= £80$$

$$\text{Total earnings} = £80 + £120$$

$$= £200$$

Salaries

People like civil servants, teachers, secretaries and company managers are paid a fixed amount each year which is called their salary. They are rarely paid overtime or commission. Salaries are usually paid monthly.

Example 6

A secretary is paid £8400 per annum. What is her monthly salary?

$$\text{Monthly salary} = £8400 \div 12$$

$$= £700$$

Exercise 8.1

1. A woman works a basic week of 35 hours for which she is paid £140. Calculate her hourly rate of pay.

2. A man works a basic week of 38 hours. His basic rate is £6 per hour. Calculate his weekly wage.

3. If the basic rate is £4.92 and the basic week is 36 hours, work out the weekly wage.

4. A building worker is paid an hourly rate of £5. Work out his overtime rate when this is paid at
   (a) time-and-a-quarter
   (b) time-and-a-half
   (c) double-time.

5. Gwen Evans is paid £100.80 for a basic week of 40 hours.
   (a) What is her basic hourly rate?
   (b) Overtime is paid at time-and-a-quarter. What is her overtime rate of pay?
   (c) She works 12 hours' overtime in a certain week. How much was she paid in overtime?
   (d) What was her total pay for the week?

6. A shop assistant works a basic week of 36 hours for which she is paid £126. During a certain week she works 4 hours overtime which is paid at time-and-a-half. How much does she earn during that week?

7. A woman is paid 5p for each bag of sweets she packs up to a limit of 350. After that she is paid a bonus of 1p per bag. Work out the amount she earns if she packs 450 bags in a day.

8. A man is paid 8p for each spot weld he makes up to 300 per day. After that he is paid a bonus of 2p per spot weld. If he makes 450 spot welds in a day, calculate the amount he earns in the day.

9. A representative sells farm machinery for £8000. If he is paid 3% commission on his sales how much commission will he get on this sale?

10. A sales assistant is paid a basic wage of £94 per week. In addition she is paid a commission of 4% on her sales. Work out her total earnings for a week in which she made sales worth £5000.

11. The following are the salaries of four people. How much is each paid monthly?
    (a) £5320   (b) £6000
    (c) £9600   (d) £6096

Deductions from Earnings

The wages (and salaries) so far discussed are not the take-home pay or salary. A number of deductions are made, the most important being

(1) Income Tax

(2) National Insurance

(3) Pension fund payments.

After the deductions have been made the wage or salary is called the **net** wage or salary. It is also called the **take-home pay.**

National Insurance

Both employees and employers pay National Insurance. Employees pay a contribution based on a certain percentage of their gross wage or salary (in 1986 this was 9%). Married women and widows can pay at a reduced rate (3.25% in 1986). Men over 65, women over 60 and employees earning less than £34 per week do not pay National Insurance.

The rates of National Insurance are fixed by the Chancellor of the Exchequer and they can vary from year to year. National Insurance pays for such things as hospitals, doctors, sick pay and unemployment benefit.

Example 7

A man earns an annual salary of £12 000. His deductions for National Insurance are 9% of his gross salary. Work out the amount paid annually in National Insurance.

$$\text{Amount paid} = 9\% \text{ of } £12\,000$$
$$= £1080$$

Income Tax

Taxes are levied by the Chancellor of the Exchequer to produce money to pay for such items as the armed services, the Civil Service and motorways. The largest producer of revenue is income tax.

Every person who earns more than a certain amount pays income tax. Tax is not paid on the entire income. Certain allowances are made. In 1988-9 these were as follows:

(1)   Single person's allowance £2605

(2)   Married couples allowance £4095

(3)   Wife's earned income allowance £2605

In certain cases other allowances such as an additional personal allowance for one parent families and blind persons are available.

The residue of the income left over after all the allowances have been deducted is called the taxable income.

Taxable income = Gross income − allowances

The amount of income tax payable per annum

$$= \text{Taxable income} \times \frac{\text{Rate of taxation}}{100}$$

In 1988-9 for taxable incomes up to £19 300 the rate of taxation was 25% but for any taxable income above £19 301 the rate of taxation was 40%.

Example 8

A woman's annual salary is £12 000. Her taxable income is calculated by deducting a personal allowance of £2605 and pension fund payments of £700 from her annual salary. She then pays income tax at 25% on her taxable income. How much does she pay in income tax per annum?

$$\text{Taxable income} = £12\,000 - £2605 - £700$$
$$= £8695$$

$$\text{Tax payable} = 25\% \text{ of } £8695$$
$$= 2173.75$$

Example 9

A man has a taxable income of £40 000. Work out the amount of income tax payable if the tax bands are 25% up to £19 300 and 40% above £19 301.

$$25\% \text{ of } £19\,300 = £4825$$

$$\text{Residue of taxable income} = £40\,000 - £19\,300$$
$$= £20\,700$$

$$40\% \text{ of } £20\,700 = £8280$$

$$\text{Total tax payable p.a.} = £4825 + £8280$$
$$= £13\,105$$

PAYE

Most people pay their income tax by a method known as **Pay-As-You-Earn** or **PAYE** for short. The tax is deducted from their wage or salary before they receive it. The taxpayer and his or her employer receive a notice of coding which gives the allowances to which the employee is entitled (based upon information given on tax forms previously completed). The notice of coding sets a code number. The employer then knows from tax tables the amount of tax to deduct from the wages of an employee.

Pension Fund

Many companies operate their own private pension scheme to provide a pension in addition to that provided by the state. Usually both employer and employee contribute. The employee's share is usually of the order of 5 or 6% of the gross annual salary.

The amount of pension received depends upon the length of service and earnings at the time of retirement.

Example 10

An employee earns £11 000 per annum. His pension fund contributions amount to 6% of his annual salary. How much are the pension fund payments per annum?

$$\text{Pension fund payments} = 6\% \text{ of } £11\,000$$
$$= £660$$

Exercise 8.2

1. A woman earns £8500 per annum. She pays 9% of her salary in National Insurance contributions. How much does she pay per annum.

2. Work out the annual amount paid in pension fund contributions when these are made at 5% of a gross wage of £9000 per annum.

3. A young woman earns £95 per week. Her deductions are National Insurance £8.55, income tax £15.78 and pension fund payments £5.70. Work out her take-home pay.

4. A man's taxable income is £3200. If tax is paid at 29% of taxable income, calculate the amount of income tax paid.

5. When income tax was levied at 30% a man paid £240 in tax. What was his taxable income?

6. A man's salary is £14 000 per annum and his total tax-free allowances are £4500.
   (a) Work out his taxable income.
   (b) If income tax is levied at 25%, calculate the amount of tax payable.

7. A single woman earns £9000 per annum. Her allowances to be set against her gross salary are as follows:

   A personal allowance of £2205

   Expenses of £35 for belonging to a learned society

   Pension fund payments of £450

   (a) Work out her total allowances to be set against her gross income.
   (b) If income tax is levied at 30%, calculate the amount payable in tax per annum.

8. In 1988 the rates of tax on taxable income were as follows:

   | Up to £19 300 | 25% |
   | From £19 301 | 40% |

   A man earned a gross salary of £32 000 per annum and his tax-free allowances amounted to £7385. Calculate the amount of income tax that he paid.

9. In a Commonwealth country a man earns $14 000 per annum and his wife earns $9600 per annum. They have four children. Allowances and taxes are calculated on their combined salaries as follows:

Husband's personal allowance	$2795
Wife's personal allowance	$1785
Pension fund payments of	$2220
For each child	$500

The rates at which income tax is levied on taxable income are as follows:

10% on the first $2000

20% on the next $2000

30% on the next $4000

40% on the remainder

Calculate:

(a) the total allowances

(b) the taxable income

(c) the total tax payable for the whole year.

10. The first £16 200 of a man's taxable income is taxed at 30%. The remainder, if any, is taxed at 50%. Work out:

(a) the amount of tax payable on a taxable income of £21 000

(b) the taxable income on which the tax payable is £8860.

Miscellaneous Exercise 8

Section A

1. A woman's gross income is £14 000 per annum. Her tax-free allowances amount to £6000.

   (a) Calculate her taxable income.

   (b) If tax is levied at 30%, work out the amount of tax payable in a year.

2. A man is paid £240 for a 40-hour week. Overtime is paid at time-and-a-half.

   (a) Calculate the man's basic hourly rate of pay.

   (b) Work out his overtime rate of pay.

   (c) If he works 7 hours' overtime during a certain week, calculate his total wages for that week.

3. A salesman is paid a basic wage of £1100 per month. On top of this he is paid commission of 2% on goods that he sells. During one certain month he sells goods to the value of £20 000. Work out his total wage for that month.

4. A pieceworker is paid 8p for every article she makes up to a limit of 150 per day. For each article she makes over this limit she is paid a bonus of 2p per article. In one day she made 220 articles. How much did she earn?

5. A woman earns £8000 per annum. Deductions from her gross annual pay are £720 for National Insurance, £1850 for income tax and £450 for pension fund payments.

   (a) What are her total deductions for the year?

   (b) What is her net annual pay?

   (c) What is her monthly take-home pay?

6. A man has a taxable income of £23 000. For up to £16 200 of taxable income tax is levied at 30%. The next £3000 of taxable income is taxed at 40%, whilst the residue is taxed at 50%. Work out the amount of income tax payable.

7. A woman is paid £3.50 per hour for a 40-hour week.

   (a) Calculate her gross weekly wage.

   (b) She pays National Insurance at 3.85% of her gross wage. How much per week does she pay in National Insurance contributions?

   (c) If, in addition, she pays £18.30 per week in income tax, work out her net weekly wage.

8. A man works a basic week of 40 hours at £5 per hour. He then earns overtime at time-and-a-quarter. Calculate his gross wage for a 44-hour week.

Section B

1. The first £10 000 of a sum of money is taxed at 30% and the remainder, if any, is taxed at 40%. Calculate:

   (a) the tax on £13 610

   (b) the sum on which the tax is £2220

   (c) the sum on which the tax is £3420.

2. A salesman is paid a basic wage of £120 per week. On top of this he is paid commission on the value of the goods he sells. During a certain week he sold goods to the value of £5000 and his total wage for that week was £195. What is the percentage rate of his commission?

3. A pieceworker is paid 25p for each toy she completes up to a limit of 60. After this she is paid a bonus of 5p for each additional toy she makes. On one day she earned £27. How many toys did she complete during that day?

4. A mechanic is paid a basic weekly wage of £210 per week for a 35-hour week. Overtime on Sunday is paid at double time. During a certain week he worked overtime on Sunday and his total weekly wage was £270. How many hours of overtime did he work?

5. A man earns £25 000 per annum and his wife has an annual salary of £12 000. Allowances and income tax are calculated on their combined salaries. The information below shows the tax-free allowances:

Husband's personal allowance	£2795
Allowance on wife's earnings	£1785
Pension fund payments	£2220

The rates of tax are as follows:

On first £14 000	30%
On next £2600	40%
On next £4600	45%
Remainder if any	50%

Calculate:

(a) their total allowances

(b) their taxable income

(c) the amount of tax payable.

Multi-Choice Questions 8

1. An agent is paid a commission of 12.5% on each sale. If goods to the value of £6400 are sold, how much is the commission?

   A £192   B £224

   C £768   D £800

A young man works a 40-hour week and he is paid £1.50 per hour. Overtime is paid at time-and-a-half. Use this information to answer questions 2, 3, 4 and 5.

2. How much money does he earn in a week when he works 52 hours?

   A £58   B £60   C £78   D £87

3. In another week his earnings are £73.50. How many hours has he worked?

   A 46   B 47   C 48   D 49

4. In a week when his wages were £78 he paid £3.90 in income tax. What percentage of his wage was paid in income tax?

   A 2%   B 5%   C 20%   D 50%

5. If the young man gains an 8% rise in wages, how much would he then earn for a 40-hour week?

   A £60   B £64.80

   C £72   D £72.50

6. In October a salesman was receiving a commission of £600. In November this was reduced by 10% but in December he got an increase of 10%. What was his commission in December?

    A £594   B £600
    C £660   D £720

7.

Salary	Tax
£3001–£6000	20% of salary
£6001–£8000	30% of salary
£8001–£10 000	35% of salary
£10 001–£12 000	40% of salary

From the above table how much will a woman earning £9000 pay in income tax?

    A £1575   B £1800
    C £3150   D £4500

8. A woman's basic pay for a 40-hour week is £160. Overtime is paid for at time-and-a-quarter. In a certain week she worked overtime and her total wage was £200. How many hours' overtime did she work?

    A 5   B 8   C 9   D 10

Mental Test 8

Try to answer the following questions without writing anything down except the answer.

1. A man's basic weekly wage is £200 for a 40-hour week. What is his hourly rate of pay?

2. A man earns £5 per hour. How much will he earn for a 40-hour week?

3. A young woman earns £3 per hour. Overtime is paid at time-and-a-half. What is her overtime rate?

4. A jobbing gardener is paid £4 per hour. How much will he be paid for 5 hours' work?

5. A pieceworker is paid 25 p for each article she completes. How much will she be paid for completing 120 articles?

6. A man has a taxable income of £500. If tax is levied at 30%, how much tax will he pay?

7. Commission is paid at 5% to a salesman. How much commission will he earn for selling goods worth £2000?

8. An employee is paid an annual salary of £6000. What is her monthly salary?

Interest

Introduction

If money is invested with a bank or a building society, interest is paid to the investor for lending the money. If money is borrowed then the borrower will be charged interest on the loan.

The money which is invested or borrowed is called the **principal**. The percentage return is called the **annual percentage rate (APR** for short).

Simple Interest

With simple interest the amount of interest earned is the same every year. If the money is invested for 2 years the amount of interest will be doubled; for 3 years the amount of interest will be trebled, i.e. multiplied by 3; for six months (i.e. half a year) the amount of interest will be halved.

Example 1

£1500 is invested at 12% per annum.

(a) Work out the interest at the end of 1 year.

(b) How much interest will be earned at the end of 5 years.

$$\text{(a)} \quad \text{Amount of interest} = 12\% \text{ of } £1500$$
$$= £180$$

(b) Amount of interest
after 5 years
$$= 5 \times £180$$
$$= £900$$

The Simple Interest Formula

From Example 1 we see that the amount of simple interest can be calculated from:

$$I = PRT$$

where P is the principal, R is the rate of interest expressed as a decimal and T is the time in years.

Since R can be expressed as a fraction with a denominator of 100

$$I = \frac{PRT}{100}$$

Example 2

A person borrows £4000 for a period of 6 years at 20% simple interest per annum. Calculate:

(a) the amount of interest payable on the loan

(b) the total amount to be repaid.

(a) We are given that $P = £4000$, $R = 20\%$ and $T = 6$ years.

$$I = \frac{4000 \times 20 \times 6}{100}$$
$$= £4800$$

(b) Total amount
repayable $\quad = £4000 + £4800$

$\qquad\qquad = £8800$

If we know any three of the quantities I, P, R and T then it is possible to find the fourth quantity. To do this we can either

(1) transpose the simple interest formula to give

$$T = \frac{100I}{PR}, \quad R = \frac{100I}{PT} \quad \text{or} \quad P = \frac{100I}{RT}$$

or

(2) solve a simple equation.

Example 3

£700 is invested for 5 years. At the end of this time the simple interest amounts to £420. Work out the rate of simple interest.

We are given $P = £700$, $T = 5$ years and $I = £420$. We have to find R.

Method 1 (using the formula)

$$R = \frac{100I}{PT}$$

$$= \frac{100 \times 420}{700 \times 5}$$

$$= 12\%$$

Hence the rate of interest is 12% p.a.

Method 2 (solving a simple equation)

$$I = \frac{PRT}{100}$$

$$420 = \frac{700 \times R \times 5}{100}$$

$$420 = 35R$$

$$R = 12\%$$

Hence, as before, the rate of interest is 12% p.a.

In modern business practice, simple interest is unusual for periods greater than one year.

Exercise 9.1

Work out the amount of simple interest for each of the following:

1. £800 invested for 1 year at 6% p.a.

2. £1500 lent for 6 years at 11% p.a.

3. £2000 invested for 7 years at 14% p.a.

Find the length of time for

4. £1000 to be the interest on £5000 invested at 5% p.a.

5. £480 to be the interest on £2000 invested at 8% p.a.

6. £1200 to be the interest on £3000 invested at 10% p.a.

Find the rate per cent per annum simple interest for

7. £420 to be the interest on £1200 invested for 5 years.

8. £72 to be the interest on £200 invested for 3 years.

9. £1200 to be the interest on £3000 invested for 4 years.

Find the principal required for

10. The simple interest to be £600 on money invested for 3 years at 5% p.a.

11. The simple interest to be £40 on money invested for 2 years at 10% p.a.

12. The simple interest to be £432 on money invested for 4 years at 9% p.a.

13. £500 is invested for 7 years at 10% p.a. simple interest. How much will the investment be worth after this period?

14. £2000 was invested at 12% p.a. simple interest. The investment is now worth £2960. How many years ago was the money invested?

Compound Interest

With compound interest the interest is added to the principal, thereby increasing it. So, by the end of the second year, the larger principal has attracted more interest. In the business world, interest is almost always in the form of compound interest.

Example 4

£5000 is invested for 2 years at 15% compound interest. Calculate the value of the investment at the end of this period.

	£
Principal	5000
1st year's interest	750 (15% of £5000)
Value of investment after 1 year	5750 (£5000 + £750)
2nd year's interest	862.50 (15% of £5750)
Value of investment after 2 years	£6612.50

The value of an investment after a period of time is usually called the amount. Thus in Example 4, the amount after 2 years is £6612.50.

Although all problems on compound interest can be worked out in the way shown in Example 4, the work takes a long time, particularly if the period is lengthy.

The following formula will allow you to calculate the compound interest but you will need a scientific calculator.

$$A = P\left(1 + \frac{R}{100}\right)^T$$

where A stands for the amount after T years, P stands for the principal, R stands for the rate per cent p.a.

Example 5

£950 is invested for 7 years at 9% p.a. compound interest.

(a) Calculate the amount accruing at the end of the 7 years.

(b) How much interest was earned during the 7 years?

We are given $P = £950$, $R = 9\%$ and $T = 7$ years.

(a)
$$A = P\left(1 + \frac{R}{100}\right)^T$$

$$= 950\left(1 + \frac{9}{100}\right)^7$$

$$= 950 \times 1.09^7$$

$$= £1736.64$$

(b) Interest earned $= £1736.64$
$$- £950$$
$$= £786.64$$

The program when using a scientific calculator is as follows:

Input	Display
1	1.
+	1.
0.09	0.09
=	1.09
x^y	1.09
7	7.
=	1.828 ...
×	1.828 ...
950	950.
=	1736.6372
−	1736.6372
950	950.
=	786.6372

Compound Interest Tables

In business, compound interest tables are used to find the amount of interest due at the end of a given period of time. Part of such a table is shown overleaf.

Years	5%	6%	7%	8%	9%
1	1.050	1.060	1.070	1.080	1.090
2	1.103	1.124	1.145	1.166	1.188
3	1.158	1.191	1.225	1.260	1.295
4	1.216	1.262	1.311	1.360	1.412
5	1.276	1.338	1.403	1.469	1.539
6	1.340	1.419	1.501	1.587	1.677
7	1.407	1.504	1.606	1.714	1.828
8	1.477	1.594	1.718	1.851	1.993
9	1.551	1.689	1.838	1.999	2.172
10	1.629	1.791	1.967	2.159	2.367

Years	10%	11%	12%	13%	14%
1	1.100	1.110	1.120	1.130	1.140
2	1.210	1.232	1.254	1.277	1.300
3	1.331	1.368	1.405	1.443	1.482
4	1.464	1.518	1.574	1.603	1.689
5	1.611	1.685	1.762	1.842	1.925
6	1.772	1.870	1.974	2.082	2.195
7	1.949	2.076	2.211	2.353	2.502
8	2.144	2.304	2.476	2.658	2.853
9	2.358	2.558	2.773	3.004	3.252
10	2.594	2.839	3.106	3.395	3.707

The table shows the appreciation (the increase in value) of £1. For instance, £1 invested for 5 years at 9% p.a. will become £1.539.

Example 6

£750 is invested for 6 years at 5% p.a. compound interest. Using the compound interest table shown above find:

(a) the amount accruing at the end of the period

(b) the amount earned in interest during this period.

From the table, in 6 years at 5% p.a. £1 becomes £1.340.

(a)   Amount accruing = £750 × 1.340

= £1005

(b)   Interest earned = £1005 − £750

= £255

Exercise 9.2

Using the compound interest formula calculate the amount accruing after the stated period:

1. £250 invested for 5 years at 8% p.a.

2. £4000 invested for 7 years at 9% p.a.

3. £1200 invested for 12 years at 10% p.a.

4. £2500 invested for 15 years at 11.5% p.a.

Using the compound interest formula, work out the amount of interest earned at the end of the stated period:

5. £5000 invested for 6 years at 7% p.a.

6. £3500 invested for 20 years at 9% p.a.

7. £8000 invested for 12 years at 12% p.a.

Using the compound interest table, calculate the amount accruing at the end of the stated period:

8. £4000 for 7 years at 8% p.a.

9. £15 000 for 9 years at 13% p.a.

10. £500 invested for 3 years at 9% p.a.

Using the compound interest table, work out the compound interest earned at the end of the stated period:

11. £350 invested for 6 years at 7% p.a.

12. £5000 invested for 5 years at 11% p.a.

Depreciation

A business will own a number of assets such as typewriters, machinery, motor transport, etc. These assets reduce in value as time goes on, i.e. they depreciate in value.

The depreciation formula is very similar to the compound interest formula and it is

$$A = P\left(1 - \frac{R}{100}\right)^T$$

where A stands for the value after T years, P stands for the initial cost, R stands for the rate of depreciation p.a.

Example 7

A company buys office machinery which costs £8000. The depreciation is calculated at an annual rate of 15%. What will be the value of the machinery at the end of 7 years?

We are given $P = £8000$, $R = 15\%$ and $T = 7$ years.

$$A = P\left(1 - \frac{R}{100}\right)^T$$

$$= 8000 \times \left(1 - \frac{15}{100}\right)^7$$

$$= 8000 \times (1 - 0.15)^7$$

$$= 8000 \times 0.85^7$$

$$= 2565$$

So the machinery will be worth £2565 at the end of 7 years.

Exercise 9.3

1. A business buys new machinery at a cost of £9000. It decides to calculate the depreciation at an annual rate of 25%. Work out the value of the machinery at the end of 3 years.

2. The value of a lathe depreciates at 18% p.a. If it cost £5000 when new, calculate its value after 5 years.

3. A car depreciates in value at an annual rate of 10%. If it cost £7000 when new, calculate its value after 7 years.

4. A lorry cost £20 000 when new. It depreciates in value at an annual rate of 12%. Estimate its value after 6 years.

5. A machine costs £18 000 when new but it depreciates at an annual rate of 15%. Work out its value at the end of 10 years.

Miscellaneous Exercise 9

Section A

1. Work out the simple interest on £3000 invested for 1 year at 11% p.a.

2. A man invests £8000 for 5 years at 10% p.a. compound interest. How much will the investment be worth at the end of this period?

3. A sum of money was invested at 12% p.a. simple interest. After 1 year the amount of interest was £600. How much money was invested?

4. It is estimated that a machine costing £20 000 has a life of 10 years. It is decided to calculate the depreciation at 12% p.a. How much is the machine worth at the end of the 10 years?

5. The value of a car, after each year's use decreases in value by a fixed percentage of its value at the beginning of the year. If a car costs £5000 when new and its value after one year was £4160, by what percentage has the value decreased? Calculate its value after a further year's use.

Section B

1. My grandfather left me £20 000 to be kept in trust for 5 years at 11% simple interest p.a. What will the money be worth at the end of the 5 years?

2. A person invests £6690 at 9% simple interest per year. How long will it be before investment is worth £9000 or more?

3. A woman invested £7000 in a building society at 8% p.a. compound interest. How much will her investment be worth at the end of 6 years?

4. A car, which cost £8000 when new, depreciates annually by 20% of its value at the beginning of the year. Estimate the value of the car at the end of 4 years.

5. A woman invested £5000 for one year in each of the following:

   (a) a building society paying 8.25% p.a. free of income tax

   (b) a municipal bond paying interest of 11% p.a. on which income tax at 30% is payable.

   Work out the net income from both investments and find which investment realises the greater net income.

Multi-Choice Questions 9

1. £500 invested for 2 years at 10% compound interest p.a. becomes

   A £600   B £605

   C £665.50   D £700

2. The value of a machine depreciates in value by 10% of its cost price each year. It cost £4000 when new. Its value after 4 years will be

   A nothing   B £1067

   C £2624   D £6034

3. A sum of £800 amounted to £815 after 9 months at simple interest. What rate of interest was earned?

   A $\frac{5}{24}$%   B $2\frac{1}{2}$%   C $13\frac{1}{3}$%   D 20%

4. The value of a bicycle depreciates by 10% of its value each year. If the price was $500 when new, what will be its value at the end of 2 years?

   A $490   B $480

   C $405   D $400

5. £2000 invested for 2 years at 10% compound interest becomes

   A £2400   B £2420

   C £2662   D £6000

Business Calculations

Profit and Loss

When a dealer buys or sells goods the cost price is the price at which the goods are bought. The selling price is the price for which the goods are sold.

$$\text{Profit} = \text{Selling price} - \text{Cost price}$$

The percentage profit is usually calculated on the cost price.

$$\text{Profit \%} = \frac{\text{Profit}}{\text{Cost price}} \times 100$$

If a loss is made the cost price is greater than the selling price. That is

$$\text{Loss} = \text{Cost price} - \text{Selling price}$$

As with the profit per cent, the loss per cent is usually calculated on the cost price.

$$\text{Loss \%} = \frac{\text{Loss}}{\text{Cost price}} \times 100$$

Example 1

(a) A shopkeeper buys goods for £30 and sells them for £36. Calculate his percentage profit.

We are given that the cost price = £30 whilst the selling price is £36.

$$\text{Profit \%} = \frac{36 - 30}{30} \times 100$$

$$= \frac{6}{30} \times 100$$

$$= 20\%$$

(b) A lady buys a car for £4000 and sells it for £2500. What is her percentage loss?

We are given that the cost price is £4000 whilst the selling price is £2500.

$$\text{Loss \%} = \frac{4000 - 2500}{4000} \times 100$$

$$= \frac{1500}{4000} \times 100$$

$$= 37.5\%$$

A dealer usually wants to make a certain percentage profit. To do this the selling price must be calculated. He can do this by letting the cost price be 100%. Then to find the selling price the percentage profit is added on.

Example 2

(a) A dealer buys an article for £50. He wishes to make a profit of 30%. Work out his selling price.

$$\text{Cost price} = £50$$

$$= 100\%$$

$$\text{Selling price} = 130\%$$

$$= \frac{130}{100} \times £50$$

$$= £65$$

(b) A greengrocer sells a bag of carrots for £14 thereby making a profit of 40%. Calculate the cost price of the carrots.

$$\text{Selling price} = 140\%$$

$$= £14$$

$$\text{Cost price} = 100\%$$

$$= \frac{100}{140} \times £14$$

$$= £10$$

Discount

Sometimes a dealer will deduct a percentage of the selling price if the customer is prepared to pay cash. This is called discount.

Example 3

A television set is offered for sale at £360. A customer is offered 5% discount for cash. How much does the customer actually pay?

$$\text{Discount} = 5\% \text{ of } £360$$

$$= £18$$

$$\text{Amount paid} = £360 - £18$$

$$= £342$$

Alternatively

Since only 95% of the selling price is paid,

$$\text{Amount paid} = \frac{95}{100} \times £360$$

$$= £342$$

Sometimes discounts are quoted as so much in the pound, for instance, 8p in the £1. Of course, 8p in the pound is the same as 8% discount. The calculation is then the same as that in Example 3.

Example 4

During a sale a shop offers a discount of 8p in the £1. How much will a young woman pay for a dress marked for sale at £25?

Since 8p in the £1 is the same as 8% discount,

$$\text{Discount} = 8\% \text{ of } £25$$

$$= £2$$

$$\text{Amount paid} = £25 - £2$$

$$= £23$$

Alternatively

Since only 92% of the selling price is paid,

$$\text{Amount paid} = \frac{92}{100} \times £25$$

$$= £23$$

Value Added Tax

Value added tax or VAT for short is a tax on goods and services which are purchased. Some goods and services bear no tax, for instance, food, books and insurance. The rate varies from time to time but at the time of writing it is 15%.

Example 5

(a) A woman buys a wardrobe which is priced at £200 plus VAT at 15%. How much does she actually pay for the wardrobe?

$$\text{VAT} = 15\% \text{ of } £200$$

$$- £30$$

$$\text{Amount paid} = £200 + £30$$

$$= £230$$

(b) A woman buys a table for £138, the price including VAT at 15%. Work out the price of the table excluding VAT.

Let   100% = price exclusive of VAT

then   115% = price inclusive of VAT

$$\text{Price exclusive of VAT} = \frac{100}{115} \times £138$$

$$= £120$$

Exercise 10.1

1. A dealer buys an article for £8 and sells it for £10. Calculate the percentage profit.

2. A shopkeeper buys ballpoint pens for 30p each and sells them for 40p. Calculate his percentage profit.

3. In a clearance sale, a coat bought for £75 is sold for £60. What is the percentage loss?

4. A greengrocer buys potatoes at £5.90 per 50 kg bag. He sells them for 14p per kilogram. What is his percentage profit?

5. A shopkeeper buys an article for £80. He wishes to make a profit of 25%. What should his selling price be?

6. A furniture store sells a table for £75, thereby making a profit of 25%. Work out the cost price of the table.

7. During a sale a dress shop offers a discount of 10% for cash. How much will a customer pay for a coat marked at £50?

8. A multiple store offers a discount of 5p in the £1 on all their furniture. How much will a customer pay for a chair marked for sale at £110?

9. A shop selling electrical goods offers a discount of 7p in the £1. A cash paying customer buys a vacuum cleaner marked for sale at £65. How much discount will be allowed on this item?

10. A furniture store offers 7% discount for cash. How much will a cash-paying customer actually pay for a suite of furniture marked for sale at £1100?

11. A householder buys a washing machine. The price is £240 plus VAT at 15%. Work out the price the householder actually pays.

12. An armchair is priced at £120 exclusive of VAT. If VAT is levied at 15%, find the amount a customer will actually pay for the chair.

13. A lawnmower is priced at £60 inclusive of VAT at 15%. What is its price exclusive of VAT?

14. A telephone bill including VAT at 15% is £57.50. How much was the bill before VAT was added?

Rates

Every property in a town or city is given a rateable value which is fixed by the local district valuer. The rateable value depends upon the size, condition and location of the property.

The rates of a district are levied at so much in the pound of rateable value, for example, 95p in the £1. The money raised by the rates is used to pay for such things as libraries, the police, education, etc.

Annual rates = Rateable value × Rate in £1

Example 6

(a) A house has a rateable value of £300 and the rates are levied at 130p in the £1. What are the annual rates?

Annual rates = 300 × 130p

= £390

(b) A householder pays £270 per year in rates on a property which has a rateable value of £200. What is the local rate?

$$\text{Local rates} = \frac{\text{Annual rates}}{\text{Rateable value}}$$

$$= \frac{270}{200}$$

$$= \text{£}1.35$$

Hence the local rates are £1.35 in the £1.

Most local authorities state the product of a penny rate on their rates demand. This is the amount of money that would be raised if the rates were levied at 1p in the £1.

Example 7

The total rateable value for a city is £24 000 000.

(a) Calculate the product of a penny rate.

(b) If the city council wish to raise £36 000 000 what should be charged in rates?

(a) Product of
penny rate $= 24\,000\,000 \times 1\text{p}$

$$= 24\,000\,000 \times \text{£}0.01$$

$$= \text{£}240\,000$$

(b) Rate in £1 $= \dfrac{\text{Amount to be raised}}{\text{Product of penny rate}}$

$$= \frac{36\,000\,000}{240\,000}$$

$$= 150\text{p}$$

Hence the rates must be levied at 150p in the £1, i.e. at £1.50 in the £1.

Note that the same result could have been obtained by dividing the amount to be raised by the total rateable value, i.e.

$$\text{Rate in £1} = \frac{36\,000\,000}{24\,000\,000}$$

$$= \text{£}1.50$$

Exercise 10.2

1. The rateable value of a house is £200. If the rate is levied at 95p in the £1, how much must the householder pay in rates for the year?

2. The rateable value of a small house is £185 and the rate is £1.30 in the £1. Calculate the annual rates.

3. A householder pays £200 in rates per annum. If the rateable value of her house is £160, work out the rate in the £1.

4. The total rateable value of all the property in a town is £8 500 000.

   (a) Work out the product of a penny rate.

   (b) If the town council need to raise £11 900 000, find the rates which must be levied by the council.

5. In a city the cost of highways is covered by a rate of 19.3p in the £1. If the rateable value of all the property in the town is £20 000 000, find the amount of money available for spending on highways.

6. The total rateable value for a town is £5 600 000. Find the cost of the libraries if a rate of 5.6p in the £1 must be levied for this purpose.

7. Work out the total income from the rates of a town of rateable value £4 300 000 when the rates are levied at £1.08 in the £1.

8. The expenditure of a town is £9 000 000 and its rates are 87p in the £1. The cost of its public park is £300 000. What rate in the £1 is needed for the upkeep of the park?

Insurance

Our future is something which is far from certain. We could become too ill to work or we could be badly injured or even killed in an accident. Our house could be burgled or burnt down. We might be involved in a car accident and be liable for injuries and damage. How do we take care of such eventualities?

The answer is to take out insurance policies. The insurance company collects premiums from thousands of people who wish to insure themselves. It invests the money to earn interest and this money (i.e. premiums + interest) is then available to pay policy holders who make a claim.

Example 8

A householder values his house and its contents at £35 000. His insurance company charges a premium of £3 per £1000 insured. How much is the annual premium?

$$\text{Annual premium} = \frac{3 \times 35\,000}{1000}$$

$$= £105$$

Car Insurance

By law a vehicle must be insured, and the owner can be prosecuted if this is not done. There are two ways of insuring a vehicle:

(1) By third-party insurance. This covers only the other person who is involved in an accident which is the fault of the policy holder. It does not cover the policy holder who will have to pay himself for any damage to his car.

(2) By a fully comprehensive policy. This covers damage to the policy holder's car plus damage to other people and their property.

If a car owner makes no claims during a year he gets a bonus (called a no-claims bonus) which means that he or she will pay a smaller premium next year.

Example 9

The premium for a fully-comprehensive policy on a car is £180. The owner is allowed a no-claims bonus of 40%. How much is the premium for the year?

$$\text{No-claims bonus} = 40\% \text{ of } £180$$

$$= £72$$

$$\text{Premium payable} = £180 - £72$$

$$= £108$$

Cost of Running a Car

When calculating the cost of running a car or motor cycle, the costs of tax, insurance, petrol, maintenance and depreciation should be taken into account.

Every motor vehicle requires a Road Fund Licence which has to be paid when the vehicle is first registered and renewed periodically. The vehicle will need insuring and maintaining and its value will depreciate with age.

Example 10

A car is bought for £7200 and used for one year. It is then sold for £5200. During this year it did 20 000 km averaging 12 km per litre of petrol which cost 45p per litre. Insurance cost £87, tax £100 and maintenance £350. Calculate the cost of a year's motoring and the cost per kilometre of running the car.

$$\text{Cost of petrol} = \pounds\frac{20\,000}{12} \times \frac{45}{100}$$

$$= \pounds750$$

$$\text{Depreciation} = \pounds2000$$

$$\text{Insurance} = \pounds87$$

$$\text{Tax} = \pounds100$$

$$\text{Maintenance} = \pounds350$$

$$\text{Total cost for the year} = \pounds3287$$

$$\text{Cost per kilometre} = \frac{3287 \times 100}{20\,000}$$

$$= 16.4\text{p}$$

Life Assurance

With this type of assurance a sum of money, depending upon the size of the premium etc., is paid to the dependants (wife or husband and children) of the policy holder upon his or her death. The size of the premium depends upon:

(1) the age of the person (the younger the policy holder is, the less he or she pays because there is less risk of him or her dying soon)

(2) the amount of money the policy holder wants his dependants to receive (the greater the amount the larger the premiums).

Example 11

A man aged 35 years wishes to assure his life for £7500. The insurance company quotes an annual premium of £13.90 per £1000 assured. Work out the amount of the monthly premiums.

$$\text{Annual premium} = \pounds13.90 \times \frac{7500}{1000}$$

$$= \pounds104.25$$

$$\text{Monthly premium} = \pounds104.25 \div 12$$

$$= \pounds8.69$$

Endowment Assurance

This is very similar to life assurance but the person can decide for how long he or she is going to pay the premiums. At the end of the chosen period a certain sum of money will be paid to the policy holder. If, however, the policy holder dies before the end of the chosen period, the assured sum of money is paid to his or her dependants.

Some endowment and life policies are 'with profits', which means that the sum assured may increase over the period depending upon the profits made by the insurance company.

Example 12

A man aged 30 years wishes to buy an endowment 'with profits' assurance policy. He is quoted a price of £48.50 per £1000 assured over a period of 20 years. Calculate the monthly premiums if he wishes to assure himself for £6000.

$$\text{Annual premiums} = \pounds48.50 \times \frac{6000}{1000}$$

$$= \pounds291.00$$

$$\text{Monthly premiums} = \pounds291.00 \div 12$$

$$= \pounds24.25$$

Exercise 10.3

1. A householder wishes to insure his house for £30 000. The insurance company charges a premium of 15 p per £100 insured. Calculate the householder's annual premium.

2. An insurance company offers the following rates to customers: buildings £1.25 per £1000 insured; contents 25 p per £100 insured. Calculate the annual premiums to be paid by a householder who insured his house for £35 000 and the contents for £8500.

3. The insurance on a car costs £160 but there is a 20% no-claims bonus. How much is the annual premium?

4. A man aged 25 years wishes to assure his life for £7000. He is quoted a rate of £9.00 per £1000 assured. How much in premiums will he pay monthly?

5. A car owner is quoted an annual premium of £180 but she is allowed a no-claims bonus of 35%. Calculate the amount of her annual premium.

6. A person aged 36 is quoted a rate of £13.90 per £1000 assured per annum. He wishes to assure his life for £12 000 but wishes to pay monthly. The company states that for monthly payments the premiums will be increased by 3%. How much per month will the policy holder actually pay?

7. For a monthly premium of £50 a woman aged 40 is guaranteed £6624 at the end of ten years. However, the insurance company states that the maturity value of the policy is likely to be £11 241.

   (a) Work out how much the woman will pay in premiums over the ten-year period.

   (b) Calculate the amount of profit she is likely to make on the policy.

8. A man drives 15 000 km in a year. If he averages 10 km per litre of petrol costing 40 p per litre, how much does the petrol cost him?

9. A second-hand car is bought for £1800. It is run for a year and sold for £1280. The cost of insurance is £126.40, tax £100, repairs £152 and maintenance £32. It is driven 12 000 km in the year and it averages 11 km per litre of petrol costing 42 p per litre.

   (a) What is the total cost of a year's motoring?

   (b) How much is the cost per kilometre?

10. A new car is bought for £6000 and it is sold two years later for £3800. During this time it travelled 22 000 miles and it averages 35 miles per gallon of petrol. If petrol costs £1.76 per gallon and other expenses were £420, calculate the cost per mile of running the car.

Miscellaneous Exercise 10

Section A

1. The rateable value of Mr Richards' house is £280. He pays rates at 93 p in the £1. How much does Mr Richards pay in rates for the year?

2. A car owner is quoted a premium of £120 per annum but he is allowed a no-claims bonus of 40%. How much does he actually pay?

3. Given that the value added tax on goods bought is 15% of their value, calculate the value added tax payable on goods valued at £90.

4. By selling a chair for £31.05 a shop-keeper makes a profit of 15% on his cost price. Calculate his profit.

5. By selling an article for £18 a dealer makes a profit of 44% on his cost price. At what price must it be sold if the profit is to be 40%?

6. A woman sells her car for £2430 and as a result loses 10% of the price she paid for it. How much did she pay for the car?

7. A city treasurer estimated that the city would need a rate of 82.5 p in the £1 in order to provide revenue of £68 887 500. He estimated that the cost to the rates of family and community services would be 8.6 p in the £1. Calculate:

   (a) the product of a penny rate

   (b) the amount of money to be raised for family and community services.

8. Find the rateable value of a house on which the amount paid in rates was £192 when the rate was £1.60 in the £1.

9. The price of a coat was £68.25 but a 5% discount was allowed during a sale. How much will the coat actually cost a customer?

10. A man aged 45 years takes out a 'with profits' insurance policy for £5000. He is charged a premium of 45p per £100 assured. He pays the premiums monthly. How much does he pay per month?

Section B

1. An agent buys 3600 articles for £2000. He sells 3060 of them at a profit of 30% and the remainder at a loss of 10%. Find the amount of his profit and express it as a percentage of his outlay.

2. Articles are purchased at £5 per 100 and sold at 7p each. Calculate the profit as a percentage of the cost price.

3. Given that the value added tax on goods supplied is 15% of their value, calculate the value added tax on goods valued at £128.

4. In a sale, a discount of 17% was allowed on the marked price of a bed. A customer paid £49.80 for the bed. What was its price before the reduction?

5. A man sells his car for £1620 and as a result loses 10% of the price he paid for it. What price did he pay for it?

6. (a) A town council decided to increase its rates by 3p in the pound. If the total rateable value of the town is £820 000 find the extra amount raised by the increase.

(b) A householder, living in the town, finds that as a result of the increase her rates go up from £110.88 to £115.20. Calculate:

   (i) the rateable value of her house.
   (ii) the number of pence in the £1 at which the new rate is charged.

7. The premium charged by an insurance company is at the following rates:

   Buildings — 15p per £100 p.a.

   Contents — 25p per £100 p.a.

How much is the premium paid by a man who insures his house for £33 600 and the contents for £8400?

8. A man bought a car for £5200 and sold it one year later for £4000. During that year he travelled 18 000 miles in it.

(a) If its fuel consumption was 28 miles per gallon, work out the amount of petrol, to the nearest number of gallons, used during the year.

(b) If the petrol cost £1.84 per gallon, how much did the petrol cost?

(c) If the other expenses of running the car were £740, work out the cost per mile of running the car.

9. A greengrocer buys 500 lb of strawberries for £150. He begins selling them at 60p per pound but when 300 lb have been sold he reduces the price to 40p per pound. He sells another 150 lb at this price. The remainder go bad and have to be thrown away.

(a) Work out the amount of money obtained by selling the strawberries.

(b) Calculate the percentage profit on his outlay.

10. A radio is advertised in a shop as having a list price of £70 plus VAT at 15%. The shopkeeper offers a discount of 18% before adding on the VAT.

Calculate:

(a) the list price including VAT

(b) the amount of discount before VAT is added

(c) the reduced final price of the radio to the nearest penny.

Multi-Choice Questions 10

1. The rateable value of a house is £150. If the rates are 80p in the £1 how much is paid in rates for the year?

    A £12   B £120

    C £140   D £150

2. In a town the total rateable value of all the property is £8 000 000. What is the product of a penny rate?

    A £8000   B £16 000

    C £20 000   D £80 000

3. The cost of highways in a city is equivalent to 6.2p in the £1. If the rateable value of all the property in the town is £3 500 000, what is the cost of highways?

    A £21 700   B £22 000

    C £217 000   D £220 000

4. The expenditure of a town is £300 000 and its rates are levied at 75p in the £1. The cost of the library is £20 000 per year. The rate needed for the upkeep of the library is

    A 4p in the £1   B 5p in the £1

    C 7.5p in the £1   D 8p in the £1

5. A car owner is quoted a premium of £99.60 to insure her car. If her no-claims bonus is 40%, how much does she pay?

    A £39.84   B £59.76

    C £99.60   D £139.44

6. When a dealer sells an article for £18 he makes a profit of £3. His percentage profit is

    A $14\frac{1}{2}$ B $16\frac{2}{3}$ C 20   D 25

7. When an article is sold for £180 a loss of 10% on its cost price is recorded. What is its cost price?

    A £162 B £190 C £198 D £200

8. During a sale a shop reduced the price of everything by 10%. What was the sale price of an article originally priced at £4.30?

    A £3.40   B £3.87

    C £3.97   D £4.73

Mental Test 10

Try to answer the following questions without writing anything down except the answer.

1. The cost price of an article was £18 and the selling price was £20. What was the profit?

2. The cost of a car was £5000 and it was sold for £4500. Calculate the loss.

3. If the profit on an article was £10 and the cost price was £50, what was the percentage profit?

4. A shopkeeper offers a discount of 20% on a suit of clothes. A suit is marked at £50. How much is the discount?

5. VAT is charged at 15%. How much VAT is payable on goods valued at £200?

6. A house has a rateable value of £200 and the rate is levied at 80p in the £1. How much must the householder pay in rates?

7. All the property in a small town has a rateable value of £500 000. What is the product of a penny rate?

8. A woman wishes to insure the contents of her house for £5000. The insurance company quotes a premium of £4 per £1000 insured. How much is her annual premium?

9. A woman is quoted a premium of £200 to insure her car but she is allowed a 20% no-claims bonus. How much does she actually pay?

10. VAT is charged at 15%. The price of a washing machine is £200 plus VAT. How much will a customer actually pay?

11 **Household Finance**

Rent

Rent is a charge for accommodation. It is usually paid weekly or monthly to the landlord who owns the property. The biggest owners of rented property are the local authorities.

The landlord may include the rates in his charge for accommodation but if he does not it is the tenants' responsibility to see that these are paid.

Example 1

A landlord charges rent of £17.50 per week and the rates are £101.40 per annum. What weekly inclusive amount for rent and rates should the landlord charge?

$$\text{Rates per week} = £101.40 \div 52$$
$$= £1.95$$
$$\text{Inclusive charge} = £17.50 + £1.95$$
$$= £19.45$$

Mortgages

If someone buying a house cannot pay outright, he or she arranges a loan, called a mortgage, from a building society. The society generally requires a deposit of 5% or 10% of the purchase price of the property, although sometimes 100% mortgages are available.

The balance of the loan plus interest is paid back over a number of years, perhaps as many as 20 or 25. The interest rates charged by the building society vary from time to time. Prior to 1983, the borrower paid the building society the full amount of interest. He then received tax relief on this interest from the Inland Revenue. Now, however, under a scheme called Mortgage Interest Relief At Source (MIRAS for short) the building society deducts the tax relief from the monthly repayments. So the borrower pays less to the building society but more in income tax, the decrease in mortgage repayments and the extra income tax being equal.

Example 2

(a) A man wishes to buy a house costing £32 000. He pays a deposit of 10% to the building society. How much mortgage will he require?

Since a deposit of 10% is paid the mortgage required is 90% of the purchase price of the house.

$$\text{Mortgage required} = 90\% \text{ of } £32\,000$$
$$= £28\,800$$

(b) A building society quotes the repayment on a mortgage as £12.96 per month per £1000 borrowed. Work out the monthly repayments on a mortgage of £25 000.

$$\text{Monthly repayments} = \frac{25\,000}{1000} \times £12.96$$
$$= £324.00$$

Sometimes a combined mortgage and endowment assurance can be arranged. An endowment policy (see Chapter 10) is taken out for the period of the loan. Interest is paid on the amount of the loan for the whole period of the endowment, after which the money received from the insurance policy is used to pay off the loan.

Example 3

A man wishes to raise a mortgage of £20 000 to buy a house. The loan is to be covered by an endowment assurance policy for a 20-year period which costs £3.88 per £1000 assured per month. The building society charges interest at 13% but tax relief under MIRAS at 30% is deducted by the society.

Calculate the monthly repayments to be paid by the borrower.

$$\text{Annual interest} = 13\% \text{ of } £20\,000$$

$$= £2600$$

$$\text{Tax relief on interest} = 30\% \text{ of } £2600$$

$$= £780$$

$$\begin{aligned}\text{Annual amount paid} \\ \text{to society}\end{aligned} = £2600 - £780$$

$$= £1820$$

$$\begin{aligned}\text{Monthly repayments} \\ \text{to society}\end{aligned} = £1820 \div 12$$

$$= £151.67$$

$$\text{Insurance premiums} = \frac{20\,000}{1000} \times £3.88$$

$$= £77.60 \text{ per month}$$

$$\begin{aligned}\text{Total monthly} \\ \text{outgoings}\end{aligned} = £151.67 + £77.60$$

$$= £229.27$$

Exercise 11.1

1. A house is rented at £29.44 per week and the annual rates are £240. Work out the weekly charge for rent and rates.

2. A house is rented for £25.50 per week. The rateable value is £348 and the rates are levied at £1.08 in the £1. Calculate

   (a) the annual amount to be paid in rates

   (b) the weekly charge for rent and rates.

3. A flat is rented for £39.50 per week. The rates are £194.40 p.a. Work out the weekly charge for rent and rates.

4. A man borrows £24 000 from a building society in order to buy a house. The society charges £12.20 per month per £1000 borrowed. How much are the monthly repayments?

5. A person borrows £16 000 from a building society to buy a bungalow. The loan is covered by an insurance policy for the ten-year period of the loan, the premiums being £7.95 per month per £1000 assured. The building society charges interest at 12% p.a. but deducts under MIRAS tax relief at 30%. Work out the total monthly outgoings of the borrower.

6. A woman wants to buy a house costing £35 000. She pays a deposit of 20% to the building society.

   (a) What mortgage does she need?

   (b) If the building society charges £15.60 per month per £1000 borrowed, work out the amount of her monthly repayments.

7. A man took out a mortgage for £24 000 when the building society interest rate was 8.5% p.a. If the rate is now 11% p.a. work out the increase in his annual interest payments to the building society.

8. A person borrows £20 000 from a building society to buy a house, the loan being covered by an endowment policy for the ten-year period which costs £7.95 per month per £1000 insured. The building society charges interest at 11% per annum but deducts, under MIRAS, tax relief at 29% on the interest. Work out the monthly outgoings of the borrower.

Hire-Purchase

When goods are purchased and paid for by instalments, they are said to be purchased by hire-purchase.

Usually the purchaser pays a deposit. The remainder of the purchase price (called the balance) plus interest is repaid in a number of instalments.

Example 4

A woman buys furniture for £280. She pays a deposit of 25% and interest at 20% is paid on the outstanding balance. She pays back the balance plus interest in twelve monthly instalments. Calculate the amount of each instalment.

$$\text{Deposit} = 25\% \text{ of } £280$$
$$= £70$$
$$\text{Balance} = £280 - £70$$
$$= £210$$
$$\text{Interest} = 20\% \text{ of } £210$$
$$= £42$$
$$\text{Total to be repaid} = £210 + £42$$
$$= £252$$
$$\text{Monthly repayments} = £252 \div 12$$
$$= £21$$

In Example 4, 20% would only be the true rate of interest if all of the outstanding balance was paid at the end of the year. However, as each instalment is paid the amount outstanding is reduced and hence a large proportion of each successive payment is interest. The true rate of interest, called the annual percentage rate or APR for short, is in fact about 38%.

Example 5

A woman buys a table for £64 on hire-purchase. She pays no deposit but repays the loan plus interest at 10% for the whole period of the loan in four quarterly instalments. What is the true rate of interest (i.e. what is the APR)?

$$\text{Interest} = 10\% \text{ of } £64$$
$$= £6.40$$
$$\text{Total to be repaid} = £64 + £6.40$$
$$= £70.40$$
$$\text{Amount of each instalment} = £70.40 \div 4$$
$$= £17.60$$

£64.00 is the balance outstanding for the first 3 months

£48.00 is the balance outstanding for the next 3 months

£32.00 is the balance outstanding for the next 3 months

£16.00 is the balance outstanding for the final 3 months

$$\begin{aligned}
\text{Average amount} \\
\text{of loan for} \\
\text{entire year} &= £(64 + 48 + 32 + 16) \div 4 \\
&= £40
\end{aligned}$$

As £6.40 is the interest paid on an actual loan of £40, the true rate of interest may be calculated by using the simple interest formula:

$$R = \frac{100I}{PT}$$
$$= \frac{100 \times 6.40}{40 \times 1}$$
$$= 16\%$$

Thus APR = 16%.

Note that in calculating outstanding balance we have said that the interest payable every 3 months is £6.40 ÷ 4 = £1.60. Hence the amount actually paid off the balance is £17.60 − £1.60 = £16. The method shown in Example 5 is called the average loan method.

Bank Loans

Many people take out personal loans from a bank. The bank will calculate the interest for the whole period of the loan. The loan plus the interest is usually repaid in equal monthly instalments.

Example 6

A man borrows £300 from a bank. The bank charges 18% interest for the whole period of the loan. If the repayments are in twelve equal monthly instalments, calculate the amount of each instalment.

$$\text{Interest} = 18\% \text{ of } £300$$
$$= £54$$

$$\text{Total amount to be repaid} = £300 + £54$$
$$= £354$$

$$\text{Amount of each instalment} = £354 ÷ 12$$
$$= £29.50$$

Exercise 11.2

1. The hire-purchase repayments on a record player are £5 per month. How much will be paid at the end of two years.

2. A woman buys a washing machine for £150. She pays a deposit of £30 and she is charged 10% interest on the balance.

   (a) How much is the balance?

   (b) Calculate the amount of interest payable.

   (c) Work out the balance plus interest.

   (d) If the balance plus interest is to be repaid in ten monthly instalments, work out the amount of each instalment.

3. A young man buys a motor cycle for £640. A deposit of 20% is paid and interest at 12% p.a. is charged on the balance for the full period of the loan. The balance plus interest is to be repaid in four quarterly instalments.

   (a) Work out the amount of the deposit.

   (b) How much is the balance?

   (c) Calculate the interest payable.

   (d) Work out the total amount to be repaid.

   (e) What is the amount of each instalment?

4. The cash price of a table is £200. To buy it on hire-purchase a deposit of £50 is payable plus ten equal instalments of £21.

   (a) Work out the amount paid for the table when it is bought on hire-purchase.

   (b) Calculate the difference between the cash and hire-purchase prices.

5. A woman buys a refrigerator for £240. She pays no deposit and is charged interest at 10% p.a. for the full period of the loan. The loan plus interest is to be repaid in four equal instalments. Work out

   (a) the amount of interest payable

   (b) the total amount to be repaid

   (c) the amount of each quarterly instalment

   (d) the average amount of the loan for the whole year

   (e) the true rate of interest charged on the loan.

6. A woman buys a chest of drawers priced at £100. She pays no deposit but agrees to repay the £100 plus interest at 20% over the entire period of the loan. If she pays two half-yearly instalments, find the amount of each instalment and the true rate of interest.

7. A man borrows £2500 from a bank at 15% interest over the whole period of the loan. If the loan plus interest is to be repaid in twelve equal monthly instalments, calculate the amount of each instalment to the nearest £1.

8. A bank lends a man £5000. They charge simple interest at 12% p.a. on the loan of £5000.

   (a) Work out the amount of interest payable.

   (b) The loan plus interest is to be repaid in 24 equal monthly instalments. Work out the amount of each instalment.

Gas Bills

Gas is charged according to the number of therms used (1 therm = 100 cubic feet approximately, the exact value depending on the calorific (heating) value of the gas). In addition a standing charge may be made.

> Amount
> of gas bill = Standing charge
> + Number of therms used
> × Charge per therm

Example 7

A customer uses 90 therms of gas during a certain quarter. She is charged 27.216 p per therm plus a standing charge of £9.00. How much is her gas bill?

> Cost of gas used = 90 × 27.216 p
>
> = £24.49
>
> Total gas bill = £24.29 + £9.00
>
> = £29.50

Reading Gas Meter Dials

The amount of gas used is always measured in cubic feet (often in hundreds of cubic feet) and is then changed into therms. In the ordinary way your gas meter will be read for you at regular intervals by a representative of the gas board. However, there may be occasions when you wish to read it yourself to check your gas consumption.

Fig. 11.1 shows the dials on a gas meter. To read the dials copy down the readings in the order in which they appear. Note that the dials read alternately in opposite directions. Where the hand stands between two figures, write down the lower figure. However if the hand stands between 9 and 0 write down 9.

Fig. 11.1

Dial 1 pointer has passed 7, reading is 7   Dial 2 pointer has passed 5, reading is 5   Dial 3 pointer has passed 1, reading is 1   Dial 4 pointer has passed 9, reading is 9

The dials in the diagram show 7519. If the previous reading was 7491, the difference would be 7519 − 7491 = 28 hundred cubic feet = 28 therms approx. This is the gas used since the meter was last read.

Electricity Bills

Electricity is charged for according to the number of units used (1 unit = 1 kilowatt hour). In addition there is usually a fixed charge.

Example 8

A customer uses 1765 units of electricity during a quarter. If the standing charge is £6.84 and the price per unit is 5.37 p, work out the amount of the quarterly electricity bill.

1765 units @ 5.37 p per unit $= 1765 \times 5.37\,\text{p}$

$$= £94.78$$

Quarterly electricity bill $= £94.78 + £6.84$

$$= £101.62$$

Reading Electricity Meters

When reading an electricity meter, remember that alternate dials revolve in opposite directions. The five dials are read from left to right. The method of reading a meter is shown in Fig. 11.2.

Fig. 11.2

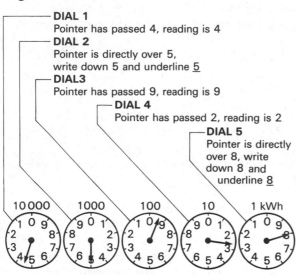

(1) Always note the number the pointer has passed. In Fig. 11.2, on the dial labelled 10 000 the pointer is between 4 and 5 so write down 4.

(2) If the pointer is directly over a figure, write down that figure and underline it. On the dial marked 1000 in the diagram, the pointer is over 5 so write 5.

(3) The reading of the dials in the diagram give 45 928.

(4) Now look at the underlined figures. If any of these is followed by a 9 reduce the underlined figure by 1. So the corrected reading is 44 928.

(5) When the meter reading has been worked out, subtract the previous reading to find the electricity consumption.

Example 9

Fig. 11.3 shows the dials of an electricity meter.

(a) Read the meter.

(b) If the previous meter reading was 53 432, work out the number of units used.

(c) If the charge per unit is 5.93 p and the standing charge is £9.80 per quarter, find the amount of the electricity bill.

Fig. 11.3

(a) The meter reads 56 378.

(b) Number of
units used $= 56\,378 - 53\,432$

$$= 2946$$

(c) Cost of
electricity $= 2946 \times £0.0593$
$$+ £9.80$$

$$= £184.49$$

Telephone Bills

Calls dialled are timed and charged in whole units of 5 p. The time allowed for each unit depends upon the destination dialled and the time of day. The table below will give some idea of how the units are timed.

		Time for 1 unit
Local calls	Cheap rate	8 min
	Standard rate	2 min
	Peak rate	1 min 30 s
Calls up to 35 km	Cheap rate	2 min
	Standard rate	40 s
	Peak rate	30 s
Calls up to 56 km	Cheap rate	1 min
	Standard rate	30 s
	Peak rate	22.5 s
Calls over 56 km	Cheap rate	48 s
	Standard rate	22.5 s
	Peak rate	17.1 s

The time when:

cheap rates apply is 6.00 p.m. to 8.00 a.m.

standard rates apply is 1.00 p.m. to 6.00 p.m.

peak rates apply is 8.00 a.m. to 1.00 p.m.

A charge is also made for the rent of the system, and VAT is charged on the total.

Example 10

(a) A telephone bill showed that a householder dialled 600 units in one quarter. Each unit was charged at 5p. The rental charge was £17.26 and VAT is levied at 15%. Work out the amount of the householder's telephone bill.

$$\text{Cost of 600 units at 5p} = £0.05 \times 600$$
$$= £30$$
$$\text{Total cost exclusive of VAT} = £30 + £17.26$$
$$= £47.26$$
$$\text{VAT at 15\%} = 15\% \text{ of } £47.26$$
$$= £7.09$$
$$\text{Total payable} = £47.26 + £7.09$$
$$= £54.35$$

(b) A telephone user made a local call lasting 6 min at peak period time. If the time for 1 unit was 1 min 30 s, find:

(i) the number of units used
(ii) the cost if each unit used is charged at 5p.

$$\text{(i)} \quad \text{Number of units used} = 6 \div 1\tfrac{1}{2}$$
$$= 4$$
$$\text{(ii)} \quad \text{Cost of the call} = 4 \times 5p$$
$$= 20p$$

Exercise 11.3

1. A customer uses 90 therms of gas during a quarter. She is charged 32.18p per therm plus a standing charge of £6.34. How much is her gas bill?

2. A householder receives a gas bill for £36 when gas is charged at 30p per therm. If his standing charge is £7.50, calculate the number of therms he used.

3. The present reading of a householder's gas meter is 1119 and the previous reading was 796 hundreds of cubic feet.

   (a) How many hundreds of cubic feet of gas have been used?

   (b) If 1 therm = 100 cubic feet, how many therms have been used?

   (c) If gas is charged at 39.0p per therm and the standing charge is £8.75, work out the householder's gas bill.

4. Copy the dials shown in Fig. 11.4 and put on hands to show:

   (a) 1700   (b) 9425
   (c) 4561

Fig. 11.4

5. Write down the readings on the gas meter dials shown in Fig. 11.5.

Fig. 11.5

6. Fig. 11.6 shows the dials of a gas meter.

    (a) Read the dials.

    (b) If the previous reading was 4392, find the number of hundreds of cubic feet.

    (c) Gas is charged at 38.3p per therm plus a standing charge of £7.90. Work out the amount of the gas bill.

Fig. 11.6

7. In a certain area electricity is charged at a fixed rate of 9.80p per unit of electricity used. A householder uses 780 units. How much will his electricity bill be?

8. A householder uses 2500 units of electricity in a quarter. Each unit used costs 5.72p and there is a standing charge of £9.75 per quarter. Work out the amount that the householder will be charged for that quarter.

9. The electricity meter reading on 15th February was 009 870 and on 17th May it was 009 986 units.

    (a) How many units of electricity have been used?

    (b) Each unit costs 5.93p and there is a standing charge of £8.70. Work out the amount of the electricity bill.

10. An economy tariff charges for electricity as follows: day rate 5.70p per unit, night rate 2.04p per unit; standing charge £2.66. 1325 units were used during the day and 932 units during the night. Calculate how much the electricity bill will be.

11. Copy the dials of the electricity meter shown in Fig. 11.7 and draw hands to show readings of:

    (a) 81 732   (b) 56 325 units.

Fig. 11.7

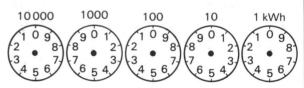

12. Read the electricity meter dials shown in Fig. 11.8.

Fig. 11.8

13. A householder receives an electricity bill for £77.60. Electricity is charged for at 5.70 p per unit and the standing charge is £9.20. Find the number of units of electricity used by the householder.

14. A telephone bill is made up of the following charges:

     Rental charge — £13.50 per quarter

     Dialled units — 603 at 5 p each

   In addition VAT is charged at 15%. Calculate the total amount of the telephone bill.

15. A telephone subscriber dialled a call to the Channel Islands which took 6 minutes to complete. If the time for 1 unit was 22.5 seconds, find:

     (a) the number of dialled units

     (b) the cost of the call if 1 unit is charged at 5 p.

16. A telephone meter was read on 17th February as 007 963 and on 15th May as 009 215.

     (a) How many units have been used during this period?

     (b) If each unit is charged at 5 p, calculate the cost of the units used.

     (c) The standing charge is £17.29. Work out the total bill, exclusive of VAT.

     (d) VAT is charged at 15% on the total bill. How much VAT is payable?

     (e) Calculate the total bill including VAT.

17. The telephone bill for a quarter was £47.00. Dialled units were charged at 5 p each and the standing charge was £18.40. In addition VAT at 15% was levied. How many dialled units were used?

Miscellaneous Exercise 11

Section A

1. A person borrows £5000 from a bank which charges interest at 15% p.a. for the whole period of the loan. If the loan plus interest is to be repaid in twelve equal monthly instalments, find:

     (a) the amount of interest payable

     (b) the amount of each instalment.

2. A man purchases a house using a combined endowment assurance and mortgage.

     (a) He borrows £18 000 from the building society which charges interest at 13% p.a. Calculate the amount of interest payable per annum.

     (b) The building society deducts, under MIRAS, tax relief at 30% on the interest. Work out the amount the borrower will actually pay to the society per annum.

     (c) The insurance premiums at £2.95 per month per £1000 assured. Work out the monthly premium.

     (d) Calculate the total monthly outgoings of the borrower.

3. A woman wants to buy a house for £30 000. She pays a deposit of 10% to the building society.

     (a) What mortgage does she require?

     (b) The monthly mortgage repayments are £11.60 per £1000 borrowed. How much will her monthly repayments be?

4. A suite of furniture is priced at £700. Hire-purchase terms are available which are: deposit 25% of cash price and 18 monthly instalments of £48.75.

     (a) Work out the amount of the deposit.

     (b) Calculate the total hire-purchase price.

     (c) Find the difference between the hire-purchase price and the cash price.

5. A small computer is priced at £300 plus VAT at 15%.

    (a) What is the total cost of the computer plus VAT?

    (b) Tindalls offer the computer at £30 off but Lyons offer it at 20% discount. Find the difference between these two offers.

6. Copy and complete the following telephone bill:

    Quarterly rental charge   £13.75

    Dialled units: 600 at 5p per unit   ....

    Total cost (excluding VAT)   ....

    VAT at 15%   ....

    Total payable   ....

7. **(a)** Fig. 11.9 shows the positions of the dials of an electricity meter at the beginning of a quarter. Write down the meter reading.

    (b) During the quarter 754 units of electricity were used. Make a sketch similar to Fig. 11.9 to show the meter reading at the end of the quarter.

Fig. 11.9

Section B

1. Calculate the monthly cost of owning a house given that the expenditure is as follows: mortgage repayments £80 per month; insurance £96 per year; rates £180 per half-year.

2. The price of electricity includes a standing quarterly charge of £8.00 plus a charge of 4.8p per unit used.

    (a) Find the cost of using 100 units per quarter. Work out the average cost per unit, i.e. (cost of 100 units + standing charge) ÷ 100.

    (b) Work out the cost of using 700 units per quarter and work out the average cost per unit when this number of units are used.

    (c) The electricity board reduces the standing charge to £6.00 but increases the cost of a unit to 5.6p per unit. A customer uses 700 units in a quarter. What is now the average price per unit and how much is the increase in the average price per unit?

3. A lady buys a suite of furniture for £2560. She pays a deposit of £320. Interest at 24% p.a. is to be paid on the outstanding balance for the full period of the agreement. If the loan plus interest is to be paid in four equal quarterly instalments, find the true rate of interest.

4. In a Commonwealth country electricity is charged over a 3 month period as follows:

Fixed charge	$9
First 50 units	11c per unit
Next 250 units	9c per unit
Over 300 units	7c per unit

    A discount of 10% is allowed on bills paid within 15 days.

    A householder uses 560 units of electricity in 3 months. If he pays the bill within 15 days, how much will he have to pay?

5. A user of electricity can use either Tariff 1 or Tariff 2 which are as follows:

   Tariff 1 (Two Part)

   A quarterly fixed charge of £8.76.

   Cost per unit 4.80 p.

   Tariff 2 (Night rate)

   A quarterly fixed charge of £10.35.

   Cost per unit used between 10.30 p.m. and 7.30 a.m. 2.40 p.

   Cost per unit used at other times 4.95 p.

   He estimates that he will use 2500 units in a quarter, 1200 of which are used at night. Which tariff should he use and how much money will he save by using the correct tariff?

6. A householder receives a gas bill for £136.44. She is charged 36.28 p per therm plus a fixed quarterly charge of £27.60. How many therms has she used?

7. A set of saucepans costs £69, the price including VAT which is levied at 15% of the value of the goods. Find the price before the VAT was added.

Multi-Choice Questions 11

1. A householder is charged 8p per unit of electricity used. He receives a bill for £48. How many units did he use?

   A 48   B 384   C 600   D 4800

2. A debt of £321.60 is to be paid in equal instalments of £40.20. How many instalments are needed?

   A 4   B 8   C 10   D 16

3. A telephone subscriber is charged a rental of £7.50 and 5p for each dialled unit used. If her telephone bill is £25, how many dialled units did she use?

   A 48   B 350   C 600   D 4800

4. Gas is charged at 57p per therm. If 560 therms are used in a quarter, how much is the quarterly bill?

   A £5.60   B £57
   C £319.20   D £560

5. Mortgages cost £15 per month per £1000 borrowed. The annual repayments on a mortgage of £20 000 are

   A £180   B £300
   C £2000   D £3600

6. A woman wishes to buy a house for £30 000. She pays a 10% deposit. What mortgage does she need?

   A £2700   B £3000
   C £27 000   D £30 000

7. A workbench is priced at £88. It can be bought on hire-purchase, the terms being £20 deposit plus 12 equal monthly payments of £6.40. The annual rate of interest is

   A 3.4%   B 5%   C 10%   D 12.9%

Mental Test 11

Try to answer the following questions without writing anything down except the answer.

1. A landlord charges an annual rent of £600. What is the monthly rent?

2. A building society charges £20 per month per £1000 borrowed. A person borrowed £20 000. What are her monthly repayments?

3. A woman buys furniture priced at £2000. She pays a deposit of 10%. What is her outstanding balance?

4. A man buys a motor car on hire-purchase. He pays a deposit of £100 and twenty instalments of £50. How much does he pay for his car?

5. A man borrows £500 from a bank which charges interest at 20% p.a. How much interest is charged if the loan is repaid in one year?

6. A housewife uses 100 therms of gas in a quarter which is charged at 36.2p per therm. Work out the amount of her gas bill.

7. A householder uses 1000 units of electricity during a quarter. The electricity board charges 6p per unit used. Find the cost of the electricity used.

8. A telephone call is made to New York. The time for 1 unit is 5 seconds. How many units are used if the call lasts 2 minutes?

Time, Distance and Speed

Measurement of Time

The units of time are:

$$60 \text{ seconds (s)} = 1 \text{ minute (min)}$$
$$60 \text{ minutes (min)} = 1 \text{ hour (h)}$$
$$24 \text{ hours (h)} = 1 \text{ day (d)}$$

Example 1

(a) Change 840 s into minutes.

$$840 \text{ s} = 840 \div 60 \text{ min}$$
$$= 14 \text{ min}$$

(b) Change 8 h into minutes.

$$8 \text{ h} = 8 \times 60 \text{ min}$$
$$= 480 \text{ min}$$

(c) How many seconds are there in 1 hour?

$$1 \text{ h} = 60 \times 60 \text{ s}$$
$$= 3600 \text{ s}$$

The Clock

There are two ways of showing the time:

(1) With the 12-hour clock (Fig. 12.1), in which there are two periods each of 12 hours' duration during each day.

The period between midnight and noon is called a.m. and the period between noon and midnight is called p.m. Thus 8.45 a.m. is a time in the morning whilst 8.45 p.m. is a time in the evening.

Fig. 12.1

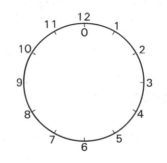

Example 2

Find the length of time between 2.15 a.m. and 8.30 p.m.

2.15 a.m. to 12 noon is	9 h 45 min
12 noon to 8.30 p.m. is	8 h 30 min
Total length of time is	18 h 15 min

(2) With a 24-hour clock (Fig. 12.2) in which there is one period of 24 hours. This clock is used for railway and airline timetables. Times between midnight and noon are stated as 0000 hours and 1200 hours whilst times between noon and midnight are stated between 1200 hours and 2400 hours.

Fig. 12.2

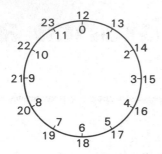

3.30 a.m. is written 0330 hours whilst 3.30 p.m. is written 1530 hours and this avoids confusion between the two times.

Note that the 24-hour clock times are written with four figures. 1528 hours should be read as 'fifteen twenty-eight hours' and 1300 as 'thirteen hundred hours'.

Example 3

Find the length of time between 0425 hours and 1812 hours.

1812 hours is	18 h 12 min
0425 hours is	4 h 25 min
Total length of time is	13 h 47 min

Exercise 12.1

1. How many minutes are there in
   (a) 4 hours   (b) 3.5 hours?

2. How many seconds are there in
   (a) 9 min   (b) 2.25 min
   (c) 3 hours?

3. Change 350 min into hours and minutes.

   Find the length of time, in hours and minutes, between

4. 2.39 a.m. and 8.54 p.m.

5. 2.15 a.m. and 7.30 p.m.

6. 0.36 a.m. and 1.15 p.m.

7. 0049 hours and 1236 hours.

8. 0242 hours and 1448 hours.

9. 0836 hours and 2115 hours.

10. 0425 hours and 1712 hours.

Timetables

Timetables are usually written using the 24-hour clock. The timetable on p. 97 is a typical bus timetable.

Example 4

If I catch the 0955 bus from Cheltenham, at what time do I arrive at Gotherington and how long does the journey take?

According to the timetable the 0955 bus arrives at Gotherington at 1018 hours.

1018 hours is	10 h 18 min a.m.
0955 hours is	9 h 55 min a.m.
Time taken is	0 h 23 min

Route Maps

Route maps are often associated with bus, rail and airline timetables. The one shown in Fig. 12.3 is typical for a rural bus service.

Exercise 12.2

Use the table opposite to answer Questions 1 to 5.

1. I can catch either the 0705 or the 0830 bus from Cheltenham to Alderton. Which is the faster bus? What is the difference in the time taken for the journey?

2. There is only one bus which leaves Woodmancote before 0900 hours for Teddington Kiosk. At what time does it leave Woodmancote and how long does the journey take?

3. I want to travel from Bishops Cleeve to Alstone.

   (a) How long does the 1315 bus take for the journey?

   (b) How long does the 1625 bus take?

   (c) What is the difference in the times taken?

4. I want to travel from Cleeve Estate to Alderton to arrive just before 1810 hours.

   (a) Which bus should I get from Cleeve?

   (b) At what time does it arrive at Alderton?

   (c) How long does the journey take?

5. (a) Which is the slowest bus from Cheltenham to Alderton?

   (b) How long does this journey take?

For questions 6, 7 and 8 use Fig. 12.3.

6. What is the number of the bus service from Stroud to Eastcombe?

7. I want to travel from Uplands to Chalford Hill. Write down the number of the bus services I must use.

8. A passenger wants to travel from East-combe to Aston Down. Write down the number of the services she must use. Where should she change buses?

CHELTENHAM · ALDERTON VIA BISHOPS CLEEVE

MONDAYS TO SATURDAYS

Cheltenham, Royal Well	0705	0735	0830	0930	0955	1020	1125	1200	1230	
Cleeve Estate	0717	0747	0842	0942	1007	1032	1137	1212	1242	
Woodmancote	0720	0750	0845	0945	1010	1035	1140	1215	1245	
Bishops Cleeve	0725	0755	0850	0950	1015	1040	1145	1220	1250	
Gotherington	0728	0758	0853	—	1018	1043	1148	—	1253	
Prescott	0732	—	—	—	—	1047	—	—	—	
Gretton	0735	—	—	—	—	1050	—	—	—	
Alderton	0740	—	0903	—	—	1055	—	—	—	
Alstone, Shelter	—	—	0909	—	—	—	—	—	—	
Teddington, Kiosk	—	—	0914	—	—	—	—	—	—	
Cheltenham, Royal Well	1255	1335	1455	1455	1605	1700	1730	1800	1930	2230
Cleeve Estate	1307	1347	1507	1507	1617	1712	1742	1812	1942	2242
Woodmancote	1310	1350	1510	1510	1620	1715	1745	1815	1945	2245
Bishops Cleeve	1315	1355	1515	1515	1625	1720	1750	1820	1950	2250
Gotherington	1318	1358	1518	1518	1628	1723	1753	1823	1953	2253
Prescott	—	1402	—	1521	1631	1726	1756	1826	—	
Gretton	1345	1405	—	1525	1635	1730	1800	1830	—	
Alderton	1340	—	—	1530	1640	1735	1805	1835	—	
Alstone, Shelter	1330	—	—	—	1645	—	—	1840	—	—
Teddington, Kiosk	1325	—	—	—	1650	—	—	1845	—	—

Fig. 12.3

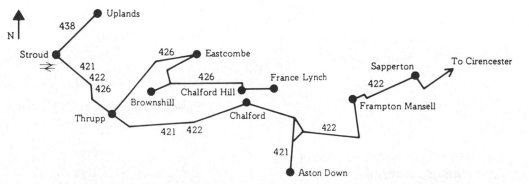

Average Speed

The road from Templeton to Myford is 50 miles long. It takes me two hours to drive from Templeton to Myford. Clearly I do not drive at the same speed all the time. Sometimes I drive slowly and sometimes I have to stop at crossroads and traffic lights.

However, the time it takes me is exactly the same as if I were driving along a straight road at 25 miles per hour, i.e. 50 miles ÷ 2 hours. So we say that my **average speed** is 25 miles per hour.

$$\text{Average speed} = \frac{\text{Distance travelled}}{\text{Time taken}}$$

The unit of speed depends upon the unit of distance and the unit of time. For instance, if the distance is measured in kilometres and the time in hours, the average speed will be measured in kilometres per hour. If the distance is measured in feet and the time in seconds then the average speed will be measured in feet per second.

The abbreviations for speed are as follows:

kilometres per hour: km/h or km h^{-1}
miles per hour: mile/h or m.p.h.
metres per second: m/s or m s^{-1}
feet per second: ft/s or ft s^{-1}

Note carefully that:

$$\text{Distance travelled} = \text{Average speed} \times \text{Time taken}$$

$$\text{Time taken} = \frac{\text{Distance travelled}}{\text{Average speed}}$$

Example 5

(a) A car travels a total distance of 200 km in 4 hours. What is its average speed?

$$\text{Average speed} = \frac{\text{Distance travelled}}{\text{Time taken}}$$

$$= \frac{200}{4}\text{km/h}$$

$$= 50\,\text{km/h}$$

(b) A lorry travels for 3 hours at an average speed of 35 miles per hour. How far does it travel in this time?

$$\text{Distance travelled} = \text{Average speed} \times \text{Time taken}$$

$$= 35 \times 3 \text{ miles}$$

$$= 105 \text{ miles}$$

(c) A train travels 120 km at 40 km/h. How long does the journey take?

$$\text{Time taken} = \frac{\text{Distance travelled}}{\text{Average speed}}$$

$$= \frac{120}{4}\text{ hours}$$

$$= 3 \text{ hours}$$

Remember that average speed is defined as total distance travelled divided by total time taken.

Example 6

(a) A car travels 60 km at an average speed of 30 km/h and for 120 km at an average speed of 40 km/h. Work out its average speed for the entire journey.

Time taken to travel 60 km at 30 km/h

$$= \frac{60}{30}\text{ hours}$$

$$= 2 \text{ hours}$$

Time taken to travel 120 km at 40 km/h

$$= \frac{120}{40}$$

$$= 3 \text{ hours}$$

Total time taken = 2 hours + 3 hours

= 5 hours

Total distance travelled = 60 km + 120 km

= 180 km

$$\text{Average speed} = \frac{180}{5}\,\text{km/h}$$

= 36 km/h

(b) A train travels for 4 hours at an average speed of 64 km/h. For the first two hours its average speed is 50 km/h. What is its average speed for the last 2 hours?

Total distance travelled in 4 hours

= 64 × 4 km

= 256 km

Distance travelled in first 2 hours

= 50 × 2 km

= 100 km

Distance travelled in last 2 hours

= 256 km − 100 km

= 156 km

Average speed in last 2 hours

$$= \frac{\text{Distance travelled}}{\text{Time taken}}$$

$$= \frac{156}{2}\,\text{km/h}$$

= 78 km/h

Exercise 12.3

1. A train travels 300 km in 4 hours. Work out its average speed.

2. A car travels 200 km at an average speed of 50 km/h. How long does the journey take?

3. A car travels for 5 hours at an average speed of 70 km/h. How far has it travelled?

4. For the first 3 hours of a journey of 182 km the average speed was 30 km/h. If the average speed for the remainder of the journey was 23 km/h, calculate the average speed for the whole journey.

5. A boy walks 3 km at a speed of 6 km/h. He then cycles 6 km at a speed of 12 km/h. What was his average speed for the entire journey?

6. A motorist travelling a steady speed of 90 km/h covers a section of motorway in 25 minutes. After a speed limit is imposed he finds that, when travelling at the maximum speed allowed, he takes 5 minutes longer than before to cover the same section. Calculate the speed limit.

7. A car travels 136 miles at an average speed of 32 miles per hour. On the return journey the speed is increased to 39 miles per hour. Calculate the average speed for the complete journey.

8. In winter a train travels between two towns 264 km apart at an average speed of 72 km/h. In summer the journey takes 22 minutes less than in winter. Work out the average speed in summer.

Miscellaneous Exercise 12

Section A

1. **(a)** A clock at a bus station shows the departure time of an evening bus to be 9.25. How would this time be written in a bus timetable which uses the 24-hour clock system?

   (b) The bus arrived at its destination at 2220 hours. How long did the journey take?

   (c) The length of the journey is 30 miles. What is the average speed of the bus?

2. The distance between two towns is 80 km.

   (a) How long does it take a motor cyclist to do the journey if his average speed is 60 km/h? Give the time in hours and minutes.

   (b) On another occasion the same journey took him 1 h 15 min. What was his average speed?

3. A car leaves London at 11.55 a.m. and arrives in Cardiff, 155 miles away, at 3.01 p.m.

   (a) How long, in hours and minutes, did the journey take?

   (b) What was the average speed of the car?

4. On a train journey of 117 km the average speed for the first 27 km was 45 km/h.

   (a) Work out the time taken, in minutes, for this part of the journey.

   (b) For the remainder of the journey the average speed was 37.5 km/h. What is the time taken, in hours and minutes, for this part of the journey?

   (c) Work out the total time for the entire journey of 117 km.

   (d) Calculate the average speed of the train for the entire journey.

5. A greyhound runs 200 metres in 50 seconds.

   (a) Calculate its average speed in metres per second.

   (b) What is its average speed in kilometres per hour?

Section B

1. A particle travels at a uniform speed of 4 m/s.

   (a) How long does it take to travel 25 m?

   (b) How far does it travel in 3.25 seconds?

2. A motorway journey takes 3 hours at an average speed of 70 miles per hour. How long will the journey take if the average speed is reduced to 50 miles per hour?

3. A car travels 20 km at an average speed of 48 km/h and then for a further 15 minutes at an average speed of 80 km/h. Calculate the average speed for the entire journey.

4. A train is travelling at a uniform speed of 90 km/h. Calculate the time, in seconds, for it to travel 600 m.

5. A man wishes to make a journey from his home to a meeting near Melton. He finds that the journey to Melton will consist of three parts:

   (1) A car journey from his home to Tranton, a distance of 48 km which he hopes will take him 45 minutes.

   (2) A journey on the 9.10 a.m. train from Tranton to Melton a distance of 315 km over which the train normally averaged 90 km/h.

   (3) A 24-minute taxi ride from Melton railway station to the meeting place over which he hopes to average 40 km/h.

   Work out:

   (a) the average speed of his car

   (b) the time at which the train would arrive at Melton

   (c) the distance between Melton railway station and the meeting place

   (d) the average speed over the journey from his home to the meeting place. Give your answer to the nearest whole number.

Multi-Choice Questions 12

1. An aeroplane flies non-stop for 2 h 15 min and travels 1620 km. What is its speed in kilometres per hour?

   A 364.5   B 720

   C 800   D 3645

2. A car travels 50 km at 50 km/h and 70 km at 70 km/h. Its average speed, in kilometres per hour, is

   A 58   B 60   C 62   D 65

3. A car travels for 3 hours at a speed of 45 mile/h and for 4 hours at a speed of 50 mile/h. How far has the car travelled?

   A 27.5 miles   B 95 miles

   C 335 miles   D 353 miles

4. A car travels 540 miles at an average speed of 30 mile/h. On the return journey the average speed is doubled. What is the average speed for the whole journey?

   A 45 mile/h   B 42 mile/h

   C 40 mile/h   D 35 mile/h

5. A car travels between two towns 270 miles apart in 9 hours. On the return journey the speed is increased by 10 mile/h. What is the time taken for the return journey?

   A 2.7 hours   B 4.5 hours

   C 6.5 hours   D 6.75 hours

6. A boy takes 15 min to walk to school at 5 km/h. How long would he take if he cycled at 15 km/h?

   A 35 min   B 25 min

   C 10 min   D 5 min

7. John lives 5 km from school. He walks 1 km at 4 km/h and travels the rest of the way by bus at 16 km/h. What is John's average speed for the whole journey?

   A 10 km/h   B 12 km/h

   C 16.25 km/h   D 20 km/h

8. A motorist made a journey of 192 km. He covered 48 km in 1 hour and for the remainder of the journey his speed was 72 km/h. Calculate his average speed for the whole journey.

   A 36 km/h   B 48 km/h

   C 54 km/h   D 64 km/h

Mental Test 12

Try to answer the following questions without writing anything down except the answer.

1. Change 480 seconds into minutes.

2. Change 7 minutes into seconds.

3. Change 240 minutes into hours.

4. Change 3 hours into minutes.

5. How many days are there in 360 hours?

6. Find the length of time between 9.30 a.m. and 12.45 a.m.

7. Work out the length of time between 1340 hours and 1750 hours.

8. A train leaves Sheffield at 1430 hours and arrives in London at 1820 hours. How long does the journey take?

9. A motorist travels at 60 miles per hour. How long did it take him to travel 240 miles?

10. A motorist travels 320 miles in 8 hours. What is his average speed?

11. A car travels at an average speed of 50 km/h for 4 hours. How far does it travel?

12. An aeroplane flies non-stop for 3 hours at an average speed of 500 miles per hour. How far does it fly in this time?

GLOUCESTER · ROSS-ON-WYE · HEREFORD

Service 38 (Summer 1987)

Mondays to Saturdays

Service No	M-F	M-F	S	X49 M-F											
GLOUCESTER (Bus Station) . . .		...	...	...	...	0930	1030	1130	1230	1330	1430	1530	1630	1735	2030
Oakle Street . . .		...	...	...	...	0943	1043	1143	1243	1343	1443	1543	1643	1748	2043
Huntley (Red Lion) . . .		...	...	...	...	0948	1048	1148	1248	1348	1448	1548	1648	1753	2048
Longhope (Nag's Head) . . .		...	...	...	...	0954	1054	1154	1254	1354	1454	1554	1654	1759	...
Lea (Crown Inn) . . .		0722	...	...	...	0958	1058	1158	1258	1358	1458	1558	1658	1803	...
Weston-under-Penyard . . .		0728	...	...	...	1004	1104	1204	1304	1404	1504	1604	1704	1809	...
ROSS-ON-WYE (Cantilupe Rd.) arr. . . .		0734	...	...	...	1010	1110	1210	1310	1410	1510	1610	1710	1815	...
ROSS-ON-WYE (Cantilupe Rd.) dep. . .	0642	0737	0752	...	0912	1012	1112	1212	1312	1412	1512	1612	1712	1817	...
Peterstow (Post Office) . . .	0648	0747	0758	...	0918	1018	1118	1218	1318	1418	1518	1618	1718	1823	...
St. Owen's Cross . . .		0751	0801	...	0921	...	...	...	...	1421	...	...	1721	...	...
Harewood End (Inn) . . .	0654	0755	0804	...	0924	1024	1124	1224	1324	1424	1524	1624	1724	1829	...
Much Birch (Church) . . .	0700	0803	0810	0829	0930	1030	1130	1230	1330	1430	1530	1630	1730	1835	...
Kingsthorne Village . . .	0702	0807	0812	0830	0932	1032	1132	1232	1332	1432	1532	1632	1732	1837	...
HEREFORD (Bus Station) . . .	0720	0830	0830	0845	0950	1050	1150	1250	1350	1450	1550	1650	1750	1855	...

Fig. 12.4

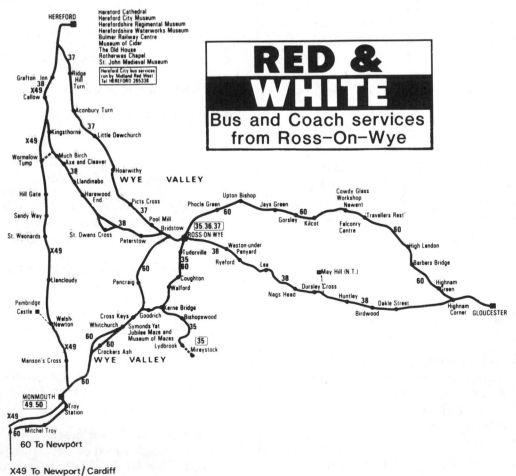

RED & WHITE
Bus and Coach services
from Ross-On-Wye

Use the bus timetable and the route map opposite (Fig. 12.4) to answer the following questions:

13. How long does the bus take to travel from Gloucester to Hereford?

14. What time must I arrive at Huntley (Red Lion) to make sure of arriving at Peterstow (Post Office) by 1125 hours?

15. What is the number of the bus service from Hereford to

(a) Picts Cross

(b) Welsh Newton?

16. Is there a bus which leaves Oakle Street for St Owen's Cross at 17 minutes to four in the afternoon?

17. A passenger wants to travel from Ross-on-Wye to Manson's Cross. Which services should she use and where should she change buses?

Basic Algebra

Introduction

The methods of algebra are an extension of those used in arithmetic. In algebra we use letters and symbols as well as numbers to represent quantities. When we write that a sum of money is £50 we are making a **particular statement** but if we write that a sum of money is £*P* we are making a **general statement**. This general statement will cover any number we care to substitute for *P*.

Use of Symbols

The following examples will show how verbal statements can be translated into algebraic symbols. Notice that we can choose any symbol we like to represent the quantities concerned.

(1)   The sum of two numbers.
      Let the two numbers be x and y.
      Sum of the two numbers = $x + y$.

(2)   Three times a number.
      Let the number be N.
      Three times the number = $3 \times N$.

(3)   One number divided by another number.
      Let one number be a and the other number be b.
      One number divided by another number
      $= \dfrac{a}{b}$.

(4)   Five times the product of two numbers.
      Let the two numbers be m and n. 5 times the product of the two numbers
      $= 5 \times m \times n$.

Exercise 13.1

Translate the following into algebraic symbols:

1.   Seven times a number x.

2.   Four times a number x minus three.

3.   Five times a number x plus a second number y.

4.   The sum of two numbers x and y divided by a third number z.

5.   Half of a number x.

6.   Eight times the product of three numbers x, y and z.

7.   The product of two numbers x and y divided by a third number z.

8.   Three times a number x minus four times a second number y.

Substitution

The process of finding the numerical value of an algebraic expression for given values of the symbols that appear in it is called **substitution**.

Example 1

If $x = 3$, $y = 4$ and $z = 5$ find the values of:

(a) $2y + 4$ (b) $3y + 5z$

(c) $8 - x$ (d) $\dfrac{y}{x}$

(e) $\dfrac{3y + 2z}{x + z}$

Note that multiplication signs are often missed out when writing algebraic expressions so that, for instance, $2y$ means $2 \times y$. These missed multiplication signs must reappear when the numbers are substituted for the symbols.

(a)
$$2y + 4 = 2 \times 4 + 4$$
$$= 8 + 4$$
$$= 12$$

(b)
$$3y + 5z = 3 \times 4 + 5 \times 5$$
$$= 12 + 25$$
$$= 37$$

(c)
$$8 - x = 8 - 3$$
$$= 5$$

(d)
$$\frac{y}{x} = \frac{4}{3}$$
$$= 1\frac{1}{3}$$

(e)
$$\frac{3y + 2z}{x + z} = \frac{3 \times 4 + 2 \times 5}{3 + 5}$$
$$= \frac{12 + 10}{8}$$
$$= \frac{22}{8}$$
$$= 2\frac{3}{4}$$

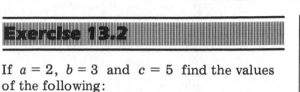
Exercise 13.2

If $a = 2$, $b = 3$ and $c = 5$ find the values of the following:

1. $a + 7$ 2. $c - 2$

3. $6 - b$ 4. $6b$

5. $9c$ 6. ab

7. $3bc$ 8. abc

9. $5c - 2$ 10. $4c + 6b$

11. $8c - 7$ 12. $a + 2b + 5c$

13. $8c - 4b$ 14. $2 \div a$

15. $\dfrac{ab}{8}$ 16. $\dfrac{abc}{6}$

17. $\dfrac{2c}{a}$ 18. $\dfrac{5a + 9b + 8c}{a + b + c}$

Powers

The quantity $a \times a \times a$ or aaa is usually written as a^3, which is called the third power of a. The number 3 which indicates the number of a's to be multiplied together is called the **index** (plural: **indices**).

$$2^4 = 2 \times 2 \times 2 \times 2$$
$$= 16$$
$$y^5 = y \times y \times y \times y \times y$$

Example 2

Find the value of b^3 when $b = 5$.
$$b^3 = 5^3$$
$$= 5 \times 5 \times 5$$
$$= 125$$

When dealing with expressions like $8mn^4$ note that it is only the symbol n which is raised to the fourth power. Thus

$$8mn^4 = 8 \times m \times n \times n \times n \times n$$

Example 3

Find the value of $7p^2q^3$ when $p = 5$ and $q = 4$.

$$7p^2q^3 = 7 \times 5^2 \times 4^3$$
$$= 7 \times 25 \times 64$$
$$= 11\,200$$

Exercise 13.3

If $a = 2$, $b = 3$ and $c = 4$ find the values of the following:

1. a^2
2. b^4
3. ab^3
4. $2a^2c$
5. ab^2c^3
6. $5a^2 + 6b^2$
7. $a^2 + c^2$
8. $7b^3c^2$
9. $3a^4$
10. $\dfrac{c^5}{ab^3}$

Addition of Algebraic Terms

Like terms are numerical multiples of the same algebraic quantity. Thus:

$$7x, 5x \text{ and } -3x$$

are three like terms.

An expression consisting of like terms can be reduced to a single term by adding the numerical coefficients together. Thus:

$$7x - 5x + 3x = (7 - 5 + 3)x$$
$$= 5x$$
$$3b^2 + 7b^2 = (3 + 7)b^2$$
$$= 10b^2$$
$$-3y - 5y = (-3 - 5)y$$
$$= -8y$$
$$q - 3q = (1 - 3)q$$
$$= -2q$$

Only like terms can be added or subtracted. Thus $7a + 3b - 2c$ is an expression containing three unlike terms and it cannot be simplified any further. Similarly with $8a^2b + 7ab^3 + 6a^2b^2$ which has three unlike terms.

It is possible to have several sets of like terms in an expression and each set can then be simplified:

$$8x + 3y - 4z - 5x + 7z - 2y + 2z$$
$$= (8 - 5)x + (3 - 2)y + (-4 + 7 + 2)z$$
$$= 3x + y + 5z$$

Multiplication and Division of Algebraic Quantities

The rules are exactly the same as those used with directed numbers:

$$(+x)(+y) = +(xy)$$
$$= +xy$$
$$= xy$$
$$5x \times 3y = 5 \times 3 \times x \times y$$
$$= 15xy$$
$$(x)(-y) = -(xy)$$
$$= -xy$$
$$(2x)(-3y) = -(2x)(3y)$$
$$= -6xy$$
$$(-4x)(2y) = -(4x)(2y)$$
$$= -8xy$$
$$(-3x)(-2y) = +(3x)(2y)$$
$$= 6xy$$
$$\frac{+x}{+y} = +\frac{x}{y}$$
$$= \frac{x}{y}$$
$$\frac{-3x}{2y} = -\frac{3x}{2y}$$

$$\frac{-5x}{-6y} = +\frac{5x}{6y}$$

$$= \frac{5x}{6y}$$

$$\frac{4x}{-3y} = -\frac{4x}{3y}$$

When **multiplying** expressions containing the same symbols, indices are used:

$$m \times m = m^2$$

$$3m \times 5m = 3 \times m \times 5 \times m$$

$$= 15m^2$$

$$(-m) \times m^2 = (-m) \times m \times m$$

$$= -m^3$$

$$5m^2n \times 3mn^3 = 5 \times m \times m \times n \times 3$$

$$\times m \times n \times n \times n$$

$$= 15m^3n^4$$

$$3mn \times (-2n^2) = 3 \times m \times n \times (-2) \times n \times n$$

$$= -6mn^3$$

When **dividing** algebraic expressions, cancellation between numerator and denominator is often possible. Cancelling is equivalent to dividing both numerator and denominator by the same quantity:

$$\frac{pq}{p} = \frac{\not{p} \times q}{\not{p}}$$

$$= q$$

$$\frac{3p^2q}{6pq^2} = \frac{3 \times \not{p} \times p \times \not{q}}{6 \times \not{p} \times q \times \not{q}}$$

$$= \frac{3p}{6q}$$

$$= \frac{p}{2q}$$

$$\frac{18x^2y^2z}{6xyz} = \frac{18 \times \not{x} \times x \times \not{y} \times y \times \not{z}}{6 \times \not{x} \times \not{y} \times \not{z}}$$

$$= 3xy$$

Exercise 13.4

Simplify the following:

1. $7x + 11x$
2. $7x - 5x$
3. $3x - 6x$
4. $-2x - 4x$
5. $-8x + 3x$
6. $-2x + 7x$
7. $8a - 6a - 7a$
8. $5m + 13m - 6m$
9. $6b^2 - 4b^2 + 3b^2$
10. $6ab - 3ab - 2ab$
11. $14xy + 5xy - 7xy + 2xy$
12. $-5x + 7x - 3x - 2x$
13. $-4x^2 - 3x^2 + 2x^2 - x^2$
14. $3x - 2y + 4z - 2x - 3y + 5z$
    $+ 6x + 2y - 3z$
15. $3a^2b + 2ab^3 + 4a^2b^2 - 5ab^3$
    $+ 11b^4 + 6a^2b$
16. $1.2x^3 - 3.4x^2 + 2.6x + 3.7x^2$
    $+ 3.6x - 2.8$
17. $pq + 2.1qr - 2.2rq + 8qp$
18. $2.6a^2b^2 - 3.4b^3 - 2.7a^3 - 3a^2b^2$
    $- 2.1b^3 + 1.5a^3$
19. $2x \times 5y$
20. $3a \times 4b$
21. $3 \times 4m$
22. $\frac{1}{4}q \times 16p$
23. $x \times (-y)$
24. $(-3a) \times (-2b)$
25. $8m \times (-3n)$
26. $(-4a) \times 3b$
27. $8p \times (-q) \times (-3r)$
28. $3a \times (-4b) \times (-c) \times 5d$
29. $12x \div 6$
30. $4a \div (-7b)$
31. $(-5a) \div 8b$
32. $(-3a) \div (-3b)$
33. $4a \div 2b$
34. $4ab \div 2a$
35. $12x^2yz^2 \div 4xz^2$
36. $(-12a^2b) \div 6a$
37. $8a^2bc^2 \div 4ac^2$
38. $7a^2b^2 \div 3ab$
39. $a \times a$
40. $b \times (-b)$

41. $(-m) \times m$ **42.** $(-p) \times (-p)$

43. $3a \times 2a$ **44.** $5X \times X$

45. $5q \times (-3q)$ **46.** $3m \times (-3m)$

47. $(-3pq) \times (-3q)$

48. $8mn \times (-3m^2n^3)$

49. $7ab \times (-3a^2)$ **50.** $2q^3r^4 \times 5qr^2$

51. $(-3m) \times 2n \times (-5p)$

52. $5a^2 \times (-3b) \times 5ab$

53. $m^2n \times (-mn) \times 5m^2n^2$

Brackets

Brackets are used for convenience in grouping terms together. When removing brackets each term within the bracket is multiplied by the quantity outside the bracket:

$$3(x + y) = 3x + 3y$$

$$5(2x + 3y) = 5 \times 2x + 5 \times 3y$$

$$= 10x + 15y$$

$$4(a - 2b) = 4 \times a - 4 \times 2b$$

$$= 4a - 8b$$

$$m(a + b) = ma + mb$$

$$3x(2p + 3q) = 3x \times 2p + 3x \times 3q$$

$$= 6px + 9qx$$

$$4a(2a + b) = 4a \times 2a + 4a \times b$$

$$= 8a^2 + 4ab$$

When a bracket has a minus sign in front of it, the signs of all the terms inside the bracket are changed when the bracket is removed. The reason for this rule may be seen from the following example:

$$-3(2x - 5y) = (-3) \times 2x + (-3) \times (-5y)$$

$$= -6x + 15y$$

$$-(m + n) = -m - n$$

$$-(p - q) = -p + q$$

$$-2(p + 3q) = -2p - 6q$$

When simplifying expressions containing brackets first remove the brackets and then add the like terms together:

$$(3x + 7y) - (4x + 3y) = 3x + 7y - 4x - 3y$$

$$= -x + 4y$$

$$3(2x + 3y) - (x + 5y) = 6x + 9y - x - 5y$$

$$= 5x + 4y$$

$$x(a + b) - x(a + 3b) = ax + bx - ax - 3bx$$

$$= -2bx$$

$$2(5a + 3b) + 3(a - 2b) = 10a + 6b + 3a$$

$$- 6b$$

$$= 13a$$

Exercise 13.5

Remove the brackets in the following:

1. $3(x + 4)$ **2.** $2(a + b)$

3. $3(3x + 2y)$ **4.** $\frac{1}{2}(x - 1)$

5. $5(2p - 3q)$ **6.** $7(a - 3m)$

7. $-(a + b)$ **8.** $-(a - 2b)$

9. $-(3p - 3q)$ **10.** $-(7m - 6)$

11. $-4(x + 3)$ **12.** $-2(2x - 5)$

13. $-5(4 - 3x)$ **14.** $2k(k - 5)$

15. $-3y(3x + 4)$ **16.** $a(p - q - r)$

17. $4xy(ab - ac + d)$

18. $3x^2(x^2 - 2xy + y^2)$

19. $-7P(2P^2 - P + 1)$

20. $-2m(-1 + 3m - 2n)$

Remove the brackets and simplify:

21. $3(x + 1) + 2(x + 4)$

22. $5(2a + 4) - 3(4a + 2)$

23. $3(x + 4) - (2x + 5)$

24. $4(1 - 2x) - 3(3x - 4)$

25. $5(2x - y) - 3(x + 2y)$

26. $\frac{1}{2}(y - 1) + \frac{1}{3}(2y - 3)$

27. $-(4a + 5b - 3c) - 2(2a + 3b - 4c)$

28. $2x(x - 5) - x(x - 2) - 3x(x - 5)$

29. $3(a - b) - 2(2a - 3b) + 4(a - 3b)$

30. $3x(x^2 + 7x - 1) - 2x(2x^2 + 3)$
    $\qquad - 3(x^2 + 5)$

Miscellaneous Exercise 13

Section A

1. Find the value of $2x^2 - 3x + 7$ when $x = 5$.

2. Express $\dfrac{a - b}{a + b}$ as a fraction in its lowest terms given that $a = 5$ and $b = 3$.

3. Make into single terms
   (a) $8a + 3a$ (b) $7a - 2a$
   (c) $3a \times 2a$ (d) $8a \div 2a$

4. Remove the brackets and simplify:
   (a) $5(2x + 3y) + 3(4x - 7y)$
   (b) $2(x + y) - (3x - 4y)$
   (c) $2x - y - (2x + y)$

5. Make each of the following into single terms:
   (a) $2a \times 3a \times 5a$
   (b) $9x - 6x - 5x$
   (c) $12a^3 \div 4a$

Section B

1. If $p = 2$, $q = 0$ and $r = -3$ find the value of $\dfrac{pq - r^2}{p^2 - qr}$.

2. Remove the brackets and simplify:
   $2(5x - 2y) + 3(3x - y - 3z)$
   $\qquad - (7x - 2y + 5z)$

3. If $y = 2(x - 1)(x - 3)$ find the value of y when
   (a) $x = 5$ (b) $x = 1$
   (c) $x = 0$

4. Find the value of $p^2 + q^2 + r^2 - 2qr$ when $p = 2$, $q = 3$ and $r = -4$.

5. If $p = -2$, $q = 3$, $r = 7$ and $s = 0$, find the value of
   (a) $(q + 2p)^3$ (b) $qr + ps$
   (c) $p(q - r)$

6. Find the value of $(2p - 3q)^2 + 12pq$ when $p = 4$ and $q = -1$.

7. $\dfrac{a}{b^2} + \dfrac{b}{c^2} - \dfrac{c}{a^2}$. What is the value of this expression when $a = 0.5$, $b = -3$ and $c = 3$?

8. Find the value of $ab(b - c) - 2abc$ when $a = 2$, $b = -5$ and $c = 1$.

Multi-Choice Questions 13

1. Five times a certain number plus six is equal to seven times the number plus one. What is the correct algebraic expression?
   A $5m + 6 = 7 + m + 1$
   B $5 + m + 6 = 7m + 1$
   C $5 + m + 6 = 7 + m + 1$
   D $5m + 6 = 7m + 1$

2. $(2x + y) - (x - 2y)$ is equal to
   A $3x - y$ B $x + 3y$
   C $x - 3y$ D $x - y$

3. Which of the following is not equal to $\frac{1}{2}pq$?
   A $\dfrac{pq}{2}$ B $q \times \dfrac{p}{2}$
   C $\dfrac{1}{2}qp$ D $\dfrac{1}{2p} \times q$

4. If $x = 5$ and $y = -3$ then $x^2 - y^3$ equals

   A -2 B 16 C 34 D 52

5. In working a problem a student writes $(a - b)^2$ as $a^2 - b^2$. If $a = 6$ and $b = -4$, what is the difference between the correct result and the student's result?

   A 80 B 48 C 20 D 16

6. The value of $3ab^2 - 3a^2b^2$ when $a = -2$ and $b = 3$ is

   A 306 B 54 C -54 D -162

7. What is the value of $\dfrac{x^2}{2} + 2$ when $x = 4$?

   A 8.5 B 10 C 12.5 D 18

8. What is the value of $x^2 - 3x + 5$ when $x = 2$?

   A 0 B 2 C 3 D 5

Mental Test 13

Try to answer the following questions without writing anything down except the answer.

1. Make $8a + 3a$ into a single term.

2. Subtract $3x$ from $5x$.

3. Express 5 times a number y plus 8 as an algebraic expression.

4. Make $6p - p$ into a single term.

5. Simplify $3a \times 4b$.

6. Express in index form $2a \times 3a \times 5a$.

7. Divide $4a^2$ by $2a$.

8. Multiply $3a^2$ by $2a$.

9. Remove the brackets from $3(2x + 3y)$.

10. Remove the brackets from $-(3x - 4y)$.

11. Remove the brackets from $-3(x - 2y)$.

12. Make into a single term $5x - 3x - 4x$.

Factorisation

Factors

The expression $3x + 3y$ has the number 3 common to both terms.

$$3x + 3y = 3(x + y)$$

3 and $(x + y)$ are said to be the factors of $3x + 3y$.

Example 1

(a) Factorise $2x + 6$.

We note that the number 2 is common to both terms so we place 2 outside the bracket. To find the terms inside the bracket we divide each of the terms making up the expression by 2. Hence

$$2x + 6 = 2(x + 3)$$

(b) Factorise $7x^2 - 14x$.

We note that $7x$ is common to both the terms making up the expression. So

$$7x^2 - 14x = 7x(x - 2)$$

(c) Factorise $p(x - y) - q(x - y)$.

Here we see that $(x - y)$ is common to both terms making up the expression. So

$$p(x - y) - q(x - y) = (x - y)(p - q)$$

Factorise the following:

1. $5x + 5y$
2. $5p - 5q$
3. $6x + 18y$
4. $ax - ay$
5. $4p - 12q$
6. $4x - 6xy$
7. $8x^2 - 4x$
8. $ax^2 - bx$
9. $x(a - b) + 2(a - b)$
10. $m(p + q) - n(p + q)$
11. $5a - 10b + 15c$

The Product of Two Binomial Expressions

A binomial expression consists of two terms. Thus $3x + 5$, $a + b$ and $4p - q$ are all binomial expressions.

To find the expansion of $(a + b)(c + d)$ we use Fig. 14.1.

Fig. 14.1

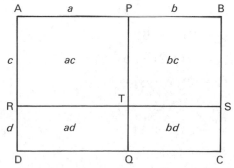

The rectangular area ABCD is made up as follows

ABCD = APTR + TQDR + PBST + STQC

i.e. $(a + b)(c + d) = ac + ad + bc + bd$

It will be seen that to find the product of two binomial expressions we multiply the terms connected by a line as shown below:

$$(a + b)(c + d) = ac + ad + bc + bd$$

Example 2

(a) $(3x + 2)(x + 4) = (3x)(x) + (3x)(4)$
$$+ (2)(x) + (2)(4)$$
$$= 3x^2 + 12x + 2x + 8$$
$$= 3x^2 + 14x + 8$$

(b) $(2x - 3)(3x - 4) = (2x)(3x)$
$$+ (2x)(-4)$$
$$+ (-3)(3x)$$
$$+ (-3)(-4)$$
$$= 6x^2 - 8x - 9x + 12$$
$$= 6x^2 - 17x + 12$$

Exercise 14.2

Remove the brackets from the following:

1. $(x + 1)(x + 3)$

2. $(2x + 5)(x + 3)$

3. $(5x - 2)(2x + 4)$

4. $(a + 3)(a - 6)$

5. $(3x + 1)(2x - 5)$

6. $(x - 2)(x - 3)$

7. $(4x - 1)(2x - 3)$

8. $(x - 3)(x + 1)$

9. $(2x - 3)(3x + 2)$

10. $(x - 4)(2x + 5)$

Factorising by Grouping

To factorise the expression $ax + ay + bx + by$ first group the terms in pairs so that each pair of terms has a common factor. Thus

$$ax + ay + bx + by = (ax + ay) + (bx + by)$$
$$= a(x + y) + b(x + y)$$
$$= (x + y)(a + b)$$

Example 3

Factorise $mp + np - mq - nq$.

$$mp + np - mq - nq = (mp + np)$$
$$- (qm + qn)$$
$$= p(m + n)$$
$$- q(m + n)$$
$$= (m + n)(p - q)$$

Exercise 14.3

Factorise each of the following:

1. $ax + ay + bx + by$

2. $ab + ac - bd - cd$

3. $2pr - 4ps + qr - 2qs$

4. $4ax + 6ay - 4bx - 6by$

5. $3mx + 2nx - 3my - 2yn$

6. $ab(p + q) - cd(p + q)$

7. $mn(3x - 1) - pq(3x - 1)$

8. $k^2l^2 - mnl - k^2l + mn$

Factors of Quadratic Expressions

An expression of the type $ax^2 + bx + c$ is called a quadratic expression. Thus

$$x^2 - 5x + 1 \quad \text{and} \quad 3p^2 + 12p - 5$$

are both quadratic expressions.

(i) *When* $a = 1$

In an expression like $x^2 + 7x + 12$, the factors will be of the type $(x + m)(x + n)$ where m and n are two numbers whose sum is 7 and whose product is 12. These two numbers must be 3 and 4. So

$$x^2 + 7x + 12 = (x + 3)(x + 4)$$

(ii) *When* $a \neq 1$

This technique is also useful when factorising expressions such as $12x^2 + 23x + 10$ in which $a = 12$, $b = 23$ and $c = 10$. We can factorise if we can find two integers whose product is ac and whose sum is b. Thus $ac = 12 \times 10 = 120$ and $b = 23$.

Now $120 = 2^3 \times 3 \times 5$. It can now be seen that the two integers are 8 and 15. Since $23x = 15x + 8x$

$$12x^2 + 23x + 10 = 12x^2 + 15x + 8x + 10$$
$$= 3x(4x + 5) + 2(4x + 5)$$
$$= (4x + 5)(3x + 2)$$

Example 4

Factorise $8x^2 - 34x + 21$.

Here $a = 8$, $b = -34$ and $c = 21$.

$ac = 8 \times 21 = 168 = 2^3 \times 3 \times 7$

Here the two integers whose sum is -34 and whose product is 168 are -6 and -28.

$$8x^2 - 34x + 21 = 8x^2 - 6x - 28x + 21$$
$$= 2x(4x - 3) - 7(4x - 3)$$
$$= (4x - 3)(2x - 7)$$

Exercise 14.4

Factorise the following:

1. $x^2 + 5x + 6$
2. $x^2 - 8x + 15$
3. $x^2 - 5x - 6$
4. $x^2 - 5x + 6$
5. $x^2 - 7x - 18$
6. $x^2 + 2x - 15$
7. $2x^2 + 7x + 5$
8. $2x^2 + 13x + 15$
9. $3x^2 + x - 2$
10. $3x^2 - 8x - 28$
11. $2x^2 - 5x - 3$
12. $10x^2 + 19x - 15$
13. $6x^2 + x - 35$
14. $5x^2 - 11x + 2$
15. $2x^2 - 11x + 5$
16. $6x^2 - 7x - 5$

Sometimes the factors form a perfect square.

$$(a + b)^2 = (a + b)(a + b)$$
$$= a^2 + 2ab + b^2$$
$$(a - b)^2 = (a - b)(a - b)$$
$$= a^2 - 2ab - b^2$$

Therefore the square of a binomial expression consists of

(First term)^2 + 2 $\times$ (First term) $\times$ (Second term) + (Second term)^2

Example 5

(a) Remove the brackets from $(3x + 5)^2$.

$$(3x + 5)^2 = (3x)^2 + 2(3x)(5) + 5^2$$
$$= 9x^2 + 30x + 25$$

(b) Remove the brackets from $(5x - 7)^2$.

$$(5x - 7)^2 = (5x)^2 + 2(5)(-7) + (-7)^2$$
$$= 25x^2 - 70x + 49$$

(c) Factorise $4x^2 + 12x + 9$.

$$4x^2 = (2x)^2; \quad 9 = 3^2; \quad 12x = 2(2x)(3)$$
$$4x^2 + 12x + 9 = (2x + 3)^2$$

The expression $a^2 - b^2$ is called the difference of two squares.

$$a^2 - b^2 = (a + b)(a - b)$$

Example 6

(a) Remove the brackets from $(3x + 5)(3x - 5)$.

$$(3x + 5)(3x - 5) = (3x)^2 - 5^2$$
$$= 9x^2 - 25$$

(b) Factorise $4x^2 - 1$.

$$4x^2 - 1 = (2x + 1)(2x - 1)$$

Exercise 14.5

Remove the brackets from the following:

1. $(x + 4)^2$ 2. $(x - 7)^2$

3. $(2x + 5)^2$ 4. $(3x - 4)^2$

5. $(x + 3)(x - 3)$ 6. $(1 - x)(1 + x)$

7. $(2x + 5)(2x - 5)$

8. $(3x + 2y)(3x - 2y)$

Factorise the following:

9. $x^2 + 6x + 9$ 10. $x^2 + 8x + 16$

11. $9x^2 - 12x + 4$ 12. $25x^2 - 10x + 1$

13. $x^2 - 1$ 14. $x^2 - 16$

15. $4x^2 - 9$ 16. $25x^2 - 36$

Harder Factorisation

Some algebraic expressions need a combination of the methods of factorisation previously discussed.

Example 7

(a) Factorise $a^2 - b^2 + 2a + 2b$.

$$a^2 - b^2 + 2a + 2b = (a^2 - b^2) + 2a + 2b$$
$$= (a + b)(a - b)$$
$$\quad + 2(a + b)$$
$$= (a + b)(a - b + 2)$$

(b) Factorise $2x^3 - 50xy^2$.

$$2x^3 - 50xy^2 = 2x(x^2 - 25y^2)$$
$$= 2x(x + 5y)(x - 5y)$$

Exercise 14.6

Factorise the following:

1. $(a - b)^2 - 2x(a - b)$

2. $3(x + y) + (x + y)^2$

3. $x^2 - y^2 - x - y$

4. $a^2 - b^2 + 3a + 3b$

5. $aR^2 - ar^2$

6. $4y^3 - 9a^2y$

7. $\frac{2}{3}\pi r^3 + \frac{1}{3}\pi r^2 h$

8. $(x - 1)^2 - 4y^2$

Miscellaneous Exercise 14

Section A

1. Factorise completely:
    (a) $2abc - 6abd$
    (b) $4(a - b) - c(a - b)$

2. Factorise $x^2 - 5x + 4$.

3. Factorise $4x^2 - 36$.

4. Factorise $x^2 - 4x + 4$.

5. Factorise completely:
    (a) $10 - 15x^2$
    (b) $4x - 8xy$

Section B

1. Factorise
    (a) $9x^2 - 16$ (b) $9x^2 - 16x$
    (c) $x^2 - 9x - 22$

2. Factorise
   (a) $4x^2 + 12x + 9$
   (b) $6x^2 - 7x - 20$
   (c) $9p^2 - 4q^2$

3. Factorise $3a^2 + 2ab - 12ac - 8bc$.

4. Factorise $7 - 63a^2$.

5. Copy and complete the following:
   $$25x^2 - 70x + \quad = (5x \quad)^2$$

6. Factorise completely: $2x^3 - 2x^2 - 4x$.

7. Factorise $ac - 2ab + 3bc - 6b^2$.

8. Factorise:
   (a) $12x^2 - 3y^2$
   (b) $b(a - 2c) + ad - 2cd$

Multi-Choice Questions 14

1. Which of the following is equal to $(2x + 1)(x - 3)$?
   A $2x^2 - x - 3$ B $2x^2 - 5x - 3$
   C $2x^2 + 5x - 3$ D $2x^2 - x - 3$

2. The difference between the factors of $x^2 + x - 6$ is
   A -3 B -2 C 3   D 5

3. If $a^2 + b^2 = 41$ and $ab = 9$, $(a + b)^2 =$
   A 41   B 50   C 59   D 81

4. What is the value of r which makes $x^2 + 10x + r$ a perfect square?
   A 1   B 5   C 10   D 25

5. When factorised $9x^2t^2 - 1$ is equal to
   A $(3xt - 1)^2$
   B $(3xt + 1)^2$
   C $(3xt^2 + 1)(3xt^2 - 1)$
   D $(3xt + 1)(3xt - 1)$

6. $(2a - 3b)^2 - (a - b)^2$ is equal to
   A $3a^2 + 8b^2$
   B $3a^2 - 8b^2$
   C $3a^2 - 14ab + 8b^2$
   D $3a^2 - 10ab + 8b^2$

15 Algebraic Fractions

Multiplication and Division of Fractions

As with ordinary arithmetic fractions, numerators can be multiplied together, as can denominators, in order to form a single fraction.

Example 1

(a)
$$\frac{a}{b} \times \frac{c}{d} = \frac{a \times c}{b \times d}$$

$$= \frac{ac}{bd}$$

(b)
$$\frac{3x}{2y} \times \frac{p}{4q} \times \frac{r^2}{s} = \frac{3x \times p \times r^2}{2y \times 4q \times s}$$

$$= \frac{3xpr^2}{8yqs}$$

Factors which are common to both numerator and denominator may be cancelled. It is important to realise that this cancelling means dividing the numerator and denominator by the same quantity.

Example 2

(a)
$$\frac{8ab}{3mn} \times \frac{9n^2m}{4ab^2}$$

$$= \frac{\cancel{8}^2 \times \cancel{a} \times \cancel{b} \times \cancel{9}^3 \times \cancel{n} \times n \times \cancel{m}}{\cancel{3} \times \cancel{m} \times \cancel{n} \times \cancel{4} \times \cancel{a} \times \cancel{b} \times b}$$

$$= \frac{6n}{b}$$

(b)
$$\frac{5x^2y}{8ab^3} \div \frac{10xy}{4a^2b}$$

$$= \frac{5x^2y}{8ab^3} \times \frac{4a^2b}{10xy}$$

$$= \frac{\cancel{5} \times \cancel{x} \times x \times \cancel{y} \times \cancel{4} \times \cancel{a} \times a \times \cancel{b}}{\cancel{8}_2 \times \cancel{a} \times \cancel{b} \times b \times b \times \cancel{10}_2 \times \cancel{x} \times \cancel{y}}$$

$$= \frac{ax}{4b^2}$$

Exercise 15.1

Simplify the following:

1. $\dfrac{6a}{b^2} \times \dfrac{b}{3a^2}$

2. $\dfrac{9x^2}{6y^2} \times \dfrac{y^3}{x^3}$

3. $\dfrac{6pq}{4rs} \times \dfrac{8s^2}{3p}$

4. $\dfrac{6ab}{c} \times \dfrac{ad}{2b} \times \dfrac{8cd^2}{4bc}$

5. $\dfrac{a}{bc^2} \times \dfrac{b^2c}{a}$

6. $\dfrac{3pq}{r} \times \dfrac{qs}{2t} \times \dfrac{3rs}{pq^2}$

7. $\dfrac{2z^2y}{3ac^2} \times \dfrac{6a^2}{5zy^2} \times \dfrac{10c^3}{3y^2}$

8. $\dfrac{ab^2}{bc^2} \div \dfrac{a^2}{bc^3}$

9. $\dfrac{6ab}{5cd} \div \dfrac{4a^2}{7bd}$

10. $\dfrac{3pq}{5rs} \div \dfrac{p^2}{15s^2}$

Addition and Subtraction of Fractions

The method for algebraic fractions is the same as for arithmetical fractions, that is:

(1) Find the LCM of the denominators.

(2) Express each fraction with the common denominators.

(3) Add or subtract the fractions.

Example 3

(a) Simplify $\dfrac{a}{2} + \dfrac{b}{3} - \dfrac{c}{4}$.

The LCM of 2, 3, and 4 is 12.

$$\frac{a}{2} + \frac{b}{3} - \frac{c}{4} = \frac{6a}{12} + \frac{4b}{12} - \frac{3c}{12}$$

$$= \frac{6a + 4b - 3c}{12}$$

(b) Simplify $\dfrac{2}{x} + \dfrac{3}{2x} + \dfrac{4}{3x}$.

The LCM of x, $2x$ and $3x$ is $6x$.

$$\frac{2}{x} + \frac{3}{2x} + \frac{4}{3x} = \frac{12 + 9 + 8}{6x}$$

$$= \frac{29}{6x}$$

The sign in front of a fraction applies to the fraction as a whole. The line which separates the numerator and denominator acts as a bracket.

Example 4

(a) Simplify $\dfrac{m}{12} + \dfrac{2m + n}{4} - \dfrac{m - 2n}{3}$.

The LCM of 12, 4 and 3 is 12.

$$\frac{m}{12} + \frac{2m + n}{4} - \frac{m - 2n}{3}$$

$$= \frac{m + 3(2m + n) - 4(m - 2n)}{12}$$

$$= \frac{m + 6m + 3n - 4m + 8n}{12}$$

$$= \frac{3m + 11n}{12}$$

(b) Simplify $\dfrac{4x}{3} - \dfrac{x - 4}{2}$.

$$\frac{4x}{3} - \frac{x - 4}{2} = \frac{2 \times 4x - 3(x - 4)}{6}$$

$$= \frac{8x - 3x + 12}{6}$$

$$= \frac{5x + 12}{6}$$

Exercise 15.2

Simplify the following:

1. $\dfrac{x}{3} + \dfrac{x}{4} + \dfrac{x}{5}$

2. $\dfrac{5a}{12} - \dfrac{7a}{18}$

3. $\dfrac{2}{q} - \dfrac{3}{2q}$

4. $\dfrac{3}{y} - \dfrac{5}{3y} + \dfrac{4}{5y}$

5. $\dfrac{3}{5p} - \dfrac{2}{3q}$

6. $\dfrac{3x}{2y} - \dfrac{5y}{6x}$

7. $3x - \dfrac{4y}{5z}$

8. $1 - \dfrac{2x}{5} + \dfrac{x}{8}$

9. $3m - \dfrac{2m + n}{7}$

10. $\dfrac{3a + 5b}{4} - \dfrac{a - 3b}{2}$

11. $\dfrac{3m - 5n}{6} - \dfrac{3m - 7n}{2}$

12. $\dfrac{x - 2}{4} + \dfrac{2}{5}$

13. $\dfrac{x - 5}{3} - \dfrac{x - 2}{4}$

14. $\dfrac{3x - 5}{10} + \dfrac{2x - 3}{15}$

Miscellaneous Exercise 15

Section A

1. Simplify $\dfrac{6a}{5} \div \dfrac{12ab}{15}$.

2. Simplify $\dfrac{9p}{4} \times \dfrac{12}{5pq}$.

3. Express $\dfrac{3x}{4} - \dfrac{x}{6}$ as a single fraction.

4. Express $\dfrac{x}{3} + \dfrac{x - 2}{2}$ as a single fraction.

5. Simplify $\dfrac{1}{2x - 5} \div \dfrac{1}{x - 4}$.

Section B

1. Simplify $\dfrac{x - 3}{3} - \dfrac{x - 7}{6}$.

2. Express $\dfrac{a + 1}{3a} + \dfrac{a + 3}{4a} + \dfrac{a - 13}{12a}$ as a single fraction in its lowest terms.

3. Express $\dfrac{2 - p}{2p} - \dfrac{3 - 2p}{3p} - \dfrac{p + 2}{6p}$ as a single fraction in its lowest terms.

4. Simplify $\dfrac{3a^2 b^3}{2cd} \times \dfrac{c^3 d^2}{ab}$.

5. Simplify $\dfrac{6pq}{5rs} \div \dfrac{4p^2}{7qs}$.

6. Express $\dfrac{2x - 5}{5x} - \dfrac{3x + 2}{4x} + \dfrac{7x + 15}{10x}$ as a single fraction in its lowest terms.

7. Make $3(x + 2) - \dfrac{4x - 5}{4}$ into a single fraction.

Multi-Choice Questions 15

1. $\dfrac{2}{x} + \dfrac{3}{2x}$ expressed as a single fraction is

   A $\dfrac{5}{3x}$ B $\dfrac{7}{2x}$ C $\dfrac{5}{2x^2}$ D $\dfrac{7}{2x^2}$

2. $\dfrac{a^2 - a}{a - 1}$ is equal to

   A a B $a^2 - 1$ C a^2 D $1 - a^2$

3. $\dfrac{6m - 2n}{2}$ is equal to

   A $\dfrac{3m - 2n}{2}$ B $6m - n$

   C $3m - 2n$ D $3m - n$

4. $t - \dfrac{3 - t}{2} + \dfrac{3 + t}{2}$ is equal to

   A t B $2t$ C $4t$ D $3 + 2t$

5. $\dfrac{3x - 7}{3} - \dfrac{2x - 5}{2}$ is equal to

   A -12 B -2 C $\dfrac{1}{6}$ D $-\dfrac{29}{6}$

Linear Equations

Introduction

Linear equations contain only the first power of the unknown quantity. Thus

$$5x - 3 = 7 \quad \text{and} \quad \frac{x}{4} = 9$$

are both linear equations.

The **solution** of an equation is that value of the unknown which, when substituted into the equation, makes the left-hand side equal to the right-hand side. Linear equations (sometimes called simple equations) have only one solution.

Solving Equations

In solving equations:

(1) The same quantity may be added or subtracted from both sides.

(2) Each side may be multiplied or divided by the same quantity.

Example 1

(a) Solve $\dfrac{x}{5} = 3$.

Multiplying each side by 5 gives

$$5 \times \frac{x}{5} = 5 \times 3$$

$$x = 15$$

(Check: substituting for x in the LHS, $\dfrac{15}{5} = 3$. Hence LHS = RHS and the solution is therefore correct.)

(b) Solve $4x = 8$.

Dividing both sides by 4 gives

$$\frac{4x}{4} = \frac{8}{4}$$

$$x = 2$$

(c) Solve $3x + 7 = 19$.

Subtract 7 from both sides then

$$3x + 7 - 7 = 19 - 7$$

$$3x = 12$$

Dividing both sides by 3 gives

$$\frac{3x}{3} = \frac{12}{3}$$

$$x = 4$$

Short cuts may be taken when you fully understand the processes outlined in Example 1. For instance:

The operation of subtracting 7 from both sides can be quickly accomplished by taking the 7 to the RHS and changing its sign.

$$3x + 7 = 19$$

$$3x = 19 - 7$$

$$3x = 12$$

$$x = \frac{12}{3}$$

$$x = 4$$

Equations Containing the Unknown on Both Sides

When an equation contains the unknown on both sides, group all the terms containing the unknown on one side of the equation and all the other terms on the other side.

Example 2

(a) Solve $5x - 20 = 3x - 8$.

$$5x - 20 = 3x - 8$$
$$5x - 3x = 20 - 8$$
$$2x = 12$$
$$x = 6$$

(b) Solve $6 - 5x = 3x + 22$.

$$6 - 5x = 3x + 22$$
$$6 - 22 = 3x + 5x$$
$$-16 = 8x$$
$$-2 = x$$
$$x = -2$$

(Note that we have taken the terms containing x to the RHS rather than the LHS. If we had used the LHS we would get $-8x = 16$ which is more difficult to manipulate.)

Equations Containing Brackets

In solving equations of this type, remove the brackets first and then solve the resulting equation using the methods previously shown.

Example 3

Solve $2(x - 3) - (x + 2) = 5$.

Removing the brackets we have
$$2x - 6 - x - 2 = 5$$
$$x - 8 = 5$$
$$x = 5 + 8$$
$$x = 13$$

Equations Containing Fractions

The first step is to get rid of the denominators by multiplying each term by the LCM of the denominators. The resulting equation can then be solved by the methods previously shown.

Example 4

(a) Solve $\dfrac{x}{2} + \dfrac{x}{4} = \dfrac{3}{5}$.

The LCM of the denominators 2, 4 and 5 is 20. Multiplying each term by 20 gives

$$\frac{x}{2} \times 20 + \frac{x}{4} \times 20 = \frac{3}{5} \times 20$$
$$10x + 5x = 12$$
$$15x = 12$$
$$x = \frac{12}{15}$$
$$x = \frac{4}{5}$$

(b) Solve $\dfrac{x - 4}{3} - \dfrac{2x - 3}{5} = 2$.

The LCM of the denominators 3 and 5 is 15. Multiplying each term by 15 gives

$$\frac{x - 4}{3} \times 15 - \frac{2x - 3}{5} \times 15 = 2 \times 15$$
$$5(x - 4) - 3(2x - 3) = 30$$
$$5x - 20 - 6x + 9 = 30$$
$$-x - 11 = 30$$
$$-x = 30 + 11$$
$$-x = 41$$
$$x = -41$$

(Note carefully that the line separating the numerator and the denominator acts as a bracket.)

Exercise 16.1

Solve the following equations:

1. $x + 3 = 8$
2. $x - 4 = 6$
3. $2x = 8$
4. $\dfrac{x}{3} = 5$
5. $2x - 7 = 9$
6. $5x + 3 = 18$
7. $3x - 7 = x - 5$
8. $9 - 2x = 3x + 7$
9. $4x - 3 = 6x - 9$
10. $5x - 8 = 3x + 2$
11. $6x + 11 = 25 - 2x$
12. $2(x + 1) = 9$
13. $5(x - 3) = 12$
14. $3(2x - 1) + 4(2x + 5) = 40$
15. $4(x - 5) = 14 - 5(2 - x)$
16. $7(2 - 3x) = 3(5x - 1)$
17. $\dfrac{x}{2} + \dfrac{x}{3} = 10$
18. $3x + \dfrac{3}{8} = 2 + \dfrac{2x}{3}$
19. $\dfrac{2x}{5} = \dfrac{x}{8} + \dfrac{1}{2}$
20. $\dfrac{6}{x} = 3$
21. $\dfrac{x + 3}{2} = \dfrac{x - 3}{3}$
22. $\dfrac{x + 3}{5} - \dfrac{2x + 1}{4} = 2$
23. $\dfrac{2x}{5} - \dfrac{x - 6}{2} - \dfrac{5x}{3} = \dfrac{3}{4}$
24. $\dfrac{3x - 5}{4} - \dfrac{9 - 2x}{3} = \dfrac{x - 3}{2}$
25. $\dfrac{1}{2x} + \dfrac{1}{3x} = \dfrac{3}{5}$
26. $\dfrac{10}{x} + 5 = 3$
27. $\dfrac{3x + 2}{2} - \dfrac{3x - 2}{3} = \dfrac{11}{4}$

Making Algebraic Expressions

Algebraic expressions can be used to translate information into symbols.

Example 5

(a) Express algebraically: the sum of two numbers a and b multiplied by a third number c.

The sum of a and $b = a + b$

The sum of a and b multiplied by $c = c(a + b)$

(b) Express as an algebraic expression: the total cost, in pounds, of a record player costing £P and n records costing Q pence each.

Cost of n records at Q pence each

$= nQ$ pence

$= £\dfrac{nQ}{100}$

Total cost of record player and records

$= £\left(P + \dfrac{nQ}{100}\right)$

Exercise 16.2

1. A girl is m years old now. How old was she 3 years ago?

2. Find the total cost, in pounds, of 8 pencils costing a pence each and b pens costing 60p each.

3. m articles cost a total of £x. Find the cost of buying n articles at the same rate.

4. In one innings a batsman scored p sixes, q fours and r single runs. Make up an algebraic expression giving the total number of runs scored.

5. Express algebraically: five times a number x minus three times a second number y.

6. If x and y represent two numbers write down algebraic expressions to represent:

   (a) 4 times the sum of the two numbers

   (b) 3 times the product of the two numbers.

7. I have a five pence coins, b ten pence coins and c fifty pence pieces. Make up an expression showing the amount of money that I have.

8. 7 times a certain number x is divided by 3 times a second number y. To the result is added a third number z. Write down an algebraic expression to represent this information.

9. If a oranges cost £b:

   (a) How much, in pence, does each orange cost?

   (b) How much, in pounds, do c oranges cost?

10. If plums cost x pence per kilogram find the weight of plums that can be bought for £y.

Construction of Simple Equations

The following notes will help you to construct simple (or linear) equations:

(1) Represent the quantity to be found by a symbol (x is usually used).

(2) Make up the equation which conforms to the given information.

(3) Make sure that both sides of the equation are in the same units.

Example 6

(a) A library buys 50 books. Some cost £5 and the others cost £7. If the library spent £290 in all, how many books costing £5 did it buy?

Let x be the number of books bought for £5 each.

Then $(50 - x)$ is the number of books bought for £7 each.

The total cost of the 50 books is

$$£5x + 7(50 - x) = £(350 - 2x)$$

But the total cost is £290. Hence

$$350 - 2x = 290$$
$$60 = 2x$$
$$x = 30$$

Hence 30 books were bought at £5 each.

(b) Find the number which when added to the numerator and denominator of the fraction $\frac{1}{3}$ makes $\frac{3}{4}$.

Let the number be x. Then

$$\frac{1 + x}{3 + x} = \frac{3}{4}$$
$$4(1 + x) = 3(3 + x)$$
$$4 + 4x = 9 + 3x$$
$$4x - 3x = 9 - 4$$
$$x = 5$$

Exercise 16.3

1. 25 articles are bought. Some cost 8p each and the others cost 7p each. If the total cost is £1.95, how many of each are bought?

2. I think of a number. If I subtract 9 from it and multiply this difference by 4 the result is 32. What is the number I thought of?

3. A rectangular room is 2 metres longer than its width. If the perimeter of the room is 24 metres, form an equation and hence find the dimensions of the room.

4. Find the number which, when added to the numerator and denominator of the fraction $\frac{5}{9}$ makes the new fraction $\frac{2}{3}$.

5. Find three consecutive whole numbers whose sum is 54.

6. I have 27 coins, some 5p and some 10p. The total value of these coins is £2.35. Form an equation and solve it to find the number of 5p coins.

7. 4 times a certain number plus 9 is equal to 3 times the number plus 20. Find the number.

8. The sides of a triangle are x cm, $(x - 5)$ cm and $(x + 3)$ cm long. If the perimeter is 25 cm, find the length of the three sides.

Miscellaneous Exercise 16

Section A

1. Solve the following equations:
   (a) $2(x - 3) + 5x = 22$
   (b) $3(x + 5) = 2(x - 5) + 20$

2. I have 25 coins. Some are ten pence and some are five pence pieces. How many of each kind do I have if their value is £1.75?

3. The sum of three successive whole numbers is 60. What are the numbers?

4. When I double a number and add 6 the result is 14. What is the number?

5. A shopping bag weighs 6 ounces. If w pounds of groceries and z pounds of meat are put in it, what is the total weight in ounces?

Section B

1. 18 books are to be bought for a library. Some cost £6 each and the remainder cost £7.50 each. If £120 is to be spent, how many of each type of book can be bought?

2. Find the number which, when added to the numerator and denominator of the fraction $\frac{3}{5}$, gives a new fraction which is equal to $\frac{4}{5}$.

3. Solve the equation $3x + \dfrac{3}{4} = 2 + \dfrac{2x}{3}$.

4. The real numbers x and y are related by
$$y = \frac{4}{5}(x - 18)$$
   (a) When $x = 4$, find the value of y.
   (b) Find the value of x when $y = 6$.

5. Solve the equation $\dfrac{3x + 2}{x + 2} = \dfrac{17}{7}$.

6. A shopkeeper buys n boxes of tea which costs him £a. He empties all the packets and divides the tea he obtains into y packets, each having a weight of p grams.
   (a) How much tea, in kilograms, did the shopkeeper buy?
   (b) What is the weight, in kilograms, of the tea in each of the original boxes?
   (c) How much has the tea in each packet cost him?

7. Solve the equations for p:
   (a) $0.2(p - 15) + 0.3(p + 20) = 8$
   (b) $\frac{1}{5}(p - 2) - \frac{1}{7}(p + 3) = 1$

Multi-Choice Questions 16

1. The value of y which satisfies the equation $4(y - 4) = 20$ is

   A 1   B 6   C 9   D 24

2. Consider the equation $y = 3x - 2$. The value of x when $y = 7$ is

   A $1\frac{2}{3}$ B 3   C 6   D 27

3. If $\dfrac{5x}{6} - \dfrac{3x}{4} = \dfrac{1}{12}$, then the value of x is

   A -8 B -2 C -1 D 1

4. When $a + 3 = \dfrac{5a}{2}$, the value of a is

   A 0   B 2   C 6   D 30

5. If $a = x - 1$ and $b = 2x - 3$, when $a = b$ the value of x is

   A -4 B -2 C $-\frac{2}{3}$ D 2

6. £$3x$ and y pence may be expressed in pence as

   A $(3x + y)$ B $3(x + y)$

   C $(300x + 100y)$ D $(300x + y)$

7. If $3(2x - 5) - 2(x - 3) = 3$, then x is equal to

   A $\frac{5}{4}$ B $\frac{11}{4}$ C 3   D 6

Mental Test 16

Try to answer the following questions without writing anything down except the answer.

Solve the following equations:

1. $3x = 12$

2. $x + 4 = 7$

3. $x + 9 = 3$

4. $\dfrac{x}{4} = 2$

5. $\dfrac{9}{x} = 3$

6. $3x = 2x - 6$

7. $2x + 3 = 3x$

8. $5x + 1 = 16$

9. A boy is k years old. His teacher is twice as old. How old is the teacher?

10. A basket contains a apples and a second basket contains b apples. The contents of the two baskets are shared equally between c boys. How many apples does each boy receive?

11. A pail holds x pints. How many times can it be filled from a tank containing k gallons?

12. The length of a playing field is x yards. Express the length in

    (a) inches   (b) feet

Formulae

Evaluating Formulae

A formula is an equation which describes the relationship between two or more quantities. The statement that $I = PRT$ is a formula for I in terms of P, R and T. The value of I may be found by substituting the values of P, R and T.

Example 1

(a) If $I = PRT$ find the value of I when $P = 20$, $R = 2$ and $T = 5$.

Substituting the given values of P, R and T and remembering that multiplication signs are omitted in formulae, we have

$$I = 20 \times 2 \times 5$$
$$= 200$$

(b) The formula $v = u + at$ is used in physics. Find the value of v when $u = 8$, $a = 3$ and $t = 2$.

$$v = 8 + 3 \times 2$$
$$= 8 + 6$$
$$= 14$$

Exercise 17.1

1. If $V = Ah$, find the value of V when $A = 6$ and $h = 3$.

2. The formula $P = \dfrac{RT}{V}$ is used in connection with the expansion of gases. Find the value of P when $R = 25$, $T = 230$ and $V = 5$.

3. If $a = b - cx$, find the value of a when $b = 32$, $c = 3$ and $x = 7$.

4. The formula $V = \sqrt{2gh}$ is used in physics. Calculate the value of V when $g = 9.8$ and $h = 7$.

5. Calculate d from the formula $d = \dfrac{2(S - an)}{n(n - p)}$ when $S = 12$, $a = 2$, $n = 5$ and $p = 3$.

Transposing Formulae

The formula $y = ax + b$ has y as its **subject**. By rearranging this formula we could make x the subject. We are then said to have transposed the formula to make x the subject.

The rules for transforming a formula are:

(1) Remove square roots or other roots.

(2) Get rid of fractions.

(3) Clear brackets.

(4) Collect together the terms containing the required subject.

(5)   Factorise if necessary.

(6)   Isolate the required subject.

These steps should be performed in the order given.

Example 2

(a) Transpose the formula $V = \dfrac{2R}{R-r}$ to make R the subject.

Step 1   Since there are no roots get rid of the fraction by multiplying both sides of the equation by $(R-r)$.

$$V(R-r) = 2R$$

Step 2   Clear the bracket.

$$VR - Vr = 2R$$

Step 3   Collect the terms containing R on the LHS.

$$VR - 2R = Vr$$

Step 4   Factorise the LHS.

$$R(V-2) = Vr$$

Step 5   Isolate R by dividing both sides of the equation by $(V-2)$.

$$R = \frac{Vr}{V-2}$$

Although we used five steps to obtain the required subject, in very many cases far fewer steps are needed. Nevertheless, you should work through the steps in the order given.

(b) Transpose $d = \sqrt{2hr}$ to make h the subject.

Step 1   Remove the square root by squaring both sides.

$$d^2 = 2hr$$

Step 2   Since there are no fractions or brackets and factorisation is not needed we can now isolate h by dividing both sides of the equation by $2r$.

$$\frac{d^2}{2r} = h$$

or $$h = \frac{d^2}{2r}$$

since it is usual to position the subject on the LHS.

Exercise 17.2

Transpose each of the following formulae:

1. $C = ad$ for d.

2. $PV = c$ for P.

3. $x = \dfrac{b}{y}$ for y.

4. $I = \dfrac{E}{R}$ for E.

5. $S = \dfrac{ta}{p}$ for a.

6. $a = b + 8$ for b.

7. $y = \dfrac{7}{4+x}$ for x.

8. $3k = kx + 5$ for k.

9. $E = \frac{1}{2}mv^2$ for m.

10. $y = mx + c$ for x.

11. $v = u + at$ for t.

12. $V = \dfrac{abh}{3}$ for h.

13. $M = 5(x + y)$ for y.

14. $C = \dfrac{N-n}{2p}$ for n.

15. $S = ar(r + h)$ for h.

16. $t = a + (n-1)d$ for n.

17. $A = 3(x - y)$ for y.

18. $d = \dfrac{v^2}{200} + k$ for k.

19. $6x + 2y = 8$ for y.

20. $y = \dfrac{2 - 5x}{2 + 3x}$ for x.

21. $k = \dfrac{3n + 2}{n + 3}$ for n.

22. $T = 2\pi \sqrt{\dfrac{R - H}{g}}$ for R.

23. $a = \sqrt{\dfrac{b}{b + c}}$ for b.

24. $K = \dfrac{mv^2}{2g}$ for v.

25. $r = \sqrt{\dfrac{A}{4\pi}}$ for A.

26. $q = \dfrac{m}{\sqrt{p}}$ for p.

27. $x = (x - a)(x + b)$ for a.

28. $y = \dfrac{a + x}{5 - bx}$ for x.

29. $x = \dfrac{5 - 4y}{3y + 2}$ for y.

30. $T = 2\sqrt{\dfrac{k^2 + h^2}{gh}}$ for k.

Miscellaneous Exercise 17

Section A

1. The volume of a cuboid is given by the formula $V = Ah$, where A is the area of the cross-section and h is the length.

    (a) Transpose the formula to make h the subject.

    (b) Find the length of a cuboid whose volume is 50 cm^3 and whose cross-sectional area is 10 cm^2.

2. The sum of the angles of a polygon is given by the formula $S = 90(2n - 4)$ where n is the number of sides possessed by the polygon. Transpose the formula to make n the subject and use the transposed formula to find n when $S = 720°$.

3. Transpose the formula $my = b + d$ to make y the subject. Find the value of y when $m = 3$, $b = 4$ and $d = 6$.

4. A formula used in geometry is $C = 180 - A - B$. Transpose the formula to express A in terms of B and C.

5. The formula $v = at + 16t^2$ is used in mechanics. Rewrite the formula to make a the subject and find its value when $v = 300$ and $t = 4$.

Section B

1. The positive numbers I, n, E, R and r are connected by the formula
$$I = \dfrac{nE}{nR + r}.$$ Express n in terms of I, E, R and r.

2. Make K the subject of the formula
$$p = \sqrt{\dfrac{K - 1}{K + 1}}.$$

3. Make x the subject of the formula
$$p = \dfrac{k}{\sqrt{x}}.$$

4. Make x the subject of the formula
$$y = \dfrac{p + x}{1 - px}.$$

5. Use the formula $T - mg = ma$ to find

    (a) the value of T when $m = 8$, $g = 32$ and $a = 12$

    (b) the value of a when $T = 200$, $g = 32$ and $m = 5$.

Multi-Choice Questions 17

Use the formula $100I = PTR$ to answer questions 1, 2 and 3.

1. When $P = 5$, $T = 6$ and $R = 7$, the value of I is

    A 2100   B 210   C 21   D 2.1

2. When $I = 42$, $P = 360$ and $R = 2\frac{1}{2}$, the value of T is

    A $\frac{1}{3}$ B $2\frac{1}{2}$ C 3   D $4\frac{2}{3}$

3. R is equal to

    A $100IPT$ B $\dfrac{100I}{PT}$

    C $100I - PT$ D $\dfrac{PTI}{100}$

4. If $X = Pr + Q$ then r is equal to

    A $\dfrac{X - Q}{P}$ B $X - Q - P$

    C $\dfrac{X}{P + Q}$ D $\dfrac{X - P}{Q}$

5. If $T = a + (n - 1)d$ then n is equal to

    A $\dfrac{T}{d} - a$ B $\dfrac{T - a}{d} + 1$

    C $\dfrac{T - a + 1}{d}$ D $\dfrac{T}{d} - a - 1$

Simultaneous Equations

Introduction

Consider the equations

$$3x + 5y = 21$$
$$2x + 3y = 13$$

Each equation contains the unknown quantities x and y. The solutions are the values of x and y which satisfy both equations simultaneously. Equations like these are called simultaneous equations.

The Method of Elimination

This method is often used to solve simultaneous equations.

Example 1

(a) Solve the simultaneous equations

$$x + 2y = 11 \qquad (1)$$
$$3x - 2y = 1 \qquad (2)$$

We can eliminate y from the equations by adding the two equations.

$$(x + 3x) + (2y - 2y) = 11 + 1$$
$$4x = 12$$
$$x = 3$$

To find the value of y we substitute for x in either equation (1) or in equation (2). Substituting for x in equation (1),

$$3 + 2y = 11$$
$$2y = 8$$
$$y = 4$$

To check the solutions substitute for x and y in equation (2) and see if the LHS equals the RHS.

$$LHS = 3 \times 3 - 2 \times 4$$
$$= 9 - 8$$
$$= 1$$
$$= RHS$$

Hence the solutions $x = 3$ and $y = 4$ are correct.

(Note that there would be no point in checking in equation (1) because this equation was used to find the value of y.)

(b) Solve the simultaneous equations

$$3x + 5y = 21 \qquad (1)$$
$$2x + 3y = 13 \qquad (2)$$

If we multiply equation (1) by 2 and equation (2) by 3 we will get the same coefficient of x in both equations so that we can eliminate x from the equations.

$$6x + 10y = 42 \qquad (3)$$
$$6x + 9y = 39 \qquad (4)$$

Subtracting equation (4) from equation (3),

$$(6x - 6x) + (10y - 9y) = 42 - 39$$
$$y = 3$$

To find the value of x we substitute for y in equation (1).

$$3x + 5 \times 3 = 21$$

$$3x + 15 = 21$$

$$3x = 6$$

$$x = 2$$

Using equation (2) for the check,

$$\text{LHS} = 2 \times 2 + 3 \times 3$$

$$= 4 + 9$$

$$= 13$$

$$= \text{RHS}$$

Hence the solutions $x = 2$ and $y = 3$ are correct.

(c) Solve the equations

$$\frac{x-2}{3} + \frac{y-1}{4} = \frac{13}{12} \qquad (1)$$

$$\frac{2-x}{2} + \frac{3+y}{3} = \frac{11}{6} \qquad (2)$$

The first step is to get rid of the fractions.

In equation (1) the LCM of the denominators is 12. Multiplying each term by 12,

$$4(x-2) + 3(y-1) = 13$$

$$4x - 8 + 3y - 3 = 13$$

$$4x + 3y = 24 \qquad (3)$$

In equation (2) the LCM of the denominators is 6. Multiplying each term by 6,

$$3(2-x) + 2(3+y) = 11$$

$$6 - 3x + 6 + 2y = 11$$

$$-3x + 2y = -1 \qquad (4)$$

To eliminate y we multiply equation (3) by 2 and equation (4) by 3.

$$8x + 6y = 48 \qquad (5)$$

$$-9x + 6y = -3 \qquad (6)$$

Subtracting equation (6) from equation (5),

$$[8x - (-9x)] + (6y - 6y) = [48 - (-3)]$$

$$17x = 51$$

$$x = 3$$

To find the value of y substitute for x in equation (5).

$$8 \times 3 + 6y = 48$$

$$24 + 6y = 48$$

$$6y = 24$$

$$y = 4$$

Since equation (5) came from equation (1) we must do the check in equation (2).

$$\text{LHS} = \frac{2-3}{2} + \frac{3+4}{3}$$

$$= -\frac{1}{2} + \frac{7}{3}$$

$$= \frac{3 \times (-1) + 2 \times 7}{6}$$

$$= \frac{-3 + 14}{6}$$

$$= \frac{11}{6}$$

$$= \text{RHS}$$

Hence the solutions $x = 3$ and $y = 4$ are correct.

Exercise 18.1

Solve the following simultaneous equations for x and y and check the solutions:

1. $2x - 3y = -8$
    $x + 3y = 14$

2. $3x + 5y = 17$
    $4x + 5y = 21$

3. $3x + 4y = 26$
    $x + y = 7$

4. $5x - 7y = 1$
    $2x + 5y = 16$

5. $\dfrac{x}{2} + \dfrac{y}{5} = 5$

 $\dfrac{2x}{3} + \dfrac{3y}{2} = 19$

6. $\dfrac{x}{4} - y = 0$

 $2x + \dfrac{y}{2} = 25\tfrac{1}{2}$

7. $\dfrac{x-3}{4} - \dfrac{y-2}{5} = \dfrac{13}{20}$

 $\dfrac{4-x}{3} + \dfrac{3-y}{2} = -\dfrac{7}{3}$

8. $x - y = 2$

 $\dfrac{3x}{2} + 4y = 19\tfrac{1}{2}$

Problems Involving Simultaneous Equations

When a problem involves two unknowns we form two separate equations from the given information.

Example 2

(a) Find two numbers x and y such that their sum is 24 and their difference is 6.

Let the two numbers be x and y.
Their sum is $x + y$ and their difference is $x - y$. The two equations are

$$x + y = 24 \qquad (1)$$
$$x - y = 6 \qquad (2)$$

To eliminate y we add equation (2) to equation (1).

$$2x = 30$$
$$x = 15$$

To find the value of y substitute for x in equation (1).

$$15 + y = 24$$
$$y = 9$$

Hence the two numbers are 15 and 9.

(b) 500 tickets were sold for a rock and roll concert. Some cost £5 and some cost £8. The cash received for the dearer tickets was £100 more than for the cheaper tickets. Find the number of each kind of ticket which were sold.

Let x be the number of dearer tickets sold and y be the number of cheaper tickets sold

Cash received for dearer tickets
$= £8x$

Cash received for the cheaper tickets
$= £5y$

The two equations are

$$x + y = 500 \qquad (1)$$
$$8x - 5y = 100 \qquad (2)$$

Multiplying equation (1) by 5,

$$5x + 5y = 2500 \qquad (3)$$

Adding equations (2) and (3),

$$13x = 2600$$
$$x = 200$$

To find the value of y substitute for x in equation (1).

$$200 + y = 500$$
$$y = 300$$

Hence 200 dearer tickets and 300 cheaper tickets were sold.

Exercise 18.2

1. Find two numbers whose sum is 42 and whose difference is 12.

2. A bill for £117 was paid using £1 and £5 notes, a total of 45 notes being used. How many £1 notes were used?

3. 800 tickets were sold for a 'pop' concert some costing £9 and some costing £12. The total cash received was £8550. Find the number of cheaper tickets sold.

4. Choosing x and y to represent two numbers write down equations to represent each of the following statements:

   (a) The sum of the two numbers is 42.

   (b) The number, x, is six times the other number.

   (c) Find values of x and y which will satisfy both statements.

5. A householder installed x radiators which cost £30 each and y electric fires which cost £50 each. His total expenditure was £490 and he spent £10 more on the fires than he did on the radiators. Form two equations and solve them.

Miscellaneous Exercise 18

1. Solve the simultaneous equations

   $$3x + 2y = 4 \qquad (1)$$
   $$x + 2y = 0 \qquad (2)$$

2. Find the value of q which satisfies the equation

   $$3p - 4q - 11 = 0 \qquad (1)$$
   $$5p + 9q + 13 = 0 \qquad (2)$$

3. Solve the simultaneous equations

   $$y = 5x + 4 \qquad (1)$$
   $$y = 4x + 6 \qquad (2)$$

4. A bill for £70 was paid with £5 notes and £10 notes, a total of 11 notes being used. How many £5 notes were used?

5. Find two numbers such that their sum is 108 and their difference is 54.

6. Given the two equations

   $$3x + 2y = 9 \qquad (1)$$
   $$2x + 3y = 16 \qquad (2)$$

   (a) by adding the two equations find the value of $x + y$,

   (b) by subtracting the two equations find the value of $x - y$,

   (c) find the values of x and y.

7. The rate of pay in a factory is £x per hour for the basic pay and £y per hour for overtime. One man works a basic week of 38 hours and 6 hours' overtime whilst a second man works a basic week of 40 hours plus 4 hours' overtime. If the first man is paid £235 and the second man £230, write down two equations in x and y and solve them simultaneously to find the basic rate of pay.

Multi-Choice Questions 18

1. The solution of the simultaneous equations

   $$2x - y = 2 \qquad (1)$$

   and $\qquad x + y = -5 \quad$ is $\qquad (2)$

   A $x = 1, y = 4$ **B** $x = -1, y = -4$

   C $x = -1, y = 4$ **D** $x = 1, y = -4$

2. Consider the equations

$$3x - y = 2 \qquad (1)$$
$$3x - 2y = 1 \qquad (2)$$

The result of subtracting equation (2) from equation (1) is

A $-3y = 1$ B $6x - 3y = 1$

C $y = 1$ D $6x + y = 1$

3. If $x + y = 5$ and $2x - y = 7$, what is the value of x which satisfies both these equations?

A 1   B 2   C 3   D 4

4. Which pair of values satisfies the equations

$$4x + y = 23 \qquad (1)$$
$$3x - y = 12 \qquad (2)$$

A $x = 6, y = 6$ B $x = 4, y = 7$

C $x = 5, y = 3$ D $x = 11, y = 21$

5. Find the value of x which satisfies the equations $x + y = 7$ and $x - y = 1$.

A 3   B 4   C 5   D 6

19 Quadratic Equations

Introduction

An equation which can be written in the form $ax^2 + bx + c = 0$ is called a **quadratic equation**. The constants a, b and c can take any numerical value. The following are all examples of quadratic equations:

$$x^2 - 9 = 0$$

in which $a = 1$, $b = 0$ and $c = -9$

$$3x^2 + 8x + 5 = 0$$

in which $a = 3$, $b = 8$ and $c = 5$

$$4x^2 - 7x = 0$$

in which $a = 4$, $b = -7$ and $c = 0$.

A quadratic equation always has two solutions (often called the roots of the equation). It is possible for one of the roots to be zero or for the two solutions to be the same.

Solving Quadratic Equations

A quadratic equation can be solved by factorisation. We make use of the fact that if the product of two factors is zero, then one of those factors must be zero. Thus if $mn = 0$ then either $m = 0$ or $n = 0$. To solve a quadratic equation by this method the expression $ax^2 + bx + c$ is written as the product of two factors.

Example 1

(a) Solve $x^2 = 9$.

Writing the equation as

$$x^2 - 9 = 0$$

and factorising the LHS

$$(x + 3)(x - 3) = 0$$

Either $x + 3 = 0$, giving $x = -3$

or $x - 3 = 0$, giving $x = 3$.

The roots are $x = -3$ and $x = 3$, often written as $x = \pm 3$.

(b) Solve $x^2 + 7x + 12 = 0$.

Factorising,

$$(x + 4)(x + 3) = 0$$

Either $x + 4 = 0$, giving $x = -4$

or $x + 3 = 0$, giving $x = -3$

The roots are $x = -4$ and $x = -3$.

(c) Solve $x^2 - 6x = 0$.

Factorising,

$$x(x - 6) = 0$$

Either $x = 0$ or $x - 6 = 0$, giving $x = 6$.

The two roots are $x = 0$ and $x = 6$.

(Note that it is incorrect to say that the solution is $x = 6$. The solution $x = 0$ must also be stated.)

(d) Solve $4x^2 - 20x + 25 = 0$.

Factorising,

$$(2x - 5)(2x - 5) = 0$$

or $\qquad\qquad (2x - 5)^2 = 0$

For both factors, $2x - 5 = 0$ giving $x = 2.5$.

With this equation both roots are the same and we say that the equation has equal roots. This always happens when the expression $ax^2 + bx + c$ forms a perfect square.

(e) Solve the equation $x - 10 + \dfrac{9}{x} = 0$.

Multiplying each term by x to clear the fraction,

$$x^2 - 10x + 9 = 0$$

Factorising,

$$(x - 9)(x - 1) = 0$$

Either $x - 9 = 0$, giving $x = 9$

or $x - 1 = 0$, giving $x = 1$.

The roots are $x = 9$ and $x = 1$.

Exercise 19.1

Solve each of the following quadratic equations using factorisation:

1. $x^2 - 4 = 0$
2. $x^2 - 16 = 0$
3. $3x^2 - 27 = 0$
4. $5x^2 - 125 = 0$
5. $(x - 7)(x + 3) = 0$
6. $x(x + 5) = 0$
7. $(3x - 5)(2x + 9) = 0$
8. $x^2 - 5x + 6 = 0$
9. $x^2 + 3x - 10 = 0$

10. $6x^2 - 11x - 35 = 0$
11. $8x^2 - 10x + 3 = 0$
12. $x^2 - 4x = 0$
13. $x^2 + 6x + 9 = 0$
14. $9x^2 + 30x + 25 = 0$
15. $4x^2 - 12x + 9 = 0$

Equations which will not Factorise

Example 2

(a) Solve the equation $2x^2 - 12 = 0$.

Since $\qquad 2x^2 - 12 = 0$

$$2x^2 = 12$$

$$x^2 = 6$$

$$x = \pm\sqrt{6}$$

and $\qquad\qquad x = \pm 2.45$

(by using a calculator to find $\sqrt{6}$)

(b) Solve the equation $2x^2 + 18 = 0$.

Since $\qquad 2x^2 + 18 = 0$

$$2x^2 = -18$$

$$x^2 = -9$$

and $\qquad\qquad x = \pm\sqrt{-9}$

The square root of a negative quantity has no arithmetic meaning and it is called an **imaginary number**. The reason is as follows:

$$(-3)^2 = 9$$

$$(+3)^2 = 9$$

$$\sqrt{9} = \pm 3$$

Hence it is not possible to give a meaning to $\sqrt{-9}$. The equation $2x^2 + 18 = 0$ is said, therefore, to have imaginary roots.

Completing the Square

It will be recalled that

$$(x + h)^2 = x^2 + 2hx + h^2$$

The quadratic expression $x^2 + 2hx + h^2$ is said to be a **perfect square**. To factorise it we note that the last term (it is h^2) is (half the coefficient of x)^2, i.e.

$$(\tfrac{1}{2} \times 2h)^2 = h^2$$

To factorise $x^2 + 8x + 16$ we note that

$$\begin{aligned}
(\tfrac{1}{2} \text{ coefficient of } x)^2 &= (\tfrac{1}{2} \times 8)^2 \\
&= 4^2 \\
&= 16
\end{aligned}$$

Since this equals the last term,

$$x^2 + 8x + 16 = (x + 4)^2$$

Similarly, because

$$(x - h)^2 = x^2 - 2hx + h^2$$
$$x^2 - 6x + 9 = (x - 3)^2$$

Many quadratic expressions do not form a perfect square. For such expressions, it is often useful to complete the square so that

$$ax^2 + bx + c = a(x + h)^2 + k$$

The problem is to find the numerical values of h and k.

Example 3

Write $3x^2 + 8x + 5$ in the form $a(x + h)^2 + k$.

$$3x^2 + 8x + 5 = 3(x^2 + \tfrac{8}{3}x) + 5$$

Adding (half the coefficient of x)^2 to the expression in the bracket, we have

$$\begin{aligned}
&3x^2 + 8x + 5 \\
&= 3[x^2 + \tfrac{8}{3}x + (\tfrac{4}{3})^2] + 5 - (\tfrac{4}{3})^2 \times 3 \\
&= 3(x + \tfrac{4}{3})^2 - \tfrac{1}{3}
\end{aligned}$$

Example 4

Write $-5x^2 + 6x + 2$ in the form $k - a(x + h)^2$.

$$\begin{aligned}
&-5x^2 + 6x + 2 \\
&= 2 - 5(x^2 - \tfrac{6}{5}x) \\
&= 2 - 5[x^2 - \tfrac{6}{5}x + (\tfrac{3}{5})^2] + (\tfrac{3}{5})^2 \times 5 \\
&= \tfrac{19}{5} - 5(x - \tfrac{3}{5})^2
\end{aligned}$$

Solving Quadratic Equations by Completing the Square

Any quadratic equation may be solved by completing the square as shown in Examples 3 and 4.

Example 5

Solve the equation $4x^2 - 8x + 1 = 0$.

Completing the square of $4x^2 - 8x + 1$, we have

$$4(x - 1)^2 - 3 = 0$$
$$4(x - 1)^2 = 3$$
$$(x - 1)^2 = \frac{3}{4}$$
$$\begin{aligned}
x - 1 &= \pm\sqrt{\frac{3}{4}} \\
&= \pm\frac{\sqrt{3}}{2} \\
&= \pm 0.866
\end{aligned}$$
$$\begin{aligned}
x &= 1 \pm 0.866 \\
&= 1.866 \quad \text{or} \quad 0.134
\end{aligned}$$

Example 6

Solve the equation $-3x^2 + 6x + 5 = 0$.

Multiplying through by -1, we have

$$3x^2 - 6x - 5 = 0$$

On completing the square of $3x^2 - 6x - 5$, we have

$$3(x-1)^2 - 8 = 0$$

$$(x-1)^2 = \frac{8}{3}$$

$$x - 1 = \pm\sqrt{\frac{8}{3}}$$

$$= \pm 1.633$$

$$x = 1 \pm 1.633$$

$$= 2.633 \quad \text{or} \quad -0.633$$

Using the Formula to Solve Quadratic Equations

Although any quadratic equation may be solved by completing the square, it is useful to derive a formula for solving such equations.

$$ax^2 + bx + c = 0$$

$\therefore \qquad ax^2 + bx = -c$

Multiply by $4a$

$\therefore \qquad 4a^2x^2 + 4abx = -4ac$

Completing the square

$$4a^2x^2 + 4abx + b^2 = b^2 - 4ac$$

$\therefore \qquad (2ax + b)^2 = b^2 - 4ac$

$\therefore \qquad 2ax + b = \pm\sqrt{b^2 - 4ac}$

$$2ax = -b \pm \sqrt{b^2 - 4ac}$$

$$x = \frac{-b \pm \sqrt{b^2 - 4ac}}{2a}$$

This equation is called the quadratic formula. Note that the whole of the numerator, including $-b$, is divided by $2a$. The formula is used when factorisation is not possible, although it may be used to solve any quadratic equation.

Example 7

(a) Solve the equation

$$3x^2 - 8x + 2 = 0$$

Comparing with $ax^2 + bx + c = 0$ we have $a = 3$, $b = -8$ and $x = 2$. Substituting these values in the formula gives

$$x = \frac{-(-8) \pm \sqrt{(-8)^2 - 4 \times 3 \times 2}}{2 \times 3}$$

$$= \frac{8 \pm \sqrt{64 - 24}}{6}$$

$$= \frac{8 \pm \sqrt{40}}{6}$$

$$= \frac{8 \pm 6.325}{6}$$

Either

$$x = \frac{8 + 6.325}{6} \quad \text{or} \quad x = \frac{8 - 6.325}{6}$$

$$= \frac{14.325}{6} \quad \text{or} \quad \frac{1.675}{6}$$

$$= 2.39 \quad \text{or} \quad 0.28$$

(b) Solve the equation

$$-2x^2 + 3x + 7 = 0$$

Where the coefficient of x^2 is negative it is best to make it positive by multiplying both sides of the equation by (-1). This is equivalent to changing the sign of each of the terms. Thus,

$$2x^2 - 3x - 7 = 0$$

This gives $a = 2$, $b = -3$ and $c = -7$.

$$x = \frac{-(-3) \pm \sqrt{(-3)^2 - 4 \times 2 \times (-7)}}{2 \times 2}$$

$$= \frac{3 \pm \sqrt{9 + 56}}{4}$$

$$= \frac{3 \pm \sqrt{65}}{4}$$

$$= \frac{3 \pm 8.062}{4}$$

Either

$$x = \frac{11.062}{4} \quad \text{or} \quad x = \frac{-5.062}{4}$$

$$= 2.766 \quad \text{or} \quad -1.266$$

Exercise 19.2

Solve the following quadratic equations, stating the answer correct to 3 significant figures:

1. $3x^2 - 42 = 0$ 2. $5x^2 - 29 = 0$

3. $7x^2 - 21 = 0$

Factorise the following:

4. $x^2 + 4x + 4$ 5. $9x^2 + 12x + 4$

6. $4x^2 - 12x + 9$ 7. $25x^2 - 10x + 1$

Express each of the following in the form $a(x + h)^2 + k$ or $k - a(x + h)^2$:

8. $x^2 + 4x + 3$ 9. $x^2 + 6x - 2$

10. $x^2 - 8x + 3$ 11. $x^2 - 8x - 2$

12. $4x^2 + 8x + 1$ 13. $3x^2 + 4x - 5$

14. $10x^2 - 20x + 9$ 15. $-x^2 + 4x - 2$

16. $-2x^2 + 4x + 7$ 17. $-4x^2 + 2x + 3$

Solve the following equations by completing the square or by using the quadratic formula, stating each root correct to 2 decimal places:

18. $4x^2 - 3x - 2 = 0$

19. $x^2 - x - 1 = 0$

20. $3x^2 + 7x - 5 = 0$

21. $7x^2 + 8x - 2 = 0$

22. $5x^2 - 4x - 1 = 0$

23. $2x^2 - 7x = 3$

24. $x(x + 4) + 2x(x + 3) = 5$

25. $5x(x + 1) - 2x(2x - 1) = 20$

26. $x(x + 5) = 66$

27. $(2x - 3)^2 = 13$

Equations Containing Fractions

When an equation contains fractions, first multiply each term in the equation by the LCM of the denominators. This will clear the fractions. Next simplify the equation and solve it.

Example 8

(a) Solve the equation $\dfrac{8}{2x} + 5 = \dfrac{x + 3}{4}$

The LCM of the denominators is $4x$. Multiplying each term by $4x$ gives

$$16 + 20x = x(x + 3)$$

$$16 + 20x = x^2 + 3x$$

$$x^2 - 17x - 16 = 0$$

Using the quadratic formula with $a = 1$, $b = -17$ and $c = -16$ we have

$$x = \frac{17 \pm \sqrt{(-17)^2 - 4 \times 1 \times (-16)}}{2 \times 1}$$

$$= \frac{17 \pm \sqrt{289 + 64}}{2}$$

$$= \frac{17 \pm \sqrt{353}}{2}$$

$$= \frac{17 \pm 18.79}{2}$$

$$= \frac{35.79}{2} \quad \text{or} \quad -\frac{1.79}{2}$$

$$= 17.89 \quad \text{or} \quad -0.89$$

(correct to 2 d.p.)

(b) Solve the equation $\dfrac{1}{x - 2} - \dfrac{1}{x^2 - 4} = \dfrac{4}{5}$

Because $x^2 - 4 = (x + 2)(x - 2)$, the LCM of the denominators is $5(x + 2)(x - 2)$. Multiplying each term by this common denominator gives:

$$5(x + 2) - 5 = 4(x^2 - 4)$$

$$5x + 10 - 5 = 4x^2 - 16$$

$$4x^2 - 5x - 21 = 0$$

$$(x - 3)(4x + 7) = 0$$

Either $x - 3 = 0$ giving $x = 3$

on $4x + 7 = 0$ giving $x = -\dfrac{7}{4}$

Exercise 19.3

Solve the following equations, stating the answer correct to 2 decimal places:

1. $\dfrac{3}{x} + 2x = 7$

2. $x + \dfrac{1}{2x} = \dfrac{9}{5}$

3. $x = \dfrac{10}{3 + x}$

4. $x = \dfrac{6}{x - 5}$

5. $\dfrac{1}{x - 1} - \dfrac{1}{x^2 - 1} = \dfrac{1}{6}$

6. $\dfrac{1}{x + 2} + \dfrac{2}{(x + 2)^2} = 2$

7. $\dfrac{1}{x - 1} - \dfrac{1}{3} = \dfrac{x}{x^2 - 1}$

Equations giving rise to Quadratic Equations

Example 9

(a) Solve the simultaneous equations

$$3x - y = 4 \tag{1}$$

$$x^2 - 3xy + 8 = 0 \tag{2}$$

From equation (1),

$$y = 3x - 4$$

Substituting for y in equation (2) gives

$$x^2 - 3x(3x - 4) + 8 = 0$$

$$x^2 - 9x^2 + 12x + 8 = 0$$

$$-8x^2 + 12x + 8 = 0$$

$$8x^2 - 12x - 8 = 0$$

Dividing throughout by 4 gives

$$2x^2 - 3x - 2 = 0$$

$$(2x + 1)(x - 2) = 0$$

Either $2x + 1 = 0$, giving $x = -\dfrac{1}{2}$

or $x - 2 = 0$, giving $x = 2$.

When $x = -\dfrac{1}{2}$, $y = 3 \times (-\dfrac{1}{2}) - 4$

$$= -5\dfrac{1}{2}$$

When $x = 2$, $y = 3 \times 2 - 4 = 2$

Thus the solutions are

$$x = -\tfrac{1}{2}, \; y = -5\tfrac{1}{2} \quad \text{or}$$

$$x = 2, \quad y = 2$$

(b) Solve the simultaneous equations

$$x - 6y - 5 = 0 \qquad (1)$$
$$xy - 6 = 0 \qquad (2)$$

From equation (1),

$$x = 6y + 5$$

Substituting for x in equation (2) gives

$$(6y + 5)y - 6 = 0$$
$$6y^2 + 5y - 6 = 0$$
$$(3y - 2)(2y + 3) = 0$$

Either $3y - 2 = 0$, giving $y = \frac{2}{3}$

or $2y + 3 = 0$, giving $y = -\frac{3}{2}$

When $y = \frac{2}{3}$, $x = 6 \times \frac{2}{3} + 5 = 9$

When $y = -\frac{3}{2}$, $x = 6 \times (-\frac{3}{2}) + 5 = -4$

Thus the solutions are

$$x = 9, \quad y = \frac{2}{3} \quad \text{or}$$
$$x = -4, \quad y = -\frac{3}{2}$$

Exercise 19.4

Solve the following simultaneous equations:

1. $x + y = 3$
   $xy = 2$

2. $ x - y = 3$
   $xy + 10x + y = 150$

3. $x^2 + y^2 - 6x + 5y = 21$
   $ x + y = 9$

4. $ x + y = 12$
   $2x^2 + 3y^2 = 7xy$

5. $2x^2 - 3y^2 = 20$
   $ 2x + y = 6$

6. $x^2 + y^2 = 34$
   $x + 2y = 13$

7. $3x + 2y = 13$
   $ xy = 2$

8. $-3x + y + 15 = 0$
   $2x^2 + 4x + y = 0$

Problems Involving Quadratic Equations

Example 10

(a) The area of a rectangle is 6 square metres. If the length is 1 metre longer than the width find the dimensions of the rectangle.

Let x metres be the width of the rectangle. Then the length of the rectangle is $(x + 1)$ metres. Since the area is length $\times$ breadth, then

$$x(x + 1) = 6$$
$$x^2 + x - 6 = 0$$
$$(x + 3)(x - 2) = 0$$

so either $x + 3 = 0$, giving $x = -3$

or $x - 2 = 0$, giving $x = 2$

The solution cannot be negative and hence $x = 2$. Hence the width is 2 metres and the length is $(2 + 1) = 3$ metres.

(b) A rectangular room is 4 metres wider than it is high and it is 8 metres longer than it is wide. The total area of the walls is 512 square metres. Find the width of the room.

Let the height of the room be x metres. Then the width of the room is $(x + 4)$ metres and the length of the room is $x + 4 + 8 = (x + 12)$ metres.

Fig. 19.1

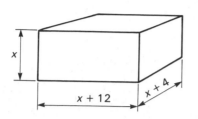

These dimensions are shown in Fig. 19.1.

The total wall area is
$2x(x + 12) + 2x(x + 4)$.

Hence $2x(x + 12) + 2x(x + 4) = 512$

Dividing both sides of the equation by 2,

$$x(x + 12) + x(x + 4) = 256$$
$$x^2 + 12x + x^2 + 4x = 256$$
$$2x^2 + 16x = 256$$

Dividing both sides of the equation by 2 again,

$$x^2 + 8x = 128$$
$$x^2 + 8x - 128 = 0$$
$$(x + 16)(x - 8) = 0$$

Either $x + 16 = 0$, giving $x = -16$

or $x - 8 = 0$, giving $x = 8$

Thus the height of the room is 8 metres. Its width is $(x + 4) = 8 + 4 = 12$ metres.

Exercise 19.5

In most of the following questions there are two answers. If one of these is not feasible say why.

1. Find the number which when added to its square gives a total of 42.

2. A rectangle is 72 square metres in area and its perimeter is 34 metres. Find its length and breadth.

3. Fig. 19.2 shows a template whose area is 50 square centimetres. Find the total length of the template.

Fig. 19.2

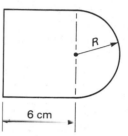

4. Two squares have a total area of 274 square centimetres and the sum of their perimeters is 88 centimetres. Find the side of the larger square.

5. The area of a rectangle is 4 square metres and its length is 3 metres longer than its width. Find the dimensions of the rectangle.

6. Part of a garden consists of a square lawn with a path 1.5 metres wide around its perimeter. If the lawn area is two-thirds of the total area find the length of a side of the lawn.

7. The largest of three consecutive positive numbers is n. The square of this number exceeds the sum of the other two numbers by 38. Find the three numbers.

8. The length of a rectangle exceeds its breadth by 4 centimetres. If the length were halved and the breadth decreased by 5 cm the area would be decreased by 55 square centimetres. Find the length of the rectangle.

9. In a certain fraction the denominator is greater than the numerator by 3. If 2 is added to both the numerator and denominator, the fraction is increased by $\frac{6}{35}$. Find the fraction.

10. A piece of wire which is 18 metres long is cut into two parts. The first part is bent to form the four sides of a square. The second part is bent to form the four sides of a rectangle. The breadth of the rectangle is 1 metre and its length is x metres. If the sum of the areas of the square and rectangle is A square metres show that:

$$A = 16 - 3x + \frac{x^2}{4}$$

If $A = 9$, calculate the value of x.

11. One side of a rectangle is d cm long. The other side is 2 cm shorter. The side of a square is 2 cm shorter still. The sum of the areas of the square and rectangle is 148 square centimetres. Find an equation for d and solve it.

Miscellaneous Exercise 19

1. Solve the equations
   (a) $x^2 + 4x = 0$
   (b) $y^2 - 4 = 0$

2. Solve the equation $9x^2 - 4 = 0$.

3. If x is a real number, find correct to 2 decimal places the positive value of x which satisfies the equation

   $$2x^2 - x - 4 = 0$$

4. (a) Write down the expansion of $(x + y)^2$.
   (b) Given that $x^2 + y^2 = 37$ and $x + y = 7$, find the value of xy.

5. Solve for x, the simultaneous equations

   $$x - 2y = 1 \qquad (1)$$
   $$xy - y = 8 \qquad (2)$$

6. Solve the simultaneous equations

   $$xy = 42 \qquad (1)$$
   $$x - 4y = 17 \qquad (2)$$

7. Solve the equation

   $$\frac{1}{x + 2} + \frac{1}{x^2 - 4} = \frac{2}{5}$$

Multi-Choice Questions 19

1. If $(x + 3)(x - 1) = 0$ then the value of x is
   A 3 or -1 B 3 or 1
   C -3 or -1 D -3 or 1

2. If $(2x - 3)(3x + 4) = 0$ then the value of x is
   A $\frac{2}{3}$ or $-\frac{3}{4}$ B $-\frac{3}{2}$ or $\frac{4}{3}$
   C $-\frac{2}{3}$ or $\frac{3}{4}$ D $\frac{3}{2}$ or $-\frac{4}{3}$

3. If $x(2x - 5) = 0$ then x is equal to
   A $-\frac{5}{2}$ B $\frac{5}{2}$
   C 0 or $\frac{5}{2}$ D 0 or $-\frac{5}{2}$

4. If $x^2 - 25 = 0$ then x is equal to
   A 0   B 5   C ± 5 D 25

5. If $3x^2 - 27 = 0$ the value of x is
   A 9   B ± 3 C 3   D 0

6. When $x^2 - 5x - 2 = 0$, the value of x is
   A $\dfrac{-5 \pm \sqrt{33}}{2}$ B $\dfrac{5 \pm \sqrt{33}}{2}$
   C $\dfrac{-5 \pm \sqrt{17}}{2}$ D $\dfrac{5 \pm \sqrt{17}}{2}$

7. If $\dfrac{3}{x-3} - \dfrac{2}{x-1} = 5$ then

A $5x^2 - 21x + 12 = 0$

B $-3x^2 - 2x - 11 = 0$

C $3x^2 - 4x - 1 = 0$

D $-3x^2 + 4x + 1 = 0$

8. When $x^2 + 9x + 7 = 0$, the value of x is given by

A $\dfrac{9 \pm \sqrt{109}}{2}$ B $\dfrac{9 \pm \sqrt{53}}{2}$

C $\dfrac{-9 \pm \sqrt{109}}{2}$ D $\dfrac{-9 \pm \sqrt{53}}{2}$

Indices

Introduction

$a^4 = a \times a \times a \times a$. The figure 4 which gives the number of a's to be multiplied together is called the **index** (plural indices). a^4 is called the **fourth power of the base** a.

Rules for Indices

(1) When **multiplying** powers of the same base **add** the indices.

$$a^m \times a^n = a^{m+n}$$
$$y^3 \times y^5 = y^{3+5}$$
$$= y^8$$
$$2^4 \times 2^6 \times 2^7 = 2^{4+6+7}$$
$$= 2^{17}$$

(2) To **divide** powers of the same base **subtract** the indices.

$$a^m \div a^n = a^{m-n}$$
$$a^5 \div a^2 = a^{5-2}$$
$$= a^3$$
$$\frac{x^3 \times x^5 \times x^6}{x^2 \times x^4} = \frac{x^{3+5+6}}{x^{2+4}}$$
$$= \frac{x^{14}}{x^6}$$
$$= x^{14-6}$$
$$= x^8$$

(3) When raising the **power of a quantity to a power multiply** the indices.

$$(a^m)^n = a^{mn}$$
$$(b^2)^3 = b^{2 \times 3}$$
$$= b^6$$
$$(x^3 y^4)^2 = x^{3 \times 2} y^{4 \times 2}$$
$$= x^6 y^8$$
$$\left(\frac{2x}{y^3}\right)^4 = \frac{2^{1 \times 4} \times x^{1 \times 4}}{y^{3 \times 4}}$$
$$= \frac{2^4 x^4}{y^{12}}$$
$$= \frac{16x^4}{y^{12}}$$

(4) A **negative index** indicates the **reciprocal** of the quantity.

$$a^{-m} = \frac{1}{a^m}$$
$$3^{-2} = \frac{1}{3^2}$$
$$= \frac{1}{9}$$
$$7x^{-3} = \frac{7}{x^3}$$
$$a^2 b^{-3} c^{-1} = \frac{a^2}{b^3 c}$$

(5) To find the **nth root** of a quantity divide the index of the quantity by n.

$$\sqrt[n]{a^m} = a^{m/n}$$

$$\sqrt[5]{x^2} = x^{2/5}$$

For a square root the number indicating the root is frequently omitted.

$$\sqrt{16x^6} = \sqrt{4^2x^6}$$

$$= 4^{2/2}x^{6/2}$$

$$= 4^1x^3$$

$$= 4x^3$$

(6) Any quantity raised to the **power of 0** equals 1.

$$a^0 = 1, \quad 8^0 = 1 \quad \text{and} \quad 395^0 = 1$$

Example 1

(a) Find the value of 2^{-4}.

$$2^{-4} = \frac{1}{2^4}$$

$$= \frac{1}{16}$$

(b) What is the value of $81^{1/4}$?

$$81^{1/4} = (3^4)^{1/4}$$

$$= 3^{4 \times 1/4}$$

$$= 3^1$$

$$= 3$$

(c) Express 8^4 as a power of 2.

$$8^4 = (2^3)^4$$

$$= 2^{3 \times 4}$$

$$= 2^{12}$$

Example 2

If $5^{2x+3} = 125^{x+5}$, find x.

$$125 = 5^3$$

$$5^{2x+3} = (5^3)^{x+5}$$

$$5^{2x+3} = 5^{3x+15}$$

Since both sides of the equation are raised to powers of 5 the indices can be equated. Hence

$$2x + 3 = 3x + 15$$

$$2x - 3x = 15 - 3$$

$$-x = 12$$

$$x = -12$$

Exercise 20.1

Simplify each of the following:

1. $c^2 \times c^5$ 2. $b^3 \times b^4 \times b^5$

3. $\dfrac{m^7}{m^3}$ 4. $\dfrac{a^2 \times a^4 \times a^5}{a^3 \times a^6}$

5. $(a^2)^3$ 6. $(5a^3bc^4)^2$

7. $\left(\dfrac{b^2}{c^4}\right)^3$

8. $p^{-3}q^2$ as a fraction.

Find the values of each of the following:

9. $2^3 \times 2^5$ 10. 5^{-2} 11. $\dfrac{5^7}{5^4}$

12. 2^{-5} 13. $8^{1/3}$ 14. $32^{1/5}$

15. $\sqrt[3]{64}$ 16. $(0.4)^0$ 17. $5^{-1} \times 10$

18. Give the answer to the following as a decimal number: $\dfrac{2^2 \times 2^3 \times 2^5}{10 \times 10^2}$.

19. Evaluate $16^{3/4}$.

20. If $2^x = 16$, find the value of x.

21. $3^{x+1} = 9^x$. Find the value of x.

22. $2^{3x+2} = 4^{x+5}$. Find x.

23. Find the value of m for which $3^{2m-1} = 243$.

24. Calculate the value of $\dfrac{p \times p^{-1/2}}{p^{1/3}}$ when $p = 64$.

25. Find the value of p for which $5^{p+3} = 25^p$.

Numbers in Standard Form

Any number can be expressed as a value between 1 and 10 multiplied by a power of 10. A number expressed in this way is said to be in **standard form**.

Note that $100 = 10^2$, $1000 = 10^3$ and $1\,000\,000 = 10^6$. The index is found by counting the number of zeros to the left of the decimal point.

Example 3

(a) $563 = 5.63 \times 100 = 5.63 \times 10^2$

(b) $86\,972 = 8.6972 \times 10\,000$
$\qquad\qquad = 8.6972 \times 10^4$

Also $0.1 = \dfrac{1}{10} = 10^{-1}$ and $0.001 = \dfrac{1}{1000}$
$= 10^{-3}$. The negative index is found by adding 1 to the number of zeros following the decimal point.

Example 4

(a) $0.063 = 6.3 \times 10^{-2}$

(b) $0.000\,0567 = 5.67 \times 10^{-5}$

When multiplying or dividing numbers in standard form the ordinary rules for indices are used.

Example 5

(a) Find the value of 3×10^4 multiplied by 2×10^5.

$$3 \times 10^4 \times 2 \times 10^5 = (3 \times 2) \times (10^4 \times 10^5)$$
$$= 6 \times 10^{4+5}$$
$$= 6 \times 10^9$$

(b) Divide 8×10^{-2} by 4×10^{-5}.

$$8 \times 10^{-2} \div 4 \times 10^{-5} = (8 \div 4) \times (10^{-2} \div 10^{-5})$$
$$= 2 \times 10^{-2-(-5)}$$
$$= 2 \times 10^3$$

When adding or subtracting numbers in standard form these should be converted to ordinary numbers before adding or subtracting.

Example 6

(a) Add 4×10^3 to 5×10^2.

$$4 \times 10^3 = 4000$$
$$5 \times 10^2 = 500$$
$$10^3 + 5 \times 10^2 = 4000 + 500$$
$$= 4500$$
$$= 4.5 \times 10^3$$

(b) Subtract 7×10^{-4} from 3×10^{-2}.

$$3 \times 10^{-2} = 0.03$$
$$7 \times 10^{-4} = 0.0007$$
$$3 \times 10^{-2} - 7 \times 10^{-4} = 0.03 - 0.0007$$
$$= 0.0293$$
$$= 2.93 \times 10^{-2}$$

A scientific calculator may be used when dealing with numbers in standard form. The program detailed in Example 7 is typical for many calculators. If this program does not work on your calculator then consult the handbook.

Example 7

Using a calculator work out the value of $3.96 \times 10^4 \div 8.53 \times 10^6$.

Input	Display
3.96	3.96
EXP	3.96 00
4	3.96 04
$\div$	39 600
8.53	8.53
EXP	8.53 00
6	8.53 06
=	4.64 −03

Hence

$$3.96 \times 10^4 \div 8.53 \times 10^6 = 4.64 \times 10^{-3}$$

(correct to 3 s.f.)

Exercise 20.2

Write the following in standard form:

1. 359
2. 7280
3. 19 400
4. 8 036 000
5. 0.06
6. 0.0056
7. 0.000 009
8. 0.000 725

Write the following as ordinary numbers:

9. 3.2×10^2
10. 9.45×10^3
11. 1.87×10^6
12. 2×10^{-3}
13. 5.67×10^{-1}
14. 3.26×10^{-2}
15. 5.6×10^{-4}
16. Add 3×10^2 and 5×10^3.
17. Subtract 5×10^2 from 8×10^3.
18. Multiply 3×10^4 by 2×10^3.

19. Divide 6×10^4 by 2×10^2.
20. Add 5×10^{-2} and 3×10^{-4}.
21. Subtract 2×10^{-3} from 5×10^{-2}.
22. Multiply 4×10^{-4} by 2×10^{-3}.
23. Divide 8×10^{-2} by 2×10^{-4}.

Using a calculator, find values for the following, stating your answer correct to 3 significant figures:

24. 3.26×10^3 multiplied by 8.17×10^5
25. 1.98×10^5 multiplied by 4.65×10^3
26. 8.17×10^{-3} multiplied by 3.52×10^{-2}
27. 1.59×10^3 multiplied by 8.63×10^{-4}
28. 6.89×10^5 divided by 4.73×10^3
29. 1.87×10^2 divided by 7.59×10^5
30. 2.65×10^3 divided by 8.17×10^{-3}

Use a calculator to work out the values of the following:

31. $8.27 \times 10^3 + 9.57 \times 10^2$
32. $5.93 \times 10^5 + 3.29 \times 10^4$
33. $6.76 \times 10^2 + 3.25 \times 10^{-2}$
34. $5.89 \times 10^{-3} + 2.68 \times 10^{-2}$
35. $6.89 \times 10^5 - 7.36 \times 10^4$
36. $8.29 \times 10^3 - 6.36 \times 10^2$
37. $4.17 \times 10^{-3} - 8.29 \times 10^{-5}$
38. $9.18 \times 10^{-4} - 7.38 \times 10^{-6}$

Miscellaneous Exercise 20

Section A

1. Rewrite the number 1 200 000 in standard form.

2. Find the value of $4.62 \times 18.93 \times 275.16$ writing the answer in standard form correct to 3 significant figures.

3. Write 5.01×10^{-3} as a decimal number, i.e. not in standard form.

4. If $p = 3 \times 10^3$ and $q = 2 \times 10^2$, work out the value of

   (a) pq (b) $p + q$

   (c) $p - q$ (d) $p \div q$

   giving the answers as ordinary decimal numbers.

5. Find the values of

   (a) $(-3)^2$ (b) 3^0 (c) 10^{-1}

6. Which is the greater, 8.73×10^2 or 1.2×10^3, and by how much?

7. Write $0.000\,893$ in standard form.

8. Find the value of n in each of the following:

   (a) $12^6 \times 12^n = 12^9$

   (b) $2^8 \div 2^n = 2^5$

   (c) $37\,500 = 3.75 \times 10^n$

   (d) $\frac{1}{8} = 2^n$

Section B

1. Find the values of $(\frac{1}{5})^0$, $125^{-1/3}$, $(1\,000\,000)^{1/3}$.

2. Work out the value of $32^{1/5} \times 25^{1/2} \times 27^{1/3}$.

3. Find the value of $\sqrt{\dfrac{p}{q}}$ when $p = 64^{2/3}$ and $q = 3^{-2}$.

4. If $2^{x+1} = 4^x$, find the value of x.

5. Calculate the value of $\dfrac{y \div y^{-1/2}}{y^{1/3}}$ when $y = 64$.

6. Find the value of p for which $3^{p+3} = 9^p$.

7. Calculate the exact value of $\dfrac{0.185\,82}{3.26}$, expressing the answer in standard form.

8. Without using tables or a calculator, evaluate

   (a) $8^{4/3}$ (b) $(-\frac{1}{2})^{-2}$ (c) $(3\frac{1}{2})^0$

Multi-Choice Questions 20

1. $(3a^2)^3$ is equal to

   A $18a$ B $27a^6$ C $27a^3$ D $9a^6$

2. The value of $64^{1/3}$ is

   A 2 B 4 C 8 D $21\frac{1}{3}$

3. The number $36\,700$ when written in standard form is

   A 367×10^2 B 36.7×10^3

   C 3.67×10^4 D 3.67×10^5

4. 4.05×10^6 is equivalent to

   A $4\,050\,000$ B $405\,000$

   C $400\,005$ D 405.000

5. When written in standard form 0.0063 is

   A 63×10^{-4} B 6.3×10^3

   C 6.3×10^{-3} D 0.63×10^{-2}

6. Find the value of $(\frac{1}{2})^{-3}$.

   A $\frac{1}{64}$ B $\frac{1}{8}$ C 8 D 64

7. Find the value of $2a^3$ when $a = -2$.

   A 16 B -12 C -16 D -64

Areas and Volumes

Units of Area

The area of a plane figure is measured by seeing how many square units it contains. 1 square metre is the area contained in a square having a side of 1 metre; 1 square centimetre is the area contained in a square having a side of 1 centimetre, etc. The standard abbreviations are

$$1 \text{ square metre } - 1 \text{ m}^2$$

$$1 \text{ square centimetre } = 1 \text{ cm}^2$$

$$1 \text{ square millimetre } = 1 \text{ mm}^2$$

$$1 \text{ square inch } = 1 \text{ in}^2$$

$$1 \text{ square foot } = 1 \text{ ft}^2$$

$$1 \text{ square yard } = 1 \text{ yd}^2$$

Table 21.1 gives the formulae for areas and perimeters of some simple geometrical shapes.

Example 1

(a) Find the area of a triangle whose sides measure 8 cm, 10 cm and 15 cm.

Using the formula,

$$A = \sqrt{s(s-a)(s-b)(s-c)}$$

$$s = \frac{8 + 10 + 15}{2}$$

$$= 16.5 \text{ cm}$$

$$A = \sqrt{16.5 \times (16.5 - 8) \times (16.5 - 10) \times \\ \times (16.5 - 15)}$$

$$= \sqrt{16.5 \times 8.5 \times 6.5 \times 1.5}$$

$$= \sqrt{1367}$$

$$= 36.98 \text{ cm}^2$$

Hence the triangle has an area of 36.98 cm².

(b) A trapezium (Fig. 21.1) has parallel sides whose lengths are 20 in and 30 in. The distance between these sides is 25 in. Find the area of the trapezium.

Fig. 21.1

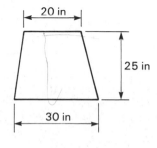

$$\text{Area } = \frac{1}{2} \times 25 \times (20 + 30)$$

$$= \frac{1}{2} \times 25 \times 50$$

$$= 625 \text{ in}^2$$

TABLE 21.1

Figure	Diagram	Formulae
Rectangle		Area $= l \times b$ Perimeter $= 2l + 2b$
Parallelogram		Area $= b \times h$
Triangle		Area $= \frac{1}{2} \times b \times h$ Area $= \sqrt{s(s-a)(s-b)(s-c)}$ where $s = \dfrac{a+b+c}{2}$
Trapezium		Area $= \frac{1}{2} \times h \times (a+b)$
Circle		Area $= \pi r^2$ Circumference $= 2\pi r = \pi d$ $\left(\pi = 3.142 \text{ or } \dfrac{22}{7}\right)$
Sector of a circle		Area $= \pi r^2 \times \dfrac{\theta}{360}$ Perimeter $= 2r + \dfrac{\pi r \theta}{180}$

(c) Fig. 21.2 shows an annulus. Calculate its area. (Take $\pi = 3.14$.)

Fig. 21.2

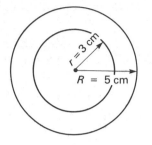

$$\begin{aligned}
\text{Area} &= \pi R^2 - \pi r^2 \\
&= \pi(R^2 - r^2) \\
&= 3.14 \times (5^2 - 3^2) \\
&= 3.14 \times (25 - 9) \\
&= 3.14 \times 16 \\
&= 50.24 \text{ cm}^2
\end{aligned}$$

(d) Find the area and length of arc of the sector of a circle shown in Fig. 21.3. (Take $\pi = \frac{22}{7}$.)

Fig. 21.3

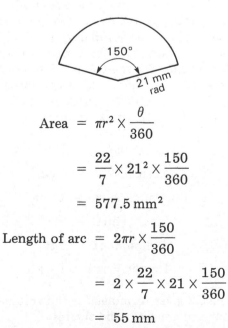

$$\begin{aligned}
\text{Area} &= \pi r^2 \times \frac{\theta}{360} \\
&= \frac{22}{7} \times 21^2 \times \frac{150}{360} \\
&= 577.5 \text{ mm}^2
\end{aligned}$$

$$\begin{aligned}
\text{Length of arc} &= 2\pi r \times \frac{150}{360} \\
&= 2 \times \frac{22}{7} \times 21 \times \frac{150}{360} \\
&= 55 \text{ mm}
\end{aligned}$$

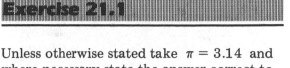

Exercise 21.1

Unless otherwise stated take $\pi = 3.14$ and where necessary state the answer correct to 3 significant figures.

1. Find the area and perimeter of a rectangle whose length is 12 in and whose width is 7 in.

2. A rectangle has a length of 5 m and an area of 15 m². What is its width?

3. A rectangular lawn is 35 m long and 18 m wide. A path $1\frac{1}{2}$ m wide is made round the lawn. Calculate the area of the path.

4. A carpet has an area of 36 m². If it is square what length of side has the carpet?

5. Calculate the area of a parallelogram which has a base 8 yd long and a vertical height of 5 yd.

6. A triangle has a base of 7 cm and an altitude of 3 cm. Calculate its area.

7. A triangle has sides which are 8 cm, 12 cm and 14 cm long. Determine the area of the triangle.

8. The area of a triangle is 40 ft². Its base is 8 ft long. Calculate its vertical height.

9. Find the area of a trapezium whose parallel sides are 45 mm and 73 mm long if the distance between them is 42 mm.

10. The parallel sides of a trapezium are 6 ft and 8 ft long. If the area of the trapezium is 105 ft², what is the distance between them?

11. Find the area of the quadrilateral shown in Fig. 21.4.

Fig. 21.4

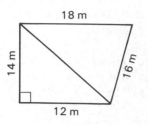

12. What is the area of the quadrilateral shown in Fig. 21.5?

Fig. 21.5

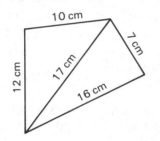

13. Find the circumference of the circles whose radii are

    (a) 3.5 cm   (b) 14 cm

    (c) 21 cm

    (Take $\pi = \frac{22}{7}$.)

14. Find the areas of the circles whose radii are

    (a) 11 cm   (b) 18 cm

    (c) 23 cm

15. A pipe has an outside diameter of 42 mm and a bore of 26 mm. Find its cross-sectional area.

16. A hollow shaft has a cross-sectional area of 4.34 in². If its inside diameter is 3.75 in, find its outside diameter in inches.

17. If r is the radius and θ is the angle subtended at the centre by an arc, find the length of arc when

    (a) $r = 4$ cm and $\theta = 60°$

    (b) $r = 9.62$ cm and $\theta = 138°$

18. If L is the length of arc and r is the radius find the angle θ subtended by the arc when

    (a) $L = 4.7$ cm and $r = 2.5$ cm

    (b) $L = 5$ cm and $r = 4$ cm

19. Find the area of the sector of a circle whose radius is 8 in if it subtends an angle of 196° at the centre.

20. Find the length of arc and the area of a sector of a circle with a radius of 15 cm if the arc subtends an angle of 258° at the centre.

Units of Volume

The volume of a solid figure is found by seeing how many cubic units it contains. 1 cubic metre is the volume contained inside a cube having an edge 1 metre long; 1 cubic centimetre is the volume contained inside a cube having an edge 1 centimetre long, etc. The standard abbreviations for the units of volume are as follows:

$$1 \text{ cubic metre} = 1 \text{ m}^3$$
$$1 \text{ cubic centimetre} = 1 \text{ cm}^3$$
$$1 \text{ cubic millimetre} = 1 \text{ mm}^3$$
$$1 \text{ cubic inch} = 1 \text{ in}^3$$
$$1 \text{ cubic foot} = 1 \text{ ft}^3$$
$$1 \text{ cubic yard} = 1 \text{ yd}^3$$

Table 21.2 gives formulae for the volumes and surface areas of solid figures.

TABLE 21.2

Solid	Volume	Surface area
Any solid having a uniform cross-section	Cross-sectional area × Length of solid	Lateral surface + Ends, i.e. (Perimeter of cross-section × Length of solid) + (Total area of ends)
Cylinder	$\pi r^2 h$	$2\pi r(h + r)$
Cone	$\frac{1}{3}\pi r^2 h$ (*h* is the vertical height)	$\pi r l$ (*l* is the slant height)
Frustum of a cone	$\frac{1}{3}\pi h(R^2 + Rr + r^2)$ (*h* is the vertical height)	Curved surface area $= \pi l(R + r)$ Total surface area $= \pi l(R + r) + \pi R^2 + \pi r^2$ (*l* is the slant height)
Sphere	$\frac{4}{3}\pi r^3$	$4\pi r^2$
Pyramid	$\frac{1}{3}Ah$	Sum of the areas of the triangles forming the sides plus the area of the base (*A* = Area of base)

Example 2

The solid prism shown in Fig. 21.6 has a constant cross-section which is a triangle with sides 8 cm, 7 cm and 7 cm long respectively.

If the length of the prism is 9 cm, calculate

(i)   the lateral surface area

(ii) the total surface area

(iii) the volume of the prism.

Fig. 21.6

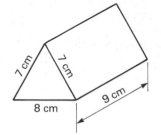

(i)   Lateral surface area = Perimeter of
                                      cross-section
                                      $\times$ Length

$$= (7 + 7 + 8) \times 9$$

$$= 22 \times 9$$

$$= 198 \, \text{cm}^2$$

(ii)

Area of
cross-section $= \sqrt{s(s - a)(s - b)(s - c)}$

$$s = \frac{7 + 7 + 8}{2}$$

$$= 11$$

Area of
cross-section

$$= \sqrt{11 \times (11 - 7) \times (11 - 7) \times (11 - 8)}$$

$$= \sqrt{11 \times 4 \times 4 \times 3}$$

$$= \sqrt{528}$$

$$= 22.98 \, \text{cm}^2$$

Total
surface area = Lateral surface area
                            + Area of the ends

$$= 198 + 2 \times 22.98$$

$$= 244 \, \text{cm}^2$$

(iii) Volume of prism = Area of cross-
                                            section
                                            $\times$ Length

$$= 22.98 \times 9$$

$$= 207 \, \text{cm}^3$$

Units of Capacity

The capacity of a container is the amount of liquid it will hold when full. In the metric system it is measured in litres or fractions of a litre.

$$1 \text{ litre } (\ell) = 100 \text{ centilitres } (c\ell)$$

$$= 1000 \text{ millilitres } (m\ell)$$

$$1 \text{ millilitre } = 1 \text{ cubic centimetre}$$

$$1000 \text{ litres } = 1 \text{ cubic metre}$$

In the imperial system capacities are measured in gallons, pints and fluid ounces.

$$20 \text{ fluid ounces (fl oz) } = 1 \text{ pint (pt)}$$

$$8 \text{ pints (pt) } = 1 \text{ gallon (gal)}$$

Example 3

(a) A cylindrical water tank has a diameter of 3 metres and a height of 4 metres. How many litres of water does it hold when full?

Volume of tank $= \pi \times 1.5^2 \times 4$

$$= 28.26 \, \text{m}^3$$

Capacity of tank $= 28.26 \times 1000$

$$= 28\,260 \text{ litres}$$

(b) A conical glass is 4.5 cm in diameter and 6 cm high.

(i) What is its capacity in millilitres?

(ii) How many glasses could be filled from a bottle of port containing 75 centilitres?

(i) Volume of wine glass $= \frac{1}{3} \times \pi \times 2.25^2$
$$\times 6$$

Since 1 millilitre $= 1 \text{ cm}^3$,

Capacity of wine glass $= 31.8$ millilitres

(ii) Amount of wine $= 75$ centilitres

$$= 750 \text{ millilitres}$$

Number of glasses $= 750 \div 31.8$

$$= 23.6 \text{ (i.e. 23 full}$$
$$\text{glasses)}$$

Exercise 21.2

Take $\pi = 3.14$ unless otherwise stated and where necessary work out the answers correct to 3 significant figures.

1. Fig. 21.7 shows the cross-section of a metal bar. If it is 12 cm long, find its volume.

Fig. 21.7

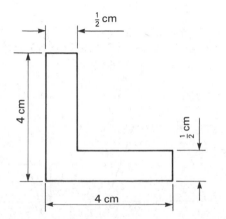

2. A block of wood has the cross-section shown in Fig. 21.8. It is 15 cm long. Find:

   (a) the lateral surface area

   (b) the area of the cross-section

   (c) the total surface area

   (d) the volume of the block.

Fig. 21.8

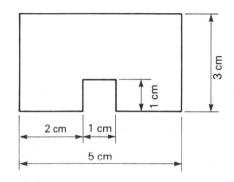

3. Find the volume of a cylinder whose radius is 14 in and whose length is 8 in. What is its total surface area? (Take $\pi = \frac{22}{7}$.)

4. Calculate the volume of a metal pipe whose inside diameter is 6 cm and whose outside diameter is 8 cm, if it is 20 cm long. (Take $\pi = \frac{22}{7}$.)

5. A sphere has a diameter of 8 in. Calculate its volume and surface area.

6. Find the volume of a cone with a diameter of 7 cm and a vertical height of 4 cm. (Take $\pi = \frac{22}{7}$.)

7. A pyramid has a square base of side 12 cm and an altitude of 15 cm. Calculate its volume.

8. Fig. 21.9 shows a laboratory flask which may be considered to be a sphere with a cylindrical neck. Calculate the volume of the flask.

Fig. 21.9

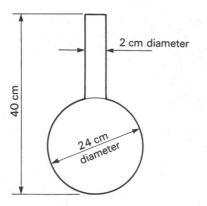

9. A rectangular tank is 2.7 m long, 1.8 m wide and 3.2 m high. How many litres of water will it hold when full?

10. A cylindrical garden pool has a diameter of 6 m. How many litres of water are needed to fill it to a depth of 120 cm?

11. A rectangular medicine bottle is 8 cm wide, 5 cm wide and 15 cm high. Calculate its capacity in millilitres and find how many 5 mℓ doses can be obtained from a full bottle.

12. An ice-cream carton is cylindrical in shape. It has a diameter of 5 cm and a height of 9 cm. How many litres of ice-cream are needed to fill 30 such cartons?

13. A block of lead is hammered out to form a square sheet 10 mm thick. The original dimensions of the block were 1.8 m × 1.2 m × 3 m. Find the dimensions of the square.

14. Calculate the diameter of a cylinder whose volume is 440 in³ if its height and diameter are equal.

15. A cone has a diameter of 7 cm and a volume of 308 cm³. Taking $\pi = \frac{22}{7}$, calculate the vertical height of the cone.

Similar Figures

Two plane figures are similar if they are the same shape. This means that:

(1) All the corresponding angles are equal.

(2) The ratios of corresponding sides are equal.

The two parallelograms shown in Fig. 21.10 are similar because:

(1) The angles C and D equal the angles R and S respectively.

(2) $\dfrac{AB}{AD} = \dfrac{PQ}{PS}$ (i.e. the ratios of corresponding sides are equal).

Fig. 21.10

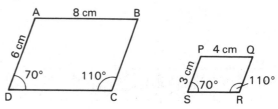

The ratio of the areas of similar figures is equal to the ratio of the squares on corresponding sides.

Thus in Fig. 21.10

$$\frac{\text{Area of ABCD}}{\text{Area of PQRS}} = \left(\frac{AB}{PQ}\right)^2$$

Example 4

The two trapeziums shown in Fig. 21.11 are similar. The area of ABCD is 30 cm². What is the area of WXYZ?

Fig. 21.11

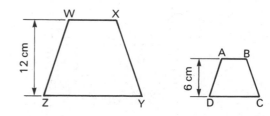

$$\frac{\text{Area of WXYZ}}{\text{Area of ABCD}} = \frac{12^2}{6^2}$$

$$= 2^2$$

$$= 4$$

$$\text{Area of WXYZ} = 4 \times \text{Area of ABCD}$$

$$= 4 \times 30$$

$$= 120 \text{ cm}^2$$

Similar Solids

Two solids are similar if they are the same shape and the ratios of their corresponding linear dimensions are equal.

Thus the two cylinders shown in Fig. 21.12 are similar because:

(1)   They are the same shape (i.e. they are both cylinders).

(2) $\dfrac{\text{Height of cylinder A}}{\text{Height of cylinder B}}$

$$= \frac{\text{Diameter of cylinder A}}{\text{Diameter of cylinder B}}.$$

Fig. 21.12

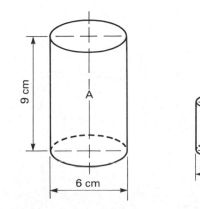

The **surface areas** of similar solids are proportional to the **squares of their linear dimensions.**

Thus in Fig. 21.12

$$\frac{\text{Surface area of cylinder A}}{\text{Surface area of cylinder B}}$$

$$= \frac{(\text{Height of cylinder A})^2}{(\text{Height of cylinder B})^2}$$

$$= \frac{(\text{Diameter of cylinder A})^2}{(\text{Diameter of cylinder B})^2}$$

The **volumes** of similar solids are proportional to the **cubes of their linear dimensions.**

Thus in Fig. 21.12

$$\frac{\text{Volume of cylinder A}}{\text{Volume of cylinder B}}$$

$$= \frac{(\text{Height of cylinder A})^3}{(\text{Height of cylinder B})^3}$$

$$= \frac{(\text{Diameter of cylinder A})^3}{(\text{Diameter of cylinder B})^3}$$

Example 5

A sphere X has a diameter of 12 cm, a surface area of 450 cm^2 and a volume of 900 cm^3. A second sphere Y has a diameter of 6 cm. What is its surface area and volume?

$$\frac{\text{Surface area of Y}}{\text{Surface area of X}} = \frac{(\text{Diameter of Y})^2}{(\text{Diameter of X})^2}$$

$$\frac{\text{Surface area of Y}}{450} = \frac{6^2}{12^2}$$

$$= \frac{36}{144}$$

$$= \frac{1}{4}$$

$$\text{Surface area of Y} = \frac{450}{4}$$

$$= 112.5 \text{ cm}^2$$

$$\frac{\text{Volume of Y}}{\text{Volume of X}} = \frac{(\text{Diameter of Y})^3}{(\text{Diameter of X})^3}$$

$$\frac{\text{Volume of Y}}{900} = \frac{6^3}{12^3}$$

$$= \frac{216}{1728}$$

$$= \frac{1}{8}$$

$$\text{Volume of Y} = \frac{900}{8}$$

$$= 112.5 \, \text{cm}^3$$

Exercise 21.3

1. A circle of 14 cm radius has an area of 616 cm². What is the area of a circle with a radius of 7 cm?

2. Two triangles are similar. One has a base 12 in long and the other has a base 8 in long. The area of the second triangle is 20 in². What is the area of the first triangle?

3. Two parallelograms A and B are similar. A has a base 10 cm long and an area of 50 cm². B has an area of 12.5 cm². Calculate the length of the base of B.

4. Two spheres have radii of 6 cm and 9 cm respectively. Calculate their surface areas and volumes. (Take $\pi = 3.14$.)

5. A spherical cap has a height of 4 cm and a volume of 64 cm³. A similar cap has a volume of 512 cm³. Find its height.

6. The volume of a cone of height 14.2 in is 210 in³. Find the height of a similar cone whose volume is 60 in³.

7. Find the total surface area and volume of a hemisphere having a diameter of 5 cm. Hence calculate the surface area and volume of a hemisphere having a diameter of 15 cm.

8. A pyramid A has a square base of side 4 cm and a volume of 16 cm³.

   (a) Calculate the height of A.

   (b) A similar pyramid B has a volume of 1024 cm³. Calculate the side of the base and the height of pyramid B.

Miscellaneous Exercise 21

Section A

1. A rectangular piece of cardboard 10 in by 6 in has equal squares of side 2 in cut from its corners. The final shape is shown in Fig. 21.13. Calculate:

   (a) the perimeter

   (b) the area of the shape.

Fig. 21.13

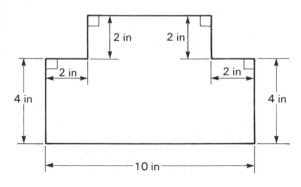

2. Refer to Fig. 21.14.

   (a) Calculate the length of AB.

   (b) Find the area of the triangle AED.

Fig. 21.14

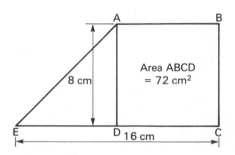

3. Work out the shaded area of the shape shown in Fig. 21.15.

Fig. 21.15

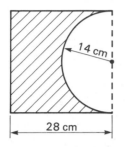

4. A rectangular tank is 4 m long, 2 m wide and 3 m high. How many litres of water can it hold when full?

5. How many doses each of 5 millilitres can be obtained from a cylindrical medicine bottle which has a diameter of 5 cm and a height of 12 cm?

Section B

1. A rectangular tank open at the top and constructed of thin sheet metal is 90 cm high. Its base measures 104 cm by 85 cm. Assuming that there is no overlap at the edges, calculate:

   (a) the area in square centimetres of sheet metal used in its construction

   (b) the capacity of the tank in litres, expressed to the nearest litre.

2. The diameter of a garden roller is 0.77 m. Work out the number of complete revolutions it makes in rolling a stretch of ground 84.7 m long. If the roller is 1.33 m wide, find the area of ground rolled for every 100 revolutions.

3. Four circles of equal area are cut from a square piece of wood of side 14 in. If the area of wood left over is $109\frac{3}{8}$ in², calculate the diameter of each circle and its circumference. (Take $\pi = 3\frac{1}{7}$.)

4. Two similar polygons have areas of 160 cm² and 360 cm² respectively. The shortest side of the larger polygon is 18 cm. Calculate the length of the shortest side of the smaller polygon.

5. A swimming bath has vertical sides and the floor slopes uniformly so that, when the bath is full, the water is 1 m deep at the shallow end and 4 m deep at the other end. If the bath is 25 m long and 12 m wide, calculate the volume of water in the bath when it is full.

Multi-Choice Questions 21

1. How many square centimetres are there in a square metre?

   A 100   B 1000
   C 10 000   D 100 000

2. If the perimeter of a square is 36 cm, then the area of the square, in square centimetres, is

   A 6   B 9   C 36   D 81

3. In Fig. 21.16, ABCD is a rectangle. The triangle BCE is removed. What is the area of ABED?

   A 132 cm²   B 162 cm²
   C 192 cm²   D 208 cm²

Fig. 21.16

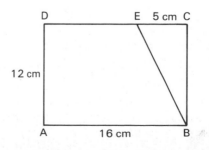

4. Triangle ABC (Fig. 21.17) is right-angled at B. What is its area in square centimetres?

   A 8.5   B 30   C 32.5   D 78

Fig. 21.17

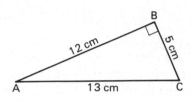

5. Soup is sold in a closed cylindrical can whose diameter is 7 cm and whose height is 10 cm. How much soup does the can hold? (Take $\pi = \frac{22}{7}$.)

   A less than 0.4 litre

   B $\frac{1}{2}$ litre

   C 1 litre

   D more than 1 litre

6. The fraction of the circle which has been shaded in Fig. 21.18 is

   A $\frac{5}{24}$ B $\frac{1}{4}$ C $\frac{1}{2}$ D $\frac{19}{24}$

Fig. 21.18

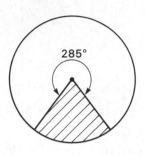

7. A cone has a base diameter of 6 cm and a vertical height of 8 cm. Calculate the volume of the cone in cubic centimetres, leaving the answer as a multiple of π.

   A 8π B 24π C 72π D 96π

Scales

Scales

Scales are used on maps, drawings of buildings, etc.

Scales may be expressed in one of two ways:

(1) As a ratio, for example 1 : 50 000. No units are involved when a scale is expressed in this way. Any distance measured on the map represents 50 000 times this distance on the ground. The scale can be written as a representative fraction, i.e. $\dfrac{1}{50\,000}$.

(2) By stating the distance represented by a distance on the map, for example 5 cm = 1 km. This means that 5 cm on the map represents 1 km on the ground.

Example 1

(a) The scale of a map is 1 : 25 000. Measured on the map, the length of a road is 20 cm. What is the actual length of the road?

20 cm on the map represents
20 × 25 000 cm on the ground.

Actual length of road =

$$\frac{20 \times 25\,000}{100} \text{ metres} = 5000 \text{ metres}$$

(b) The scale of a map is 1 : 800 000. A motorway is 120 km long. What distance will this be on the map?

$$120 \text{ km} = 120 \times 1000 \text{ m}$$
$$= 120 \times 1000 \times 100 \text{ cm}$$

$$\text{Distance on the map} = \frac{120 \times 1000 \times 100}{800\,000} \text{cm}$$

$$= 15 \text{ cm}$$

(c) On the plan of a house drawn to a scale of 1 cm = 5 m, the length of the kitchen is 1.3 cm. What is the actual length of the kitchen?

$$1 \text{ cm} = 5 \text{ m}$$
$$1.3 \text{ cm} = 5 \times 1.3 \text{ m}$$
$$= 6.5 \text{ m}$$

Length of kitchen is 6.5 m.

Areas and Scales

Since areas are proportional to the squares of linear dimensions,

$$\frac{\text{Actual area}}{\text{Area on map}} = (\text{Map scale})^2$$

For large areas such as fields, farms, parks, etc. the hectare (ha) is used such that

$$1 \text{ hectare} = 10\,000 \text{ m}^2$$

Example 2

(a) A map is drawn to a scale of $1 : 20\,000$. Measured on the map a field has an area of $6\,cm^2$. Calculate the actual area of the field in hectares.

$$\frac{\text{Actual area}}{\text{Area on map}} = (\text{Map scale})^2$$

$$\begin{aligned}
\text{Actual area} &= \text{Area on map} \times (\text{Map scale})^2 \\
&= 6 \times (20\,000)^2 \text{ square} \\
&\qquad\qquad \text{centimetres} \\
&= \frac{6 \times (20\,000)^2}{100 \times 100} \text{ square metres} \\
&= 240\,000 \text{ square metres} \\
&= \frac{240\,000}{10\,000} \text{ hectares} \\
&= 24 \text{ hectares}
\end{aligned}$$

(b) A drawing is made to a scale of $2\,cm = 5\,m$. Measured on the drawing a room has an area of $4.8\,cm^2$. Calculate the actual area of the room.

$$\begin{aligned}
\text{Since} \qquad 2\,cm &= 5\,m \\
1\,cm &= 2.5\,m \\
1\,cm^2 &= (2.5)^2\,m^2 \\
&= 6.25\,m^2 \\
4.8\,cm^2 &= 4.8 \times 6.25 \\
&= 30\,m^2
\end{aligned}$$

Hence the actual area of the room is $30\,m^2$.

Volumes and Scales

Since volumes are proportional to the cubes of linear dimensions

$$\frac{\text{Actual volume}}{\text{Volume in model}} = (\text{Model scale})^3$$

Example 3

A model aircraft is built to a scale of $1 : 25$. The volume of the hold in the actual aircraft is $350\,m^3$. What is the volume of the hold in the model?

$$\begin{aligned}
\text{Volume in model} &= \frac{\text{Actual volume}}{(\text{Model scale})^3} \\
&= \frac{350}{25^3} \\
&= 0.0224\,m^3 \\
&= 0.0224 \times 100 \times 100 \\
&\qquad \times 100 \text{ cubic centimetres} \\
&= 22\,400\,cm^3
\end{aligned}$$

Exercise 22.1

1. The scale of a map is $1 : 20\,000$. What distance in metres does $4\,cm$ on the map represent?

2. The scale of a map is $1\,cm = 5\,m$. Express this scale as a ratio.

3. The scale of a map is $1 : 50\,000$. What distance, in centimetres, represents $12\,km$?

4. A drawing is made to a scale of $1\,cm = 10\,m$. The actual length of a garden is $150\,m$. What length, in centimetres, will it be on the drawing?

5. The scale of a map is $2\,cm = 1\,km$. On the map two towns are $30.4\,cm$ apart. What distance, in kilometres, separates the towns?

6. The scale of a map is $1 : 20\,000$. What area, in square metres, does $3\,cm^2$ represent?

7. A small farm is represented on a map by an area of $12\,cm^2$. If the map scale is $1 : 50\,000$, what is the actual area, in hectares, of the farm?

8. The scale of a map is 5 cm to 1 km. A field is represented by an area of 260 cm². Calculate the actual area of the field in hectares.

9. A model of an aircraft is made to $\frac{1}{10}$ scale. On the aircraft itself, the wing area is 240 m². What area, in square centimetres, will represent the wing area on the model?

10. A model of a building is made to $\frac{1}{10}$ scale. Fig. 22.1 shows a pictorial view of the garage. Calculate:

    (a) the actual area of the end of the garage

    (b) the area, in square centimetres, of the end of the garage on the model

    (c) the actual volume of the garage

    (d) the volume, in cubic centimetres, of the model.

Fig. 22.1

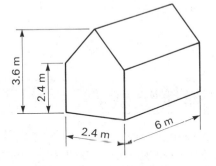

11. A concert hall is in the shape of a cuboid. A model of it with a scale of 1 : 50 has length 2 m, width 1.2 m and height 0.8 m. Calculate:

    (a) the volume, in cubic centimetres, of the model of the hall

    (b) the actual volume of the hall in cubic metres

    (c) the floor area of the hall in square metres.

12. A model of an aircraft is made to a scale of 1 : 10. The volume of the hold on the model is 20 000 cm³. Calculate the actual volume, in cubic metres, of the hold.

Gradient

When used in connection with roads and railways the gradient denotes the ratio of the vertical distance to the corresponding distance measured along the slope. Thus in Fig. 22.2,

$$\text{Gradient} = \frac{\text{BC}}{\text{AB}}$$

Fig. 22.2

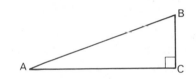

The gradient is expressed as either a ratio or a percentage. If a road has a gradient of 1 in 20 it means that for every 20 m along the slope the road rises or falls 1 m vertically. This gradient can also be stated as a percentage, i.e. $\frac{1}{20} \times 100 = 5\%$.

Example 4

(a) Two points P and Q are linked by a straight road with a uniform gradient. If Q is 50 m higher than P and the distance PQ measured along the road is 800 m, what is the gradient of the road?

Fig. 22.3

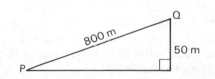

From Fig. 22.3 we see that

$$\text{Gradient} = \frac{50}{800}$$

$$= \frac{1}{16}$$

$$= \frac{1}{16} \times 100\%$$

$$= 6\tfrac{1}{4}\%$$

The gradient of the road is 1 in 16 or $6\tfrac{1}{4}\%$.

(b) A man walks 5 km up a road whose gradient is 8%. How much higher, in metres, is he than he was when he started?

From Fig. 22.4,

$$\text{Gradient of road} = \frac{x}{5000}$$

$$\frac{x}{5000} \times 100 = 8$$

$$\frac{x}{50} = 8$$

$$x = 50 \times 8$$

$$= 400$$

The man is 400 m higher than when he started.

Fig. 22.4

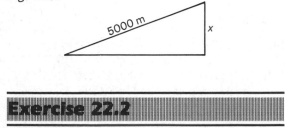

 Exercise 22.2

1. Two points A and B are connected by a straight road with a uniform gradient. If A is 40 m higher than B and the distance AB (measured along the road) is 1 km, what is the gradient of the road? State the answer as a ratio.

2. Two points P and Q whose heights differ by 50 m are connected by a straight road 2 km long. What is the gradient of the road? State the answer as a ratio.

3. A road has a gradient of 1 in 15. A man walks $1\tfrac{1}{2}$ km up the road. How much higher is he than he was when he first started?

4. A man walks up a road whose gradient is 1 in 30. If he walks a distance of 900 m how much higher is he than he was when he first started?

5. A road has a gradient of $12\tfrac{1}{2}\%$. Express this in the form $1:n$.

6. A mountain track has a uniform gradient and rises 20 ft in a distance of 400 ft (measured along the road). Express the gradient of the track as $1:n$.

7. A man walks 100 m up a slope whose gradient is 5% and then a further 80 m up a slope whose gradient is $12\tfrac{1}{2}\%$. How much higher is he than he was when he first started?

 **Miscellaneous Exercise 22**

1. A map is drawn to a scale of $1:25\,000$. What length of road, in metres, is represented by 2 cm on the map?

2. A rectangular kitchen is $3\tfrac{1}{2}$ m wide, $4\tfrac{1}{2}$ m long and $2\tfrac{1}{2}$ m high. A model of the house is made to a scale of 1 cm = 2 m. Work out the dimensions of the kitchen on the model.

3. The building plans of a house are drawn to a scale of $1:50$.

   (a) Find the length of the house if it is represented by a length of 20 cm on the plan.

(b) The actual height of the house is to be 7 m. Find the length in centimetres which represents this dimension on the plan.

(c) On the plan the area of the kitchen floor is 80 cm². Calculate the actual area of the kitchen floor.

4. A model of a public building is made to a scale of 1 : 20.

(a) The model has a length of 4 m. What is the actual length of the building?

(b) The actual area of one of the rooms is 2000 m². What is the area on the model?

(c) The volume of the model is 36 m³. What is the actual volume of the building?

5. The plan of a house and garden is drawn to a scale of 1 : 60.

(a) Find the distance on the plan which represents an actual distance of 30 m.

(b) On the plan the garden measures 60 cm by 30 cm. What is the actual area of the garden?

(c) On the plan the volume of the house is shown to be 90 cm³. What is the actual volume of the house?

6. A model of a lorry is made to a scale of 1 in 10.

(a) The windscreen of the model has an area of 100 cm². Work out the area, in square centimetres, of the windscreen on the lorry.

(b) The fuel tank of the lorry, when full, holds 100 litres. Find the capacity, in cubic centimetres, of the fuel tank of the model.

7. A model aircraft, similar to the full-size aircraft, is made to a scale of 1 in 20. The model has a tail 6 inches high, a wing area of 420 in² and a cabin volume of 950 in³. Calculate the corresponding values for the full-size aircraft.

8. Two points A and B whose heights differ by 10 metres are shown 1.2 cm apart on a map whose scale is 1 : 10 000.

(a) Work out the horizontal distance between A and B.

(b) If the road connecting A and B slopes uniformly, work out the distance between A and B along the slope.

(c) Calculate the gradient of the road and express it as a percentage correct to 1 decimal place.

Tables, Charts and Diagrams

Tables

Conversion tables of various kinds are used in business, science and engineering.

Ready reckoners provide a quick way of determining bank loan and hire-purchase repayments. Table 1 shows an extract from hire-purchase payment tables put out by a finance house.

TABLE 23.1

Outstanding Balance	Monthly instalments		
	36 months	24 months	12 months
£5	£0.23	£0.30	£0.51
£10	£0.45	£0.59	£1.01
£20	£0.90	£1.18	£2.02
£30	£1.35	£1.77	£3.02
£40	£1.80	£2.36	£4.03
£50	£2.25	£2.95	£5.04
£60	£2.70	£3.55	£6.05
£70	£3.17	£4.14	£7.05
£80	£3.62	£4.73	£8.06
£90	£4.07	£5.32	£9.07
£100	£4.52	£5.91	£10.08
£200	£9.04	£11.82	£20.16
£300	£13.56	£17.73	£30.24
£400	£18.08	£23.64	£40.32
£500	£22.60	£29.55	£50.40
£600	£27.12	£35.46	£60.48
£700	£31.64	£41.37	£70.56
£800	£36.16	£47.28	£80.64
£900	£40.68	£53.19	£90.72

Example 1

Use Table 1 to find the monthly instalments for outstanding balances of

(a) £70 over 36 months

(b) £95 over 24 months

(c) £785 over 12 months.

    (a) Directly from the table:

        Monthly instalment = £3.17

    (b)   £95 = £90 + £5

    Monthly instalment = £5.32 + £0.30

               = £5.62

    (c)   £785 = £700 + £80 + £5

    Monthly instalment = £70.56 + £8.06

               + £0.51

               = £79.13

Table 23.2 shows part of a temperature conversion table.

TABLE 23.2

TEMPERATURE	Degrees Fahrenheit to degrees Celsius									
°F	0	1	2	3	4	5	6	7	8	9
0	−17.8	−17.2	−16.7	−16.1	−15.6	−15.0	−14.4	−13.9	−13.3	−12.8
10	−12.2	−11.7	−11.1	−10.6	−10.0	−9.4	−8.9	−8.3	−7.8	−7.2
20	−6.7	−6.1	−5.6	−5.0	−4.4	−3.9	−3.3	−2.8	−2.2	−1.7
30	−1.1	−0.6	0	0.6	1.1	1.7	2.2	2.8	3.3	3.9
40	4.4	5.0	5.6	6.1	6.7	7.2	7.8	8.3	8.9	9.4
50	10.0	10.6	11.1	11.7	12.2	12.8	13.3	13.9	14.4	15.0
60	15.6	16.1	16.7	17.2	17.8	18.3	18.9	19.4	20.0	20.6
70	21.1	21.7	22.2	22.8	23.3	23.9	24.4	25.0	25.6	26.1
80	26.7	27.2	27.8	28.3	28.9	29.4	30.0	30.6	31.1	31.7
90	32.2	32.8	33.3	33.9	34.4	35.0	35.6	36.1	36.7	37.2
100	37.8	38.3	38.9	39.4	40.0	40.6	41.1	41.7	42.2	42.8
110	43.3	43.9	44.4	45.0	45.6	46.1	46.7	47.2	47.8	48.3
120	48.9	49.4	50.0	50.6	51.1	51.7	52.2	52.8	53.8	53.9
130	54.4	55.0	55.6	56.1	56.7	57.2	57.8	58.3	58.9	59.4
140	60.0	60.6	61.1	61.7	62.2	62.8	63.3	63.9	64.4	65.0
150	65.6	66.1	66.7	67.2	67.8	68.3	68.9	69.4	70.0	70.6
160	71.1	71.7	72.2	72.8	73.3	73.9	74.4	75.0	75.6	76.1
170	76.7	77.2	77.8	78.3	78.9	79.4	80.0	80.6	81.1	81.7
180	82.2	82.8	83.3	83.9	84.4	85.0	85.6	86.1	86.7	87.2
190	87.8	88.3	88.9	89.4	90.0	90.6	91.1	91.7	92.2	92.8
Interpolation:	deg F: 0.1	0.2	0.3	0.4	0.5	0.6	0.7	0.8	0.9	
	deg C: 0.1	0.1	0.2	0.2	0.3	0.3	0.4	0.4	0.5	

Example 2

Convert:

(a) 54°F to degrees Celsius

(b) 122.7°F to degrees Celsius

(c) 35°C to degrees Fahrenheit

(d) 62.7°C to degrees Fahrenheit.

(a) We first find 50 in the first column and move along this row until we find the column headed 4. We find the number 12.2 and hence 54°F = 12.2°C.

(b) To convert 122.7°F into degrees Celsius we make use of the figures given under the heading 'interpolation'. From the table we find 122°F = 50°C. Looking at the figures given at the foot of the table we can find 0.7°F and this is equivalent to 0.4°C which is shown immediately below. Thus

$$122.7°F = 50°C + 0.4°C$$
$$= 50.4°C$$

(c) To convert degrees Celsius into degrees Fahrenheit we use the table in reverse. To convert 35°C into degrees Fahrenheit we search in the body of the table until we find the figure 35. This occurs in the column headed 5 in the row starting with 90. Hence 35°C is equivalent to 95°F.

(d) To convert 62.7°C into degrees Fahrenheit we look in the body of the table for a number as close to 62.7 as possible, but less than 62.7. This number is 62.2 corresponding to 144°F. Now 62.2°C is 0.5°C less than 62.7°C. Using the interpolation figures we see that 0.5°C corresponds to 0.9°F. Therefore

$$62.7°C = 62.2°C + 0.5°C$$
$$= 144°F + 0.9°F$$
$$= 144.9°F$$

Mileage charts are often set out like Table 23.3.

TABLE 23.3

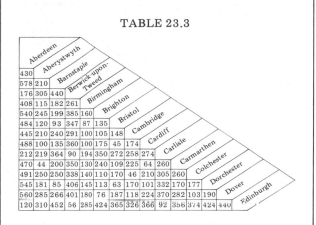

Example 3

Use Table 23.3 to find the distance between

(a) Aberdeen and Carlisle

(b) Barnstaple and Cambridge.

(a) To find the distance from Aberdeen to Carlisle, find Aberdeen and Carlisle in the sloping list. Move vertically down the column for Aberdeen and horizontally along the row for Carlisle until the two movements coincide. The figure given in the table at this point is 212 and thus the distance between Aberdeen and Carlisle is 212 miles.

(b) Find Barnstaple in the sloping list and move down the third column. The row for Cambridge is the seventh row from the top. In the third column of the seventh row we find the figure 240 and hence the distance between Barnstaple and Cambridge is 240 miles.

Parallel Scale Conversion Charts

A system of parallel scales may be used when we wish to convert from one set of units to another related set. Fig. 23.1 is a chart relating degrees Fahrenheit and degrees Celsius.

Fig. 23.1

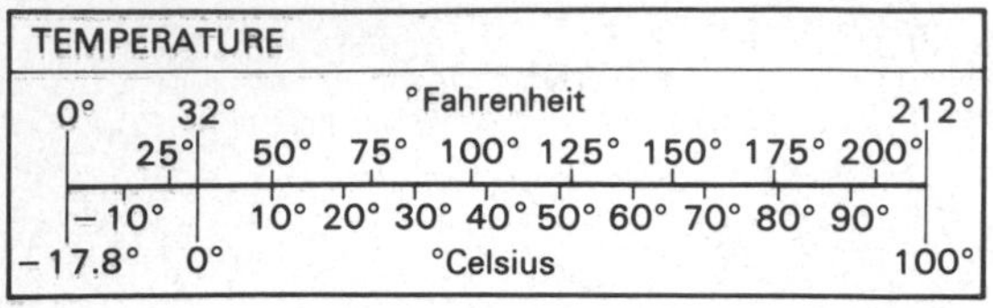

```
TEMPERATURE
 0°     32°           °Fahrenheit                    212°
        25°    50°  75°  100° 125° 150° 175° 200°
 -10°          10° 20° 30° 40° 50° 60° 70° 80° 90°
-17.8°   0°               °Celsius                   100°
```

Example 4

Using the chart in Fig. 23.1 convert
(a) 50 °F to degrees Celsius
(b) 75 °F to degrees Celsius
(c) 20 °C to degrees Fahrenheit.

(a) From the chart we see that 50 °F corresponds with 10 °C and therefore 50 °F is equivalent to 10 °C.

(b) From the chart we see that 75 °F corresponds roughly with 24 °C. Note carefully that when using the chart we cannot expect accuracy greater than 1°.

(c) From the chart we see that 20 °C is about 68 °F.

Exercise 23.1

1. Use Table 23.1 to find the monthly instalments for outstanding balances of
    (a) £90 over 36 months
    (b) £700 over 12 months
    (c) £65 over 24 months
    (d) £805 over 12 months
    (e) £540 over 36 months
    (f) £735 over 24 months.

2. Use the temperature conversion table (Table 23.2) to convert
    (a) 83 °F to degrees Celsius
    (b) 114.6 °F to degrees Celsius
    (c) 55 °C to degrees Fahrenheit
    (d) 68.7 °C to degrees Fahrenheit.

3. Use the mileage chart (Table 23.3) to find the distances between
    (a) Aberdeen and Cardiff
    (b) Birmingham and Carmarthen
    (c) Bristol and Dorchester.

4. Table 23.4 shows a comparison between gradients expressed as a ratio and gradients expressed as a percentage. Use the table to convert
    (a) a gradient of 1 : 7 to a percentage
    (b) a gradient of 7.2% to a ratio.

TABLE 23.4

Gradients	
Ratio	%
1 : 3	33.3
1 : 4	25
1 : 5	20
1 : 6	16.4
1 : 7	14.2
1 : 8	12.4
1 : 9	11.1
1 : 10	10
1 : 11	9.1
1 : 12	8.4
1 : 13	7.9
1 : 14	7.2
1 : 15	6.6
1 : 16	6.4
1 : 17	5.8
1 : 18	5.5
1 : 19	5.2
1 : 20	5

5. Table 23.5 allows conversion from miles to kilometres (and also miles per hour to kilometres per hour) to be made. Use the table to convert
    (a) 15 miles to kilometres
    (b) 48.27 km/h to miles per hour
    (c) 58 miles to kilometres
    (d) 568 miles per hour to kilometres per hour.

TABLE 23.5

Distance and speed					
Miles to kilometres					
Miles per hour to kilometres per hour					
1	1.60	20	32.18	75	120.7
2	3.21	25	40.23	80	128.7
3	4.82	30	48.27	85	136.8
4	6.43	35	56.32	90	144.8
5	8.04	40	64.37	95	152.9
6	9.65	45	72.41	100	160.9
7	11.26	50	80.46	200	321.9
8	12.87	55	88.51	300	482.8
9	14.48	60	96.55	400	643.7
10	16.09	65	104.60	500	804.7
15	24.13	70	112.70	1000	1609.3

1 mile = 1.609 344 km
1 kilometre = 0.621 371 miles

6. Use the chart (Fig. 23.1) relating temperatures in degrees Fahrenheit to degrees Celsius to convert

    (a) 100 °F to degrees Celsius

    (b) 80 °C to degrees Fahrenheit

    (c) 64 °F to degrees Celsius

    (d) 48 °C to degrees Fahrenheit.

7. Fig. 23.2 is a chart relating pounds avoirdupois to kilograms. Use the chart to convert

    (a) 3.4 lb to kilograms

    (b) 568 lb to kilograms

    (c) 1.8 kg to pounds.

Fig. 23.2

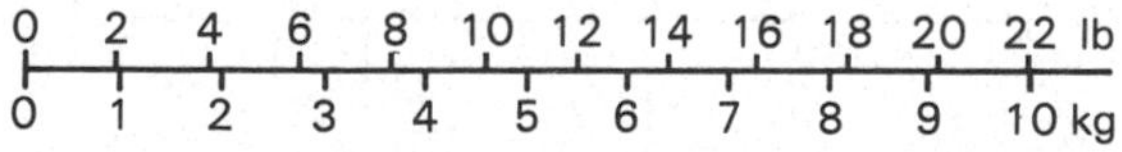

```
0   2   4   6   8  10  12  14  16  18  20  22 lb
+---+---+---+---+---+---+---+---+---+---+---+
0   1   2   3   4   5   6   7   8   9  10 kg
```

8. Table 23.6 shows a comparison of tyre pressures. Convert:

    (a) 22 lb/in^2 into kilograms per square centimetre

(b) 2.10 kg/cm^2 into pounds per square inch.

(c) Estimate the tyre pressure in pounds per square inch corresponding to 3 kg/cm^2.

TABLE 23.6

Tyre pressures					
Pounds per sq in to kg per sq cm					
16	1.12	26	1.83	40	2.80
18	1.26	28	1.96	50	3.50
20	1.40	30	2.10	55	3.85
22	1.54	32	2.24	60	4.20
24	1.68	36	2.52	65	4.55

The Proportionate Bar Chart

The proportionate bar chart (Fig. 23.3) relies on heights (or areas) to convey the proportions of a whole. The bar should be of the same width throughout its length or height. This diagram is accurate, quick and easy to construct and it can show quite a large number of components without confusion. Although Fig. 23.3 shows the bar drawn vertically it may also be drawn horizontally if desired.

Fig. 23.3

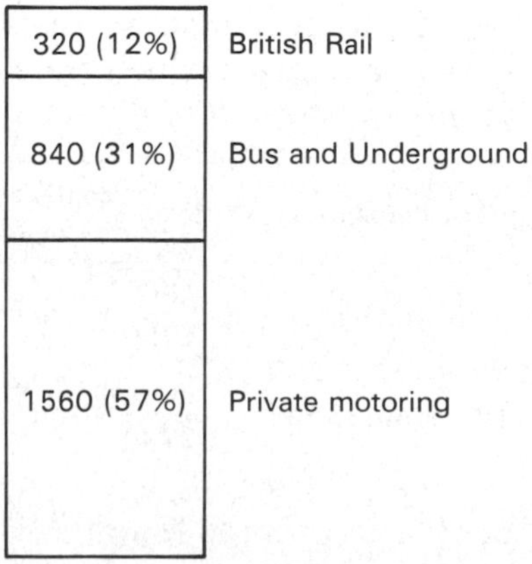

320 (12%)	British Rail
840 (31%)	Bus and Underground
1560 (57%)	Private motoring

Example 5

Draw a proportionate bar chart for the figures below which show the way commuters in the South-east region travelled to the London area.

Type of transport	Numbers using
Private motoring	1560
Bus and underground	840
British Rail	320

The easiest way is to draw the chart on graph paper. However, if plain paper is used, the lengths of the component parts must be calculated and then drawn accurately using a rule (Fig. 23.3).

Total number $= 1560 + 840 + 320$

$= 2720$

Suppose that the total height of the diagram is to be 6 cm. Then

1560 commuters are represented by

$$\frac{1560}{2720} \times 6 = 3.44 \text{ cm}$$

840 commuters are represented by

$$\frac{840}{2720} \times 6 = 1.85 \text{ cm}$$

320 commuters are represented by

$$\frac{320}{2720} \times 6 = 0.71 \text{ cm}$$

Alternatively, the proportions can be expressed as percentages which are calculated as shown below.

Type of transport	Percentage of commuters using
Private motoring	$\frac{1560}{2720} \times 100 = 57\%$
Bus and underground	$\frac{840}{2720} \times 100 = 31\%$
British Rail	$\frac{320}{2720} \times 100 = 12\%$

Simple Bar Charts

In these charts the information is represented by a series of bars all of the same width. The height or the length of each bar represents the magnitude of the figures. The bars may be drawn vertically or horizontally as shown in Figs. 23.4 and 23.5 which present the information given in Example 5.

Fig. 23.4

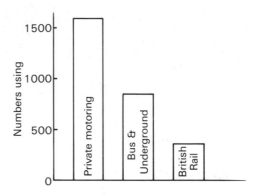

Fig. 23.5

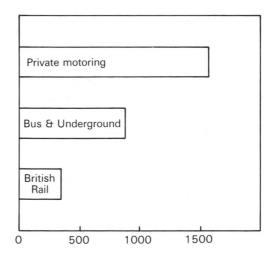

Chronological Bar Charts

This type of chart compares quantities over periods of time. It is very similar to the vertical bar chart and is basically the same in construction as a graph.

Example 6

The information below gives the number of colour television sets sold in Southern England during the period 1970-5.

Year	Number of sets sold (thousands)
1970	77.2
1971	84.0
1972	91.3
1973	114.6
1974	130.9
1975	142.5

Draw a chronological bar chart to represent this information.

When drawing a chronological bar chart, time is always marked off along the horizontal axis. The chart is drawn in Fig. 23.6 and it clearly shows how the sales of TV sets have increased over the period illustrated.

Fig. 23.6

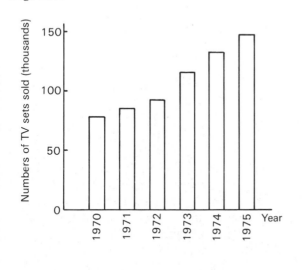

Pie Charts

A pie chart displays the proportions of a whole as sector angles or sector areas. The circle as a whole represents the total of the component parts.

Example 7

Represent the information given in Example 5 in the form of a pie chart.

The first step is to calculate the sector angles. Remembering that a circle contains $360°$ the sector angles are calculated as shown below:

Type of transport	Sector angle (degrees)
Private motoring	$\frac{1560}{2720} \times 360 = 206°$
Bus and underground	$\frac{840}{2720} \times 360 = 111°$
British Rail	$\frac{320}{2720} \times 360 = 43°$

Using a protractor the pie chart (Fig. 23.7) can now be drawn. If desired percentages can be displayed on the diagram.

Fig. 23.7

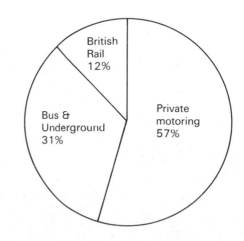

Pie charts are very useful when component parts of a whole are to be represented. Up to eight component parts can be accommodated but above this number the chart loses its effectiveness.

Exercise 23.2

1. In Fig. 23.8 find the values of x and y.

Fig. 23.8

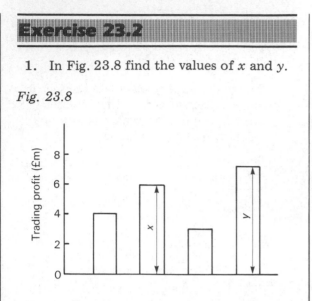

2. Draw a proportionate bar chart for the information below which relates to expenditure per head on transport. In each case find the percentage expenditure and show it on your diagram.

Type of transport	Expenditure (£)
Private motoring	1.10
Rail	2.75
Other public transport	3.15
Total	7.00

3. The table below shows the number of people employed on various kinds of work in a factory.

Type of personnel	Number employed
Unskilled workers	45
Craftsmen	25
Draughtsmen	5
Clerical staff	10
Total	85

(a) Draw a vertical bar chart to represent this information.

(b) Draw a simple horizontal bar chart to show this information.

(c) Draw a pie chart, showing percentages, for this information.

4. The information below gives details of the temperature range used when forging various metals. Draw a horizontal bar chart to represent this data.

Metal	Temperature (°C)
Carbon steel	770–1300
Wrought iron	860–1340
Brass	600– 800
Copper	500–1000

5. The figures below give the world population (in millions of people) from 1750 to 1950. Draw a chronological bar chart to represent this information.

Year	Population (millions)
1750	728
1800	906
1850	1171
1900	1608
1950	2504

6. The information in the following table gives the production of grain on a certain farm during the period from 1968–73. Draw a chronological bar chart to represent this information.

Year	Grain production (tonnes)
1968	395
1969	410
1970	495
1971	560
1972	420
1973	515

7. The table below gives the area of various regions of the world.

Region	Area (millions of square miles)
Africa	11.7
Asia	10.6
Europe	1.9
N. America	9.1
S. America	7.2
Oceania	3.3
U.S.S.R.	8.6

Draw a pie chart to depict this information.

8. Figure 23.9 is a pie chart which shows the total sales of a department store for one day. Find the correct size of each sector angle. (The diagram is NOT drawn to scale.)

Fig. 23.9

Pictograms

These are diagrams in the form of pictures which are used to present information to those who are unskilled in dealing with figures or to those who have only a limited interest in the topic depicted.

Example 8

The table below shows the output of bicycles for the years 1970 to 1974.

Year	Output
1970	2000
1971	4000
1972	7000
1973	8500
1974	9000

Represent this data in the form of a pictogram.

The pictogram is shown in Fig. 23.10. It will be seen that each bicycle in the diagram represents an output of 2000 bicycles. Part of a symbol is shown in 1972, 1973 and 1974 to represent a fraction of 2000 but clearly this is not a very precise way of representing the output.

Fig. 23.10

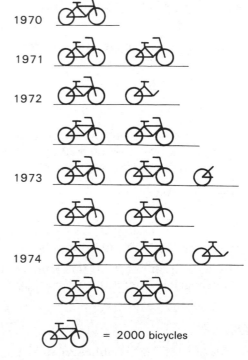

It is essential that the diagram is labelled correctly and that the source of the information is stated.

A method not recommended is shown in Fig. 23.11. Comparison is difficult because the reader is not sure whether to compare heights, areas or volumes.

Fig. 23.11

Sales of milk in 1950 and 1970

(millions of litres)

Example 9

If a square of 3 cm side represents a population of 18 000, what population is represented by a square of 4 cm side?

Here the quantities are represented by the areas of squares.

The area of the square is

3 cm × 3 cm = 9 square centimetres

Hence 9 square centimetres represents a population of 18 000 and 1 square centimetre represents a population of $\frac{18\,000}{9} = 2000$.

The area of a square of side 4 cm is 4 cm × 4 cm = 16 square centimetres.

Hence a square of 4 cm side represents a population of 2000 × 16 = 32 000.

Example 10

A production of 1000 tonnes of steel is represented by a cube of side 2 cm. Calculate the production of steel represented by a cube of side 3 cm.

Here, the quantities are represented by the volumes of cubes. Since the volume of a cube is side × side × side = side³:

Volume of a cube of 2 cm side

= 2 cm × 2 cm × 2 cm

= 8 cubic centimetres

Therefore 8 cubic centimetres represents 1000 tonnes and 1 cubic centimetre represents

$\frac{1000}{8} = 125$ tonnes

Volume of a cube of 3 cm side

= 3 cm × 3 cm × 3 cm

= 27 cubic centimetres

Hence a cube of 3 cm side represents

125 tonnes × 27 = 3375 tonnes

Exercise 23.3

1. Figure 23.12 is a pictogram showing the method by which first year boys come to school.

   (a) How many come by bus?

   (b) How many come by car?

   (c) Estimate the number of boys who live within 1 kilometre of the school (i.e. those who walk).

Fig. 23.12

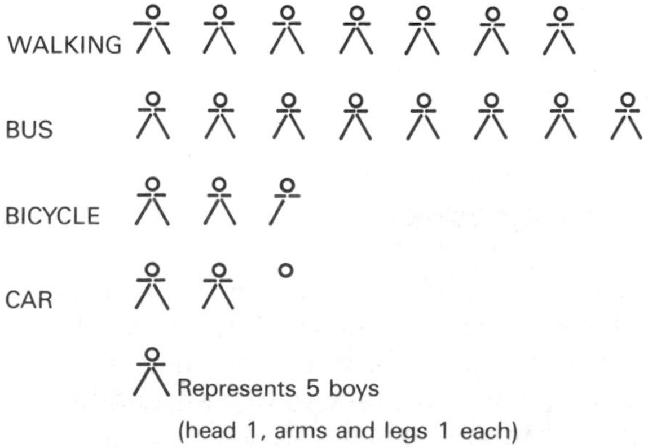

Represents 5 boys

(head 1, arms and legs 1 each)

2. The sales of motor cars by Mortimer & Co. Ltd. were as follows:

Year	Sales
1972	2000
1973	2500
1974	3200
1975	2700
1976	3000

Represent this information in a pictogram.

3. The table below gives the number of houses completed in the S.W. of England.

Year	Number (thousands)
1965	81
1967	69
1969	73
1971	84
1973	80

Draw a pictogram to represent this information.

4. The information below gives the production of tyres (in thousands) produced by Treadwell & Co. for the first six months of 1977.

Month	Production
January	40
February	43
March	39
April	38
May	37
June	45

Draw a pictogram to represent this information.

5. The pictogram (Fig. 23.13) is an attempt to show the sales of a company in the years 1972, 1974 and 1976. Why does the pictogram not give a true indication of the company's sales?

Fig. 23.13

1972
1 000 000 cornets of ice cream

1974
1 500 000 cornets of ice cream

1976
2 000 000 cornets of ice cream

6. A firm of carriers indicated their increase in parcel traffic as shown in Fig. 23.14. This indicated that they had increased eightfold. In 1970 the firm carried 1500 parcels.

(a) How many did they carry in 1972?

(b) Draw a figure in the same pattern to indicate the volume of parcel traffic in 1976 if it was 40 500 parcels.

Fig. 23.14

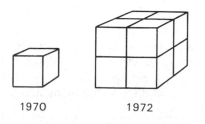

1970   1972

7. A circle of 2 cm radius represents sales of 200 items. If sales are depicted by area what does a circle of 4 cm radius represent?

8. The volume of a sphere of diameter 3 cm represents a production of 54 000 articles. What production will a sphere of 2 cm diameter represent?

Miscellaneous Exercise 23

1. Draw a proportionate bar chart for the figures shown below which relate to the way in which people travel to work in the London area.

Type of transport	Numbers using
Bus	780
Private motoring	420
Other (foot, bicycle, motor cycle, etc.)	160
Total	1360

2. Draw a pie chart to represent the information below which relates to the way in which a family spends its weekly income.

Item	Amount spent (£)
Food and drink	95
Housing	42
Transport	29
Clothing	33
Other	51
Total	250

3. The table below shows the result of a survey of the colours of doors on a housing estate. Draw a vertical bar chart to depict this information.

Colour of door	Number of houses
White	85
Red	17
Green	43
Brown	70
Blue	15

4. The table below shows the output of bicycles from a factory for the years 1982–6. Represent this information in the form of a chronological bar chart.

Year	1982	1983	1984	1985	1986
Number of bicycles	2000	4000	7000	8500	9000

5. The pie chart (Fig. 23.15) shows a local election result. There were three candidates, White, Green and Brown. The angle representing the number of votes cast for White is 144° and for Green is 90°.

    (a) Work out the angle representing the number of votes cast for Brown.

    (b) Calculate the number of votes cast for each candidate if the total number voting was 5000.

Fig. 23.15

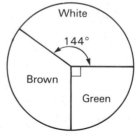

6. A pie chart was drawn to show the number of tons of various crops grown by a group of farmers. The sector angles corresponding to these crops were as follows:

Crop	Sector angle (degrees)
Vegetables	100
Grain	80
Potatoes	50
Fruit	40
Roots	30
Other	60

    Calculate the amount of each crop grown if the total weight of crops was 500 tons.

7. A factory buys various amounts of different metals. The proportionate bar chart (Fig. 23.16) shows the amount spent on each of these metals. Find the expenditure on steel, brass and bronze.

Fig. 23.16

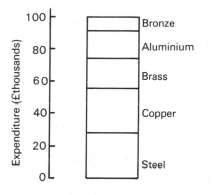

8. The chart below shows the distances (in kilometres) between five towns in Spain.

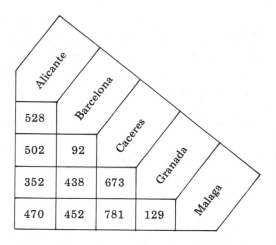

(a) What is the distance between Alicante and Barcelona?

(b) What is the distance between Granada and Alicante?

(c) Which two towns are nearest to each other?

(d) Work out the distance in miles between Alicante and Caceres given that 8 km = 5 miles approximately.

(e) A car journey between Malaga and Alicante took $5\frac{1}{2}$ hours. What was the average speed of the car in miles per hour? State the answer correct to 1 decimal place.

Multi-Choice Questions 23

1. The bar chart in Fig. 23.17 shows the distribution of children in classes 1 to 6 in a school. Each child is given three exercise books for the term. If an exercise book costs 20p how much was spent altogether on exercise books?

   A £180 B £360 C £600 D £1800

Fig. 23.17

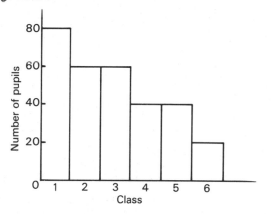

2. A student's performance in a class test was graded as 'excellent', 'very good', 'good', 'average' or 'poor'. The pie chart shown in Fig. 23.18 depicts the results of a mathematics test in which 50 students took part. How many students were graded as good?

   A 20   B 10   C 8   D 6

Fig. 23.18

3. 5000 students were interviewed to find out which of four sporting activities they liked most. The results are shown below:

Rugby	1200
Hockey	1000
Soccer	2000
Athletics	800
Total	5000

A pie chart was drawn to represent these figures. The angle of the sector representing those who like hockey was

A 95° B 72° C 65° D 60°

4. Fig. 23.19 shows the frequency distribution of the marks obtained in a test. How many candidates took the test?

A 15   B 24   C 50   D 62

Fig. 23.19

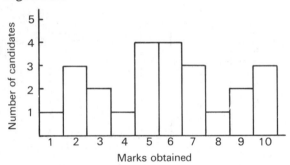

The chart below shows the distance (in kilometres) between four towns in Brittany.

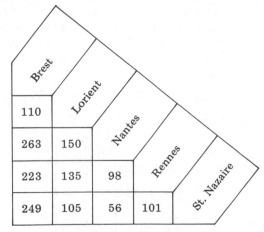

Use the chart to answer Questions 5, 6, 7 and 8.

5. What is the distance, in kilometres, between Brest and Nantes?

A 101   B 110   C 150   D 263

6. Which two towns are closest together?

A Brest and Lorient
B Nantes and St. Nazaire
C Nantes and Rennes
D Lorient and Nantes

7. Given that 8 km = 5 miles, what is the distance in miles between Brest and St. Nazaire?

A 156   B 66   C 63   D 35

8. A car travels from Brest to Rennes. Its average speed was 50 km/h. How long did the journey take?

A 2.2 hours   B 4.46 hours
C 4.98 hours   D 5.26 hours

Mental Test 23

Try to answer the following questions without writing anything down except the answer.

1. Use the chart below to find the distance from
   (a) Exeter to Gloucester
   (b) Exeter to Lincoln
   (c) Hereford to Liverpool
   (d) Kendal to Manchester.

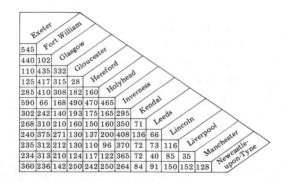

2. Fig. 23.20 shows a diagram which can be used to convert degrees Celsius into degrees Fahrenheit, and vice versa. Use the diagram to convert

   (a) $-40\,^{\circ}$C into degrees Fahrenheit

   (b) $20\,^{\circ}$C into degrees Fahrenheit

   (c) $100\,^{\circ}$F into degrees Celsius.

Fig. 23.20

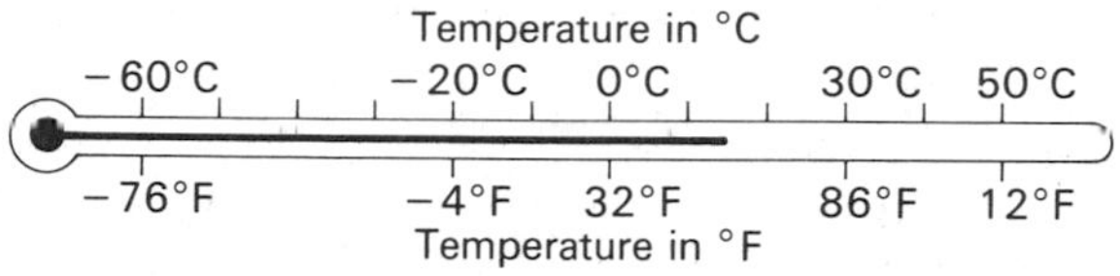

3. The bar chart (Fig. 23.21) shows the ages of children in a school.

   (a) How many 13 year olds took part in the survey?

   (b) How many children in total took part?

Fig. 23.21

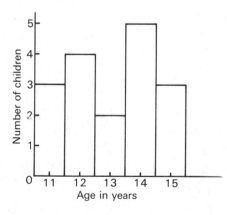

4. The pie chart (Fig. 23.22) illustrates the sports preferred by a group of male students.

   (a) What percentage preferred tennis?

   (b) What fraction preferred cricket?

Fig. 23.22

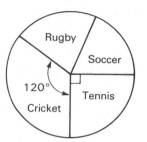

5. The main leisure interests of 100 fifth year pupils are shown below:

Soccer	40
Reading	10
Television	25
Scouts	5
Other	20

   If a pie chart is drawn for this information, find the size of the sector angles to represent

   (a) reading

   (b) television.

6. The chronological bar chart shown in Fig. 23.23 shows the number of colour television sets sold in England during the years 1970–5. Use the diagram to estimate the numbers of colour television sets sold in

   (a) 1971   (b) 1973   (c) 1975

Fig. 23.23

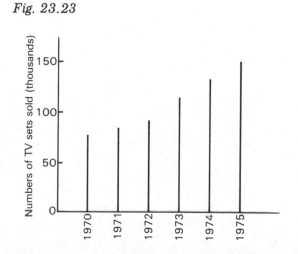

7. The pictogram (Fig. 23.24) shows the number of cars sold in the years 1975-9.

   (a) In which year were the most cars sold?

   (b) How many cars were sold in 1978?

   (c) How many cars were sold in 1976?

   (d) Estimate the total number of cars sold in the years depicted by the pictogram.

Fig. 23.24

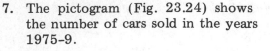

Graphs, Functions and Relations

Axes of Reference

To plot a graph the **axes of reference** are first drawn. Their intersection is called the **origin**. The vertical axis is called the y-axis and the horizontal axis is called the x-axis (Fig. 24.1).

Fig. 24.1

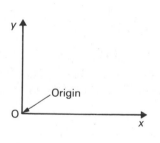

Scales

The number of units represented by a unit length along an axis is called the **scale**, for instance, $1\ cm = 5$ units.

Coordinates

Coordinates are used to mark the points on a graph. In Fig. 24.2 the point P has been plotted so that $x = 4$ and $y = 6$. The values of 4 and 6 are called the **rectangular coordinates** of P. For brevity the point P is said to be the point $(4, 6)$. Note carefully that the value of x is always given first and because the order in which the coordinates

are given is important the values $(4, 6)$ are called **an ordered pair**. Hence any pair of coordinates constitutes an ordered pair.

Fig. 24.2

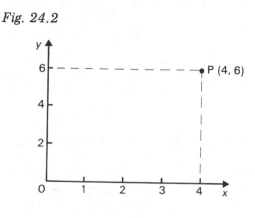

Plotting a Graph

Every graph shows a **relation** between two sets of figures.

Example 1

The information given below gives the production of car tyres (in thousands) produced at a certain factory for the first six years of production.

Year	1980	1981	1982	1983	1984	1985
Production	10	23	30	35	37	40

Draw a graph of this information.

One way of representing the information on a graph is to draw a bar chart, as shown in Fig. 24.3. It consists of a number of vertical lines whose heights represent the production of tyres during the years given on the horizontal axis.

Fig. 24.3

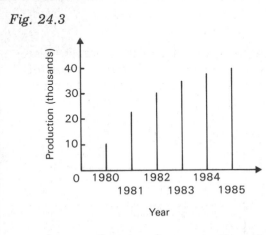

A second way of representing the information is to draw a graph of the data, as shown in Fig. 24.4. The points on the graph are plotted using the coordinates given in the table. Thus the first point has the coordinates (1980, 10), the second point has the coordinates (1981, 23), and so on. The points are then joined by a series of straight lines.

Fig. 24.4

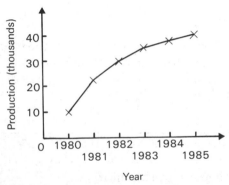

Exercise 24.1

1. The children in a class of a school are given a test periodically. One child obtained the following scores out of 50.

Number of test	1	2	3	4	5	6
Score	18	23	32	36	42	47

Draw:

(a) a bar chart

(b) a line graph

to represent this information. Plot the scores on the vertical axis.

2. The table below gives the temperature at 12.00 noon on 7 successive days. Plot:

(a) a bar chart

(b) a line graph

to depict this information. Mark off the days on the horizontal axis.

Day	1	2	3	4	5	6	7
Temperature (°C)	15	24	30	28	17	25	28

3. The information below gives the number of colour television sets sold in a Caribbean country during the years 1977 to 1982.

Year	1977	1978	1979	1980	1981	1982
Number sold (thousands)	35	43	58	67	77	92

Draw a graph, with the year on the horizontal axis, to represent this information.

4. The table below shows the amount of fruit produced by a farmer during 7 successive years.

Year	1	2	3	4	5	6	7
Fruit production (tonnes)	23	45	34	63	32	54	70

Draw a line graph, with the year on the horizontal axis, to depict this information.

5. In a primary school the ages of the pupils were as follows:

Age (years)	5	6	7	8	9	10
Number of children	21	19	36	42	28	32

Draw a line graph of this information, taking the ages on the horizontal axis.

Example 2

The values below give corresponding values of x and y. Using scales of 1 cm to 2 units on the horizontal axis and 1 cm to 5 units on the vertical axis, plot the graph.

x	0	2	4	6	8	10
y	1	7	13	19	25	31

The points are first plotted and we see that a straight line passes through all the points (Fig. 24.5).

Fig. 24.5

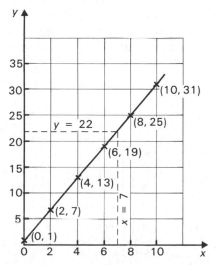

Interpolation

When a graph is a straight line or a smooth curve it can be used to deduce corresponding values of x and y not given in the original table of values. Thus to find the value of y corresponding to $x = 7$, we draw the vertical and horizontal lines shown broken in Fig. 24.5. We find that when $x = 7$, $y = 22$.

Using a graph in this way to find values not given in the original table of values is called *interpolation*.

Exercise 24.2

1. Fig. 24.6 shows a graph of y plotted against x. Find:

    (a) the values of y when $x = 4$, $x = 7$ and $x = 11$

    (b) the values of x when $y = 14$, $y = 22$ and $y = 34$.

Fig. 24.6

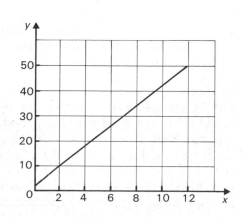

2. Fig. 24.7 shows a graph of distance plotted against time. From the graph find

(a) the distances corresponding to times of 2, 3 and 5 seconds

(b) the times corresponding to 4, 28 and 54 metres.

Fig. 24.7

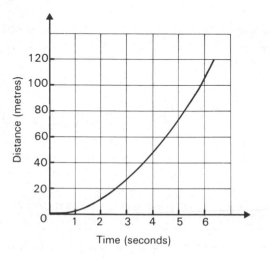

Time (seconds)

3. The values below give corresponding values of P and Q.

P	0	1	2	3	4	5
Q	0	2	8	18	32	50

Plot P horizontally and Q vertically using 1 large square to represent 1 unit horizontally and 1 large square to represent 10 units vertically. Draw the graph and use it to find the value of Q when $P = 2.5$.

4. The figures below show corresponding values for the units of electricity used and the cost. Draw a graph of this information taking units used along the horizontal axis and the cost along the vertical axis. Use scales of 1 large square to represent 100 units on the horizontal axis and 1 large square to represent £5 on the vertical axis. Use the graph to find the cost of 300 units of electricity.

Units used	100	200	400	600	700
'Cost (£)'	11	15	23	31	35

5. The table below shows corresponding values of x and y. Plot a graph using scales of 1 large square to represent 1 unit horizontally and vertically. Use your graph to find the value of y when $x = 3.5$.

x	1	2	3	4	5	6
y	8.0	4.0	2.7	2.0	1.6	1.3

Arrow Diagrams

The relation between two sets of values may also be shown by means of an arrow diagram. The arrow diagram (Fig. 24.8) shows the relation between x and y. The set of starting elements 0, 3, 6, 9 is called the **domain** and the set of finishing elements 1, 7, 13 and 19 is called the **range**. The values connected by the arrowed lines constitute ordered pairs and hence the relation between x and y may also be shown by means of a graph (Fig. 24.9).

Fig. 24.8

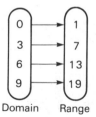

Domain   Range

Fig. 24.9

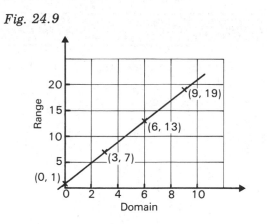

The arrow diagram (Fig. 24.8) shows a one to one relation. Other types are many to many (Fig. 24.10), one to many (Fig. 24.11) and many to one (Fig. 24.12).

Fig. 24.10

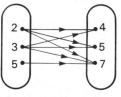

Many to many (diagram shows the relation 'is less than'.)

Fig. 24.11

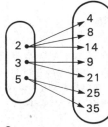

One to many (diagram shows the relation 'is a factor of'.)

Fig. 24.12

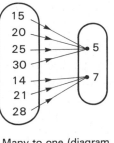

Many to one (diagram shows the relation 'is a multiple of'.)

Functions

A **function** is a relation in which each **first** element of an ordered pair has only one second element. Thus the relations shown in the arrow diagrams of Figs. 24.8 and 24.12 are functions whilst the relations shown in Figs. 24.10 and 24.11 are **not** functions.

When a relation is not a function a vertical line will pass through two or more ordered pairs when the relation is shown as a graph (Fig. 24.13).

Fig. 24.13

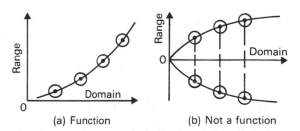

(a) Function   (b) Not a function

Function Notation

There are several ways of describing a function. They are by means of:

(1)   a mapping diagram like Fig. 24.8

(2)   an algebraic equation such as
$y = 3x + 2$

(3)   a graph such as Fig. 24.13(a)

(4)   a set of ordered pairs like $\{(0,3), (1,5),$ $(2,7), (3,9)\}$.

Consider the relation $x \to 2x + 7$ with domain $\{0,1,2,3\}$. The range is found by substituting each member of the domain into $2x + 7$.

Thus when

$$x = 0 \qquad 2x + 7 = (2 \times 0) + 7$$
$$= 7$$

$$x = 1 \qquad 2x + 7 = (2 \times 1) + 7$$
$$= 9$$

$$x = 2 \qquad 2x + 7 = (2 \times 2) + 7$$
$$= 11$$

$$x = 3 \qquad 2x + 7 = (2 \times 3) + 7$$
$$= 13$$

Hence the range is the set $\{7,9,11,13\}$.

When only one value of y corresponds to one value of x in an equation (as, for instance, above) function notation may be used. The relation $y = 2x + 7$ is a function and we may write $f : x \to 2x + 7$ or $f(x) = 2x + 7$, both expressions meaning that y is a function of x. $g(x)$ and $h(x)$ are also used to denote functions. Thus if $y = 3x^2 - 5x + 2$ then since y is a function of x we may write $g(x) = 3x^2 - 5x + 2$.

Example 3

If $f(x) = 2x^2 - 4x + 3$ find $f(-2)$ and $f(3)$.

To find $f(-2)$ substitute $x = -2$ into the expression $2x^2 - 4x + 3$. Thus

$$f(-2) = 2 \times (-2)^2 - 4 \times (-2) + 3$$
$$= 8 + 8 + 3$$
$$= 19$$

$$f(3) = 2 \times 3^2 - 4 \times 3 + 3$$
$$= 18 - 12 + 3$$
$$= 9$$

Exercise 24.3

1. Copy and complete Fig. 24.14 if the relation is

    (a) $x \to 2x + 4$ (b) $x \to 6x$

Fig. 24.14

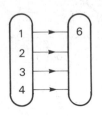

2. With $\{1,3,4,7\}$ as domain draw mapping diagrams for $f : x \to 4x - 1$ and $f : x \to x^2 + 2$.

3. If $f(x) = 7x - 9$ find:

    (a) $f(-3)$ (b) $f(0)$ (c) $f(5)$

4. Draw a mapping diagram for the function $f(x) = x^2 - 3x + 2$ with domain $\{0,4,5,7,8\}$.

5. If $f(x) = 3x^2 - 5x - 8$, write down the values of

    (a) $f(-2)$ (b) $f(0)$ (c) $f(2)$

Flow Charts

There are many times when a series of operations of instructions has to be carried out. If these are written down in sequence then this list is called a **program**.

A simple and clear way of giving a program is by using a **flow chart**. An example of a flow chart is shown in Fig. 24.15.

Fig. 24.15

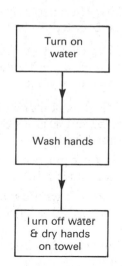

The instructions are given in rectangles which are connected by arrows showing the order in which to proceed.

The flow chart for mathematical calculations may be drawn in a similar way to Fig. 24.15.

Example 4

Draw a flow chart for the calculation of $5 \times 4 \div 2$.

The flow chart is shown in Fig. 24.16.

Frequently, algebraic expressions are represented on flow charts.

Fig. 24.16

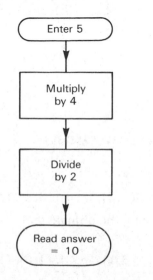

Functions may also be illustrated by flow charts.

Example 5

Draw a flow chart to represent the equation $f(x) = 3(2x + 5)$.

Fig. 24.17

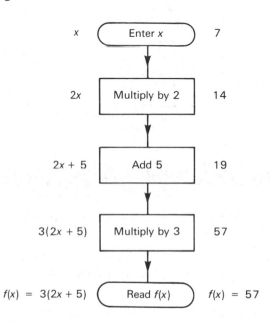

The flow chart is shown in Fig. 24.17. This flow chart allows the value of y to be calculated for any value of x. Thus if $x = 7$, the flow chart gives the results shown alongside Fig. 24.17.

Example 6

Use the flow chart shown in Fig. 24.18 to find the relationship between x and y.

The symbols shown alongside the chart show the operations performed and we see that the relationship between x and y is $y = (2x + 9)^2$.

Fig. 24.18

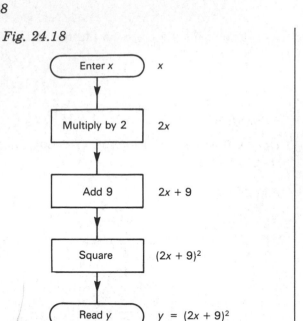

3. Draw a flow chart for the arithmetic operations $5 \times (3 + 4) - 7$.

4. Draw a flow chart to represent the equation $y = 2(3x - 5)$ and use it to find the value of y when $x = 4$.

5. The flow diagram shown in Fig. 24.20 is used to convert degrees Celsius into degrees Fahrenheit.

   (a) Draw a flow diagram for converting degrees Fahrenheit into degrees Celsius.

   (b) By using the two flow diagrams:
      (i) Convert $30°F$ into degrees Celsius.
      (ii) Convert $20°C$ into degrees Fahrenheit.

Exercise 24.4

1. Draw a flow chart for the calculation $7 + 8 \times 3$.

2. Use the flow diagram shown in Fig. 24.19 to express y in terms of x.

Fig. 24.20

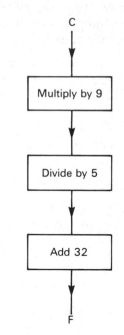

Fig. 24.19

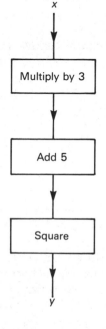

Inverse Functions

Consider the relation

$$F = \{(1, 3), (2, 6), (3, 4), (4, 12)\}$$

Interchanging the domain and the range, we have

$$G = \{(3, 1), (6, 2), (4, 3), (12, 4)\}$$

F and G are said to be **inverse relations.** G is the inverse of F which is denoted by F^{-1}. Thus

$$G = F^{-1}$$

We see that the inverse of a relation F is obtained by interchanging the elements of each ordered pair in F.

Example 7

If f is a function defined by $f(x) = \dfrac{3x - 5}{x + 2}$, write down an expression for $f^{-1}(x)$ and find the value of $f^{-1}(2)$.

Writing the equation as

$$y = \frac{3x - 5}{x + 2}$$

the defining equation for $f^{-1}(x)$ is

$$x = \frac{3y - 5}{y + 2}$$

since the inverse of a function is obtained by interchanging corresponding values of x and y.

We now transpose this equation to make y the subject. Thus

$$y = \frac{2x + 5}{3 - x}$$

$$\therefore \quad f^{-1}(x) = \frac{2x + 5}{3 - x}$$

$$f^{-1}(2) = \frac{2 \times 2 + 5}{3 - 2}$$

$$= \frac{9}{1}$$

$$= 9$$

For some kinds of functions an inverse flow diagram is useful when finding the inverse of a function.

Figure 24.21 is a flow diagram for the function $f : x \rightarrow \dfrac{x + 4}{5}$.

Fig. 24.21

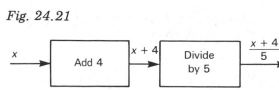

The inverse function is obtained from the reverse flow diagram (Fig. 24.22).

Fig. 24.22

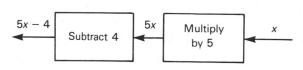

Note that in reversing the flow diagram add becomes subtract, subtract becomes add, multiply becomes divide and divide becomes multiply.

We see that when $f : x \rightarrow \dfrac{x + 4}{5}$ then $f^{-1} : x \rightarrow 5x - 4$. Alternatively we may write when $f(x) = \dfrac{x + 4}{5}$ then $f^{-1}(x) = 5x - 4$.

The inverse of a function may be used to solve equations.

Example 8

By finding the inverse of $f(x) = \dfrac{4x + 5}{x + 2}$, solve the equation $\dfrac{4x + 5}{x + 2} = 3$.

Writing the defining equation for $f(x)$ as

$$y = \frac{4x + 5}{x + 2}$$

the defining equation for $f^{-1}(x)$ is

$$x = \frac{4y + 5}{y + 2}$$

Transposing this equation to make y the subject gives

$$y = \frac{5 - 2x}{x - 4}$$

$$f^{-1}(x) = \frac{5 - 2x}{x - 4}$$

The solution of the given equation is found by finding the value of $f^{-1}(3)$. Thus

$$f^{-1}(3) = \frac{5 - 2 \times 3}{3 - 4}$$

$$= 1$$

Hence the solution of the given equation is $x = 1$.

Composite Functions

Suppose $f: x \to 2x - 1$ and $g: x \to 3x + 2$ then $gf: x \to 3(2x - 1) + 2$ or $gf: x \to 6x - 1$.

Note that gf means do f first and then g.

Composite functions are seldom commutative, that is,

$$gf \neq fg$$

For the above,

$$fg: x \to 2(3x + 2) - 1$$

or $\quad fg: x \to 6x + 3 = 3(2x + 1)$

Hence $\qquad gf \neq fg$

Example 9

If $f: x \to 2x + 5$, $g: x \to \frac{1}{2}x$ and $h: x \to 3x - 1$ find:

(a) $fg(2)$ (b) $gf(-1)$

(c) $fgh(3)$ (d) $ghf(-2)$

(a) $fg: x \to 2(\frac{1}{2}x) + 5 = x + 5$

$$fg(2) = 2 + 5$$

$$= 7$$

(b) $gf: x \to \frac{1}{2}(2x + 5) = x + \frac{5}{2}$

$$gf(-1) = -1 + \frac{5}{2}$$

$$= \frac{3}{2}$$

(c) $fgh: x \to 2[\frac{1}{2}(3x - 1)] + 5 = 3x + 4$

$$fgh(3) = 3 \times 3 + 4$$

$$= 13$$

(d) $ghf: x \to \frac{1}{2}[3(2x + 5) - 1] = 3x + 7$

$$ghf(-2) = 3 \times (-2) + 7$$

$$= 1$$

The Inverse of a Composite Function

If $f: x \to 3x - 5$ and $g: x \to 2x + 1$ then $gf: x \to 2(3x - 5) + 1 = 6x - 9$

$$(gf)^{-1}: x \to \frac{1}{6}(x + 9)$$

Now $f^{-1}: x \to \frac{1}{3}(x + 5)$ and $g^{-1}: x \to \frac{1}{2}(x - 1)$

$$f^{-1}g^{-1}: x \to \frac{1}{3}[\frac{1}{2}(x - 1)] + \frac{5}{3}$$

$$= \frac{1}{6}(x - 1) + \frac{5}{3}$$

$$= \frac{1}{6}x - \frac{1}{6} + \frac{5}{3}$$

$$= \frac{1}{6}x + \frac{9}{6}$$

$$= \frac{1}{6}(x + 9)$$

Hence $(gf)^{-1} = f^{-1}g^{-1}$.

Using this relationship often simplifies the work in finding the inverse of a composite function.

Example 10

If $f: x \rightarrow 7x + 2$ and $g: x \rightarrow 2x - 1$ find $(gf)^{-1}$.

Now $f^{-1}: x \rightarrow \frac{1}{7}(x - 2)$ and
$g^{-1}: x \rightarrow \frac{1}{2}(x + 1)$

$$(gf)^{-1} = f^{-1}g^{-1}: x \rightarrow \frac{1}{7}[\frac{1}{2}(x + 1)] - \frac{2}{7}$$

$$= \frac{1}{14}(x + 1) - \frac{2}{7}$$

$$= \frac{1}{14}x + \frac{1}{14} - \frac{2}{7}$$

$$= \frac{1}{14}(x - 3)$$

Exercise 24.5

Find the inverse functions for the following:

1. $f: x \rightarrow 3x$

2. $f: x \rightarrow 2x - 3$

3. $f: x \rightarrow \dfrac{x - 3}{2}$

4. $f: x \rightarrow 3(2x - 5)$

5. $f: x \rightarrow \dfrac{2x - 5}{x + 2}$

6. If $f: x \rightarrow 2(3x - 1)$ find
   (a) $f^{-1}(-2)$ (b) $f^{-1}(3)$

7. If $f: x \rightarrow \dfrac{x - 4}{3 - x}$ find
   (a) $f^{-1}(2)$ (b) $f^{-1}(-2)$

8. If $f: x \rightarrow 4x - 7$ find $f^{-1}(3)$ and $f^{-1}(-2)$.

9. If $f: x \rightarrow 2x - 5$, $g: x \rightarrow 3x + 1$ and $h: x \rightarrow 4x$, find
   (a) $fg(2)$ (b) $gf(2)$
   (c) $fgh(-1)$ (d) $ghf(-3)$
   (e) $hgf(2)$

10. If $f: x \rightarrow 5x - 1$ and $g: x \rightarrow 3x + 2$, find $(fg)^{-1}$ and $(gf)^{-1}$.

11. By finding the inverse of $f(x) = \dfrac{x + 3}{x - 5}$, solve the equation $\dfrac{x + 3}{x - 5} = 2$.

12. By finding the inverse of $f(x) = \dfrac{3x + 1}{4x - 3}$, solve the equation $\dfrac{3x + 1}{4x - 3} = 5$.

Miscellaneous Exercise 24

Section A

1. Fig. 24.23 shows the relation between the number of units of electricity used and the cost of a certain electricity bill. Find:

   (a) the cost when 360 units are used

   (b) the number of units used when the cost is £17.

   (c) The bill is made up of a standing charge plus so much per unit used. What is the standing charge and what is the cost per unit?

Fig. 24.23

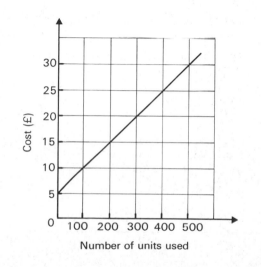

Number of units used

2. Using a scale of 1 cm to 1 unit on the x-axis and 1 cm to 2 units on the y-axis, construct a pair of axes on graph paper for values of x between −4 and 5 and for values of y between −10 and 10.

    (a) On your graph plot the points A(4,8) and B(−3,1). Join A and B with a straight line.

    (b) Use your graph to find the coordinates of the point where the line AB crosses the y-axis.

3. In Fig. 24.24, RP is the line $y = 2x + 3$. If ON is 3 units, find:

    (a) the length of PN

    (b) the area of the trapezium AONP.

Fig. 24.24

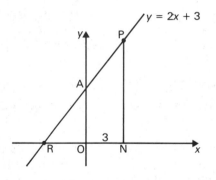

4. Fig. 24.25 shows a straight line passing through the points A and B, whose equation is $y = \frac{1}{2}x + 6$.

    (a) Write down the coordinates of the points A and B.

    (b) Calculate the area of the triangle AOB.

Fig. 24.25

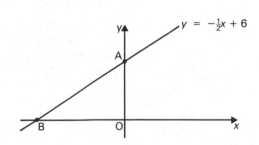

5. Fig. 24.26 shows a mapping diagram. Copy and complete it if $x \rightarrow 3x - 1$.

Fig. 24.26

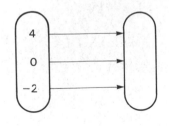

Section B

1. Draw a mapping diagram for the function $f: x \rightarrow 3^x$ defined as the domain $\{0, 1, 2\}$.

2. If $f(x) = \dfrac{3x - 4}{x + 1}$, find an expression for $f^{-1}(x)$ and work out the value of $f^{-1}(2)$.

3. By finding the inverse of $f(x) = \dfrac{x + 3}{x + 5}$ solve the equation $\dfrac{x + 3}{x + 5} = 2$.

4. If $f: x \rightarrow 2x - 5$ and $g: x \rightarrow 3x + 1$, find

    (a) $gf(2)$ (b) $fg(2)$

5. If $f(x) = 5x^2 - 7x - 8$, write down the values of $f(-2)$, $f(0)$ and $f(3)$.

6. Draw a flow chart to represent $f(x) = 5(2x - 3)$. By drawing a reverse flow diagram find an expression for $f^{-1}(x)$ and find the value of $f^{-1}(-2)$.

7. If $f: x \rightarrow 5x - 1$ and $g: x \rightarrow 3x + 2$, find expressions for $(gf)^{-1}$.

8. If $g(x) = x - 2$ and $f(x) = 2x + 1$, find expressions for
   (a) $gf(x)$ (b) $fg(x)$.

Multi-Choice Questions 24

1. What is the relation shown in the mapping diagram (Fig. 24.27)?
   A $x \rightarrow x + 1$ B $x \rightarrow x + 4$
   C $x \rightarrow x + 3$ D $x \rightarrow 2x + 1$

Fig. 24.27

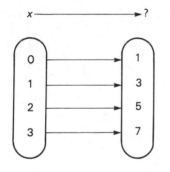

2. All the points on the line PQ (Fig. 24.28) satisfy a certain relation. What is this relation?
   A $y = x$ B $y = -3$
   C $y = x + 3$ D $y = x - 3$

Fig. 24.28

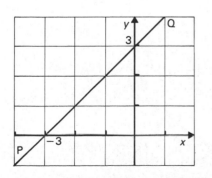

3. Which of the functions on the set $\{1, 2, 3, 4\}$ does Fig. 24.29 represent?
   A $f(x) = 2x + 2$ B $f(x) = 2x - 2$
   C $f(x) = \frac{1}{2}x + 2$ D $f(x) = \frac{1}{2}x - 2$

Fig. 24.29

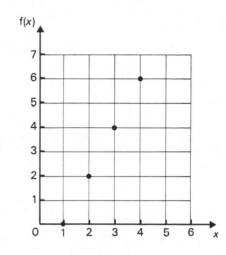

4. In Fig. 24.30, if $BM = 13$ then the length of OM is
   A $3\frac{1}{2}$ B 5   C $6\frac{1}{2}$ D 8

Fig. 24.30

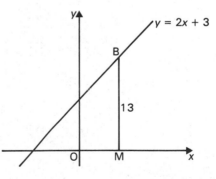

5. The straight line $y = 2x - 3$ passes through one of the following points. Which one?
   A $(0, -3)$ B $(0, 3)$
   C $(-3, 0)$ D $(3, 0)$

6. x and y are coordinates of the points on a straight line. The missing number in the table below is

x	0	?	12
y	0	4	9

A $2\frac{2}{3}$ B 3   C $5\frac{1}{3}$ D 6

7. Find f^{-1} when $f: x \rightarrow 2x + 1$.

A $2(x + 1)$ B $\dfrac{x - 1}{2}$

C $\dfrac{x + 1}{2}$ D $2(x - 1)$

25 Coordinate Geometry

Rectangular Coordinates

It has been shown on page 181 that a point can be positioned by means of rectangular coordinates. Thus in Fig. 25.1 the point P has the coordinates $x = 3$ and $y = 5$. We say that P is the point $(3, 5)$. Similarly Q is the point $(-2, -4)$.

Fig. 25.1

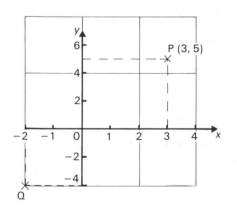

Example 1

Plot the points $A(3, 4)$, $B(5, 4)$, $C(6, 2)$ and $D(2, 2)$. Join up the points in alphabetical order to form the quadrilateral ABCD.

(a) Name the quadrilateral.

(b) Find its area.

(a) The points are plotted in Fig. 25.2 and the quadrilateral is seen to be a trapezium.

Fig. 25.2

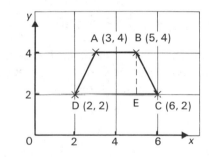

(b) To find its area we need to know the lengths of the two parallel sides and the distance between them. From the diagram:

$$AB = 5 - 3$$
$$= 2 \text{ units}$$
$$DC = 6 - 2$$
$$= 4 \text{ units}$$
$$BE = 4 - 2$$
$$= 2 \text{ units}$$

Here

$$\text{Area} = \tfrac{1}{2}(AB + DC) \times BE$$
$$= \tfrac{1}{2} \times (2 + 4) \times 2$$
$$= \tfrac{1}{2} \times 6 \times 2$$
$$= 6 \text{ square units}$$

In each of the following problems, plot the given points. Join them up in alphabetical order and name the resulting figure. Find the area of each (in square units).

1. P(3, 2), Q(6, 2) and R(6, 8).

2. A(2, 3), B(8, 3), C(8, 8) and D(2, 8).

3. E(2, 5), F(6, 5), G(7, 10) and H(3, 10).

4. W(−2, 2), X(3, 2), Y(2, 4) and Z(−1, 4).

5. A(−3, −2), B(2, 1) and C(2, −2).

The Length of a Line

Given the coordinates of its end points, the length of a line can be found by Pythagoras' theorem as shown in Example 2.

Example 2

A is the point (3, 5) and B is the point (6, 3). Find the length of AB.

The line is drawn in Fig. 25.3. Constructing the right-angled triangle ABC we see that C has the coordinates (3, 3). Hence

$$AC = (5 - 3) \text{ units}$$
$$= 2 \text{ units}$$

and $$BC = (6 - 3) \text{ units}$$
$$= 3 \text{ units}$$

Applying Pythagoras' theorem we have
$$AB^2 = AC^2 + BC^2$$
$$= 2^2 + 3^2$$
$$= 4 + 9$$
$$= 13$$
$$AB = \sqrt{13}$$
$$= 3.606 \text{ units}$$

Fig. 25.3

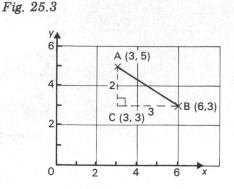

The Mid-Point of a Line

Example 3

A is the point (4, 2) and B is the point (12, 4). Find the coordinates of the mid-point of the line AB.

The line AB is drawn in Fig. 25.4. C is the mid-point of the line. From the construction shown:

$$x \text{ coordinate of C} = \tfrac{1}{2}(12 + 4)$$
$$= \tfrac{1}{2} \times 16$$
$$= 8$$

$$y \text{ coordinate of C} = \tfrac{1}{2}(2 + 4)$$
$$= \tfrac{1}{2} \times 6$$
$$= 3$$

Hence C is the point (8, 3).

Fig. 25.4

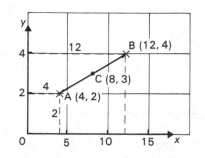

Exercise 25.2

Find the lengths of the lines AB in Questions 1 to 5.

1. $A(3, 5)$, $B(6, 8)$.

2. $A(1, 6)$, $B(3, 9)$.

3. $A(-1, 5)$, $B(4, 9)$.

4. $A(-2, -5)$, $B(3, -8)$.

5. $A(7, -2)$, $B(9, -6)$.

Find the coordinates of the mid-points of the following lines AB.

6. $A(3, 7)$, $B(0, 4)$.

7. $A(-3, 5)$, $B(-1, 8)$.

8. $A(0, 3)$, $B(5, 9)$.

Graphs of Linear Equations

Consider the equation:

$$y = 2x + 5$$

We can give x any value we please and so calculate a corresponding value for y. Thus,

when $x = 0$ $\quad y = 2 \times 0 + 5$
$$= 5$$

when $x = 1$ $\quad y = 2 \times 1 + 5$
$$= 7$$

when $x = 2$ $\quad y = 2 \times 2 + 5$
$$= 9$$

and so on.

The value of y therefore depends on the value allocated to x. We therefore call y the **dependent variable**. Since we can give x any value we please, we call x the **independent variable**. It is usual to mark the values of the independent variable along the horizontal axis and this axis is frequently called the x-axis. The values of the dependent variable are then marked off along the vertical axis which is often called the y-axis.

In plotting graphs representing equations we may have to include coordinates which are positive and negative. To represent these on a graph we make use of the number scales used in directed numbers (Fig. 25.5).

Fig. 25.5

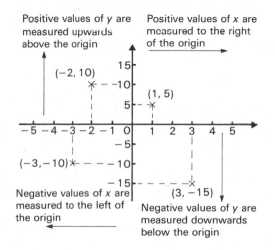

Positive values of y are measured upwards above the origin

Positive values of x are measured to the right of the origin

Negative values of x are measured to the left of the origin

Negative values of y are measured downwards below the origin

Example 4

(a) Draw the graph of $y = 2x - 5$ for values of x between -3 and 4.

Having decided on some values for x we calculate the corresponding values for y by substituting in the given equation. Thus,

when $x = -3$
$$y = 2 \times (-3) - 5$$
$$= -6 - 5$$
$$= -11$$

For convenience the calculations are tabulated as shown below.

x	-3	-2	-1	0
$2x$	-6	-4	-2	0
-5	-5	-5	-5	-5
$y = 2x - 5$	-11	-9	-7	-5
x	1	2	3	4
$2x$	2	4	6	8
-5	-5	-5	-5	-5
$y = 2x - 5$	-3	-1	1	3

A graph may now be plotted using these values of x and y (Fig. 25.6). The graph is a straight line.

Fig. 25.6

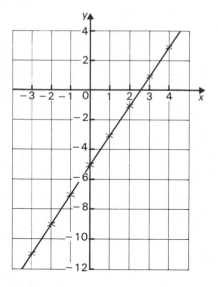

Equations of the type $y = 2x - 5$, where the highest powers of the variables, x and y, are the first are called equations of the **first degree**. All equations of this type give graphs which are straight lines and hence they are often called **linear equations**. In order to draw graphs of linear equations we need only take two points.

It is safer, however, to take three points, the third point acting as a check on the other two.

(b) By means of a graph show the relationship between x and y in the equation $y = 5x + 3$. Plot the graph between $x = -3$ and $x = 3$.

Since this is a linear equation we need only take three points.

x	-3	0	$+3$
$y = 5x + 3$	-12	3	$+18$

The graph is shown in Fig. 25.7.

Fig. 25.7

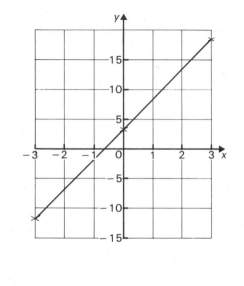

Exercise 25.3

Draw graphs of the following simple equations:

1. $y = x + 2$ taking values of x between -3 and 2.

2. $y = 2x + 5$ taking values of x between -4 and 4.

3. $y = 3x - 4$ taking values of x between -4 and 3.

4. $y = 5 - 4x$ taking values of x between -2 and 4.

The Equation of a Straight Line

Every linear equation may be written in the standard form:

$$y = mx + c$$

Hence $y = 2x - 5$ is in the standard form with $m = 2$ and $c = -5$.

The equation $y = 4 - 3x$ is in standard form if we rearrange it to give $y = -3x + 4$ so that we see $m = -3$ and $c = 4$.

The equation

$$4x + 5y = 6$$

may be written in standard form if it is rearranged to give

$$5y = -4x + 6$$

i.e.

$$y = -\frac{4}{5}x + \frac{6}{5}$$

Hence $m = -\frac{4}{5}$ and $c = \frac{6}{5}$.

The Meaning of *m* and *c* in the Equation of a Straight Line

The point B is any point on the straight line shown in Fig. 25.8 and it has the coordinates x and y. Point A is where the line cuts the y-axis and it has coordinates $x = 0$ and $y = c$.

Fig. 25.8

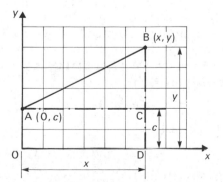

$\dfrac{BC}{AC}$ is called the gradient of the line.

Now

$$BC = \frac{BC}{AC} \times AC$$

$$= AC \times \text{Gradient of the line}$$

$$y = BC + CD$$

$$= BC + AO$$

$$= AC \times \text{Gradient of the line} + AO$$

$$= x \times \text{Gradient of the line} + c$$

But $y = mx + c$.

Hence it can be seen that

$$m = \text{Gradient of the line}$$

$$c = \text{Intercept on the } y\text{-axis}$$

Figure 25.9 shows the difference between **positive** and **negative** gradients.

(When the angle between the line and the positive direction of Ox is **acute** the gradient is **positive**. When the angle is **obtuse** the gradient is **negative**.)

Fig. 25.9

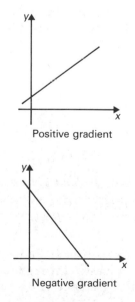

Positive gradient

Negative gradient

Example 5

(a) Find the equation of the straight line shown in Fig. 25.10.

Fig. 25.10

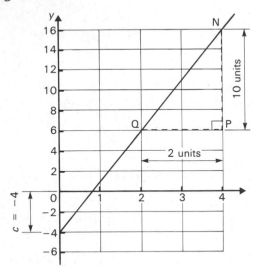

Since the origin is at the intersection of the axes, c is the intercept on the y-axis. From Fig. 25.10 it will be seen that $c = -4$. We now have to find m. Since this is the gradient of the line we draw $\triangle QNP$, making the sides reasonably long since a small triangle will give very inaccurate results. Using the scales of x and y we see that $QP = 2$ units and $PN = 10$ units.

$$\therefore \qquad m = \frac{NP}{QP}$$

$$= \frac{10}{2}$$

$$= 5$$

Therefore the standard equation of a straight line $y = mx + c$ becomes $y = 5x - 4$.

(b) Find the values of m and c if the straight line $y = mx + c$ passes through the point $(-1, 3)$ and has a gradient of 6.

Since the gradient is 6 we have $m = 6$

$$\therefore \qquad y = 6x + c$$

Since the line passes through the point $(-1, 3)$ we have $y = 3$ when $x = -1$. By substitution,

$$3 = 6 \times (-1) + c$$

$$3 = -6 + c$$

$$\therefore \qquad c = 9$$

Hence $y = 6x + 9$.

(c) Find the equation of the straight line shown in Fig. 25.11.

Fig. 25.11

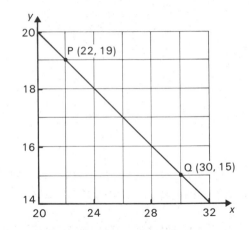

It will seem from Fig. 25.11 that the **origin is not at the intersection of the axes.** In order to determine the equation of the straight line we use two simultaneous equations as follows: Choose two convenient points P and Q and find their coordinates (these two points should be as far apart as possible to get maximum accuracy). If a point lies on a line then the x and y values of that point must satisfy the equation:

$$y = mx + c$$

At point P,

$$x = 22 \quad \text{and} \quad y = 19$$

$$\therefore \qquad 19 = 22m + c \qquad (1)$$

At point Q,

$$x = 30 \quad \text{and} \quad y = 15$$
$$15 = 30m + c \qquad (2)$$

Subtracting equation (2) from equation (1), we have

$$4 = -8m$$

$$\therefore \qquad m = \frac{4}{-8}$$

$$m = -0.5$$

Substituting $m = -0.5$ in equation (1), we have

$$19 = 22 \times (-0.5) + c$$
$$19 = -11 + c$$
$$c = 30$$

Thus the equation of the line shown in Fig. 25.11 is

$$y = -0.5x + 30$$

(d) Find the values of m and c if the straight line $y = mx + c$ passes through the points $(3, 4)$ and $(7, 10)$.

$$y = mx + c$$

The first point has coordinates $x = 3$, $y = 4$. Hence

$$4 = 3m + c \qquad (1)$$

The second point has coordinates $x = 7$, $y = 10$. Hence

$$10 = 7m + c \qquad (2)$$

Subtracting equation (1) from equation (2), we have

$$6 = 4m$$

$$\therefore \qquad m = 1.5$$

Substituting for $m = 1.5$ in equation (1), we have

$$4 = 4.5 + c$$

$$\therefore \qquad c = -0.5$$

The equation of the straight line is

$$y = 1.5x - 0.5$$

Experimental Data

One of the most important applications of the straight-line equation is the determination of an equation connecting two quantities when values have been obtained from an experiment.

Example 6

In an experiment carried out with a lifting machine the effort E and the load W were found to have the values given in the table below:

W (kg)	15	25	40	50	60
E (kg)	2.75	3.80	5.75	7.00	8.20

Plot these results and obtain the equation connecting E and W which is thought to be of the type $E = aW + b$.

If E and W are connected by an equation of the type $E = aW + b$ then the graph must be a straight line. Note that when plotting the graph, W is the independent variable and must be plotted on the horizontal axis. E is the dependent variable and must be plotted on the vertical axis.

Fig. 25.12

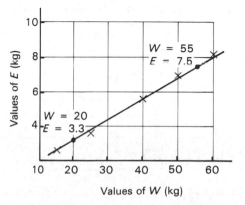

On plotting the points (Fig. 25.12) it will be noticed that they deviate only slightly from a straight line. Since the data are experimental we must expect errors in measurement and observation and hence slight deviations from a straight line must be expected. Although the straight line will not pass through some of the points, an attempt must be made to ensure an even spread of the points above and below the line.

To determine the equation we choose two points which lie on the straight line. Do not use any of the experimental results from the table unless they happen to lie exactly on the line. Choose the points as far apart as is convenient because this will help the accuracy of your result.

The point $W = 55$, $E = 7.5$ lies on the line. Hence

$$7.5 \ = \ 55a + b \qquad (1)$$

The point $W = 20$, $E = 3.3$ also lies on the line. Hence

$$3.3 \ = \ 20a + b \qquad (2)$$

Subtracting equation (2) from equation (1), we have

$$4.2 \ = \ 35a$$

$$a \ = \ 0.12$$

Substituting for $a = 0.12$ in equation (2), we have

$$3.3 \ = \ 20 \times 0.12 + b$$

$$b \ = \ 0.9$$

The required equation connecting E and W is therefore

$$E \ = \ 0.12W + 0.9$$

Exercise 25.4

The following equations represent straight lines. State in each case the gradient of the line and the intercept on the y-axis.

1. $y = x + 3$ 2. $y = -3x + 4$

3. $y = -5x - 2$ 4. $y = 4x - 3$

5. Find the values of m and c if the straight line $y = mx + c$ passes through the point $(-2, 5)$ and has a gradient of 4.

6. Find the values of m and c if the straight line $y = mx + c$ passes through the point $(3, 4)$ and the intercept on the y-axis is -2.

In the following find the values of m and c if the straight line $y = mx + c$ passes through the given points:

7. $(-2, -3)$ and $(3, 7)$

8. $(1, 1)$ and $(2, 4)$

9. $(-2, 1)$ and $(3, -9)$

10. $(-3, 13)$ and $(1, 1)$

11. $(2, 17)$ and $(4, 27)$.

12. The following table gives values of x and y which are connected by an equation of the type $y = ax + b$. Plot the graph and from it find the values of a and b.

x	2	4	6	8	10	12
y	10	16	22	28	34	40

13. The following observed values of P and Q are supposed to be related by the linear equation $P = aQ + b$, but there are experimental errors. Find by plotting the graph the most probable values of a and b.

Q	2.5	3.5	4.4	5.8	7.5	9.6	12.0	15.1
P	13.6	17.6	22.2	28.0	35.5	47.4	56.1	74.6

14. In an experiment carried out with a machine the effort E and the load W were found to have the values given in the table below. The equation connecting E and W is thought to be of the type $E = aW + b$. By plotting the graph check whether this is so and hence find a and b.

W (kg)	10	30	50	60	80	100
E (kg)	8.9	19.1	29	33	45	54

15. A test on a metal filament lamp gave the following values of resistance (R ohms) at various voltages (V volts).

V	62	75	89	100	120
R	100	117	135	149	175

These results are expected to agree with an equation of the type $R = mV + c$ where m and c are constants. Test this by drawing the graph and find suitable values for m and c.

16. During an experiment to verify Ohm's Law the following results were obtained.

E (volts)	0	1.0	2.0	2.5	3.7
I (amperes)	0	0.24	0.5	0.63	0.92

E (volts)	4.1	5.9	6.8	8.0
I (amperes)	1.05	1.48	1.70	2.05

Plot these values with I horizontal and find the equation connecting E and I.

Non-Linear Equations which can be Reduced to the Linear Form

Many non-linear equations can be reduced to the linear form by making a suitable substitution.

Consider the equation $y = ax^2 + b$. Let $z = x^2$ so that the given equation becomes $y = az + b$. If we now plot values of z against values of y we shall obtain a straight line because $y = az + b$ is of the standard linear form.

Example 7

The fusing current I amperes for wires of various diameters d mm is as shown in the table below.

d (mm)	5	10	15	20	25
I (amperes)	6.25	10	16.25	25	36.25

Construct another table showing values of d^2 against I and verify graphically that d and I are connected by a law of the form

$$I = ad^2 + b$$

where a and b are constants. Use your graph to estimate

(a) values for a and b

(b) the value of I when $d = 12.5$ mm

(c) the value of d when $I = 22$ amperes.

Drawing up the new table:

d^2	25	100	225	400	625
I	6.25	10	16.25	25	36.25

From the graph (Fig. 25.13) we see that the points lie on a straight line and hence we have verified that $I = ad^2 + b$. Note that I is the dependent variable and is plotted on the vertical axis and that d^2 is the independent variable and is plotted on the horizontal axis.

Fig. 25.13

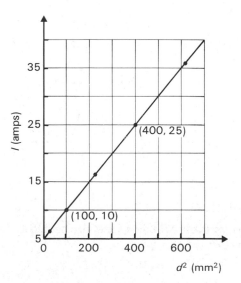

(a) To find the values of a and b choose two points which lie on the line and find their coordinates.

The point $(400, 25)$ lies on the line.

$$\therefore \qquad 25 = 400a + b \qquad (1)$$

The point $(100, 10)$ lies on the line.

$$\therefore \qquad 10 = 100a + b \qquad (2)$$

Subtracting equation (2) from equation (1),

$$15 = 300a$$

$$a = 0.05$$

Substituting $a = 0.05$ in equation (2),

$$10 = 100 \times 0.05 + b$$

$$b = 5$$

Therefore the law is:

$$I = 0.05d^2 + 5$$

(b) When $d = 12.5$,

$$I = 0.05 \times (12.5)^2 + 5$$

$$= 12.81$$

Hence when $d = 12.5\,\text{mm}$, $I = 12.81$ amperes.

(c) When $I = 22$,

$$22 = 0.05d^2 + 5$$

$$17 = 0.05d^2$$

$$d = \sqrt{\frac{17}{0.05}}$$

$$= 18.44$$

Therefore when $I = 22$ amperes, $d = 18.44\,\text{mm}$.

Example 8

Two variables P and Q are connected by an equation of the type

$$P = a\sqrt{Q} + b$$

where a and b are constants. When $Q = 9$, $P = 5$ and when $Q = 25$, $P = 8$. Draw a graph which shows the relationship between P and $\sqrt{Q}$. Use your graph to estimate:

(a) probable values of a and b

(b) the value of Q when $P = 7$

(c) the value of P when $Q = 20$.

Drawing up a table to show corresponding values of P and $\sqrt{Q}$,

$\sqrt{Q}$	3	5
P	5	8

(a) The graph is plotted in Fig. 25.14. Since $\sqrt{Q}$ is the independent variable it is plotted horizontally. The gradient may be found by drawing the right-angled triangle ABC, from which

$$a = \frac{BC}{AC}$$

$$= \frac{3}{2}$$

$$= 1.5$$

Fig. 25.14

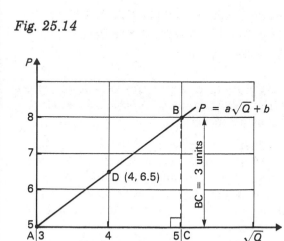

To find the value of b, take any point on the line AB and find its coordinates. Thus D$(4, 6.5)$ lies on the line AB. Now substituting these values of P and Q in the equation,

$$P = 1.5\sqrt{Q} + b$$

$$6.5 = 1.5 \times 4 + b$$

$$6.5 = 6 + b$$

$$b = 0.5$$

$$\therefore \qquad P = 1.5\sqrt{Q} + 0.5$$

(b) When $P = 7$,

$$7 = 1.5\sqrt{Q} + 0.5$$

$$\sqrt{Q} = \frac{7 - 0.5}{1.5}$$

$$= 4.333$$

$$Q = (4.333)^2$$

$$= 18.77$$

(c) When $Q = 20$,

$$P = 1.5 \times \sqrt{20} + 0.5$$

$$= 1.5 \times 4.472 + 0.5$$

$$= 7.208$$

Note that the method used here to find a and b is an alternative to the simultaneous equation method shown in Example 5(c).

Exercise 25.5

1. A car travelling along a straight road passed a post P, and T seconds after passing P its distance (D metres) from P was estimated. The following results were obtained:

T	2	3	$3\frac{1}{2}$	$4\frac{1}{2}$	5	$5\frac{1}{2}$	6
D	46	68	82	117	138	161	186

Construct another table showing values of T^2 against D, and verify graphically that T and D are connected approximately by a law of the form

$$D = aT^2 + b$$

where a and b are constants.

Use your graph to estimate

(a) values for a and b

(b) the time when the car was 175 m past P giving your answer correct to the nearest tenth of a second

(c) the distance travelled from $T = 4$ to $T = 5$.

(*Scales*: Take 2 cm to represent 20 m on the D-axis and 5 units on the T^2-axis.)

2. Two variables, X and Y, are connected by a law of the form

$$\sqrt{Y} = aX + b$$

where a and b are constants. When $X = 2.5$, $Y = 48$ and when $X = 15$, $Y = 529$.

Draw a graph of $\sqrt{Y}$ against X and use your graph to estimate values for a and b. Also use your graph to estimate

(a) the value of X when $Y = 441$

(b) the percentage increase in Y as X increases from 6 to 12.

(*Scales*: Take 2 cm to represent 2 units on the X-axis and 2 units on the $\sqrt{Y}$-axis.)

3. The prototype of a new car was tested on a straight road to find the resistance (R newtons) to the motion of the car at various speeds (V km/h). The following results were obtained:

V	13	30	42	50	60	65
R	1080	1240	1440	1580	1840	1960

Construct another table giving values of V^2 against R. Plot a graph of R against V^2 and show that V and R are connected by a law of the form

$$R = aV^2 + b$$

where a and b are constants.

(*Scales*: Take 2 cm to represent 500 units on the V^2-axis and 100 units on the R-axis.)

Use your graph to estimate

(a) probable values of a and b

(b) the speed at which the resistance was 1500 newtons

(c) the percentage increase in the resistance as the speed increased from 25 km/h to 50 km/h.

4. In a laboratory experiment a heavy spring was suspended vertically from a horizontal beam. A mass of M kilograms was hung on the lower end of the spring which was stretched and then released. The time of oscillation, T seconds, of the mass was measured and the experiment was repeated for various values of M. The following results were obtained:

M	0.13	0.25	0.37	0.50	0.63	0.71	0.85
T	1.8	2.0	2.2	2.4	2.6	2.7	2.9

Construct a new table showing M against T^2 and by plotting these new values on a graph show that M and T^2 are connected by a law of the form

$$T^2 = aM + b$$

where a and b are constants.

(*Scales*: Take 2 cm to represent 0.1 kg on the M-axis and take 2 cm to represent 1 unit on the T^2-axis.)

Use your graph to estimate

(a) values of a and b

(b) the value of M for which $T = 2.5$

(c) the percentage increase in T as M increases from 0.4 to 0.8.

Parallel Lines

Two straight lines are parallel if their gradients are the same.

It was previously shown that the general equation of a straight line is

$$y = mx + c$$

where m is the gradient of the line and c is the intercept on the y-axis.

Example 9

Show that the straight lines $2y + 4x = 7$ and $3y + 6x = 2$ are parallel.

If $2y + 4x = 7$

then $y = 3\frac{1}{2} - 2x$

If $3y + 6x = 2$

then $y = \frac{2}{3} - 2x$

For each line the gradient is -2 and hence the two straight lines are parallel.

Perpendicular Lines

Two straight lines are perpendicular if the product of their gradients is -1.

Example 10

Show that the lines $y - 3x = 2$ and $3y + x = 10$ are perpendicular.

When $y - 3x = 2$

$$y = 3x + 2 \qquad (1)$$

When $3y + x = 10$

$$y = -\frac{1}{3}x + \frac{10}{3} \qquad (2)$$

The gradient for line (1) is 3 and the gradient for line (2) is $-\frac{1}{3}$. The product of these two gradients is $3 \times (-\frac{1}{3}) = -1$ and hence the two lines are perpendicular.

The two lines are shown plotted in Fig. 25.15. Note that the scales on both axes are the same and the lines are clearly perpendicular to each other. If the scales are different on the two axes, the lines will not look as though they are perpendicular.

Fig. 25.15

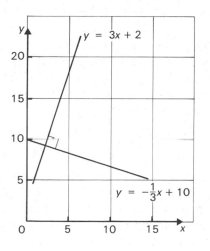

State if the following lines are

(a) parallel

(b) perpendicular to each other

(c) neither parallel nor perpendicular to each other:

1. $y = 5x - 3$ and $y = 7 + 5x$

2. $y = 2x - 5$ and $y = 5 - \frac{1}{2}x$

3. $y = 3x + 1$ and $y = 2x + 1$

4. $3y + 6x = 2$ and $5y + 10x = 8$

5. $2y + x = 7$ and $3y - 2x = 7$

6. $3y + 2x = 5$ and $2y - 3x = 5$

7. $4y + x = 4$ and $2y - x = 8$

Miscellaneous Exercise 25

Section A

1. Using a scale of 1 large square to 1 unit on the x-axis and 1 large square to 2 units on the y-axis construct a pair of axes on graph paper for values of x between -4 and 6 and values of y between -10 and 10.

    (a) On your graph plot the points A$(-3, -9)$ and B$(5, 7)$.

    (b) Join A and B with a straight line and write down its gradient.

    (c) Write down the equation of the straight line AB in the form $y = mx + c$.

2. The equation of the straight line PQ is expressed in the form $y = mx + c$. Given that P is the point $(0, 4)$ and the gradient of PQ is 5:

    (a) Write down the equation of the line PQ.

    (b) The point R has its x-coordinate equal to -3. Work out its y-coordinate.

3. The line $y = 3x + c$ passes through the point $(4, 27)$. What is the value of c?

4. A straight line passes through the points $(0, 4)$ and $(5, 5)$. Find the equation of the line.

5. Find the equation of the straight line passing through the points $(-2, 4)$ and $(4, 16)$.

Section B

1. The equation of the straight line AB (Fig. 25.16) is $y = 3x + 5$ and A is the point $(0, a)$.

   (a) Find the value of a.

   (b) PQ is parallel to AB. Write down the equation of PQ if AP = 7.

Fig. 25.16

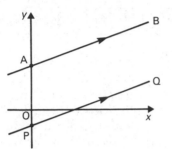

2. A straight line has a gradient of 4 and it passes through the point $(2, 5)$. Find the coordinates of the point where the line cuts the y-axis.

3. A straight line is drawn through the points $A(2, 3)$ and $B(3, 5)$. Find the equation of the line AB.

4. In Fig. 25.17, A is the point $(0, 3)$, B is the point $(3, 8)$ and M is the mid-point of AB. Find:

   (a) the gradient of the line AB

   (b) the equation of the line OB

   (c) the coordinates of M.

Fig. 25.17

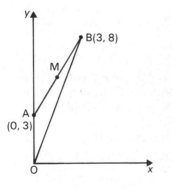

5. A machine was tested in a laboratory to find the effort, P newtons, required to lift a mass of M kilograms. The following results were obtained:

M	1	3	4	5	7	8
P	7.0	12.7	15.5	18.2	23.5	26.5

Plot these points on a graph and show that M and P could be connected by an equation of the form $P = aM + b$, where a and b are constants. Use your graph to estimate the values of a and b.

(*Scales*: Take 2 cm to represent 1 kg on the M-axis and 2 cm to represent 2 newtons on the P-axis.)

6. The total cost (P pence) of operating a machine on a workbench was compared with the time (T hours) for which the machine operated and the following results were obtained:

T	4	10	20	40	60	80
P	65	85	115	145	175	195

Construct another table that shows values of $\sqrt{T}$ against P, and by plotting these results on a graph, show that T and P are approximately connected by a law of the form $P = a\sqrt{T} + b$ where a and b are constants.

(*Scales*: Take 2 cm to represent 1 unit on the T-axis and 20 units on the P-axis.)

Use your graph to estimate

(a) the probable values of a and b

(b) the time for which the total cost of operating the machine is 125p.

Multi-Choice Questions 25

1. When the points $A(1, 2)$, $B(4, 2)$, $C(5, 4)$ and $D(2, 4)$ are joined in alphabetical order they form which figure?

   A square   **B** rectangle

   C parallelogram   **D** trapezium

2. A is the point $(1, 3)$ and B is the point $(4, 7)$. What is the length of the straight line AB?

    A 5 units   B 10 units

    C 15 units   D 25 units

3. $A(5, 2)$ and $B(9, 6)$ are joined by a straight line. What are the rectangular coordinates of the mid-point of AB?

    A $(7, 4)$ B $(14, 8)$

    C $(8, 14)$ D $(4, 7)$

4. Which of the following is a linear equation?

    A $y = 3 + \dfrac{5}{x}$ B $y = x^2 + 3$

    C $y = 3 - 8x$ D $y = 3\sqrt{x} + 7$

5. One of the following diagrams (Fig. 25.18) represents the graph of the equation $y = 5x - 2$. Which?

Fig. 25.18

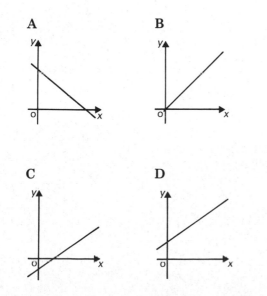

6. Which one of the diagrams in Fig. 25.19 represents the graph of $y = 5 - 2x$?

Fig. 25.19

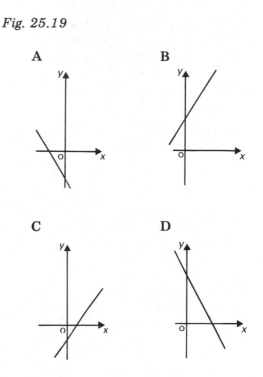

7. What is the equation of the straight line shown in Fig. 25.20?

    A $y = x + 2$ B $y = 2x + 1$

    C $y = 4x + 6$ D $y = 6x + 4$

Fig. 25.20

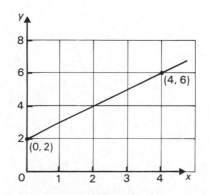

8. Four pairs of lines are given below. In which pair are the lines perpendicular?

   A $y = 3x + 2$
      $y = -3x - 2$

   B $y = 4x + 1$
      $y = \frac{1}{4}x + 1$

   C $y = 2 - 3x$
      $y = -\frac{1}{2} - 3x$

   D $y = 2x + 3$
      $y = 4 - \frac{1}{2}x$

9. In Question 8, in which pair are the lines parallel?

10. What is the gradient of the line represented by the equation $4x - 7y = 7$?

    A $\frac{4}{7}$ B $\frac{7}{4}$ C 4   D -7

11. What is the equation of the straight line which passes through the origin and whose gradient is 2?

    A $y = 0$

    B $y = 2x$

    C $y = -2x$

    D $y = \frac{1}{2}x$

12. What is the gradient of the straight line which passes through the points $(2, 5)$ and $(-2, -1)$?

    A 0   B 1   C $1\frac{1}{2}$ D 2

Further Graphical Work

Graphs of Quadratic Functions

The expression $ax^2 + bx + c$ where a, b and c are constants is called a **quadratic function** of x. When plotted, quadratic functions always give a smooth curve known as a parabola.

Example 1

Plot the graph of $y = 3x^2 + 10x - 8$ between $x = -6$ and $x = 4$.

A table may be drawn up as follows giving values of y for chosen values of x.

x	-6	-5	-4	-3	-2	-1
$3x^2$	108	75	48	27	12	3
$10x$	-60	-50	-40	-30	-20	-10
-8	-8	-8	-8	-8	-8	-8
y	40	17	0	-11	-16	-15

x	0	1	2	3	4
$3x^2$	0	3	12	27	48
$10x$	0	10	20	30	40
-8	-8	-8	-8	-8	-8
y	-8	5	24	49	80

The graph is shown in Fig. 26.1 and it is a smooth curve.

Fig. 26.1

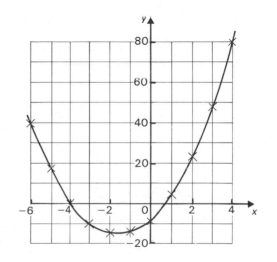

The Axis of Symmetry of a Parabola

Fig. 26.2 shows the graph of

$$y = ax^2 + bx + c$$

The curve is symmetrical about the line

$$x = -\frac{b}{2a}$$

which is the axis of symmetry. By using the axis of symmetry we can often reduce the amount of work in drawing up a table of values for a quadratic function.

211

Example 2

Plot the graph of $y = x^2 - 4x + 3$ between $x = -2$ and $x = 6$.

Fig. 26.2

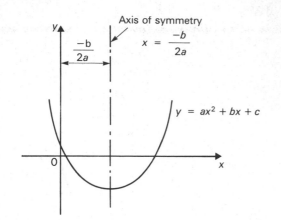

Fig. 26.3

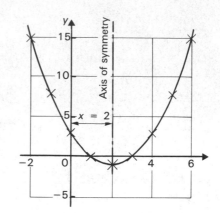

The axis of symmetry is

$$x = -\frac{(-4)}{2} = 2$$

Hence in drawing up the table of corresponding values of x and y we choose values of x so that they are symmetrical about the line $x = 2$. The graph is drawn in Fig. 26.3.

x	-2	-1	0	1	2	3	4	5	6
x^2	4	1	0	1	4				
$-4x$	8	4	0	-4	-8				
3	3	3	3	3	3				
y	15	8	3	0	-1	0	3	8	15

Note that, for example, the point $(5, 8)$ is the image of the point $(-1, 8)$ when it is reflected in the line $x = 2$ (see Chapter 31), and similarly for the points $(-2, 15)$ and $(6, 15)$, $(0, 3)$ and $(4, 3)$, $(1, 0)$ and $(3, 0)$.

Solution of Equations

An equation may be solved by means of a graph. The following example shows the method.

Example 3

(a) Plot the graph of $y = 6x^2 - 7x - 5$ between $x = -2$ and $x = 3$. Hence solve the equation $6x^2 - 7x - 5 = 0$.

A table is drawn up as follows.

x	-2	-1	0	1	2	3
y	33	8	-5	-6	5	28

The curve is shown in Fig. 26.4. To solve the equation $6x^2 - 7x - 5 = 0$ we have to find the values of x when $y = 0$. That is, we have to find the values of x where the graph cuts the x-axis. These are points A and B in Fig. 26.4 and hence the solutions are:

$$x = -0.5 \quad \text{or} \quad x = 1.67$$

Fig. 26.4

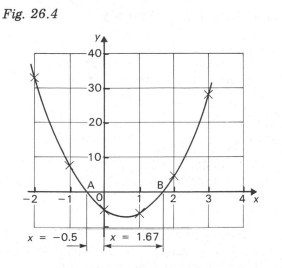

(b) Plot the graph of $y = 2x^2 - x - 6$ and hence solve the following equations:

(i) $2x^2 - x - 6 = 0$

(ii) $2x^2 - x - 4 = 0$

(iii) $2x^2 - x - 9 = 0$

Take values of x between -4 and 6.

To plot $y = 2x^2 - x - 6$ draw up table values as shown below:

x	-4	-3	-2	-1	0	
y	30	15	4	-3	-6	
x	1	2	3	4	5	6
y	-5	0	9	22	39	60

(i) The graph is plotted as shown in Fig. 26.5. The curve cuts the x-axis, i.e. where $y = 0$, at the points where $x = -1.5$ and $x = 2$. Hence the solutions of the equation $2x^2 - x - 6 = 0$ are

$$x = -1.5 \quad \text{or} \quad x = 2$$

(ii) The equation $2x^2 - x - 4 = 0$ may be written in the form

$$2x^2 - x - 6 = -2$$

Hence if we find the values of x when $y = -2$ we shall obtain the solutions required. These are where the line $y = -2$ cuts the curve (see Fig. 26.5). The solutions are therefore

$$x = -1.19 \quad \text{or} \quad 1.69$$

Fig. 26.5

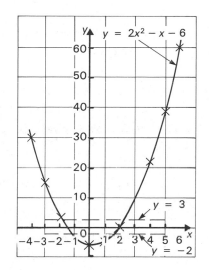

(iii) The equation $2x^2 - x - 9 = 0$ may be written in the form

$$2x^2 - x - 6 = 3$$

Hence by drawing the line $y = 3$ and finding where it cuts the curve we shall obtain the solutions. They are

$$x = -1.89 \quad \text{or} \quad 2.38$$

Exercise 26.1

Plot the graphs of the following equations:

1. $y = 2x^2 - 7x - 5$ between $x = -4$ and $x = 12$.

2. $y = x^2 - 4x + 4$ between $x = -3$ and $x = 3$.

3. $y = 6x^2 - 11x - 35$ between $x = -3$ and $x = 5$.

4. $y = 3x^2 - 5$ between $x = -2$ and $x = 4$.

5. $y = 1 + 3x - x^2$ between $x = -2$ and $x = 3$.

By plotting suitable graphs solve the following equations:

6. $x^2 - 7x + 12 = 0$ (take values of x between 0 and 6).

7. $x^2 + 16 = 8x$ (take values of x between 1 and 7).

8. $x^2 - 9 = 0$ (take values of x between -4 and 4).

9. $3x^2 + 5x = 60$ (take values of x between 0 and 4).

10. Plot the graph of $y = x^2 + 7x + 3$ taking values of x between -12 and 2. Hence solve the equations:
    (a) $x^2 + 7x + 3 = 0$
    (b) $x^2 + 7x - 2 = 0$
    (c) $x^2 + 7x + 6 = 0$

11. Draw the graph of $y = 1 - 2x - 3x^2$ between $x = -4$ and $x = 4$. Hence solve the equations:
    (a) $1 - 2x - 3x^2 = 0$
    (b) $3 - 2x - 3x^2 = 0$
    (c) $9x^2 + 6x = 6$

12. Draw the graph of $y = x^2 - 9$ taking values of x between -5 and 5. Hence solve the equations:
    (a) $x^2 - 9 = 0$
    (b) $x^2 - 5 = 0$
    (c) $x^2 + 6 = 0$

Intersecting Graphs

Equations may also be solved graphically by using intersecting graphs. The method is shown in the following example.

Example 4

Plot the graph of $y = 2x^2$ and use it to solve the equation $2x^2 - 3x - 2 = 0$. Take values of x between -2 and 4.

The equation $2x^2 - 3x - 2 = 0$ can be solved graphically by the method used in earlier examples, but the alternative method shown here is often preferable. The equation $2x^2 - 3x - 2 = 0$ may be written in the form $2x^2 = 3x + 2$. We now plot on the same axes and to the same scales the graphs

$$y = 2x^2 \quad \text{and} \quad y = 3x + 2$$

x		-2	-1	0	1	2	3	4
$y = 2x^2$		8	2	0	2	8	18	32
$y = 3x + 2$		-4		2				14

Note that to plot $y = 3x + 2$ we need only three points since this is a linear equation. The graphs are shown plotted in Fig. 26.6.

Fig. 26.6

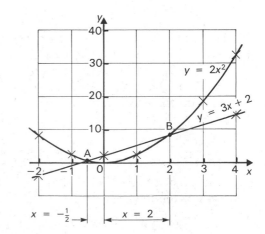

At the points of intersection of the curve and the line (points A and B in Fig. 26.6) the y value of $2x^2$ is the same as the y value of $3x^2 + 2$. Therefore at these points the equation $2x^2 = 3x + 2$ is satisfied. The required values of x may now be found by inspection of the graph. They are at A, where $x = -\frac{1}{2}$ and at B, where $x = 2$.

The required solutions are therefore

$$x = -\frac{1}{2} \quad \text{or} \quad x = 2$$

Graphical Solutions of Simultaneous Equations

The method is shown in the following examples.

Example 5

(a) Solve graphically

$$y - 2x = 2 \qquad (1)$$

$$3y + x = 20 \qquad (2)$$

Equation (1) may be written as

$$y = 2 + 2x$$

Equation (2) may be written as

$$y = \frac{20 - x}{3}$$

Drawing up the following table we can plot the two equations on the same axes.

x	-3	0	3
$y = 2 + 2x$	-4	2	8
$y = \dfrac{20 - x}{3}$		7.7	6.7	5.7

The solutions of the equations are the coordinates of the point where the two lines cross (that is, point P in Fig. 26.7). The coordinates of P are $x = 2$ and $y = 6$.

Hence the solutions of the given equations are

$$x = 2 \quad \text{and} \quad y = 6$$

Fig. 26.7

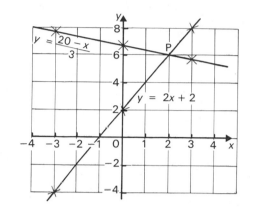

(b) Draw the graph of $y = (3 + 2x)(3 - x)$ for values of x from $-1\frac{1}{2}$ to 3. On the same axes, and with the same scales, draw the graph of $3y = 2x + 14$. From your graphs determine the values of x for which $3(3 + 2x)(3 - x) = 2x + 14$.

To plot the graph of $y = (3 + 2x)(3 - x)$ we draw up the following table.

x		$-1\frac{1}{2}$	-1	$-\frac{1}{2}$	0
$y = (3 + 2x)(3 - x)$		0	4	7	9

x	$\frac{1}{2}$	1	$1\frac{1}{2}$	2	$2\frac{1}{2}$	3
$y = (3 + 2x)(3 - x)$	10	10	9	7	4	0

The equation $3y = 2x + 14$ may be rewritten as

$$y = \frac{2x + 14}{3}$$

To draw this graph we need only take three points since it is a linear equation:

x	-1	1	3
$y = \dfrac{2x + 14}{3}$	4	$5\frac{1}{3}$	$6\frac{2}{3}$

Fig. 26.8

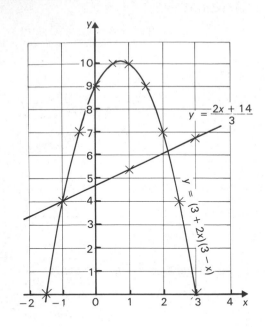

The graphs are shown in Fig. 26.8. Since the equation $3(3 + 2x)(3 - x) = 2x + 14$ may be rewritten to give

$$(3 + 2x)(3 - x) = \frac{2x + 14}{3}$$

the coordinates where the curve and the line intersect give the solutions which are

$$x = -1 \quad \text{and} \quad x = 2.17$$

Exercise 26.2

1. Plot the graph of $y = 3x^2$ taking values of x between -3 and 4. Hence solve the following equations:

   (a) $3x^2 = 4$

   (b) $3x^2 - 2x - 3 = 0$

   (c) $3x^2 - 7x = 0$

2. Plot the graph of $y = x^2 + 8x - 2$ taking values of x between -12 and 2. On the same axes, and to the same scale, plot the graph of $y = 2x - 1$. Hence find the values of x which satisfy the equation

$$x^2 + 8x - 2 = 2x - 1$$

Solve graphically the following simultaneous equations:

3. $2x - 3y = 5, \quad x - 2y = 2$

4. $7x - 4y = 37, \quad 6x + 3y = 51$

5. $\dfrac{x}{2} + \dfrac{y}{3} = \dfrac{13}{6}, \quad \dfrac{2x}{7} - \dfrac{y}{4} = \dfrac{5}{14}$

6. If $y = x^2(15 - 2x)$ construct a table of values of y for values of x from -1 to $1\frac{1}{2}$ at half-unit intervals. Hence draw the graph of this function. Using the same axes and scales draw the straight line $y = 10x + 10$. Write down and simplify an equation which is satisfied by the values of x where the two graphs intersect. From your graph find the approximate values of the two roots of this equation.

7. Write down the three values missing from the following table which gives values of $2x^3 + x + 3$ for values of x from -2 to 2.

x	-2.0	-1.5	-1.0	-0.5
$2x^3 + x + 3$	-15.0	-5.25		2.25

x	0	0.5	1.0	1.5	2.0
$2x^3 + x + 3$		3.75	6.0		21.0

Using the same axes draw the graphs of $y = 2x^3 + x + 3$ and $y = 9x + 3$. Use your graphs to write down:

(a) The range of values of x for which $2x^3 + x + 3$ is less than $9x + 3$.

(b) The solution of the equation $2x^3 + x + 3 = 5$. Write down and simplify the equation which is satisfied by the values of x at the points of intersection of the two graphs.

8. Write down the three values missing from the following table which gives values, correct to two decimal places, of $6 - \dfrac{10}{2x+1}$ for values of x from 0.25 to 5.

x	0.25	0.5	1	1.5
$6 - \dfrac{10}{2x+1}$	−0.67	1.00	2.67	3.50

x	2	3	3.5	4	4.5	5
$6 - \dfrac{10}{2x+1}$		4.57		4.89		5.09

Using the same axes draw the graphs of $y = 6 - \dfrac{10}{2x+1}$ and $y = x + 1$. Use your graphs to solve the equation $2x^2 - 9x + 5 = 0$.

9. If $y = \dfrac{x+10}{x+1}$ construct a table of values of y when $x = 0, 1, 2, 3, 4, 5$. Draw the graph of this function and also using the same axes and scales draw the graph of $y = x - 1$. Write down, and simplify, an equation which is satisfied by the value of x where the graphs intersect. From your graphs find the approximate value of the positive root of this equation.

10. Calculate the values of $\dfrac{x^2}{4} + \dfrac{24}{x} - 12$ which are omitted from the table below.

x	2	2.5	3	3.5	4
$\dfrac{x^2}{4} + \dfrac{24}{x} - 12$		−0.84		−2.08	

x	4.5	5	5.5	6
$\dfrac{x^2}{4} + \dfrac{24}{x} - 12$	−1.60	−0.95	−0.07	1.00

Draw the graph of $y = \dfrac{x^2}{4} + \dfrac{24}{x} - 12$ from $x = 2$ to $x = 6$. Using the same scales and axes draw the graph of $y = \dfrac{x}{3} - 2$. Write down, but do not simplify, an equation which is satisfied by the values of x where the graphs intersect. From your graphs find approximate values for the two roots of this equation.

The Reciprocal Function

The recriprocal of x is $\dfrac{1}{x}$. The function $f: x \rightarrow \dfrac{k}{x}$ is called the reciprocal function since the values of the range are k times the reciprocal of the domain (k being a constant).

Example 6

Draw the graph of $y = \dfrac{4}{x}$ for values of x between −4 and 4.

x	−4	−3	−2	−1	1	2	3	4
y	−1	$-1\frac{1}{3}$	−2	−4	4	2	$1\frac{1}{3}$	1

Note that the reciprocal of a negative number is a negative number, whilst the reciprocal of a positive number is a positive number. It is impossible to find a value for the reciprocal of zero (i.e., we cannot divide by zero) and, as shown in Fig. 26.9, the graph consists of two separate branches. The negative branch is the image of the positive branch reflected in the line $y = -x$.

Fig. 26.9

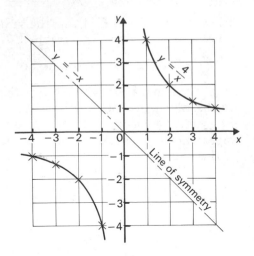

The Law of Natural Growth

Suppose the sum of $2000 is invested at 10% per annum compound interest.

To find the amount accruing after x years the formula below is used:

$$A_x = 2000 \left(1 + \frac{10}{100}\right)^x$$

$$= 2000 \times 1.1^x$$

That is

after 1 year	A_1	$= 2000 \times 1.1^1$
		$= 2200$
after 2 years	A_2	$= 2000 \times 1.1^2$
		$= 2420$
after 3 years	A_3	$= 2000 \times 1.1^3$
		$= 2662$

and so on.

The graph depicting the growth of the money invested is shown in Fig. 26.10. It will be seen that the curve rises as the number of years increase. Curves like this depict the law of natural growth. There are very many examples of natural growth such as population growth, multiplication of bacteria, etc. Expressed mathematically, the exponential or growth function is

$$f: x \rightarrow kn^x$$

Fig. 26.10

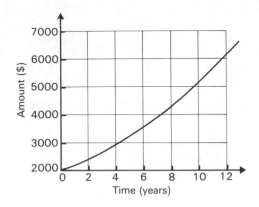

Example 7

The population of a certain country appears to be increasing at a rate of 3% per annum. If its population is 8 millions in 1978, estimate its population in the year 2000, assuming that the law of natural growth applies and that the rate of increase of 3% per annum is maintained.

We have

$$P = kR^x$$

$$= 8 \times 1.03^x$$

where P is the population after x years. Hence after 22 years

$$P = 8 \times 1.03^{22}$$

$$= 15.3 \text{ millions}$$

That is, by the year 2000, the population will very nearly double.

Example 8

The growth of a colony of bacteria is monitored regularly and recorded in the table below.

Number of bacteria (thousands)	3	6	12	24	48	96	192	384
Time (hours)		0	1	2	3	4	5	6	7

Plot a graph to show the multiplication of the bacteria, with the time along the horizontal axis. Find:

(a) the factor by which the number of bacteria is multiplied every hour, and write down an equation connecting time and number of bacteria

(b) the number of bacteria after $2\frac{1}{2}$ hours

(c) the time when the number of bacteria will be 150 000

(d) the time when the number of bacteria should reach 500 000.

Fig. 26.11

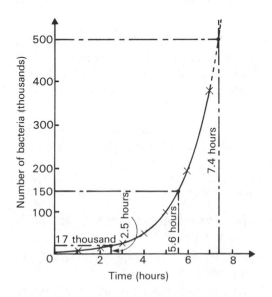

Time (hours)

The graph is shown in Fig. 26.11 and it is seen to be a growth curve.

(a) The multiplication factor is 2 because, from the table, we see that every hour the number of bacteria doubles. The equation connecting time (T hours) and number of bacteria (N) is

$$N = 3000 \times 2^T$$

(b) From the graph the number of bacteria after $2\frac{1}{2}$ hours is found to be 17 000.

(c) From the graph the time when the number of bacteria will reach 150 000 is 5.6 hours.

(d) By extending the curve, the estimated time, when the number of bacteria will reach 500 000, is 7.4 hours.

The Graph of x^3

The graph of $y = x^3$ is shown in Fig. 26.12 and the negative part will be seen to be the image of the positive part turned through $180°$ about the origin.

Fig. 26.12

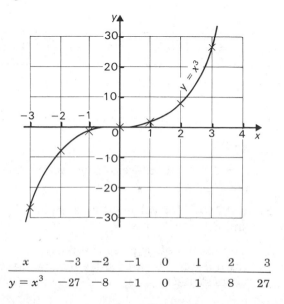

x	-3	-2	-1	0	1	2	3
$y = x^3$	-27	-8	-1	0	1	8	27

The Graph of x^{-2}

The graph of

$$y = \frac{1}{x^2} \quad \text{or} \quad y = x^{-2}$$

is shown in Fig. 26.13 and it is seen to be symmetrical about the y-axis (i.e. about the line $x = 0$).

Fig. 26.13

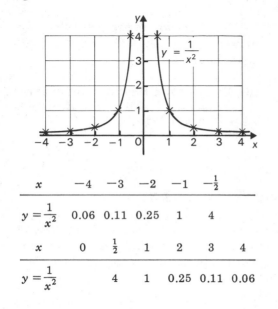

x	-4	-3	-2	-1	$-\frac{1}{2}$
$y = \dfrac{1}{x^2}$	0.06	0.11	0.25	1	4

x	0	$\frac{1}{2}$	1	2	3	4
$y = \dfrac{1}{x^2}$		4	1	0.25	0.11	0.06

Exercise 26.3

Draw graphs of the following functions:

1. $x \to \dfrac{5}{x}$ for values of x between -5 and 5

2. $x \to 3x^{-2}$ for values of x between -2 and 2

3. $x \to 10x^3$ for values of x between -3 and 3.

4. The table below gives the population (in millions) of a country during the years stated.

Year	1900	1920	1940	1950	1960	1970
Population (millions)	1.50	1.78	2.11	2.30	2.51	2.74

(a) Draw a graph of this information taking the year as the domain.

(b) Write down the factor by which the population increases in each 10 year period.

(c) From the answer to part (b) write down the equation connecting the year and the population.

(d) Use the answer to part (c) to estimate the population in the year 2000 assuming that the rate of increase remains the same.

(e) From your graph obtain the population in 1930.

5. A colony of bacteria multiplies according to the equation $N = 2^T$ where N is the number of bacteria after a time T hours.

(a) Make a table showing the number of bacteria at 1, 2, 3, 4 and 5 hours.

(b) Draw a graph of the equation $N = 2^T$ and use your graph to estimate the number of bacteria after $2\frac{1}{2}$ hours.

6. The population of a Caribbean territory is increasing at 4% per annum. In 1975 the population was 2 000 000. Estimate the population in 1990.

7. The figures below show how £5000 appreciates over a period of years.

Years	2	5	7	10	15
Amount	6050	8050	9745	12 970	20 900

(a) Draw a graph taking years as the domain.

(b) What is the rate of interest per annum?

(c) What is the amount after 8 years?

(d) What is the number of years for the amount to reach £15 000?

Miscellaneous Exercise 26

Section A

1. Copy and complete the table below given that $f(x) = 2x^2 - x - 6$.

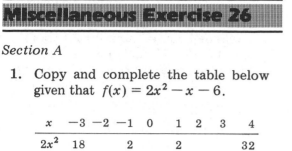

x	-3	-2	-1	0	1	2	3	4
$2x^2$	18		2		2			32
$-x$	3		1		-1			-4
-6	-6		-6		-6			-6
$f(x)$	15		-3		-5			22

Use the completed table to plot the graph of $f(x) = 2x^2 - x - 6$ taking 2 cm to represent 1 unit on the x-axis and 2 cm to represent 5 units on the y-axis. Use your graph to solve the equation $2x^2 - x - 6 = 0$.

2. (a) Copy and complete the table of values given below for the function $y = 5 - x$, where x is a real number.

x	-3	-2	0	1	2	3	4	5
y				4	3	2	1	0

(b) Plot these points and join them up to form a straight line.

(c) Copy and complete the table given below for the function $y = x^2 - 1$, where x is a real number.

x	-3	-2	-1	0	1	2	3
y							

(d) Plot these points on the same axes as for (b) and join them up to form a smooth curve.

(e) Use your graphs to obtain the solutions of the equation $x^2 + x - 6 = 0$.

(Suitable scales are 2 cm to represent 1 unit on the x-axis and 2 cm to represent 2 units on the y-axis.)

3. Using a scale of 1 cm = 1 unit on both axes and using the same axes for both graphs, draw the following relations for values of x from 0 to 10 only:

(a) $y = x + 2$ (b) $y = 10 - x$

(i) Use your graphs to find the solutions of the simultaneous equations

$$x - y = 2 \qquad (1)$$
$$x + y = 10 \qquad (2)$$

(ii) A is the point $(4, 0)$ and K is the point $(8, 4)$. Plot these points on your graph and join them with a straight line. Find the equation of the line.

4. Copy and complete the table below given that $y = x^2 + 4x$.

x	-6	-5	-4	-3	-2	-1	0	1	2
y	12		0	-3		-3			5

(a) Plot the graph of $y = x^2 + 4x$ on graph paper using scales for x between -6 and 2 and between -5 and 15 for y. Use 1 cm to represent 1 unit on both axes.

(b) Use your graph to solve the equation $x^2 + 4x = 3$.

(c) Insert on your graph the straight line $y = -2x$ to cut the curve. Find the solution of the equation $x^2 + 6x = 0$.

5. The population of a certain African country is increasing according to the law $N = 2 \times 1.05^T$ where T is the number of years after 1960 and N is the population. Copy and complete the table below.

Year	1960	1970	1980	1990	2000
N	2 000 000				

Draw a graph of the growth of the population and from it estimate the population in the year 1985.

Section B

1. Copy and complete the table below, given that $y = x^2 - 6x + 8$.

x	-1	0	1	2	2.5	3	3.5	4	5
x^2	1				6.25		12.25		25
$-6x$	6				-15.00		-21.00		-30
$+8$	8				8.00		8.00		8
y	15				-0.75		-0.75		3

(a) Use the completed table to draw the graph of $y = x^2 - 6x + 8$ taking 2 cm to represent 1 unit on the x-axis and 1 cm to represent 1 unit on the y-axis.

(b) On the same sheet of paper, using the same axes, draw a straight line whose equation is $y - 5 = 0$.

(c) Use your graphs to obtain the solutions of the equations:
    (i) $x^2 - 6x + 8 = 0$
    (ii) $x^2 - 6x + 3 = 0$

2. The table below shows part of a table of values for the graph of $y = x - 4 + \dfrac{3}{x}$.

x	0.5	1	1.5	2	2.5	3	4	5
y	2.5			-0.5	-0.3		0.75	

(a) Calculate the four missing values.

(b) Plot the graph of $y = x - 4 + \dfrac{3}{x}$ for values of x from 0.5 to 5 using a scale of 2 cm to 1 unit on the x-axis and 4 cm to 1 unit on the y-axis.

(c) Use your graph to solve the equation

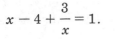

$$x - 4 + \frac{3}{x} = 1.$$

3. Draw the graph of $y = 1 - 2x - 3x^2$ for values of x between -4 and 4, using scales of 1 cm to 1 unit on the x-axis and 1 cm to 10 units on the y-axis. Use your graph to solve the equations:
(a) $1 - 2x - 3x^2 = 0$
(b) $3 - 2x - 3x^2 = 0$
(c) $9x^2 + 6x = 6$

4. Plot the graph of $y = 3x^2$ taking values of x between -3 and 4 using scales of 2 cm to 1 unit on the x-axis and 2 cm to 5 units on the y-axis.

(a) Use your graph to solve the equation $3x^2 = 12$.

(b) On the same axes plot the graph of $y = 2x + 3$ and hence solve the equation $3x^2 - 2x - 3 = 0$.

(c) On the same axes plot the graph of $y = 7x$ and hence solve the equation $3x^2 - 7x = 0$.

5. Copy and complete the table below given that $y = 2^x$.

x	0	1	2	3	4	5	6
y							

(a) Draw the graph of $y = 2^x$ for values of x between 0 and 6 using scales of 2 cm to 1 unit on the x-axis and 2 cm to 5 units on the y-axis.

(b) Use the graph to find x when $y = 25$.

(c) Find the gradient of the line joining the point on the curve where $x = 1$ to the point on the curve where $x = 5$.

Rates of Change, Travel Graphs and Variation

The Gradient of a Curve

Consider the graph $y = x^2$ part of which is shown in Fig. 27.1. As the values of x increase so do the values of y, but they do not increase at the same rate. A glance at the portion of the curve shown in Fig. 27.1 shows that the values of y increase faster when x is large because the gradient of the curve is increasing.

Fig. 27.1

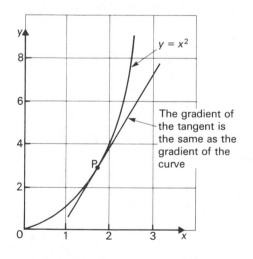

To find the rate of change of y with respect to x at a particular point we need to find the gradient of the curve at that point. If we draw a tangent to the curve at the point, the gradient of the tangent will be the same as the gradient of the curve.

Example 1

(a) Draw the curve of $y = x^2$ and find the gradient of the curve at the points where $x = 2$ and $x = -2$.

To draw the curve the table below is drawn up

x	-3	-2	-1	0	1	2	3
$y = x^2$	9	4	1	0	1	4	9

The point where $x = 2$ is the point $(2, 4)$. We draw a tangent at this point as shown in Fig. 27.2. Then by constructing a right-angled triangle the gradient is found to be $\dfrac{4}{1} = 4$.

Fig. 27.2

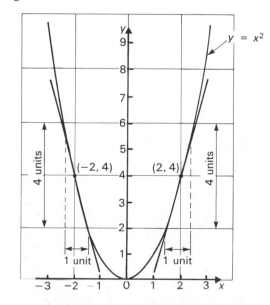

223

This gradient is positive since the tangent slopes upwards from left to right.

A positive value of the gradient indicates that y is increasing as x increases.

The point where $x = -2$ is the point $(-2, 4)$.

By drawing the tangent at this point and constructing a right-angled triangle as shown in Fig. 27.2, the gradient is found to be

$$\frac{-4}{1} = -4$$

The gradient is negative because the tangent slopes downwards from left to right.

A negative value of the gradient indicates that y is decreasing as x increases.

(b) Draw the graph of $y = x^2 - 3x + 7$ between $x = -4$ and $x = 4$ and hence find the gradients of the curve at the points $x = -3$ and $x = 2$.

To plot the curve draw up the following table.

x	-4	-3	-2	-1	0	1	2	3	4
y	35	25	17	11	7	5	5	7	11

Fig. 27.3

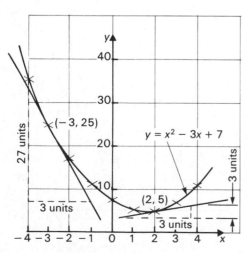

At the point where $x = -3$, $y = 25$.

At the point $(-3, 25)$ draw a tangent to the curve as shown in Fig. 27.3. The gradient is found by drawing a right-angled triangle (which should be as large as possible for accuracy) as shown and measuring its height and base.

Hence the gradient at the point $(-3, 25) =$

$$-\frac{27}{3} = -9$$

At the point where $x = 2$, $y = 5$.
Hence by drawing a tangent and a right-angled triangle at the point $(2, 5)$,

$$\text{Gradient at the point } (2, 5) = \frac{3}{3}$$

$$= 1$$

(c) Draw the graph of $y = x^2 + 3x - 2$ taking values of x between $x = -1$ and $x = 4$. Hence find the value of $y = x^2 + 3x - 2$ where the gradient of the curve is 7.

To plot the curve the following table is drawn up

x	-1	0	1	2	3	4
y	-4	-2	2	8	16	26

The curve is shown in Fig. 27.4.

Fig. 27.4

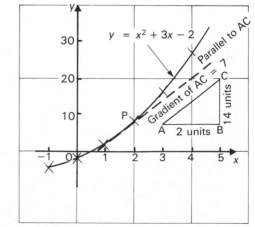

To obtain a line whose gradient is 7 we draw the $\triangle ABC$ making (for convenience) $AB = 2$ units to the scale on the x-axis and $BC = 14$ units to the scale on the y-axis. Hence

$$\text{Gradient of AC} = \frac{BC}{AB}$$

$$= \frac{14}{2}$$

$$= 7$$

Using set-squares we draw a tangent to the curve so that the tangent is parallel to AC. As can be seen this tangent touches the curve at the point P where $x = 2$. Hence the gradient of the curve is 7 at the point where $x = 2$ and $y = 8$.

Exercise 27.1

1. Draw the graph of $y = 3x^2 + 7x + 3$ for values of x between -4 and 4 and find the gradient of the curve at the points where $x = -2$ and $x = 2$.

2. Draw the graph of $y = 2x^2 - 5$ for values of x between -2 and 3. Find the gradient of the curve at the points where $x = -1$ and $x = 2$.

3. Draw the curve of $y = x^2 - 3x + 2$ from $x = 2.5$ to $x = 3.5$ and find its gradient at the point where $x = 3$.

4. For what values of x is the gradient of the curve $y = \dfrac{x^3}{3} + \dfrac{x^2}{2} - 33x + 7$ equal to 3? In drawing the curve take values of x between -8 and 6.

5. If $y = (1 + x)(5 - 2x)$ copy and complete the table below.

x	-2	$-1\frac{1}{2}$	-1	0	$\frac{1}{2}$	1	$1\frac{1}{2}$	2	3
y	-9		0	5		6	5	3	-4

Hence draw the graph of $y = (1 + x)(5 - 2x)$. Find the value of x at which the gradient of the curve is -2.

6. If $y = x^2 - 5x + 4$ find by plotting the curve between $x = 4$ and $x = 12$ the value of x at which the gradient of the curve is 11.

Rates of Change

There are many applications for rates of change, for example the decay rate of a radioactive substance, the rate of increase of the population of a country, etc.

Example 2

The number of bacteria in a colony is monitored regularly and the results are recorded in the table below:

Number of bacteria (thousands)	5.0	20.8	29.0	77.0	171	302
Time (hours)	1.0	2.5	3.0	5.0	7.5	10.0

Draw a graph of this information with time marked on the horizontal axis. Hence find the rate of increase of the bacteria after (a) 4 hours, (b) 8 hours

The graph is drawn in Fig. 27.5 and it is seen to be a smooth curve.

Fig. 27.5

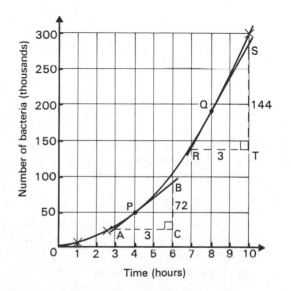

(a) To find the rate of increase of the bacteria after 4 hours we draw a tangent to the curve at P and we find its gradient by drawing the right-angled triangle ABC.

$$\text{Rate of increase} = \frac{BC}{AC}$$

$$= \frac{72}{3}$$

$$= 24$$

Therefore at a time of 4 hours the bacteria are increasing at a rate of 24 000 per hour.

(b) To find the rate of increase of the bacteria after 8 hours a tangent to the curve is drawn at Q. Using the right-angled triangle RST, we have

$$\text{Gradient} = \frac{144}{3}$$

$$= 48$$

Therefore at a time of 8 hours the bacteria are increasing at a rate of 48 000 per hour.

Example 3

The table below shows how the mass of a radioactive material varies with time.

Time (seconds)	0	1	2	3	4	5
Mass (grams)	1	0.5	0.25	0.125	0.063	0.032

Taking time on the horizontal axis, plot a graph of this information and from it determine the rate of decay after 2.5 seconds.

The graph is shown in Fig. 27.6 which shows that the mass is decreasing with time. To determine the rate of decrease of mass with time (i.e. the decay rate) we draw a tangent to the curve at P.

Fig. 27.6

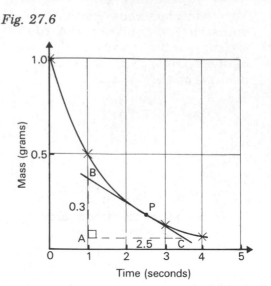

Drawing the right-angled triangle ABC,

$$\text{Gradient at P} = -\frac{AB}{AC}$$

$$= -\frac{0.3}{2.5}$$

$$= -0.12$$

Hence the rate of decay after 2.5 seconds is 0.12 gram per second. (The minus sign here indicates that the mass decreases with time.)

Mathematics in the Media

Mathematical statements are frequently made in newspapers, on television and on the radio. For instance, 'the amount of money lent by the banks is rising rapidly' or 'the amount of money deposited with the building societies is still rising but not so rapidly as in previous years'.

On graphs it is the gradient of the graph which determines the rate of increase or decrease. Words such as 'soar' and 'rise' indicate that the rate of increase is positive (i.e. the graph is moving upwards) whilst words like 'drop' and 'plummet' indicate that the graph is moving downwards. Some examples are given in Fig. 27.7.

Fig. 27.7

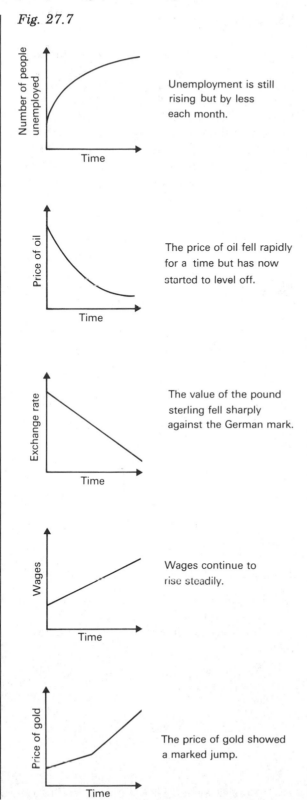

Unemployment is still rising but by less each month.

The price of oil fell rapidly for a time but has now started to level off.

The value of the pound sterling fell sharply against the German mark.

Wages continue to rise steadily.

The price of gold showed a marked jump.

Exercise 27.2

1. The growth of a bacterial colony, monitored at regular intervals, gave the following figures:

Number of bacteria (thousands)	2.75	9.50	20.8	36.5
Time (minutes)	15	30	45	60

Number of bacteria (thousands)	49.5	64.5	100
Time (minutes)	70	80	100

Plot a graph of this information with time on the horizontal axis. From your graph determine the rate of increase of the bacteria

(a) after 40 minutes

(b) after 75 minutes.

2. The table below shows corresponding values of pressure p, in millimetres, and volume v, in cubic metres, of a given mass of gas at constant temperature.

p	90	100	130	150	170	190
v	16.7	13.6	11.5	9.95	8.82	7.89

With p on the horizontal axis plot a graph of this data and from it find the rate of change of v with respect to p, when $p = 140 \,\mathrm{mm}$.

3. The way in which a radioactive substance decays with time is shown in the table below.

Time (seconds)	0	1	2	3	4
Mass (grams)	1	0.37	0.14	0.05	0.02

With time on the horizontal axis plot a graph of this data and from it determine the rate of decay after 2.5 seconds.

4. The table below shows how the population increased over a period of 50 years.

Number of years	0	10	20	30	40	50
Population (millions)	3	3.6	4.5	5.4	6.6	8.07

With time on the horizontal axis plot a graph of this information. Hence find the rate of population growth after

(a) 15 years   (b) 45 years

5. The table below shows how barometric pressure varied with time.

Time (hours)	6	8	10	12
Pressure (mm of mercury)	751.5	753.2	754.7	755.3

Time (hours)	14	16	18
Pressure (mm of mercury)	755.9	755.5	755.2

With time on the horizontal axis plot a graph of pressure against time. From it find the rate of change of pressure at

(a) 9 hours   (b) 15 hours

6. Which of the graphs shown in Fig. 27.8 best illustrates the statement 'the rate of inflation is decreasing as time goes on'?

Fig. 27.8

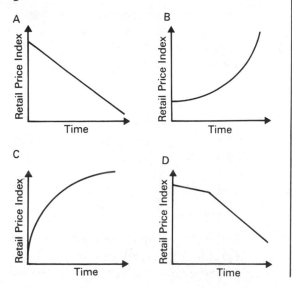

Turning Points

At the points P and Q (Fig. 27.9) the tangent to the curve is parallel to the x-axis. The points P and Q are called **turning points**. The turning point at P is called a **maximum** turning point and the turning point at Q is called a **minimum** turning point. It will be seen from Fig. 27.9 that the value of y at P is not the greatest value of y nor is the value of y at Q the least. The terms maximum and minimum values apply only to the values of y at the turning points and not to the values of y in general.

Fig. 27.9

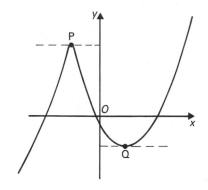

Example 4

(a) Plot the graph of $y = x^3 - 5x^2 + 2x + 8$ for values of x between -2 and 6. Hence find the maximum and minimum values of y.

To plot the graph we draw up a table in the usual way.

x	-2	-1	0	1
$y = x^3 - 5x^2 + 2x + 8$	-24	0	8	6

x	2	3	4	5	6
$y = x^3 - 5x^2 + 2x + 8$	0	-4	0	18	56

Fig. 27.10

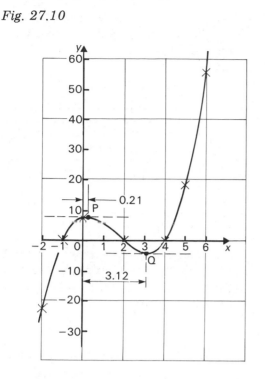

Fig. 27.11

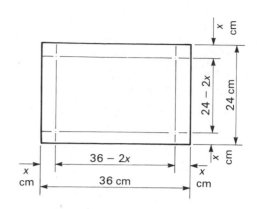

The graph is shown in Fig. 27.10. The maximum value occurs at the point P where the tangent to the curve is parallel to the x-axis. The minimum value occurs at the point Q where again the tangent to the curve is parallel to the x-axis. From the graph the maximum value of y is 8.21 and the minimum value of y is -4.06.

Notice that the value of y at P is not the greatest value of y nor is the value of y at Q the least. However, the values of y at P and Q are called the maximum and minimum values of y respectively.

(b) A small box is to be made from a rectangular sheet of metal 36 cm by 24 cm. Equal squares of side x cm are cut from each of the corners and the box is then made by folding up the sides. Prove that the volume V of the box is given by the expression $V = x(36 - 2x)(24 - 2x)$. Find the value of x so that the volume may be a maximum and find this maximum volume.

Referring to Fig. 27.11 we see that after the box has been formed

$$\text{Length} = 36 - 2x$$

$$\text{Breadth} = 24 - 2x$$

$$\text{Height} = x$$

The volume of the box is

$$V = \text{Length} \times \text{Breadth} \times \text{Height}$$

$$\therefore \quad V = x(36 - 2x)(24 - 2x)$$

We now have to plot a graph of this equation and so we draw up the table below:

x	1	2	3	4
$36 - 2x$	34	32	30	28
$24 - 2x$	22	20	18	16
V	748	1280	1620	1792

x	5	6	7	8
$36 - 2x$	26	24	22	20
$24 - 2x$	14	12	10	8
V	1820	1728	1540	1280

The graph is shown in Fig. 27.12 and it can be seen that the maximum volume is 1825 cm^3 which occurs when $x = 4.71$ cm.

Fig. 27.12

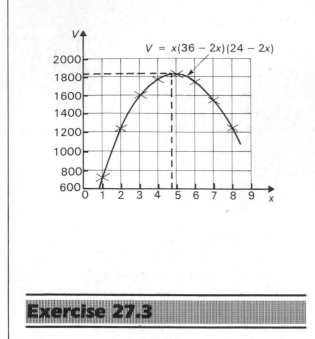

Exercise 27.3

1. Find the minimum value of the curve $y = 3x^2 + 2x - 3$. Plot the graph for values of x between -2 and 3.

2. Find the maximum value of the curve $y = -x^2 + 5x + 7$. Plot the graph for values of x between -2 and 4.

3. Plot the graph of $y = x^3 - 9x^2 + 15x + 2$ taking values of x from 0 to 7. Hence find the maximum and minimum values of y.

4. Draw the graph of $y = x^2 - 3x$ from $x = -1$ to $x = 4$ and use your graph to find:

   (a) the least value of y

   (b) the two solutions of the equation $x^2 - 3x = 1$

   (c) the two solutions of the equation $x^2 - 2x - 1 = 0$.

5. Draw the graph of $y = (x - 1)(4 - x)$ for values of x from 0 to 5. From your graph find the greatest value of $(x - 1)(4 - x)$.

6. Write down the three values missing from the following table which gives values of $\frac{1}{2}(3x^2 - 5x - 1)$ for values of x from -2 to 3.

x	-2	-1.5	-1	-0.5
$\frac{1}{2}(3x^2 - 5x - 1)$	10.50	6.63		1.13

x	0	0.5	1	1.5
$\frac{1}{2}(3x^2 - 5x - 1)$		-1.38	-1.50	-0.88

x	2	2.5	3.0
$\frac{1}{2}(3x^2 - 5x - 1)$		2.63	5.50

Draw the graph of $y = \frac{1}{2}(3x^2 - 5x - 1)$ and from it find the minimum value of $\frac{1}{2}(3x^2 - 5x - 1)$ and the value of x at which it occurs.

7. A piece of sheet metal $20\,\text{cm} \times 12\,\text{cm}$ is used to make an open box. To do this, squares of side x cm are cut from the corners and the sides and ends folded over. Show that the volume of the box is

$$V = x(20 - 2x)(12 - 2x)$$

By taking values of x from $1\,\text{cm}$ to $5\,\text{cm}$ in $0.5\,\text{cm}$ steps, plot a graph of V against x and find the value of x which gives a maximum volume. What is the maximum volume of the box?

8. An open tank which has a square base of x metres has to hold 200 cubic metres of liquid when full. Show that the height of the tank is $\dfrac{200}{x^2}$ and hence prove that the surface area of the tank is given by $A = \left(x^2 + \dfrac{800}{x}\right)$ square metres. By plotting a graph of A against x find the dimensions of the tank so that the surface area is a minimum. (Take values of x from 3 to 9.)

9. A rectangular parcel of length x metres, width k metres and height k metres is to be sent through the post. The total length and girth (i.e. the distance round) of the parcel is to be exactly 2 metres. Show that the volume of the parcel is

$$V = \frac{x}{16}(2 - x)^2$$

Draw a graph of V against x for values of x 0.3 to 1 in steps of 0.1 and hence find the dimensions of the parcel which has the greatest possible volume.

10. A farmer uses 100 m of hurdles to make a rectangular cattle pen. If he makes a pen of length x metres show that the area enclosed is $(50x - x^2)$ square metres. Draw the graph of $y = 50x - x^2$ for values of x between 0 and 50 and use your graph to find:

    (a) the greatest possible area that can be enclosed

    (b) the dimensions of the pen when the area enclosed is 450 square metres.

Maximum or Minimum Value of a Quadratic Expression

If the squared term in the quadratic expression $ax^2 + bx + c$ is positive, then the graph of $y = ax^2 + bx + c$ is similar to Fig. 27.13.

Fig. 27.13

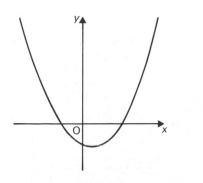

The expression $ax^2 + bx + c$ has a minimum value as shown in the diagram. Although the minimum value can be found by drawing a graph it can also be found by completing the square (see page 136).

Since, on completing the square,

$$y = a(x + h)^2 + k$$

the minimum value of y occurs when

$$(x + h) = 0$$

i.e. when $\qquad x = -h$

and its value is

$$y = k$$

Example 5

Find the minimum value of $y = 4x^2 + 8x + 1$ and the value of x where it occurs.

Completing the square of $4x^2 + 8x + 1$, we have

$$y = 4(x + 1)^2 - 3$$

The minimum value of y occurs when

$$x + 1 = 0$$

i.e. when $\qquad x = -1$

and the minimum value is

$$y = -3$$

When the squared term of a quadratic expression is negative then the graph of $y = ax^2 + bx + c$ is similar to Fig. 27.14.

Fig. 27.14

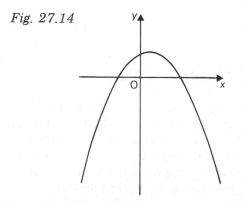

Since, on completing the square,

$$y = k - a(x + h)^2$$

the maximum value of y occurs when

$$(x + h) = 0$$

i.e. when $\qquad x = -h$

and its value is

$$y = k$$

Example 6

Find the maximum value of the quadratic expression $y = -2x^2 + 4x + 7$ and the value of x where it occurs.

On completing the square,

$$y = 9 - 2(x - 1)^2$$

The maximum value of y occurs when

$$x - 1 = 0$$

i.e. when $\qquad x = 1$

and its value is

$$y = 9$$

Exercise 27.4

Find, by completing the square, the maximum or minimum value of each of the following quadratic expressions and the value of x where it occurs.

1. $x^2 + 2x - 3$ 2. $-x^2 + 4x + 7$
3. $2x^2 - 4x + 5$ 4. $4x^2 - 8x - 1$
5. $-3x^2 + 9x - 2$ 6. $-2x^2 - 8x + 7$

Area under a Curve

The area under a graph may be found by one of several approximate methods. The simplest method is by counting the squares on the graph paper. Although it is a simple method it gives results which are as accurate as those obtained by more complicated methods.

Example 7

Plot the graph of the function $y = 2x^2 - 7x + 8$ for values of x between 0 and 8. Hence, by counting squares, find the area under the curve between $x = 2$ and $x = 6$.

x	0	1	2	3	4	5	6	7	8
y	8	3	2	5	12	23	38	57	80

The graph is drawn in Fig. 27.15. On the horizontal axis a scale of 1 large square = 2 units has been used and on the vertical axis the scale is 1 large square = 20 units. Hence, on the horizontal axis 1 small square = $\frac{2}{10}$ = 0.2 units and on the vertical axis 1 small square = $\frac{20}{10}$ = 2 units. Therefore, 1 small square represents an area = 0.2 × 2 = 0.4 square units (Fig. 27.16).

Fig. 27.15

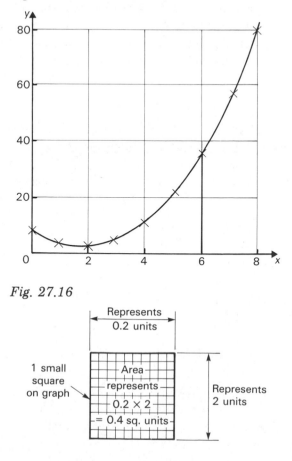

Fig. 27.16

To find the area between the graph, the x-axis and the lines $x = 2$ and $x = 6$ we count up the number of small squares in this region, judgement being exercised in the case of parts of small squares. The number of small squares is found to be about 145. Hence

$$\text{Area required} = 145 \times 0.4$$

$$= 58 \text{ square units}$$

(The exact area is $58\frac{2}{3}$ square units. Hence the method of counting squares is a very accurate method.)

The Trapezium Rule

This is the second method for finding the area under a curve. In order to find the area shown in Fig. 27.17 we divide the area up into a number of equal strips each of width b.

Fig. 27.17

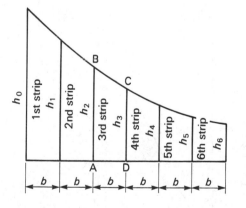

Consider the third strip. If we join BC then ABCD is a trapezium and its area is $b \times \frac{1}{2}(AB + CD)$ which very nearly equals the area of the strip.

If the ordinates at the extremes of the strips are $h_0, h_1, h_2, \ldots$ etc., then the area of the first strip is

$$A_1 = b \times \tfrac{1}{2}(h_0 + h_1)$$

$$= \tfrac{1}{2}bh_0 + \tfrac{1}{2}bh_1$$

The area of the second strip is

$$A_2 = b \times \tfrac{1}{2}(h_1 + h_2)$$

$$= \tfrac{1}{2}bh_1 + \tfrac{1}{2}bh_2$$

The area of the third strip is

$$A_3 = b \times \tfrac{1}{2}(h_2 + h_3)$$

$$= \tfrac{1}{2}bh_2 + \tfrac{1}{2}bh_3$$

and so on.

The total area is

$$A = A_1 + A_2 + A_3 + \ldots$$

$$= \tfrac{1}{2}bh_0 + \tfrac{1}{2}bh_1 + \tfrac{1}{2}bh_1$$
$$\qquad + \tfrac{1}{2}bh_2 + \tfrac{1}{2}bh_2 + \tfrac{1}{2}bh_3 + \ldots$$

$$= \tfrac{1}{2}bh_0 + bh_1 + bh_2 + \ldots$$

$$= b(\tfrac{1}{2}h_0 + h_1 + h_2 + \ldots)$$

$$= \text{width of strips} \times (\tfrac{1}{2} \text{ the sum of}$$
$$\text{first and last ordinates}$$
$$+ \text{ sum of remaining ordinates})$$

This is known as the **trapezium rule**.

Example 8

Find the area required in Example 7 by using the trapezium rule.

Using the values of y given in the table of Example 7 we have:

$$\text{Area} = 1 \times [\tfrac{1}{2}(2 + 38) + 5 + 12 + 23]$$

$$= 60 \text{ square units}$$

This compares very well with the exact area of $58\frac{2}{3}$. A more accurate estimate may be obtained by taking more strips.

Exercise 27.5

Find the areas under the following curves by

(a) counting squares

(b) using the trapezium rule.

1. Between the curve $y = x^3$, the x-axis and the lines $x = 5$ and $x = 3$.

2. Between the curve $y = 3 + 2x + 3x^2$, the x-axis and the lines $x = 1$ and $x = 2$.

3. Between the curve $y = x^2(2x - 1)$, the x-axis and the lines $x = 1$ and $x = 2$.

4. Between the curve $y = (x + 1)^2$, the x-axis and the lines $x = 1$ and $x = 3$.

5. Between the curve $y = 5x - x^3$, the x-axis and the lines $x = 1$ and $x = 2$.

Average Speed

In Chapter 12 it was shown that

$$\text{Average speed} = \frac{\text{Distance travelled}}{\text{Time taken}}$$

If the distance is measured in metres and the time in seconds, then the speed is measured in metres per second (m/s). Similarly, if the distance is measured in kilometres and the time in hours, then the speed is measured in kilometres per hour (km/h).

Example 9

(a) A car travels a distance of 100 km in 2 hours. What is the average speed?

$$\text{Average speed} = \frac{100\ \text{km}}{2\ \text{h}}$$

$$= 50\ \text{km/h}$$

(b) A body travels a distance of 80 metres in 4 seconds. What is its average speed?

$$\text{Average speed} = \frac{80\ \text{m}}{4\ \text{s}}$$

$$= 20\ \text{m/s}$$

Distance–Time Graphs

Since

$$\text{Distance} = \text{Speed} \times \text{Time}$$

If the speed is constant the distance travelled is proportional to time and a graph of distance against time will be a straight line passing through the origin. The gradient of this graph will represent the speed.

Example 10

A man travels a distance of 120 km in 2 hours by car. He then cycles 20 km in $1\frac{1}{2}$ hours and finally walks a distance of 8 km in 1 hour all at constant speed. Draw a graph to illustrate this journey and from it find the average speed for the entire journey.

Fig. 27.18

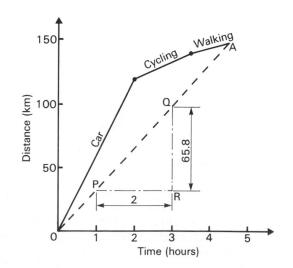

The graph is drawn in Fig. 27.18 and it consists of three straight lines. The average speed is found by drawing the straight line OA and finding its gradient.

To find the gradient of OA we draw the right-angled triangle PQR, which, for accuracy, should be as large as possible.

$$\text{Gradient of OA} = \frac{QR}{PR}$$

$$= \frac{65.8}{2}$$

$$= 32.9$$

Hence average speed = 32.9 km/h.

If the speed is not constant the distance-time graph will be a curve.

Example 11

The table shown below gives the distance travelled, *s* metres, of a vehicle after a time of *t* seconds.

t	0	1	2	3	4
s	0	1	8	27	64

Draw a smooth curve to show how *s* varies with *t* and use the graph to find the speed after 2.5 seconds.

Fig. 27.19

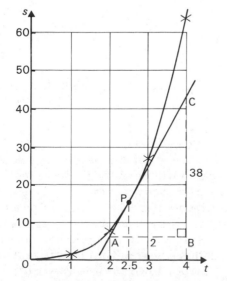

The graph is drawn in Fig. 27.19. To find the speed after 2.5 seconds draw a tangent to the curve at the point P and find its gradient. This is done by drawing the right-angled triangle ABC.

The gradient is

$$\frac{38}{2} = 19.$$

Hence the speed after 2.5 seconds is 19 m/s.

Exercise 27.6

1. A vehicle travels 300 km in 5 hours. Calculate its average speed.

2. A car travels 400 km at an average speed of 80 km/h. How long does the journey take?

3. A train travels for 5 hours at an average speed of 60 km/h. How far has it travelled?

4. A vehicle travels a distance of 250 km in a time of 5 hours. Draw a distance-time graph to depict the journey and from it find the average speed of the vehicle.

5. A car travels a distance of 120 km in 3 hours. It then changes speed and travels a further distance of 80 km in $2\frac{1}{2}$ hours. Assuming that the two speeds are constant draw a distance-time graph. From the graph find the average speed of the journey.

6. A girl cycles a distance of 20 km in 100 minutes. She then rests for 20 minutes and then cycles a further 10 km which takes her 50 minutes. Draw a distance-time graph to represent the journey and from it find the average speed for the entire journey.

7. A man travels a distance of 90 km by car which takes him $1\frac{1}{2}$ hours. He then cycles 18 km in a time of $1\frac{1}{2}$ hours. He then rests for 15 minutes before continuing on foot during which he walks 8 km in 2 hours. By drawing a distance-time graph find his average speed for the entire journey, assuming that he travels at constant speed for each of the three parts of the journey.

8. The following table gives the distance travelled by s metres, after a time of t seconds:

t	0	2	4	6	8
s	0	16	128	432	1024

Draw a graph to show how s varies with t and use your graph to find the speed after 5 seconds.

9. A body moves a distance of s metres in t seconds so that $s = t^3 - 3t^2 + 8$.

(a) Draw a graph to show how s varies with t for values of t between 1 second and 6 seconds.

(b) Find the speed of the body at the end of 4 seconds.

10. A body moves s metres in t seconds, where

$$s = \frac{1}{t^2}$$

By drawing a suitable graph find the speed of the body after 3 seconds.

11. The table below shows how the distance travelled by a body (s metres) varies with the time (t seconds).

t	0	1	2	3	4	5
s	0	1	16	63	160	325

Draw a graph of this information and hence find the speed of the body after 3 seconds.

Velocity

Velocity is speed in a given direction, e.g. 50 km/h due North.

Velocity–Time Graphs

If a velocity–time graph or speed–time graph is drawn (Fig. 27.20), the area under the graph gives the distance travelled. The gradient of the curve gives the acceleration, since acceleration is the rate of change of velocity. If the velocity is measured in metres per second (m/s or $m\,s^{-1}$, the acceleration will be measured in metres per second per second (m/s^2 or $m\,s^{-2}$).

Fig. 27.20

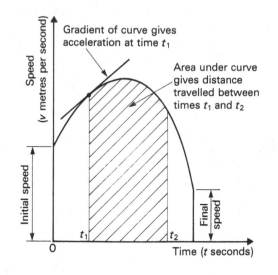

If the acceleration is constant the velocity-time graph will be a straight line (Fig. 27.21). When the velocity is increasing the graph has a positive gradient (i.e. it slopes upwards to the right). If the velocity is decreasing (i.e. there is deceleration, sometimes called retardation) the graph has a negative gradient (i.e. it slopes downwards to the right).

Fig. 27.21

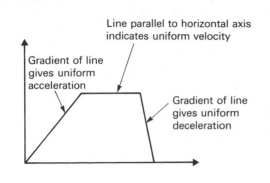

Line parallel to horizontal axis indicates uniform velocity

Gradient of line gives uniform acceleration

Gradient of line gives uniform deceleration

Example 12

A car, starting from rest, attains a velocity of 20 m/s after 5 seconds. It continues at this speed for 15 seconds and then slows down and comes to rest in a further 8 seconds. If the acceleration and retardation are constant, draw a velocity–time graph and from it, find:

(a) the acceleration of the car

(b) the retardation of the car

(c) the distance travelled in the total time of 28 seconds.

The velocity–time graph is drawn in Fig. 27.22.

Fig. 27.22

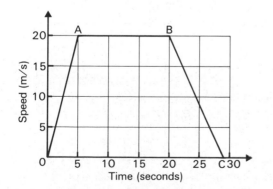

(a) The acceleration is given by the gradient of the line OA.

$$\text{Acceleration} = \frac{20}{5}$$
$$= 4 \text{ m/s}^2$$

(b) The retardation is given by the gradient of the line BC.

$$\text{Retardation} = \frac{20}{8}$$
$$= 2.5 \text{ m/s}^2$$

(c) The distance travelled in the 28 seconds the car was travelling is given by the area of the trapezium OABC.

$$\text{Distance travelled} = 20 \times \tfrac{1}{2} \times (15 + 28)$$
$$= 20 \times \tfrac{1}{2} \times 43$$
$$= 430 \text{ m}$$

Example 13

The table below gives the speed of a car, v metres per second, after a time of t seconds.

t	0	5	10	15	20	25
v	0	2.4	5.0	7.5	9.5	10.2
t	30	35	40	45	50	55
v	9.2	5.2	2.7	2.3	2.7	3.5

Draw a smooth curve to show how v varies with t. Use the graph:

(a) to find the speed of the car after 47 seconds

(b) to find the times when the speed is 4 m/s

(c) to find the acceleration after 15 seconds

(d) to find the retardation after 35 seconds.

The graph is drawn in Fig. 27.23.

Fig. 27.23

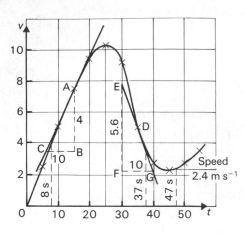

(a) The speed after 47 seconds is read directly from the graph and it is found to be $2.4\,\text{m s}^{-1}$.

(b) The times when the speed is 4 m/s are also read directly from the graph and they are found to be 8 seconds and 37 seconds.

(c) To find the acceleration at a time of 15 seconds, draw a tangent to the curve at the point A and find its gradient by constructing the right-angled triangle ABC.

The gradient $= \dfrac{4}{10} = 0.4$.

Hence the acceleration is $0.4\,\text{m s}^{-2}$.

(d) To find the retardation at a time of 35 seconds, draw a tangent to the curve at the point D. Then by constructing the right-angled triangle EFG,

$$\text{Retardation} = \frac{5.6}{10}$$

$$= 0.56\,\text{m s}^{-2}.$$

Note that the car is slowing down at this time and hence retardation occurs. This is also shown by the negative gradient of the tangent at the point D.

Exercise 27.7

1. Figure 27.24 shows a number of velocity–time diagrams. In each case state the distance travelled.

Fig. 27.24

(a)

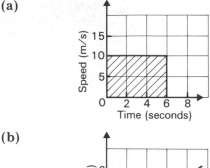

(b)

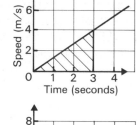

(c)

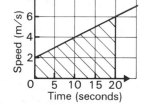

(d)

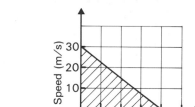

(e)

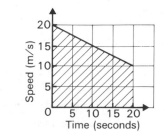

2. In Fig. 27.25 are shown some speed–time graphs. For each write down the acceleration or retardation.

Fig. 27.25

(a)

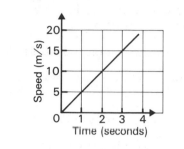

(b)

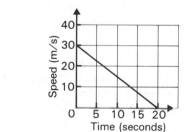

(c)

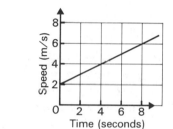

(d)

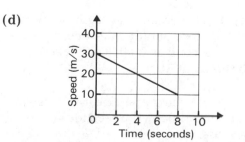

3. Figure 27.26 shows the speed–time graph for a vehicle travelling at a constant speed of 10 m/s. Find the total distance travelled in 15 s.

Fig. 27.26

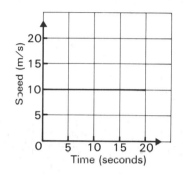

4. The speed–time graph shown in Fig. 27.27 shows a car travelling with constant acceleration.

   (a) What is the acceleration?

   (b) What is the initial speed?

   (c) What is the distance travelled in 30 seconds?

Fig. 27.27

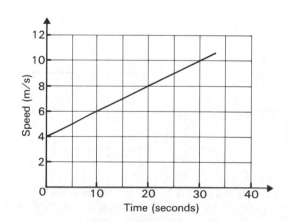

5. From the speed–time graph of Fig. 27.28, find:

(a) the acceleration

(b) the retardation

(c) the initial speed

(d) the distance travelled in the first 20 seconds.

Fig. 27.28

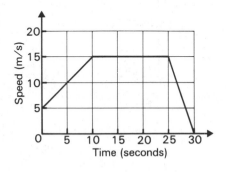

6. Figure 27.29 shows a speed–time graph. Find:

(a) the acceleration after 5 seconds

(b) the acceleration after 18 seconds

(c) the acceleration after 40 seconds

(d) the maximum speed reached

(c) the total distance travelled in the first 50 seconds.

Fig. 27.29

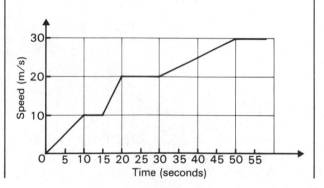

7. A vehicle starting from rest attains a velocity of 16 m/s after it has been travelling for 8 seconds with uniform acceleration. It continues at this speed for 15 seconds and then it slows down, with uniform retardation, until it finally comes to rest in a further 10 seconds.

(a) Draw the velocity–time graph.

(b) Determine the acceleration of the vehicle.

(c) What is the retardation?

(d) From the diagram, find the distance travelled during the 33 seconds represented on the graph.

8. A car has an initial velocity of 5 m/s. It then accelerates uniformly for 6 seconds at $\frac{1}{2}$ m/s^2. It then proceeds with this speed for a further 25 seconds.

(a) Draw a velocity–time graph from this information.

(b) Calculate the distance travelled by the car in the time of 31 seconds.

9. The speed of a body, v metres per second, at various times, t seconds, is shown in the following table:

t	0	1	2	3	4	5	6	7	8
v	0	1	2	6	12	20	30	42	56

(a) Draw a graph showing how v varies with t. Horizontally take 1 cm to represent 1 second and vertically take 1 cm to represent 10 m/s.

(b) From the graph find the acceleration after times (i) 2 seconds, (ii) 6 seconds.

10. The graph, Fig. 27.30, shows how the speed of a vehicle varies over a period of 20 seconds. From the graph, find:

    (a) the acceleration of the vehicle at a time of 4 seconds

    (b) the time at which the speed of the car is decreasing at the greatest rate

    (c) the distance travelled by the vehicle in the 20 seconds represented on the graph.

Fig. 27.30

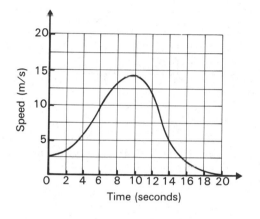

Direct Variation

The statement that y is proportional to x (often written $y \propto x$) means that the graph of y against x is a straight line passing through the origin (Fig. 27.31). If the gradient of this line is k, then $y = kx$. The value of k is called the constant of proportionality. Thus, the ratio of y to x is equal to the constant of proportionality and y is said to **vary directly** as x. Hence direct variation means that if x is doubled then y is also doubled, if x is halved then y is halved and so on. Some examples of direct variation are as follows:

(1) The circumference of a circle is directly proportional to its diameter.

(2) The volume of a cone of given radius is directly proportional to its height.

Fig. 27.31

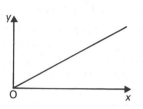

Most problems on direct variation involve first finding the constant of proportionality from information given in the problem as shown in the following example.

Example 14

If y is directly proportional to x and $y = 2$ when $x = 5$, find the value of y when $x = 6$.

Since $y \propto x$ then $y = kx$.

We are given that when $y = 2$, $x = 5$.

Hence,

$$2 = k \times 5 \quad \text{or} \quad k = \frac{2}{5}$$

$$\therefore \qquad y = \frac{2}{5}x$$

When $x = 6$,

$$y = \frac{2}{5} \times 6$$

$$= \frac{12}{5}$$

The area of a circle is given by the equation $A = \pi r^2$. From this equation we see that A varies directly as the square of r ($A \propto r^2$). and the constant of proportionality is π.

If y is proportional to x^2 ($y \propto x^2$) the graph of y against x^2 is a straight line passing through the origin. If the gradient of this line is k then $y = kx^2$.

Example 15

The surface area of a sphere, A square millimetres, varies directly as the square of its radius, r millimetres. If the surface area of a sphere of 2 mm radius is 50.24 mm^2 find the surface area of a sphere whose radius is 4 mm.

Since $A \propto r^2$ then $A = kr^2$.

We are given that $A = 50.24$ when $r = 2$.

Hence:

$$50.24 = k \times 2^2 \quad \text{or} \quad k = \frac{50.24}{2^2}$$

$$= 12.56$$

$$\therefore \qquad A = 12.56r^2$$

When $r = 4$,

$$A = 12.56 \times 4^2$$

$$= 200.96 \text{ mm}^2$$

Inverse Variation

If y is inversely proportional to x then the graph of y against $\dfrac{1}{x}$ is a straight line passing through the origin (Fig. 27.32). If the gradient of this line is k, then

$$y = k \times \frac{1}{x}$$

$$= \frac{k}{x}$$

Fig. 27.32

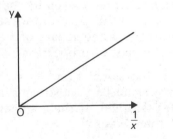

Example 16

The electrical resistance, R ohms, of a wire of given length is inversely proportional to the square of the diameter of the wire, d mm. If R is 4.25 ohms when d is 2 mm, find the value of R when $d = 3$ mm.

Fig. 27.33

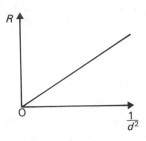

Since R is inversely proportional to d^2, (Fig. 27.33)

$$R \propto \frac{1}{d^2}$$

$$R = \frac{k}{d^2}$$

When $R = 4.25$, $d = 2$.

Hence $\quad 4.25 = \dfrac{k}{2^2}$

$$k = 4.25 \times 2^2$$

$$= 17$$

$$\therefore \qquad R = \frac{17}{d^2}$$

When $d = 3$,

$$R = \frac{17}{3^2}$$

$$= 1.9 \text{ ohms}$$

Exercise 27.8

1. Express the following with an equals sign and a constant:

   (a) y varies directly as x^2.

   (b) U varies directly as V.

   (c) S varies inversely as T.

   (d) h varies inversely as the square of m.

2. If $y = 2$ when $x = 4$ write down the value of y when $x = 9$ for the following:

   (a) y varies directly as the square of x.

   (b) y varies inversely as the square of x.

   (c) y varies inversely as x.

3. If S varies inversely as T^2 and $S = 2$ when $T = 3$ find the value of T when $S = 16$.

4. If U varies directly as V and $U = 27$ when $V = 9$ find the value of V when $U = 4$.

5. The surface area of a sphere, $V\,mm^2$, varies directly as the square of its diameter, $d\,mm$. If the surface area is to be doubled, by what ratio must the diameter be altered?

Miscellaneous Exercise 27

Section A

1. From the speed–time graph shown in Fig. 27.34 find:

   (a) the acceleration

   (b) the retardation

   (c) the initial speed

   (d) the distance travelled in the first 20 seconds.

Fig. 27.34

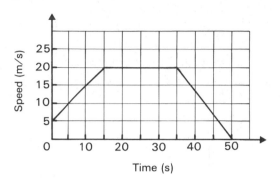

2. The speed of a body, v metres per second, at various times t seconds is shown in the table below.

t	0	1	2	3	4	5	6
v	0	2	8	18	32	50	72

   (a) Draw a graph showing how v varies with t. Horizontally take 1 cm to represent 1 second and vertically take 1 cm to represent 10 metres per second.

   (b) From the graph find the acceleration after times (i) 2 seconds, (ii) 4 seconds.

   (c) Estimate the distance travelled by the body in the first 6 seconds.

3. A ball is rolling down an inclined plane. The distance, s centimetres, travelled in a time t seconds is given in the table below.

t	0	1	2	2.5	3	3.5	4	4.5	5
s	0	2.8	11.2	17.5	25.2	34.3	44.8	56.7	70.0

   (a) Using a scale of 2 cm to represent 1 second horizontally and a scale of 2 cm to represent 10 cm vertically, plot these values of s and t.

   (b) Find the speed of the ball when $t = 3$ seconds.

   (c) If $s = kt^2$, find the value of the constant k.

4. Fig. 27.35 shows a speed–time graph for a train journey. Find:

   (a) how far the train travelled in the first 20 seconds

   (b) the total distance travelled.

Fig. 27.35

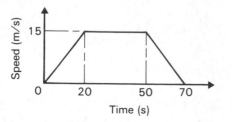

5. A man in training runs for 2 hours at 14 km/h and then for 1 hour at 12 km/h and finally for 1 hour at 10 km/h.

   (a) How far does he run in the first two hours?

   (b) How far does he run altogether?

   (c) Draw a sketch graph to show his speeds during the run.

   (d) On another day he runs the same distance at uniform speed in the same time. What was this uniform speed?

Section B

1. The shape shown in Fig. 27.36 is obtained by removing the triangle QRT from the rectangle PQRS. The angle QTR = 90° and QT = TR. Given that PQ = 18 cm and PS = 3x cm, verify that the area A, in square centimetres, of the shape PQTRS is given by the formula

$$A = 54x - \frac{9x^2}{4}$$

   Copy and complete the following table:

x	0	3	6	9	11	13	15	18
A	0	142			322		304	

Fig. 27.36

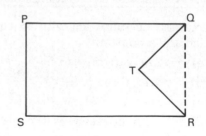

Draw a graph showing how A varies with x using a scale of 1 cm to represent 2 units on the x-axis and 1 cm to represent 50 cm² on the A-axis. Use your graph to find the maximum value of A.

2. Copy and complete the table of values given below for the function $y = x^3 - 3x^2 + 2$.

x	−2	−1	0	1	2	3	4
y	−18					2	18

   Use the completed table to draw the graph of $y = x^3 - 3x^2 + 2$ for values of x between −2 and 4 taking 1 cm to represent 1 unit on the x-axis and 2 cm to represent 5 units on the y-axis. Use your graph to find:

   (a) the gradient of the curve at the point (3, 2)

   (b) the coordinates of the points on the curve at which y has a maximum or minimum value and distinguish clearly between these points.

3. Draw a graph of the curve $y = x^2 + 3$ for values of x in the range −3 to 3 including the points for which x = −2.5 and x = 2.5. Use scales of 1 cm = 1 unit on the x-axis and the y-axis. Use the trapezium rule to find the area bounded by the curve, the x-axis and the lines x = −3 and x = 3.

4. If y varies directly as x^2 and $y = 5$ when $x = 2$, calculate the value of y when $x = 4$.

5. Given that y is inversely proportional to x and $y = -6$ when $x = -4$, find the value of x when $y = 8$.

6. 'Unemployment is still rising and at an increasing rate.' Which of the sketch graphs shown in Fig. 27.37 best represents this statement?

Fig. 27.37

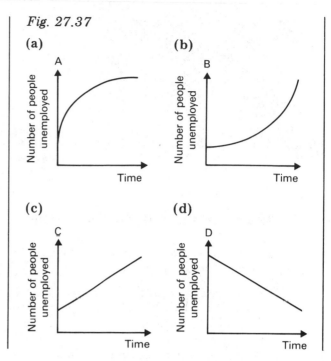

(a)

(b)

(c)

(d)

Inequalities

Symbols used for Inequalities

The following symbols are used when dealing with inequalities:

$>$ means greater than

$<$ means less than

$\geqslant$ means equal to or greater than

$\leqslant$ means equal to or less than

Note that the 'arrow' always points to the smaller quantity.

Solution of Inequalities

Fig. 28.1 shows a number line. One number is greater than another if it lies to the *right* of the other on the number line. Thus

$$5 > 3 \quad \text{and} \quad -2 > -5$$

If $x < 3$ all the possible values of x are shown in Fig. 28.1. The empty circle at the end of the arrowed line shows that $x = 3$ is not included.

Fig. 28.1

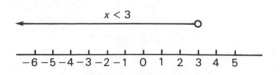

Fig. 28.2 shows all the solutions for the inequality $x \geqslant -2$. Since $x = -2$ is included, the arrowed line ends in a solid circle.

Fig. 28.2

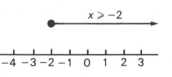

Combined Inequalities

Suppose we have a number x such that $4 < x$ and $x < 7$. The inequalities can be combined to give $4 < x < 7$, which means that the value of x lies between 4 and 7. If x is a real number the inequality $4 < x < 7$ represents a region on the number line (Fig. 28.3). The region is the set $S = \{x : 4 < x < 7\}$ (see Chapter 29).

Fig. 28.3

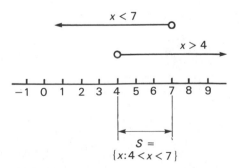

Example 1

If x is an integer find the solution set for $-3 \leqslant x \leqslant 2$.

Drawing the number line (Fig. 28.4) it can be seen that the solution set is $\{-3, -2, -1, 0, 1, 2\}$.

Fig. 28.4

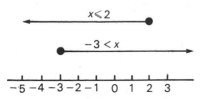

Rules for Solving Inequalities

(1) The same number may be added or subtracted from both sides.

(2) Multiplying or dividing both sides by the same positive number leaves the inequality unaltered.

(3) Multiplying or dividing both sides by the same negative number reverses the inequality sign.

Example 2

(a) Solve the inequality $4x + 5 \geqslant 2x + 9$.

$$4x + 5 \geqslant 2x + 9$$
$$4x - 2x \geqslant 9 - 5$$
$$2x \geqslant 4$$
$$x \geqslant 2$$

(b) Solve the inequality $3 - 7x < 2x + 21$.

$$3 - 7x < 2x + 21$$
$$-7x - 2x < 21 - 3$$
$$-9x < 18$$
$$-x < 2$$
$$x > -2 \quad \text{(by multiplying both sides by } -1\text{)}$$

(c) If x is an integer find the solution set for $3 \leqslant x + 1 \leqslant 5$.

Dealing with each inequality separately we have

$$3 \leqslant x + 1 \qquad x + 1 \leqslant 5$$
$$2 \leqslant x \qquad\qquad x \leqslant 4$$

So $\qquad 2 \leqslant x \leqslant 4$

Hence the solution set is $\{2, 3, 4\}$.

Exercise 28.1

Solve the following inequalities and represent the solutions on number lines:

1. $3x > 6$
2. $x - 3 < 3$
3. $x + 4 \leqslant 6$
4. $3x - 5 \geqslant 7$
5. $6x + 11 > 18 - x$
6. $3x + 25 < 8x - 10$
7. $x + 3 \geqslant 7 + 3x$
8. $2x - 7 \geqslant 5x + 8$

Use number lines to find solutions for the following pairs of inequalities. In each case x must be one of the set $\{0, 1, 2, 3, 4, 5, 6, 7, 8, 9\}$.

9. $5 < x < 2$
10. $3 \leqslant x \leqslant 9$
11. $0 < x < 7$
12. $1 \leqslant x \leqslant 5$

If x is an integer find solution sets for each of the following:

13. $-4 < x < 2$
14. $-5 < x < -1$
15. $3 \leqslant x \leqslant 9$
16. $-2 \leqslant x \leqslant 8$
17. $3 \leqslant x \leqslant 5$
18. $3 \leqslant 2x - 1 \leqslant 5$
19. $-4 \leqslant 3x + 2 \leqslant 11$
20. $-2 < 4x + 6 < 10$

Graphs of Linear Inequalities

Example 3

(a) Illustrate on a graph the solution of the inequality $2y + 3x \geqslant 8$.

First express the inequality in the form:

$$2y \geqslant 8 - 3x$$

$$y \geqslant 4 - \frac{3x}{2}$$

Next draw the graph of $y = 4 - \frac{3x}{2}$ as shown in Fig. 28.5. Since the solution set includes points on the line $y = 4 - \frac{3x}{2}$ this is shown as a full line. The solution of the inequality $2y + 3x \geqslant 8$ is the unshaded region above the straight line representing the equation $y = 4 - \frac{3x}{2}$.

Fig. 28.5

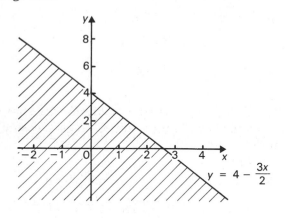

$$y = 4 - \frac{3x}{2}$$

(b) Find graphically the solution set $S = \{(x, y): -1 < y < 3, 0 < x < 3, 4y + 5x < 20, \text{ where } x \text{ and } y \text{ are integers}\}$.

In Fig. 28.6, the lines are shown broken which is the convention for strict inequalities. The solution is given by the unshaded region in the diagram. The members of S are shown by the large dots. Hence $S = \{(1, 2), (1, 1), (1, 0), (2, 2), (2, 1), (2, 0)\}$.

Fig. 28.6

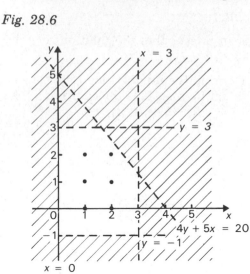

Exercise 28.2

Illustrate graphically the following inequalities:

1. $x + y \geqslant 3$

2. $y < 3x + 5$

3. $y + x - 2 < 0$

4. $3x + 2y - 6 \geqslant 0$

5. $4y + 3x - 12 \geqslant 0$

Indicate on diagrams the regions which represent the following:

6. $x \geqslant 0$, $y \geqslant 0$, $x + y \geqslant 8$

7. $x \geqslant 1$, $y \geqslant 1$, $2x + 3y \leqslant 12$, $5x + 2y \leqslant 20$

8. $x \geqslant 2$, $y \geqslant 1$, $x + y \leqslant 7$, $y - x \geqslant 0$

9. $y \geqslant 0$, $3 \leqslant x \leqslant 6$, $y \leqslant 2x - 3$

10. $0 \leqslant y \leqslant 2$, $0 \leqslant x \leqslant 3$, $2y + 3x \geqslant 6$

Miscellaneous Exercise 28

Section A

1. If x is a positive integer such that $10 < x < 100$, state the least possible value of x.

2. Use one of the symbols $< = >$ in each of the following to make true statements.
   (a) $13^2 - 12^2 \qquad 5^2$
   (b) $9^2 - 3^2 \qquad 3^2$
   (c) $\dfrac{5}{9} - \dfrac{1}{3} \qquad \dfrac{2}{3}$
   (d) $+\sqrt{14.4} \qquad 1.2$
   (e) $0.5^2 \qquad 0.5^3$

3. Which of the following symbols $<$, $=$ or $>$ can be correctly inserted in the following?
   (a) $12 \qquad 3 \times 4 \times 0$
   (b) $3.99 \times 3.97 \qquad 16$
   (c) $2^3 \times 3^3 \qquad 6^3$
   (d) $15^2 \qquad 8^2 + 7^2$

4. Find the solution sets for the following pairs of inequalities, where x is an integer:
   (a) $x < 4$ and $x > 2$
   (b) $x > 4$ and $x < 9$

5. If x is an integer find the solution sets for
   (a) $-2 \leqslant x < 3$ \qquad (b) $-4 < x < 5$

Section B

1. Find the positive integer such that $x + 1 < 19 < x + 3$.

2. Find the smallest whole number which satisfies the inequality $3x - 5 > x + 13$.

3. The cost of photocopying is 6p per copy for the first ten copies and 3p for subsequent copies. Write down an expression for the cost in pence of producing n copies where $n > 10$.

4. Draw, on graph paper, taking 2 cm to represent 1 unit on both the x and y-axes, the set of points (x, y) which obey the inequalities $y \geqslant 1$, $y \leqslant 9$, $x \leqslant 6$, $y \leqslant x + 5$, $5x + 2y \geqslant 10$.

5. In Fig. 28.7 the coordinates of all the points on the line through B and E satisfy the relationship $x + 2y = 4$.
   (a) What relationship is satisfied by all the points on the line through (i) A and D, (ii) C and E?
   (b) What are the coordinates of D?
   (c) Give three inequalities satisfied by the coordinates of all the points inside the shaded triangle OEB.

Fig. 28.7

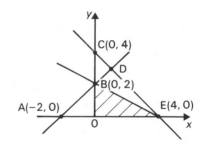

6. If x is an integer, solve the inequality $14 \leqslant (2x + 5) \leqslant 23$.

7. (a) Draw the parabola $y = x^2$ for values of x between -3 and $+3$, using a scale of 2 cm $= 1$ unit on the horizontal axis and 1 cm $= 1$ unit on the vertical axis.
   (b) Using the same axes as for part (a), draw the straight line $y = 3 - x$ for values of x between -2 and $+3$.
   (c) On the diagram shade the area where $y \leqslant 3 - x$.
   (d) Write down the coordinates of any points on the parabola $y = x^2$ such that $y \geqslant 3 - x$.

Multi-Choice Questions 28

1. If x is a real number, for what set of values is $-x + 1 < 0$?

   A $x < 1$ B $x < -1$

   C $x > 1$ D $x > -1$

2. The solution to $28 - 5x < 2x + 9$ is

   A $x > \dfrac{19}{3}$ B $x < \dfrac{19}{3}$

   C $x > \dfrac{19}{7}$ D $x < \dfrac{19}{7}$

   Fig. 28.8 shows the graphs of $x^2 - 5x + 3$ and $y = \frac{1}{4}x + 1$. Use them to answer Questions 3 and 4.

Fig. 28.8

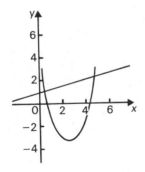

3. For what range of values of x is $x^2 - 5x + 3 < 0$?

   A between 0.4 and 4.3

   B between 0.5 and 4.5

   C between 0.7 and 4.3

   D between 1 and 5

4. For what range of values of x is $x^2 - 5x + 3 < \frac{1}{4}x + 1$?

   A between 0.4 and 4.8

   B between 0.5 and 4.5

   C between 0.7 and 4.3

   D between 1 and 5

5. The range of values of x for which $\dfrac{x}{2} > \dfrac{x}{3} + 1$ is

   A $x > 1$ B $x < 1$

   C $x > 6$ D $x < 6$

6. The smallest whole number which satisfies the inequality $9 - 2x < 5x - 12$ is

   A 1   B 2   C 3   D 4

7. Which of the following inequalities is true for the point $(4, -2)$?

   A $x + y > 3$ B $x < 3y$

   C $x - y < 10$ D $2x - 3y < 12$

Sets

Definitions

A **set** is a collection of objects, numbers, ideas, etc. The different objects etc. are called the **elements** or **members** of the set.

A set may be **defined** by using one of the following methods:

(1) By listing all of the members, for instance, $A = \{3, 5, 7, 9, 11\}$. The order in which the elements are written does not matter and each element is listed once only.

(2) By listing only enough elements to indicate the pattern and showing that the pattern continues by using dots. For instance $B = \{2, 4, 6, 8, \ldots\}$.

(3) By a description such as $S = \{\text{all odd numbers}\}$.

(4) By using an algebraic expression such as $C = \{x \colon 2 \leqslant x \leqslant 7, \; x \text{ is an integer}\}$, which means 'the set of elements, x, such that x is an integer whose value lies between 2 and 7, that is $C = \{2, 3, 4, 5, 6, 7\}$.

Types of Set

A **finite set** is one in which all the elements are listed such as $\{3, 7, 9\}$.

An **infinite set** is one in which it is impossible to list all the members. For instance $\{1, 3, 5, 7, \ldots\}$ where the dots mean 'and so on'.

The **null** or **empty** set which contains no elements. It is denoted by $\varnothing$ or by $\{\,\}$.

Membership of a Set

The symbol $\in$ means 'is a member of the set'. Thus because 7 is a member of the set $S = \{2, 5, 6, 7, 9\}$ we write $7 \in S$. The symbol $\notin$ means 'is not a member of the set'. Because 3 is not a member of S we write $3 \notin S$.

Order of a Set

The order of a set is the number of elements contained in the set. If a set A has 5 members we write $n(A) = 5$.

Subsets

If all the members of a set A are also members of a set B then A is said to be a **subset** of B. Thus if $A = \{p, q, r\}$ and $B = \{p, q, r, s\}$ we write $A \subset B$. Every set has at least two subsets, itself and the null set.

Example 1

List all the subsets of $\{a, b, c\}$.

The subsets are $\varnothing, \{a\}, \{b\}, \{c\}, \{a, b\}, \{a, c\}, \{b, c\}$ and $\{a, b, c\}$.

Every possible subset of a given set, except the set itself, is called a **proper subset**.

If there are n elements in a set then the total number of subsets is given by

$$N = 2^n$$

Example 2

A set has 5 members. How many subsets can be formed?

We have $n = 5$. Hence

$$N = 2^5$$
$$= 32$$

Therefore 32 subsets may be formed.

The Universal Set

The **universal set** for any particular problem is the set which contains all the available elements for that problem. Thus if the universal set is all the odd numbers up to and including 11, we write

$$\mathcal{E} = \{1, 3, 5, 7, 9, 11\}$$

The **complement** of a set A is the set of elements of $\mathcal{E}$ which do not belong to A. Thus if

$$A = \{2, 4, 6\}$$

and $\qquad \mathcal{E} = \{1, 2, 3, 4, 5, 6, 7\}$

the complement of A is

$$A' = \{1, 3, 5, 7\}$$

Equality and Equivalence

The order in which the elements of a set are written does not matter. Thus $\{a, b, c, d\}$ is the same as $\{c, a, d, b\}$. Two sets are said to be equal if they have exactly the same elements. Thus if $A = \{2, 3, 5, 8\}$ and $B = \{3, 8, 2, 5\}$ then $A = B$.

Two sets are said to be **equivalent** if they have exactly the same number of elements. Thus if $A = \{5, 7, 9\}$ and $B = \{a, b, c\}$ then $n(A) = n(B) = 3$ and the two sets A and B are equivalent.

Write down the members of the following sets:

1. {odd numbers from 3 to 11 inclusive}

2. {even numbers less than 10}

3. {multiples of 2 up to and including 16}

4. State which of the following sets are finite or null:
    (a) $\{1, 3, 5, 7, 9, \ldots\}$
    (b) $\{2, 4, 6, 8, 10\}$
    (c) {letters of the alphabet}
    (d) {people who have swum the Atlantic ocean}
    (e) {odd numbers which can be divided exactly by 2}

5. If $A = \{a, b, c, d, e\}$ what is $n(A)$?

6. If $B = \{2, 4, 6, 8, 10, 12, 14, 16\}$ write down $n(B)$.

7. State which of the following statements are true?
    (a) $7 \in$ {prime factors of 63}
    (b) $24 \in$ {multiples of 5}
    (c) octagon $\notin$ {polygons}

8. $A = \{2, 3, 5, 6, 8, 9, 11, 13, 14, 15\}$. List the members of the following subsets:
    (a) {odd numbers of A}
    (b) {even numbers of A}
    (c) {prime numbers of A}
    (d) {numbers divisible by 3 in A}

9. Write down all the subsets of $\{a, b, c, d\}$.

10. How many subsets has $B = \{2, 3, 4, 5, 6, 7\}$? How many of these are proper subsets?

11. Below are given eight sets. Connect appropriate sets with the symbol ⊂:

    (a) {integers between 1 and including 24}

    (b) {all cutlery}

    (c) {all footwear}

    (d) {letters of the alphabet}

    (e) {boot, shoe}

    (f) {a, e, i, o, u}

    (g) {2, 4, 6, 8}

    (h) {knife, fork, spoon}

12. A set has 8 members. How many subsets can be formed from its elements?

13. If $\mathscr{E} = \{2, 3, 5, 6, 7, 9, 11, 12, 13, 15, 16, 17\}$ write down the subsets of

    (a) {odd numbers}

    (b) {prime numbers}

    (c) {multiples of 3}

14. Show that
    $\{x : 2 \leqslant x \leqslant 5\} \subset \{x : 1 \leqslant x \leqslant 9\}$

15. If $A = \{3, 5, 7, 8, 9\}$, $B = \{5, 7, 9\}$ and $C = \{7, 10\}$ which of the following statements is/are correct?

    $A \subset B$, $B \subset C$, $B \subset A$, $C \subset B$.

16. If $\mathscr{E} = \{1, 2, 4, 5, 7, 9, 10, 12, 14, 15\}$ and $A = \{4, 5, 7, 10, 14, 15\}$ write down the elements of A'.

17. If $A = \{a, b, c, d, e\}$, $B = \{2, 3, 4, 5\}$, $C = \{b, d, a, c, e\}$ and $D = \{m, n, o, p\}$ write down the sets which are

    (a) equal   (b) equivalent

Venn Diagrams

Set problems may be solved by using Venn diagrams.

The universal set is represented by a rectangle and subsets of this set are represented by closed curves (Fig. 29.1). The shaded region of the diagram represents the complement of A, i.e. A'.

Fig. 29.1

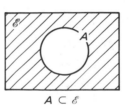

$A \subset \mathscr{E}$

Subsets are represented by a curve within a curve (Fig. 29.2).

Fig. 29.2

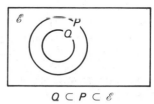

$Q \subset P \subset \mathscr{E}$

Union and Intersection

The **intersection** of two sets A and B is the set of elements which are members of both A and B. Thus if $A = \{2, 4, 7\}$ and $B = \{2, 3, 7, 8\}$ then the intersection of A and B is $\{2, 7\}$. We write $A \cap B = \{2, 7\}$.

The shaded region of the Venn diagram shown in Fig. 29.3 represents $P \cap Q$.

Fig. 29.3

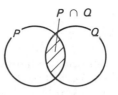

$P \cap Q$

The **union** of the sets A and B is the set of all the elements contained in A and B. Thus if $A = \{3, 4, 6\}$ and $B = \{2, 3, 4, 5, 7, 8\}$ then the union of A and B is $\{2, 3, 4, 5, 6, 7, 8\}$. We write

$$A \cup B = \{2, 3, 4, 5, 6, 7, 8\}$$

The shaded portion of the Venn diagram shown in Fig. 29.4 represents $X \cup Y$.

Fig. 29.4

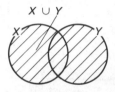

Problems with Intersections and Unions

So far we have dealt with the intersection and union of only two sets. It is, however, quite usual for there to be intersections between three or more sets. In the Venn diagram (Fig. 29.7) the shaded portion represents $A \cap B \cap C$.

Fig. 29.7

Example 3

If $A = \{3,4,5,6\}$, $B = \{2,3,5,7,9\}$ and $\mathscr{E} = \{1,2,3,4,5,6,7,8,9,10,11\}$, draw a Venn diagram to represent this information. Hence write down the elements of
(a) A' (b) $A \cap B$ (c) $A \cup B$

The Venn diagram is shown in Fig. 29.5 and from it:

(a) $A' = \{1,2,7,8,9,10,11\}$

(b) $A \cap B = \{3,5\}$

(c) $A \cup B = \{2,3,4,5,6,7,9\}$

Fig. 29.5

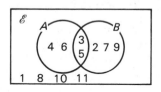

Two sets A and B are said to be **disjoint** if they have no members which are common to each other, i.e. $A \cap B = \emptyset$. Two disjoint sets X and Y are shown in the Venn diagram of Fig. 29.6.

Fig. 29.6

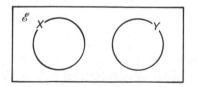

Example 4

If $A = \{2,4,6,8,10\}$, $B = \{1,2,3,4,5\}$ and $C = \{2,5,6,8\}$ determine:

(a) $A \cap (B \cap C)$ (b) $(A \cap B) \cap C$

(a) In problems of this kind always work out the brackets first. Thus

$$B \cap C = \{2,5\}$$
$$A \cap (B \cap C) = \{2\}$$

(b) $$A \cap B = \{2,4\}$$
$$(A \cap B) \cap C = \{2\}$$

It will be noticed that the way in which sets A, B and C are grouped makes no difference to the final result. Thus

$$A \cap (B \cap C) = (A \cap B) \cap C$$

This is known as the **associative law** for sets.

Problems with the Number of Elements of a Set

For two intersecting sets A and B it can be shown that

$$n(A \cup B) = n(A) + n(B) - n(A \cap B)$$

Example 5

The number of elements in sets A and B are shown in Fig. 29.8. If $n(A) = n(B)$ find

(a) x (b) $n(A)$

(c) $n(B)$ (d) $n(A \cup B)$

  (a) $2x - 4 + x = x + x + 5$

               $3x - 4 = 2x + 5$

                   $x = 9$

  (b) $n(A) = 3x - 1$

           $= 3 \times 9 - 4$

           $= 23$

  (c) Since $n(A) = n(B)$, $n(B) = 23$

  (d) $n(A \cup B) = 2x - 4 + x + x + 5$

                $= 4x + 1$

                $= 4 \times 9 + 1$

                $= 37$

Fig. 29.8

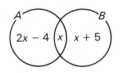

Example 6

In a class of 25 members, 15 take history, 17 take geography and 3 take neither subject. How many class members take both subjects?

If H = set of students who take history and G = set of students who take geography, then $n(H) = 15$, $n(G) = 17$, $n(H \cup G) = 22$.

The Venn diagram of Fig. 29.9 shows the number of students in each set.

Since we do not know $n(H \cap G)$ we let x represent this set. Thus

$$n(H \cup G) = 15 - x + x + 17 - x$$

$$= 32 - x$$

$$22 = 32 - x$$

$$x = 10$$

Hence the number of students taking both subjects is 10.

Alternatively, using the equation:

$$n(H \cup G) = n(H) + n(G) - n(H \cap G)$$

$$22 = 15 + 17 - x$$

$$x = 10 \quad \text{(as before)}$$

When three sets A, B and C intersect as shown in Fig. 29.10:

$$n(A \cup B \cup C) = n(A) + n(B) + n(C)$$
$$- n(A \cap B) - n(A \cap C)$$
$$- n(B \cap C)$$
$$+ n(A \cap B \cap C)$$

Fig. 29.9

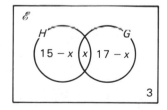

Fig. 29.10

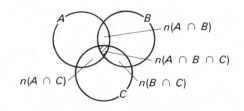

Example 7

The entries in Fig. 29.11 show the number of elements in the various regions. Find:

(a) $n(A)$ (b) $n(B)$

(c) $n(C)$ (d) $n(A \cap B)$

(e) $n(A \cap C)$ (f) $n(B \cap C)$

(g) $n(A \cap B \cap C)$ (h) $n(A \cup B \cup C)$

Fig. 29.11

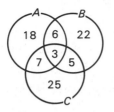

(a) $n(A) = 18 + 6 + 3 + 7$

$= 34$

(b) $n(B) = 6 + 22 + 5 + 3$

$= 36$

(c) $n(C) = 7 + 3 + 5 + 25$

$= 40$

(d) $n(A \cap B) = 6 + 3$

$= 9$

(e) $n(A \cap C) = 7 + 3$

$= 10$

(f) $n(B \cap C) = 3 + 5$

$= 8$

(g) $n(A \cap B \cap C) = 3$

(h) $n(A \cup B \cup C) = 18 + 6 + 3 + 7$

$+ 22 + 5 + 25$

$= 86$

Example 8

In an examination 60 candidates offered mathematics, 80 offered English and 50 offered chemistry. If 20 offered mathematics and English, 15 English and chemistry, and 10 offered all three subjects, how many candidates sat the examination?

We are given: $n(M) = 60$, $n(E) = 80$, $n(C) = 50$, $n(M \cap E) = 20$, $n(E \cap C) = 15$, $n(M \cap C) = 25$ and $n(M \cap E \cap C) = 10$. We have to find $n(M \cup E \cup C)$. The Venn diagram of Fig. 29.12 shows the number of elements in each region. From this diagram we have

$$n(M \cup E \cup C) = 25 + 15 + 10 + 10$$
$$+ 5 + 55 + 20$$
$$= 140$$

Hence 140 candidates took the examination.

Alternatively, using the equation:

$$n(M \cup E \cup C) = n(M) + n(E) + n(C)$$
$$- n(M \cap E)$$
$$- n(M \cap C)$$
$$- n(E \cap C)$$
$$+ n(M \cap E \cap C)$$
$$= 60 + 80 + 50 - 20$$
$$- 15 - 25 + 10$$
$$= 140 \quad \text{(as before)}$$

Fig. 29.12

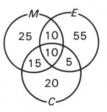

Exercise 29.2

1. Fig. 29.13 is a Venn diagram which shows the universal set & and two subsets A and B. Write down the elements of

   (a) &   (b) A

   (c) B (d) A'

   (e) B' (f) $A \cap B$

   (g) $A \cup B$ (h) $(A \cup B)'$

Fig. 29.13

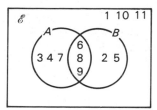

2. Draw a Venn diagram to represent
   & $= \{1,2,3,4,5,6,7,8,9,10\}$,
   $X = \{2,4,7,9\}$ and $Y = \{1,4,5,6,7,10\}$.
   Hence write down the elements of $X \cap Y$.

3. Use set notation to describe the shaded portions of the Venn diagrams of Fig. 29.14.

Fig. 29.14

(a)   (b)

(c)   (d)

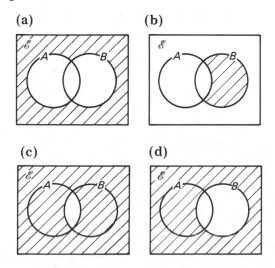

4. If $A = \{3,4,6,8\}$, $B = \{3,4,5,7\}$ and $C = \{3,6,7,8\}$, list the following sets:

   (a) $A \cap (B \cap C)$ (b) $(A \cap B) \cap C$

   (c) $B \cap (A \cup C)$ (d) $A \cup (B \cup C)$

   (e) $(A \cap B) \cup (B \cap C)$

5. Given that & $= \{x: 1 \leqslant x \leqslant 10\}$,
   $A = \{3,4,5,6\}$ and $B = \{5,6,8,9\}$,
   list the following sets:

   (a) $A \cap (A' \cap B)$ (b) $B \cup (A' \cap B)$

   (c) $A' \cap (A \cap B')$ (d) $B' \cup (A \cap B')$

   (e) $A \cup (A' \cap B')$

6. Fig. 29.15 shows two intersecting sets A and B. The entries give the number of elements in each region. Find:

   (a) $n(A)$ (b) $n(B)$

   (c) $n(A \cap B)$ (d) $n(A \cup B)$

Fig. 29.15

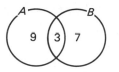

7. Fig. 29.16 is a Venn diagram showing the two intersecting sets A and B. The number of elements in each region are shown in the diagram. If $n(A) = n(B)$, find:

   (a) $n(A)$ (b) $n(B)$

   (c) $n(A \cup B)$ (d) $n(A \cap B)$

Fig. 29.16

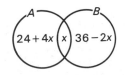

8. In a group of 75 girls, 54 like hockey and 42 like tennis. How many like both sports?

9. In a group of 90 students, 54 take physics and 51 take chemistry. 9 take neither. How many students take both subjects?

10. The entries in the Venn diagram (Fig. 29.17) show the number of elements in each region. Determine:

(a) $n(A)$ (b) $n(B)$

(c) $n(C)$ (d) $n(A \cap B)$

(e) $n(A \cap C)$ (f) $n(B \cap C)$

(g) $n(A \cap B \cap C)$ (h) $n(A \cup B \cup C)$

Fig. 29.17

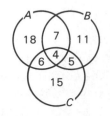

11. Fig. 29.18 shows the number of elements in the various regions of a Venn diagram which contains three intersecting sets A, B and C. If $n(A \cup B \cup C) = 127$, find:

(a) x (b) $n(A)$

(c) $n(B)$ (d) $n(C)$

(e) $n(A \cap B)$ (f) $n(A \cap C)$

(g) $n(B \cap C)$ (h) $n(A \cap B \cap C)$

Fig. 29.18

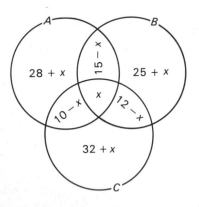

12. Out of 68 pupils in a school, 30 take English, 50 take mathematics and 24 take physics. If 10 take mathematics and physics, 14 take English and physics and 22 take English and mathematics, how many pupils take all three subjects?

13. In an examination 180 candidates offered French, 240 offered English and 150 offered German. If 60 offered French and English, 45 English and German and 75 French and German and 30 offered all three, how many candidates sat the examination?

14. Out of three sports, rugby, association football and hockey, 240 young people were asked to state their preference. 120 preferred association football, 172 preferred rugby and 128 preferred hockey. 64 liked rugby and association football, 76 liked rugby and hockey whilst 68 liked association football and hockey. How many liked all three sports?

Miscellaneous Exercise 29

1. The sets P and Q of letters from the alphabet are:

$$P = \{a, l, d, e, r, s, h, o, t\}$$

$$Q = \{e, x, a, m, s\}$$

(a) Write down the value of $n(P)$.

(b) List the elements of $P \cup Q$.

2. In a Venn diagram (Fig. 29.19), the number of elements are as shown. Given that $\mathscr{E} = A \cup B \cup C$ and $n(\mathscr{E}) = 35$ find:

(a) $n(A \cup C)$ (b) $n(A \cap C)$

(c) $n(A' \cap C')$

Fig. 29.19

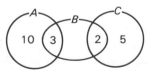

3. Fig. 29.20 shows a Venn diagram in which the number of elements are as shown. If $n(\mathscr{E}) = 25$, find the values of

(a) $n(A')$ (b) $n(A \cap B')$

(c) $n(A' \cap B')$

Fig. 29.20

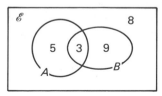

4. $\mathscr{E} = \{x : x$ is an integer and $20 \leqslant x \leqslant 80\}$.

$P = \{x : x$ is divisible by 5$\}$.

$Q = \{x : x$ is a perfect square$\}$.

$R = \{x :$ units digit of x is 7$\}$.

(a) List the elements of $P \cap Q$.

(b) Simplify $P \cap R$.

(c) List the elements of $Q \cup R$.

5. Copy Fig. 29.21 and draw and label a Venn diagram to illustrate the relationship between the sets

$\mathscr{E}$ = {natural numbers}

P = {even numbers}

Q = {multiples of 3}

Write each of the numbers 3, 4, 5 and 6 in the appropriate position on your Venn diagram and describe a set R such that $Q \cap P = R$.

Fig. 29.21

6. Fig. 29.22 represents two sets A and B which are subsets of the universal set $\mathscr{E}$. Copy and shade the region which represents $(A \cup B) \cap B'$.

Fig. 29.22

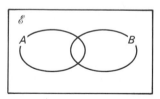

7. $\mathscr{E} = \{5, 6, 7, 8, 9, 10, 11, 12, 13\}$. List the members of the following subset of $\mathscr{E}$:

$\{x : 54 - 5x = 19\}$

8. $\mathscr{E} = \{x : x$ is a positive integer$\}$, $A = \{x : x < 10\}$ and $B = \{x : x \geqslant 3\}$.

(a) List the members of $A \cap B$.

(b) Find $n(B')$.

9. The sets A, B and C are subsets of the universal set $\{x : x$ is an integer and $10 \leqslant x \leqslant 30\}$. $A = \{x : x$ is a multiple of 5$\}$, $B = \{x : 12 \leqslant x \leqslant 18\}$ and $C = \{x : x$ is a prime number$\}$. Find:

(a) $A' \cap B$ (b) $n(C)$

10. 60 candidates in an examination offered history and geography. Every student takes at least one of these subjects and some take both. 55 take history and 45 take geography. How many take both?

11. Out of 136 students in a school, 60 take French, 100 take chemistry and 48 take physics. If 28 take French and chemistry, 44 take chemistry and physics and 20 take French and physics, how many take all three subjects?

Multi-Choice Questions 29

1. If $M_x = \{\text{multiples of } x\}$ and $F_x = \{\text{factors of } x\}$, find $M_3 \cap F_{12}$.

   A $\{3, 6, 9, 12, 15, \ldots\}$

   B $\{2, 3, 4, 6, 12\}$

   C $\{3, 6, 12\}$

   D $\varnothing$

2. Which of the following sets is equivalent to $\{\text{prime numbers}\} \cap \{\text{even numbers}\}$?

   A $\varnothing$ B $\{2\}$ C $\{3\}$ D $\{4\}$

3. $n(P) = 4$ means that there are 4 elements in the set P. Given that $n(X \cup Y) = 50$, $n(X) = 20$ and $n(Y) = 40$, $n(X \cap Y)$ equals

   A 10   B 20   C 25   D 30

4. If $X = \{x : 9 < x < 18\}$ and $Y = \{y : 10 < y < 21\}$ what is $n(X \cup Y)$, given that x and y are integers?

   A 18   B 14   C 11   D 9

5. If $X = \{\text{prime numbers less than 20}\}$ and $\{Y = \text{multiples of 3 less than 20}\}$ then $X \cap Y$ is

   A $\{3, 9, 15\}$ B $\{3\}$

   C $\{3, 9, 15, 21\}$ D $\varnothing$

6. Fig. 29.23 shows the number of men who play football or cricket. Hence 9 is the number of men who play

   A cricket only

   B football only

   C both football and cricket

   D neither football nor cricket

Fig. 29.23

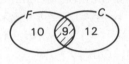

7. If the universal set is $\& = \{1, 2, 3, 4, 5, 6, 7, 8, 9\}$ and $A = \{1, 3, 5, 7, 9\}$ then A' is

   A $\{1, 3, 5, 7, 9\}$

   B $\{1, 2, 3, 4, 5, 6, 7, 8, 9\}$

   C $\{2, 4, 6, 8\}$

   D $\varnothing$

8. Which of the following pairs of sets are equal?

   (i) $\{a, b, c, d\}$ and $\{w, x, y, z\}$

   (ii) $\{1\}$ and $\varnothing$

   (iii) $\{1, 2, 3, 4\}$ and $\{2, 4, 1, 3\}$

   A (i) only   B (ii) only

   C (iii) only   D (i), (ii) and (iii)

9. If $P = \{a, b, c, d, e\}$ and $Q = \{a, q, r, s, e\}$, how many members has $P \cup Q$?

   A 10   B 9   C 8   D 7

10. The universal set is $\& = \{a, b, c, d, e\}$, P is the subset $\{a, b, d, e\}$ and Q is the subset $\{c, d, e\}$. What is the set $P \cap Q'$?

   A $\{a, b\}$ B $\{b, c, e\}$

   C $\{a, b, c, d, e\}$ D $\{d, e\}$

Matrices

Introduction

When a large amount of numerical data has to be used it is often convenient to arrange the numbers in the form of a matrix.

Suppose that a nurseryman offers collections of fruit trees in three separate collections. The table below shows the name of each collection and the number of each type of tree included in it.

	Apple	Pear	Plum	Cherry
Collection				
A	6	2	1	1
B	3	2	2	1
C	3	1	1	0

After a time the headings and titles could be removed because those concerned with the packing of the collections would know what the various numbers meant. The table could then look like this:

$$\begin{pmatrix} 6 & 2 & 1 & 1 \\ 3 & 2 & 2 & 1 \\ 3 & 1 & 1 & 0 \end{pmatrix}$$

The information has now been arranged in the form of a **matrix**, that is, in the form of an array of numbers.

A matrix is always enclosed in curved brackets. The above matrix has 3 rows and 4 columns. It is called a matrix of order 3×4. In defining the order of a matrix the number of rows is always stated first

and then the number of columns. The matrix shown below is of order 2×5 because it has 2 rows and 5 columns.

$$\begin{pmatrix} 1 & 2 & 5 & 2 & 4 \\ 3 & 0 & 3 & 1 & 2 \end{pmatrix}$$

Types of Matrix

(1) **Row matrix.** This is a matrix having only one row. Thus $(3 \quad 5)$ is a row matrix.

(2) **Column matrix.** This is a matrix having only one column. Thus $\begin{pmatrix} 1 \\ 6 \end{pmatrix}$ is a column matrix.

(3) **Null matrix.** This is a matrix with all its elements zero. Thus $\begin{pmatrix} 0 & 0 \\ 0 & 0 \end{pmatrix}$ is a null matrix.

(4) **Square matrix.** This is a matrix having the same number of rows and columns. Thus $\begin{pmatrix} 2 & 1 \\ 6 & 3 \end{pmatrix}$ is a square matrix.

(5) **Diagonal matrix.** This is a square matrix in which all the elements are zero except the diagonal elements. Thus $\begin{pmatrix} 2 & 0 \\ 0 & 3 \end{pmatrix}$ is a diagonal matrix. Note that the diagonal in a matrix always runs from upper left to lower right.

(6) **Unit matrix** or **identity matrix.** This is a diagonal matrix in which the diagonal elements equal 1. An identity matrix is usually denoted by the symbol I. Thus

$$I = \begin{pmatrix} 1 & 0 \\ 0 & 1 \end{pmatrix}$$

Addition and Subtraction of Matrices

Two matrices may be added or subtracted provided they are of the **same order**. Addition is done by adding together the corresponding elements of each of the two matrices. Thus

$$\begin{pmatrix} 3 & 5 \\ 6 & 2 \end{pmatrix} + \begin{pmatrix} 4 & 7 \\ 8 & 1 \end{pmatrix} = \begin{pmatrix} 3+4 & 5+7 \\ 6+8 & 2+1 \end{pmatrix}$$

$$= \begin{pmatrix} 7 & 12 \\ 14 & 3 \end{pmatrix}$$

Subtraction is done in a similar fashion except that the corresponding elements are subtracted. Thus

$$\begin{pmatrix} 6 & 2 \\ 1 & 8 \end{pmatrix} - \begin{pmatrix} 4 & 3 \\ 7 & 5 \end{pmatrix} = \begin{pmatrix} 6-4 & 2-3 \\ 1-7 & 8-5 \end{pmatrix}$$

$$= \begin{pmatrix} 2 & -1 \\ -6 & 3 \end{pmatrix}$$

Multiplication of Matrices

(1) **Scalar multiplication.** A matrix may be multiplied by a number as follows:

$$3\begin{pmatrix} 2 & 1 \\ 6 & 4 \end{pmatrix} = \begin{pmatrix} 3 \times 2 & 3 \times 1 \\ 3 \times 6 & 3 \times 4 \end{pmatrix}$$

$$= \begin{pmatrix} 6 & 3 \\ 18 & 12 \end{pmatrix}$$

(2) **General matrix multiplication.** Two matrices can only be multiplied together if the number of columns in the first one is equal to the number of rows in the second. The multiplication is done by multiplying a row by a column as shown below.

(a) $\begin{pmatrix} 2 & 3 \\ 4 & 5 \end{pmatrix} \times \begin{pmatrix} 5 & 2 \\ 3 & 6 \end{pmatrix}$

$$= \begin{pmatrix} 2 \times 5 + 3 \times 3 & 2 \times 2 + 3 \times 6 \\ 4 \times 5 + 5 \times 3 & 4 \times 2 + 5 \times 6 \end{pmatrix}$$

$$= \begin{pmatrix} 19 & 22 \\ 35 & 38 \end{pmatrix}$$

(b) $\begin{pmatrix} 3 & 4 \\ 2 & 5 \end{pmatrix} \times \begin{pmatrix} 6 \\ 7 \end{pmatrix}$

$$= \begin{pmatrix} 3 \times 6 + 4 \times 7 \\ 2 \times 6 + 5 \times 7 \end{pmatrix}$$

$$= \begin{pmatrix} 46 \\ 47 \end{pmatrix}$$

(c) $\begin{pmatrix} 2 & 3 & 1 \\ 5 & 4 & 6 \end{pmatrix} \begin{pmatrix} 1 & 2 \\ 3 & 7 \\ 5 & 4 \end{pmatrix}$

$$= \begin{pmatrix} 2 \times 1 + 3 \times 3 + 1 \times 5 \\ 5 \times 1 + 4 \times 3 + 6 \times 5 \end{pmatrix}$$

$$\begin{pmatrix} 2 \times 2 + 3 \times 7 + 1 \times 4 \\ 5 \times 2 + 4 \times 7 + 6 \times 4 \end{pmatrix}$$

$$= \begin{pmatrix} 14 & 29 \\ 47 & 62 \end{pmatrix}$$

Matrix Notation

It is usual to denote matrices by capital letters. Thus

$$\mathbf{A} = \begin{pmatrix} 3 & 1 \\ 7 & 4 \end{pmatrix} \quad \text{and} \quad \mathbf{B} = \begin{pmatrix} 2 \\ 3 \end{pmatrix}$$

Generally speaking matrix products are **non-commutative**, that is

$$\mathbf{A} \times \mathbf{B} \text{ does not equal } \mathbf{B} \times \mathbf{A}$$

If **A** is of order 4×3 and **B** is of order 3×2, then **AB** is of order 4×2.

Example 1

(a) Form $C = A + B$ if

$$A = \begin{pmatrix} 3 & 4 \\ 2 & 1 \end{pmatrix} \quad \text{and} \quad B = \begin{pmatrix} 2 & 3 \\ 4 & 2 \end{pmatrix}$$

$$C = \begin{pmatrix} 3 & 4 \\ 2 & 1 \end{pmatrix} + \begin{pmatrix} 2 & 3 \\ 4 & 2 \end{pmatrix}$$

$$= \begin{pmatrix} 5 & 7 \\ 6 & 3 \end{pmatrix}$$

(b) Form $Q = RS$ if

$$R = \begin{pmatrix} 1 & 2 \\ 3 & 4 \end{pmatrix} \quad \text{and} \quad S = \begin{pmatrix} 3 & 1 \\ 5 & 6 \end{pmatrix}$$

$$Q = \begin{pmatrix} 1 & 2 \\ 3 & 4 \end{pmatrix} \begin{pmatrix} 3 & 1 \\ 5 & 6 \end{pmatrix}$$

$$= \begin{pmatrix} 13 & 13 \\ 29 & 27 \end{pmatrix}$$

Note that just as in ordinary algebra the multiplication sign is omitted so we omit it in matrix algebra.

(c) Form $M = PQR$ if

$$P = \begin{pmatrix} 2 & 0 \\ 1 & 0 \end{pmatrix}, \quad Q = \begin{pmatrix} -1 & 0 \\ 0 & 1 \end{pmatrix}$$

$$\text{and } R = \begin{pmatrix} 2 & 1 \\ 3 & 0 \end{pmatrix}$$

$$PQ = \begin{pmatrix} 2 & 0 \\ 1 & 0 \end{pmatrix} \begin{pmatrix} -1 & 0 \\ 0 & 1 \end{pmatrix}$$

$$= \begin{pmatrix} -2 & 0 \\ -1 & 0 \end{pmatrix}$$

$$M = (PQ)R$$

$$= \begin{pmatrix} -2 & 0 \\ -1 & 0 \end{pmatrix} \begin{pmatrix} 2 & 1 \\ 3 & 0 \end{pmatrix}$$

$$= \begin{pmatrix} -4 & -2 \\ -2 & -1 \end{pmatrix}$$

Transposition of Matrices

When the rows of a matrix are interchanged with its column the matrix is said to be transposed. If the original matrix is **A**, the transpose is denoted by A^T. Thus

$$A = \begin{pmatrix} 3 & 4 \\ 5 & 6 \end{pmatrix}, \quad A^T = \begin{pmatrix} 3 & 5 \\ 4 & 6 \end{pmatrix}$$

Determinants

If $A = \begin{pmatrix} a & b \\ c & d \end{pmatrix}$ then its determinant is $|A| = ad - bc$ which is a numerical value.

Example 2

Find the determinant of the matrix $A = \begin{pmatrix} 5 & 2 \\ 3 & 4 \end{pmatrix}$.

$$|A| = \begin{vmatrix} 5 & 2 \\ 3 & 4 \end{vmatrix}$$

$$= 5 \times 4 - 2 \times 3$$

$$= 20 - 6$$

$$= 14$$

Inverting a Matrix

If $AB = I$ (**I** is the unit matrix) then **B** is called the **inverse** or **reciprocal** of **A**. The inverse of **A** is usually written A^{-1} and hence

$$AA^{-1} = I$$

If

$$A = \begin{pmatrix} a & b \\ c & d \end{pmatrix}$$

then

$$A^{-1} = \frac{1}{|A|} \begin{pmatrix} d & -b \\ -c & a \end{pmatrix}$$

Example 3

If $A = \begin{pmatrix} 4 & 1 \\ 2 & 3 \end{pmatrix}$ form A^{-1}.

$$|A| = \begin{vmatrix} 4 & 1 \\ 2 & 3 \end{vmatrix}$$

$$= 4 \times 3 - 1 \times 2$$

$$= 12 - 2$$

$$= 10$$

$$A^{-1} = \frac{1}{10}\begin{pmatrix} 3 & -1 \\ -2 & 4 \end{pmatrix}$$

$$= \begin{pmatrix} 0.3 & -0.1 \\ -0.2 & 0.4 \end{pmatrix}$$

To check:

$$AA^{-1} = \begin{pmatrix} 4 & 1 \\ 2 & 3 \end{pmatrix}\begin{pmatrix} 0.3 & -0.1 \\ -0.2 & 0.4 \end{pmatrix}$$

$$= \begin{pmatrix} 1 & 0 \\ 0 & 1 \end{pmatrix}$$

Singular Matrices

A singular matrix is a matrix which does not have an inverse. The determinant of a singular matrix is zero. Thus the matrix $\begin{pmatrix} 6 & 9 \\ 2 & 3 \end{pmatrix}$ is singular because its determinant is

$$\begin{vmatrix} 6 & 9 \\ 2 & 3 \end{vmatrix} = 6 \times 3 - 2 \times 9$$

$$= 18 - 18$$

$$= 0$$

Hence this matrix has no inverse.

Equality of Matrices

If two matrices are equal then their corresponding elements are equal. Thus if

$$\begin{pmatrix} a & b \\ c & d \end{pmatrix} = \begin{pmatrix} e & f \\ g & h \end{pmatrix}$$

then $a = e$, $b = f$, $c = g$ and $d = h$.

Example 4

Find the values of x and y if

$$\begin{pmatrix} 2 & 1 \\ 3 & 4 \end{pmatrix}\begin{pmatrix} x & 2 \\ 5 & y \end{pmatrix} = \begin{pmatrix} 7 & 10 \\ 23 & 30 \end{pmatrix}$$

$$\begin{pmatrix} 2x + 5 & 4 + y \\ 3x + 20 & 6 + 4y \end{pmatrix} = \begin{pmatrix} 7 & 10 \\ 23 & 30 \end{pmatrix}$$

$$\therefore \quad 2x + 5 = 7 \quad \text{and} \quad x = 1$$

$$4 + y = 10 \quad \text{and} \quad y = 6$$

(We could have used $3x + 20 = 23$ and $6 + 4y = 30$ if we had desired.)

Solution of Simultaneous Equations

Consider the two simultaneous equations

$$3x + 2y = 12 \qquad (1)$$

$$4x + 5y = 23 \qquad (2)$$

We may write these equations in matrix form as follows:

$$\begin{pmatrix} 3 & 2 \\ 4 & 5 \end{pmatrix}\begin{pmatrix} x \\ y \end{pmatrix} = \begin{pmatrix} 12 \\ 23 \end{pmatrix}$$

If we let

$$A = \begin{pmatrix} 3 & 2 \\ 4 & 5 \end{pmatrix}, \quad X = \begin{pmatrix} x \\ y \end{pmatrix}, \quad K = \begin{pmatrix} 12 \\ 23 \end{pmatrix}$$

Then $$AX = K$$

and $$X = A^{-1}K$$

$$A^{-1} = \frac{1}{3 \times 5 - 2 \times 4}\begin{pmatrix} 5 & -2 \\ -4 & 3 \end{pmatrix}$$

$$= \begin{pmatrix} \dfrac{5}{7} & -\dfrac{2}{7} \\ -\dfrac{4}{7} & \dfrac{3}{7} \end{pmatrix}$$

$$\begin{pmatrix} x \\ y \end{pmatrix} = \begin{pmatrix} \dfrac{5}{7} & -\dfrac{2}{7} \\ -\dfrac{4}{7} & \dfrac{3}{7} \end{pmatrix} \begin{pmatrix} 12 \\ 23 \end{pmatrix}$$

$$= \begin{pmatrix} 2 \\ 3 \end{pmatrix}$$

Therefore the solutions are $x = 2$ and $y = 3$.

Exercise 30.1

1. Find the values of L and M in the following matrix addition:

$$\begin{pmatrix} L & 4 \\ -3 & 1 \end{pmatrix} + \begin{pmatrix} 1 & 2 \\ -2 & 4 \end{pmatrix} + \begin{pmatrix} -1 & 2 \\ M & 4 \end{pmatrix}$$

$$= \begin{pmatrix} 3 & 8 \\ -6 & 9 \end{pmatrix}$$

2. If $A = \begin{pmatrix} 4 & 5 \\ 2 & 3 \end{pmatrix}$ find A^2.

3. If $A = \begin{pmatrix} 3 & 1 \\ 2 & 0 \end{pmatrix}$ and $B = \begin{pmatrix} 4 & -1 \\ 2 & 3 \end{pmatrix}$
   calculate the following matrices:
   (a) $A + B$ (b) $3A - 2B$
   (c) AB (d) BA

4. $P = \begin{pmatrix} 2 & 1 \\ 3 & 1 \end{pmatrix}$ $Q = \begin{pmatrix} 1 & 0 \\ 0 & 1 \end{pmatrix}$

   $R = \begin{pmatrix} 0 & 1 \\ 1 & 0 \end{pmatrix}$ $S = \begin{pmatrix} 1 & -2 \\ -6 & 3 \end{pmatrix}$

   (a) Find each of the following as a single matrix: PQ, RS, $PQRS$, $P^2 - Q^2$.

   (b) Find the values of a and b if $aP + bQ = S$.

5. A and B are two matrices.
   If $A = \begin{pmatrix} -2 & 3 \\ 4 & -1 \end{pmatrix}$ find A^2 and then use your answer to find B, given that $A^2 = A - B$.

6. Find the value of
   $$\begin{pmatrix} 2 & 3 & 1 \\ 0 & 1 & 2 \end{pmatrix} \begin{pmatrix} 1 \\ 3 \\ -2 \end{pmatrix}$$

7. If $\begin{pmatrix} 2 & 3 \\ 4 & 5 \end{pmatrix} \begin{pmatrix} p & 2 \\ 7 & q \end{pmatrix} = \begin{pmatrix} 31 & 1 \\ 55 & 3 \end{pmatrix}$
   find p and q.

8. If $A = \begin{pmatrix} 2 & -1 \\ 1 & 1 \end{pmatrix}$ and $B = \begin{pmatrix} 1 & 2 \\ 1 & 1 \end{pmatrix}$,
   write as a single matrix:
   (a) $A + B$ (b) $A \times B$
   (c) the inverse of B.

9. If matrix $A = \begin{pmatrix} 3 & 1 \\ 2 & 4 \end{pmatrix}$ and
   matrix $B = \begin{pmatrix} 4 & 2 \\ 1 & 0 \end{pmatrix}$, calculate:
   (a) $A + B$ (b) $3A - 2B$
   (c) AB

10. If $A = \begin{pmatrix} 3 & 1 \\ -2 & 0 \end{pmatrix}$ and $B = \begin{pmatrix} -1 & 3 \\ -4 & 2 \end{pmatrix}$
    calculate AB and BA.

11. Solve the equation $\begin{pmatrix} 2 & 5 \\ 1 & 3 \end{pmatrix} \begin{pmatrix} x \\ y \end{pmatrix} = \begin{pmatrix} 3 \\ 1 \end{pmatrix}$.

12. Solve the following simultaneous equations using matrices:
    (a) $x + 3y = 7$ (1)
    $2x + 5y = 12$ (2)
    (b) $4x + 3y = 24$ (1)
    $2x + 5y = 26$ (2)
    (c) $2x + 7y = 11$ (1)
    $5x + 3y = 13$ (2)

Application of Matrices

Matrices provide a way of tabulating data in a systematic way which often gives a decided advantage over other methods of solution.

Example 5

Tom and Betty buy bulbs for their window box. Tom buys 6 daffodil bulbs, no tulip bulbs and 5 crocus bulbs. Betty buys 3 daffodils, 4 tulips and 6 crocus. If daffodil bulbs cost 8p each, tulip bulbs 10p each and crocus bulbs 3p each, find the amount that each spent on bulbs.

Tabulating the information:

Daffodils	Tulips	Crocus	
6	0	5	Tom
3	4	6	Betty

In effect, the information has been written in the form of a 2×3 matrix which we will call **B**.

$$\mathbf{B} = \begin{pmatrix} 6 & 0 & 5 \\ 3 & 4 & 6 \end{pmatrix}$$

We can tabulate the cost of the bulbs in a similar way:

Cost (pence)
$$\begin{array}{c} 8 \\ 10 \\ 3 \\ \hline \end{array}$$

This information is in the form of a 3×1 matrix which we will call **C**.

$$\mathbf{C} = \begin{pmatrix} 8 \\ 10 \\ 3 \end{pmatrix}$$

To find how much each spent on bulbs we multiply **B** by **C** to give

$$\mathbf{BC} = \begin{pmatrix} 6 & 0 & 5 \\ 3 & 4 & 6 \end{pmatrix} \begin{pmatrix} 8 \\ 10 \\ 3 \end{pmatrix}$$

$$= \begin{pmatrix} 63 \\ 82 \end{pmatrix}$$

The top row of **BC** gives the amount that Tom spent; the bottom row gives the amount that Betty spent. Hence Tom spent 63p whilst Betty spent 82p.

Exercise 30.2

1. The scores of five soccer clubs in the first 12 matches of the 1985–6 season are given by the matrix:

	Won	Drawn	Lost
Aston Villa	4	4	4
Arsenal	3	2	7
Chelsea	6	3	3
Everton	5	2	5
Newcastle	2	4	6

The points awarded for a win, a draw or a lost match are given by the matrix

Won	$\begin{pmatrix} 3 \\ 1 \\ 0 \end{pmatrix}$
Drawn	
Lost	

Calculate a matrix which shows the total number of points scored by each of these clubs, showing clearly the method used.

2. (a) Eastern Airlines have 6 Tridents, 4 VC10s and 2 Jumbo Jets. Western Airlines have 3 Tridents, 6 VC10s and 1 Jumbo Jet. Write this information as a 2 by 3 matrix.

(b) Tridents carry 150 passengers, VC10s carry 120 passengers and Jumbo Jets carry 375 passengers. Write this information as a 3 by 1 matrix.

(c) Multiply these matrices to find a 2 by 1 matrix.

(d) What does the 2 by 1 matrix represent?

3. White, Green and Brown played in a cricket match. White hit 2 sixes, 3 fours and 8 singles, Green scored 1 six, 2 fours and 5 singles whilst Brown scored no sixes, 2 fours and 9 singles. Set out this information in the form of two matrices and use them to find the total number of runs scored by each batsman.

4. Susan and Margaret went shopping and their purchases were as follows:

	Meat cubes	Chicken cubes	Packets of soup
Susan	6	4	3
Margaret	8	5	2

(a) Write this information in the form of a 2×3 matrix and call it **P**.

(b) Meat cubes cost 4p each, chicken cubes cost 5p each and soup costs 12p per packet. Write this information as a 3×1 matrix and call it **C**.

(c) Work out **PC**.

(d) Pre-multiply **PC** by the row matrix (1 1) and explain what the result means.

5. In a football league, City scored 1 goal in each of 5 matches, 2 goals in each of 7 matches and 3 goals in each of 2 matches. United scored 1 goal in each of 6 matches, 2 goals in each of 5 matches and 3 goals in each of 3 matches. Set out this information in the form of two matrices and use them to work out the total number of goals scored by each team.

Miscellaneous Exercise 30

1. Write down the inverse of the matrix $\begin{pmatrix} 2 & -1 \\ 2 & 4 \end{pmatrix}$.

2. If $\mathbf{p} = \begin{pmatrix} 4 \\ 2 \end{pmatrix}$ and $\mathbf{q} = \begin{pmatrix} 2 \\ 3 \end{pmatrix}$, calculate $\mathbf{p} - 2\mathbf{q}$.

3. If $\mathbf{M} = \begin{pmatrix} 2 & 4 \\ 1 & 0 \end{pmatrix}$ and $\mathbf{N} = \begin{pmatrix} 2 & 1 \\ 0 & 3 \end{pmatrix}$ find $\mathbf{MN} - \mathbf{NM}$.

4. If $\begin{pmatrix} 2 & x \\ 4 & y \end{pmatrix} \begin{pmatrix} 5 \\ 2 \end{pmatrix} = \begin{pmatrix} 14 \\ 30 \end{pmatrix}$, find x and y.

5. Matrices **A** and **B** are given as follows:
$$\mathbf{A} = \begin{pmatrix} 1 & -1 \\ 2 & 3 \end{pmatrix} \quad \mathbf{B} = \begin{pmatrix} 3 & 1 \\ -2 & 1 \end{pmatrix}$$
(a) Find the products **BA** and **AB**.

(b) Find the matrix **X** such that **AX** = **B**.

6. Show that the matrix $\begin{pmatrix} 8 & 4 \\ 2 & 1 \end{pmatrix}$ has no inverse.

7. Given that $a > 0$ and $b > 0$ and that
$$\begin{pmatrix} a & a \\ b & b \end{pmatrix} \begin{pmatrix} a & b \\ a & b \end{pmatrix} = \begin{pmatrix} 2 & c \\ c & 8 \end{pmatrix}$$
find the values of a, b and c.

8. If $\mathbf{A} = \begin{pmatrix} 4 \\ 6 \end{pmatrix}$ and $\mathbf{B} = \begin{pmatrix} -3 \\ 3 \end{pmatrix}$ find

(a) $\frac{1}{2}\mathbf{A}$

(b) **X** if $\mathbf{A} + \mathbf{X} = 4\mathbf{B}$.

9. Find x if $(2x \quad 3) \begin{pmatrix} 11 \\ -6x \end{pmatrix} = (100)$.

10. Evaluate as a single matrix
$$2\begin{pmatrix} 3 & -1 \\ 5 & 2 \end{pmatrix} + 3\begin{pmatrix} 4 & -2 \\ 3 & -1 \end{pmatrix}$$

11. A gardener buys 2 apple trees, 3 plum trees and 1 pear tree. Apple trees cost £5 each, plum trees £4 each and pear trees £6 each. Set out this information in the form of two matrices and use them to find the total cost of purchasing the fruit trees.

Multi-Choice Questions 30

1. If $W = \begin{pmatrix} -2 \\ 3 \end{pmatrix}$ and $S = \begin{pmatrix} -1 \\ -2 \end{pmatrix}$ then $W + \frac{1}{2}S$ equals

   A $\begin{pmatrix} -2\frac{1}{2} \\ 2 \end{pmatrix}$ 　　　　 B $\begin{pmatrix} -1\frac{1}{2} \\ \frac{1}{2} \end{pmatrix}$

   C $\begin{pmatrix} -3 \\ 2 \end{pmatrix}$ 　　　　 D $\begin{pmatrix} 2\frac{1}{2} \\ 2 \end{pmatrix}$

2. If $A = \begin{pmatrix} 6 \\ -5 \end{pmatrix}$ and $B = \begin{pmatrix} -3 \\ 2 \end{pmatrix}$ what is the value of $3B - 2A$?

   A $\begin{pmatrix} -21 \\ 16 \end{pmatrix}$ 　　　　 B $\begin{pmatrix} 3 \\ 4 \end{pmatrix}$

   C $\begin{pmatrix} 24 \\ -19 \end{pmatrix}$ 　　　　 D $\begin{pmatrix} 21 \\ -16 \end{pmatrix}$

3. The inverse of the matrix $\begin{pmatrix} 6 & 10 \\ 2 & 4 \end{pmatrix}$ is

   A $\begin{pmatrix} 1 & -2\frac{1}{2} \\ -\frac{1}{2} & 1\frac{1}{2} \end{pmatrix}$ 　　 B $\begin{pmatrix} 4 & -10 \\ -2 & 6 \end{pmatrix}$

   C $\begin{pmatrix} -2 & 6 \\ -10 & 4 \end{pmatrix}$ 　　 D $\begin{pmatrix} 1 & 2\frac{1}{2} \\ \frac{1}{2} & 1\frac{1}{2} \end{pmatrix}$

4. If $\begin{pmatrix} 3 & -2 \\ 4 & 1 \end{pmatrix} \begin{pmatrix} 6 & x \\ 2 & 2 \end{pmatrix} = \begin{pmatrix} 14 & -13 \\ 26 & -10 \end{pmatrix}$ what is the value of x?

   A -3 　　 B -1 　　 C 2 　　　 D 3

5. If $M = \begin{pmatrix} 2 & 3 \\ 4 & 5 \end{pmatrix}$ and $N = \begin{pmatrix} 3 & 1 \\ 2 & 2 \end{pmatrix}$ then MN is

   A $\begin{pmatrix} 5 & 4 \\ 6 & 7 \end{pmatrix}$ 　　　　 B $\begin{pmatrix} 10 & 14 \\ 12 & 16 \end{pmatrix}$

   C $\begin{pmatrix} 6 & 3 \\ 8 & 10 \end{pmatrix}$ 　　　　 D $\begin{pmatrix} 12 & 8 \\ 22 & 14 \end{pmatrix}$

6. Express as a single matrix
   $$\begin{pmatrix} 1 & 0 \\ 3 & 2 \end{pmatrix} \begin{pmatrix} 4 & 1 \\ 1 & 0 \end{pmatrix}$$

   A $\begin{pmatrix} 5 & 1 \\ 4 & 2 \end{pmatrix}$ 　　　　 B $\begin{pmatrix} 4 & 1 \\ 14 & 3 \end{pmatrix}$

   C $\begin{pmatrix} 7 & 2 \\ 1 & 0 \end{pmatrix}$ 　　　　 D $\begin{pmatrix} 7 & 2 \\ 1 & 0 \end{pmatrix}$

7. If $A = \begin{pmatrix} a & b \\ c & d \end{pmatrix}$ then A^{-1} equals

   A $\begin{pmatrix} \dfrac{1}{a} & \dfrac{1}{b} \\ \dfrac{1}{c} & \dfrac{1}{d} \end{pmatrix}$ 　　 B $\begin{pmatrix} c & d \\ a & b \end{pmatrix}$

   C $\dfrac{1}{|A|}\begin{pmatrix} d & -b \\ -c & a \end{pmatrix}$ 　 D $\dfrac{1}{|A|}\begin{pmatrix} b & -d \\ -a & c \end{pmatrix}$

Transformation Geometry

Introduction

A transformation describes the relation between any point and its image point. Thus in Fig. 31.1, the point $P(4, 3)$ has been transformed into the point $P'(-4, 3)$. The point P is called the **object** or the **pre-image** whilst P' is called the **image**.

Fig. 31.1

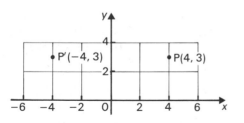

Translation

If every point in a line or a plane figure moves the same distance in the same direction, the transformation is called a **translation**. Thus in Fig. 31.2 every point in the line AB has been moved 2 units to the right and 3 units upwards. Thus

$$A(2, 1) \text{ becomes } A'(4, 4)$$
$$B(6, 3) \text{ becomes } B'(8, 6)$$

The lines AB and A'B' are parallel and equal in length. Hence the translation can be described by the **displacement vector** $\begin{pmatrix} 2 \\ 3 \end{pmatrix}$ which is, in effect, a column matrix.

Generally speaking, for this translation, a point $P(x, y)$ maps on to the point $P'(x + 2, y + 3)$. This is equivalent to

$$\begin{pmatrix} x \\ y \end{pmatrix} + \begin{pmatrix} 2 \\ 3 \end{pmatrix} = \begin{pmatrix} x + 2 \\ y + 3 \end{pmatrix}$$

Fig. 31.2

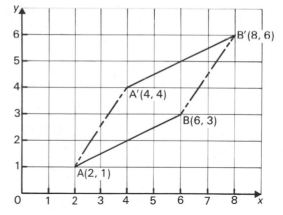

Any translation can be represented by the displacement vector $\begin{pmatrix} a \\ b \end{pmatrix}$ and its inverse by $\begin{pmatrix} -a \\ -b \end{pmatrix}$.

Size and shape do not change under a translation and the object and the image are **congruent** since the shape of the object can be fitted exactly over the shape of the image.

An **isometry** is a transformation in which the object and the image are congruent. Hence translation is an isometry.

Example 1

ABCD is a square with vertices A(1,1), B(4,1), C(4,4) and D(1,4). It is translated by the displacement vector $\begin{pmatrix} 5 \\ 2 \end{pmatrix}$. Find the coordinates of A', B', C' and D' the vertices of the image of ABCD.

The **position vector** of

$$A' = \begin{pmatrix} 1 \\ 1 \end{pmatrix} + \begin{pmatrix} 5 \\ 2 \end{pmatrix} = \begin{pmatrix} 6 \\ 3 \end{pmatrix}$$

The position vector of

$$B' = \begin{pmatrix} 4 \\ 1 \end{pmatrix} + \begin{pmatrix} 5 \\ 2 \end{pmatrix} = \begin{pmatrix} 9 \\ 3 \end{pmatrix}$$

The position vector of

$$C' = \begin{pmatrix} 4 \\ 4 \end{pmatrix} + \begin{pmatrix} 5 \\ 2 \end{pmatrix} = \begin{pmatrix} 9 \\ 6 \end{pmatrix}$$

The position vector of

$$D' = \begin{pmatrix} 1 \\ 4 \end{pmatrix} + \begin{pmatrix} 5 \\ 2 \end{pmatrix} = \begin{pmatrix} 6 \\ 6 \end{pmatrix}$$

The coordinates of the vertices of A'B'C'D' are A'(6,3), B'(9,3), C'(9,6) and D'(6,6). The object ABCD and its image A'B'C'D' are shown in Fig. 31.3.

Fig. 31.3

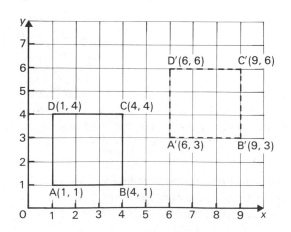

Example 2

Triangle A'B'C' is the image of triangle ABC under translation by the vector $\begin{pmatrix} -3 \\ 2 \end{pmatrix}$. A', B' and C' have the coordinates (5,3), (3,1) and (4,2) respectively. Find the coordinates of A, B and C before the translation.

The translation is reversed by the vector $\begin{pmatrix} 3 \\ -2 \end{pmatrix}$.

The position vector of A is

$$\begin{pmatrix} 5 \\ 3 \end{pmatrix} + \begin{pmatrix} 3 \\ -2 \end{pmatrix} = \begin{pmatrix} 8 \\ 1 \end{pmatrix}$$

The position vector of B is

$$\begin{pmatrix} 3 \\ 1 \end{pmatrix} + \begin{pmatrix} 3 \\ -2 \end{pmatrix} = \begin{pmatrix} 6 \\ -1 \end{pmatrix}$$

The position vector of C is

$$\begin{pmatrix} 4 \\ 2 \end{pmatrix} + \begin{pmatrix} 3 \\ -2 \end{pmatrix} = \begin{pmatrix} 7 \\ 0 \end{pmatrix}$$

Hence the coordinates of A, B and C are (8,1), (6,−1) and (7,0) respectively.

Exercise 31.1

1. Write down the coordinates of the following points under the translation $\begin{pmatrix} -2 \\ 4 \end{pmatrix}$.

   (a) (1,3)   (b) (2,1)
   (c) (−4,−3)   (d) (−2,5)
   (e) (7,0)

2. The vertices of the triangle ABC have the following coordinates: A(2,1), B(5,1) and C(5,4). Find the coordinates of A, B and C under the translation $\begin{pmatrix} -3 \\ -7 \end{pmatrix}$.

3. The rectangle ABCD is formed by joining four points A(1,5), B(5,5), C(5,3) and D(1,3). Find the image of ABCD under the translation $\begin{pmatrix} 3 \\ 4 \end{pmatrix}$.

4. The image of a rectangle ABCD is A'(8,12), B'(16,12), C'(16,16) and D'(8,16). The vector $\begin{pmatrix} -5 \\ 3 \end{pmatrix}$ effects the translation. Find the coordinates of A, B, C and D.

5. The triangle XYZ (Fig. 31.4) is translated by the vector $\begin{pmatrix} 4 \\ 2 \end{pmatrix}$. Copy the diagram and show on it triangle X'Y'Z' under this translation.

Fig. 31.4

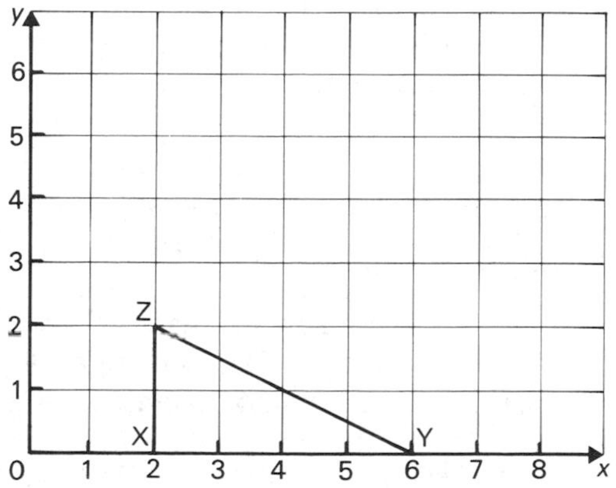

6. Triangle A'B'C' is the image of triangle ABC after translation by the vector $\begin{pmatrix} -2 \\ 3 \end{pmatrix}$. Copy the diagram (Fig. 31.5) and show the position of triangle ABC before translation.

Fig. 31.5

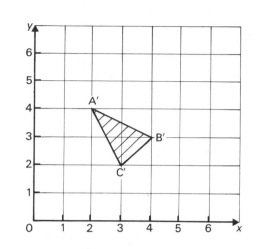

7. The triangle ABC is formed by joining the points A(−2,−2), B(−1,−4) and C(−1,−1). Find the image of ABC under the translation $\begin{pmatrix} 2 \\ -1 \end{pmatrix}$.

8. Write as column matrices, the inverses of the following translations:

(a) $\begin{pmatrix} 5 \\ -3 \end{pmatrix}$ (b) $\begin{pmatrix} 7 \\ 0 \end{pmatrix}$

(c) $\begin{pmatrix} -3 \\ 5 \end{pmatrix}$ (d) $\begin{pmatrix} -6 \\ -2 \end{pmatrix}$

(e) $\begin{pmatrix} -1 \\ 3 \end{pmatrix}$

Reflection

If a point P is reflected in a mirror so that its image is P', the mirror line or line of reflection is the perpendicular bisector of PP' (see Chapter 33) as shown in Fig. 31.6.

Fig. 31.6

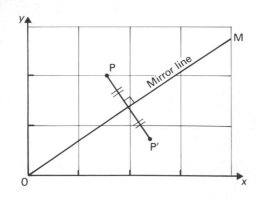

In general the point $P(x, y)$ maps to $P'(x, -y)$ when reflected in the x-axis. Writing the points as column matrices gives

$$\begin{pmatrix} x \\ y \end{pmatrix} \text{ maps to } \begin{pmatrix} x \\ -y \end{pmatrix}$$

but

$$\begin{pmatrix} 1 & 0 \\ 0 & -1 \end{pmatrix} \begin{pmatrix} x \\ y \end{pmatrix} = \begin{pmatrix} x \\ -y \end{pmatrix}$$

Hence if the position vector of $P(x, y)$ is pre-multiplied by the matrix $\begin{pmatrix} 1 & 0 \\ 0 & -1 \end{pmatrix}$ the image of P in the x-axis obtained.

Reflection in the *x*-axis

Consider the point $A(2, 4)$ in Fig. 31.7. Its distance from the x-axis is given by its y coordinate, i.e. $y = 4$. Since A is 4 units above the x-axis its image A' will be 4 units below the x-axis, i.e. $y = -4$. Hence the coordinates of A' are $(2, -4)$.

Fig. 31.7

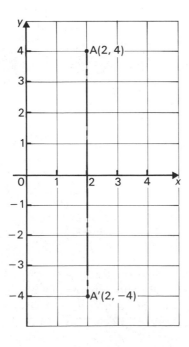

Example 3

The triangle ABC is formed by joining the three points $A(3, 2)$, $B(5, 2)$ and $C(3, 6)$. Reflect this triangle in the x-axis and state the coordinates of the transformed points A', B' and C'.

When dealing with reflections it is convenient to write down the coordinates of the points A, B and C in the form of a coordinate matrix:

$$\begin{array}{ccc} A & B & C \end{array}$$
$$\begin{pmatrix} 3 & 5 & 3 \\ 2 & 2 & 6 \end{pmatrix}$$

To find the coordinate matrix for the reflected triangle $A'B'C'$ we pre-multiply the coordinate matrix given above by the matrix $\begin{pmatrix} 1 & 0 \\ 0 & -1 \end{pmatrix}$.

$$\begin{array}{ccc} A & B & C \end{array} \qquad \begin{array}{ccc} A' & B' & C' \end{array}$$
$$\begin{pmatrix} 1 & 0 \\ 0 & -1 \end{pmatrix} \begin{pmatrix} 3 & 5 & 3 \\ 2 & 2 & 6 \end{pmatrix} = \begin{pmatrix} 3 & 5 & 3 \\ -2 & -2 & -6 \end{pmatrix}$$

The coordinates of the vertices of the reflected triangle are $A'(3, -2)$, $B'(5, -2)$ and $C'(3, -6)$.

Reflection in the *y*-axis

Reflections in other mirror lines follow the same principles as reflection in the *x*-axis. When a point P (Fig. 31.8) is reflected in the *y*-axis its image P′ lies as far behind the *y*-axis as P lies in front of it. In general,

$$P(x, y) \text{ maps to } P'(-x, y)$$

This is equivalent to the operation:

$$\begin{pmatrix} -1 & 0 \\ 0 & 1 \end{pmatrix}\begin{pmatrix} x \\ y \end{pmatrix}$$

Fig. 31.8

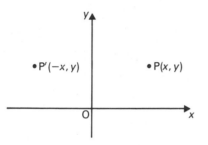

Example 4

A rectangle ABCD has the coordinate

matrix $\begin{pmatrix} A & B & C & D \\ 2 & 6 & 6 & 2 \\ 1 & 1 & 3 & 3 \end{pmatrix}$.

Find the coordinate matrix for A′B′C′D′ the image of ABCD reflected in the *y*-axis.

$$\begin{pmatrix} -1 & 0 \\ 0 & 1 \end{pmatrix}\begin{pmatrix} 2 & 6 & 6 & 2 \\ 1 & 1 & 3 & 3 \end{pmatrix}$$

$$= \begin{pmatrix} A' & B' & C' & D' \\ -2 & -6 & -6 & -2 \\ 1 & 1 & 3 & 3 \end{pmatrix}$$

The object ABCD and its image A′B′C′D′ are shown in Fig. 31.9.

Fig. 31.9

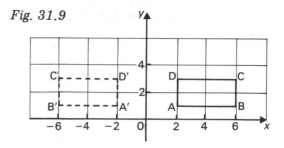

Reflection in the line *y = x*

If the scales on the *x* and *y* axes are the same the line $y = x$ will be inclined upwards at 45° to the *x*-axis. If the position vector of the point $P(x, y)$ is pre-multiplied by $\begin{pmatrix} 0 & 1 \\ 1 & 0 \end{pmatrix}$ the position vector of P′, the image of P in the line $y = x$ is obtained. The effect of this transformation is to reverse the coordinates of P.

Example 5

The point $P(3, 4)$ is reflected in the line $y = x$. Find the position vector of P′, the image of P.

$$\begin{pmatrix} 0 & 1 \\ 1 & 0 \end{pmatrix}\begin{pmatrix} 3 \\ 4 \end{pmatrix} = \begin{pmatrix} 4 \\ 3 \end{pmatrix}$$

Hence the position vector of P′ is $\begin{pmatrix} 4 \\ 3 \end{pmatrix}$ as shown in Fig. 31.10.

Fig. 31.10

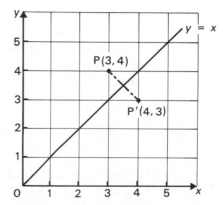

Reflection in the line $y = -x$

In Fig. 31.11, the point P has been reflected in the line $y = -x$ which is inclined downwards at $45°$ to the x-axis. The line $y = -x$ is the perpendicular bisector of PP′.

Pre-multiplying $\begin{pmatrix} x \\ y \end{pmatrix}$ by $\begin{pmatrix} 0 & -1 \\ -1 & 0 \end{pmatrix}$ transforms the point P into its reflection in the line $y = -x$.

Fig. 31.11

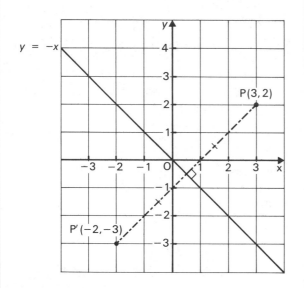

Properties of Reflections

(1)   A point and its image are the same distance from the mirror line.

(2)   The mirror line bisects the angle between a line segment and its image (Fig. 31.12).

(3)   The object and its image are the same size and shape.

(4)   Reflection is its own inverse.

Fig. 31.12

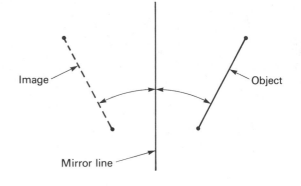

Example 6

The point P has coordinates $(3, 2)$. Find its reflection in the line $y = -x$.

To find the coordinates of the image of P its position vector is pre-multiplied by $\begin{pmatrix} 0 & -1 \\ -1 & 0 \end{pmatrix}$. Thus

$$\begin{pmatrix} 0 & -1 \\ -1 & 0 \end{pmatrix} \begin{pmatrix} 3 \\ 2 \end{pmatrix} = \begin{pmatrix} -2 \\ -3 \end{pmatrix}$$

The coordinates of the image of P are P′$(-2, -3)$.

Example 7

Triangle A′B′C′ is the image of triangle ABC after reflection in the x-axis. A′, B′ and C′ have the coordinates $(5, 3), (3, 1)$ and $(4, 2)$ respectively. Find the coordinates of A, B and C before reflection.

Since reflection is its own inverse the coordinate matrix for ABC is found by pre-multiplying the coordinate matrix for A′B′C′ by $\begin{pmatrix} 1 & 0 \\ 0 & -1 \end{pmatrix}$.

$$\begin{pmatrix} 1 & 0 \\ 0 & -1 \end{pmatrix} \overset{\displaystyle A' \quad B' \quad C'}{\begin{pmatrix} 5 & 3 & 4 \\ 3 & 1 & 2 \end{pmatrix}} = \overset{\displaystyle A \quad B \quad C}{\begin{pmatrix} 5 & 3 & 4 \\ -3 & -1 & -2 \end{pmatrix}}$$

The coordinates of A, B and C are $(5, -3), (3, -1)$ and $(4, -2)$ respectively.

Exercise 31.2

1. The vertices of the square ABCD are $(0,1)$, $(2,1)$, $(2,3)$ and $(0,3)$ respectively. Obtain the coordinate matrix for the image of ABCD after each of the following transformations:

   (a) Reflection in the x-axis.

   (b) Reflection in the y-axis.

   (c) Reflection in the line $y = -x$.

   (d) Reflection in the line $y = x$.

2. The trapezium ABCD with vertices at $A(2,1)$, $B(7,1)$, $C(6,3)$ and $D(3,3)$. Obtain the coordinate matrix after the following transformations:

   (a) Reflection in the x-axis.

   (b) Reflection in the y-axis.

   (c) Reflection in the line $y = x$.

3. WXYZ is a square such that W is the point $(2,3)$, X is the point $(4,3)$ and Y is the point $(4,5)$.

   (a) Write down the coordinates of the point Z.

   (b) Draw the square on graph paper using a scale of $2 \text{ cm} = 1$ unit on both axes.

   (c) Draw on the same axes the line $y = x$.

   (d) Reflect WXYZ in the line $y = x$.

   (e) State the coordinates of the transformed points W, X, Y and Z.

4. Triangle A'B'C' is the image of triangle ABC after reflection in the x-axis. A', B' and C' have the coordinates $(2,-3)$, $(4,-4)$ and $(3,-5)$ respectively. Give the coordinates of A, B and C before the reflection.

5. Draw on graph paper the points $A(1,1)$ and $B(4,4)$ and join them with a straight line. Reflect the line in the y-axis and find the equation of

   (a) AB

   (b) the image of AB

Symmetry

Look at the kite shown in Fig. 31.13. We see that it is symmetrical about the line AC which means that triangle ABC is a reflection of triangle ADC, AC being the mirror line. It follows that:

(1) AC is at right angles to BD and bisects it, because B is a reflection of the point D.

(2) AC bisects the angles BAD and BCD.

(3) Triangles ABC and ADC are the same size and shape.

If a plane figure contains its own reflection in a line m, then the figure is said to be symmetrical about m. The line m is the mirror line or the axis of symmetry.

Fig. 31.13

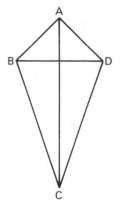

Some plane figures have more than one axis of symmetry. The rectangle shown in Fig. 31.14 has two axes of symmetry, and a square (Fig. 31.15) has four axes of symmetry. Symmetry is dealt with in more detail in Chapter 32.

Fig. 31.14

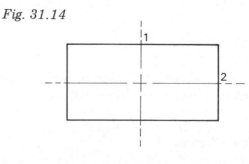

Fig. 31.15

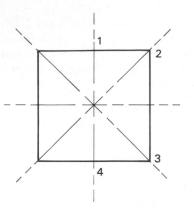

Rotation

A point which has been moved through an arc of a circle is said to have been rotated. A rotation needs the following three quantities to describe it:

(1)   The centre of rotation (O in Fig. 31.16).

(2)   The angle of rotation (θ in Fig. 31.16).

(3)   The sense of the rotation, i.e. either clockwise or anticlockwise. Anticlockwise rotation may be considered to be positive and clockwise rotation negative.

Fig. 31.16

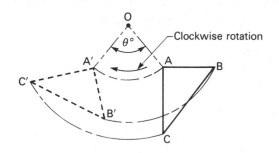

With rotations only the centre of the rotation remains invariant (i.e. it does not move). Since every point (except O) moves through the same angle the object and its image are congruent figures.

Finding the Centre of Rotation

In Fig. 31.17 the line AB has been rotated to A′B′. To find the centre of rotation:

(1)   Join AA′ and BB′.

(2)   Construct the perpendicular bisectors of AA′ and BB′.

(3)   The point of intersection of the two perpendicular bisectors (O in the diagram) is the centre of the rotation.

Fig. 31.17

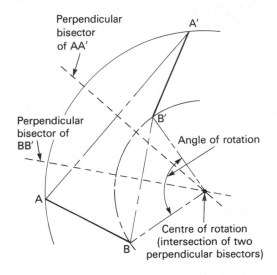

Example 8

The line AB with end points A(2, 2) and B(4, 2) is transformed into A′B′ by a rotation such that A′ is the point (8, 4) and B′ is the point (8, 2). Find, by drawing the co-ordinates of the centre of rotation, the angle of rotation and its direction.

The construction is shown in Fig. 31.18 from which the coordinates of the centre of rotation are (6, 0) and the angle of rotation is $90°$ clockwise.

Fig. 31.18

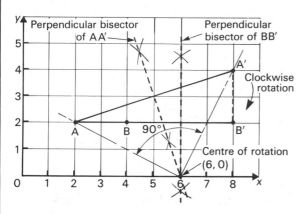

Example 9

Triangle ABC has vertices A(3, 4), B(7, 4) and C(7, 6). Plot these points on graph paper and join them to form the triangle ABC. Show the image of ABC after a clockwise rotation of 90° about the origin and mark it A'B'C'. Write down the coordinates of A', B' and C'.

Triangle ABC is drawn in Fig. 31.19. The image, triangle A'B'C' has vertices A'(4, −3), B'(4, −7) and C'(6, −7).

Fig. 31.19

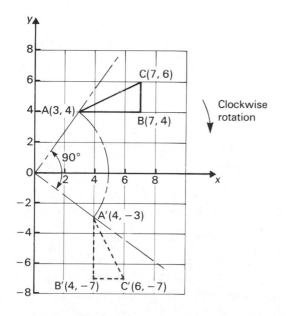

If the centre of rotation is located at the origin then the rotation can be described by a 2 × 2 matrix.

The matrix $M_1 = \begin{pmatrix} 0 & -1 \\ 1 & 0 \end{pmatrix}$ gives the point P(x, y) an anticlockwise rotation of 90° about the origin.

The matrix $M_2 = \begin{pmatrix} -1 & 0 \\ 0 & -1 \end{pmatrix}$ gives the point P(x, y) an anticlockwise rotation of 180° about the origin.

The matrix $M_3 = \begin{pmatrix} 0 & 1 \\ -1 & 0 \end{pmatrix}$ gives the point P(x, y) an anticlockwise rotation of 270° about the origin.

A rotation of 360° (a full revolution) takes the point back to its original position. The coordinates of the point P and its image P' are the same and hence a rotation of 360° may be described by the identity matrix $\begin{pmatrix} 1 & 0 \\ 0 & 1 \end{pmatrix}$.

For a clockwise rotation of 90° use the matrix M_3 because this rotation is equivalent to an anticlockwise rotation of 270°.

For a clockwise rotation of 270° use the matrix M_1 since this rotation is equivalent to an anticlockwise rotation of 90°.

Example 10

(a) The rectangle A(1, 1), B(4, 1), C(4, 3), D(1, 3) is given a clockwise rotation of 90° about the origin. Find the coordinates of A', B', C' and D', the vertices of the image of ABCD under this transformation.

$$\begin{matrix} & & A & B & C & D \\ \begin{pmatrix} 0 & 1 \\ -1 & 0 \end{pmatrix} & \begin{pmatrix} 1 & 4 & 4 & 1 \\ 1 & 1 & 3 & 3 \end{pmatrix} \end{matrix}$$

$$\begin{matrix} & A' & B' & C' & D' \\ = & \begin{pmatrix} 1 & 1 & 3 & 3 \\ -1 & -4 & -4 & -1 \end{pmatrix} \end{matrix}$$

Hence the required coordinates are A'$(1, -1)$, B'$(1, -4)$, C'$(3, -4)$ and D'$(3, -1)$. The object ABCD and its image A'B'C'D' are shown in Fig. 31.20.

Fig. 31.20

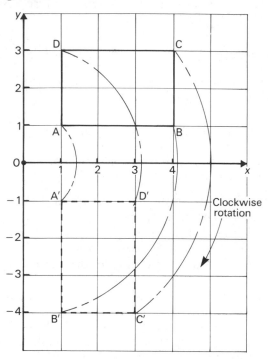

(b) The trapezium A$(2, 2)$, B$(5, 2)$, C$(5, 4)$, D$(3, 4)$ is given a rotation of $90°$ anticlockwise about the origin. Find the co-ordinate matrix for A'B'C'D'. the image of ABCD under this transformation.

$$
\begin{array}{cccc} & A & B & C & D \end{array}
$$
$$
\begin{pmatrix} 0 & -1 \\ 1 & 0 \end{pmatrix} \begin{pmatrix} 2 & 5 & 5 & 3 \\ 2 & 2 & 4 & 4 \end{pmatrix}
$$

$$
\begin{array}{cccc} A' & B' & C' & D' \end{array}
$$
$$
= \begin{pmatrix} -2 & -2 & -4 & -4 \\ 2 & 5 & 5 & 3 \end{pmatrix}
$$

The object ABCD and its image are shown in Fig. 31.21.

Fig. 31.21

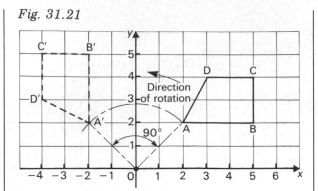

The Inverse of a Rotation

Since the transformation of P to P' is described by the matrix A the transformation of P' back to P is described by the matrix A^{-1}, i.e. the inverse of the matrix A.

Example 11

Quadrilateral A'B'C'D' is the image of ABCD after rotation through $90°$ anticlockwise about the origin. A', B', C' and D' have the coordinates $(-2, 2)$, $(-1, 4)$, $(-4, 5)$ and $(-4, 3)$ respectively. Find the coordinates of A, B, C and D before the rotation.

For a rotation of $90°$ anticlockwise the matrix $M_1 = \begin{pmatrix} 0 & -1 \\ 1 & 0 \end{pmatrix}$ transforms the point P(x, y) to its image P'. Therefore the matrix M_1^{-1} will transform P' back to P.

$$
M_1^{-1} = \begin{pmatrix} 0 & 1 \\ -1 & 0 \end{pmatrix}
$$

$$
\begin{array}{cccc} A' & B' & C' & D' \end{array}
$$
$$
\begin{pmatrix} 0 & 1 \\ -1 & 0 \end{pmatrix} \begin{pmatrix} -2 & -1 & -4 & -4 \\ 2 & 4 & 5 & 3 \end{pmatrix}
$$

$$
\begin{array}{cccc} A & B & C & D \end{array}
$$
$$
= \begin{pmatrix} 2 & 4 & 5 & 3 \\ 2 & 1 & 4 & 4 \end{pmatrix}
$$

The coordinates of A', B', C' and D' are $(2, 2)$, $(4, 1)$, $(5, 4)$ and $(3, 4)$ respectively.

Exercise 31.3

1. Map the points shown in the following coordinate matrix.

$$\begin{array}{cccc} A & B & C & D \\ \begin{pmatrix} 1 & 3 & 6 & 4 \\ 2 & 3 & 5 & 0 \end{pmatrix} \end{array}$$

    Join them to form the plane figure ABCD. Rotate ABCD through 90° anticlockwise about the origin to give the image of ABCD. Mark the image A′B′C′D′ and write down the coordinate matrix for this image.

2. Triangle ABC has vertices A(2, 3), B(4, 4) and C(3, 5). Plot these points on graph paper and join them to form ABC. Show on the same axes the image of ABC after it has been given a rotation of 90° anticlockwise about the origin. Write down the coordinates of the images of A, B and C.

3. The line AB with end points (1, 1) and (5, 1) is transformed to A′B′ by a rotation such that A′ is the point (6, 6) and B′ is the point (6, 2). Find the coordinates of the centre of rotation and write down the angle of rotation.

4. The point P(3, 3) is transformed to P′(−1, 5) by a rotation of 90° anti-clockwise. Find the coordinates of the centre of the rotation.

5. A′B′ is the image of the line AB after a rotation of 90° anticlockwise about the origin. A′ and B′ have the coordinates (−1, 2) and (−1, 4) respectively. Find the coordinates of the points A and B before the transformation.

6. Triangle XYZ has vertices X(3, 3), Y(6, 3) and Z(6, 5). Find the co-ordinates of the image of XYZ after it has been given a rotation of 90° clockwise about the origin.

7. The square A(2, 1), B(4, 1), C(4, 3), D(2, 3) is given a rotation of 270° clockwise about the origin so that its image is A′B′C′D′. Find the coordinates of the points A′, B′, C′ and D′.

Other Transformations

If a position vector is pre-multiplied by a 2 × 2 matrix it is transformed into another column matrix.

Example 12

A parallelogram has the coordinate matrix

$$\begin{array}{cccc} A & B & C & D \\ \begin{pmatrix} 1 & 3 & 4 & 2 \\ 1 & 2 & 5 & 4 \end{pmatrix} \end{array}$$

It is transformed into A′B′C′D′ by the matrix $\begin{pmatrix} 2 & 0 \\ 1 & 1 \end{pmatrix}$. Find the coordinate matrix for A′B′C′D′.

Fig. 31.22

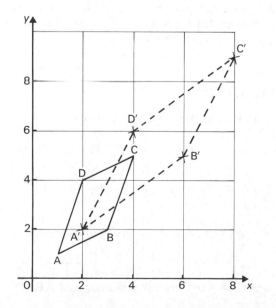

To obtain the position vectors of A',
B', C' and D' we multiply the coordinate
matrix for ABCD by $\begin{pmatrix} 2 & 0 \\ 1 & 1 \end{pmatrix}$.

$$\begin{matrix} & A & B & C & D \\ \begin{pmatrix} 2 & 0 \\ 1 & 1 \end{pmatrix} & \begin{pmatrix} 1 & 3 & 4 & 2 \\ 1 & 2 & 5 & 4 \end{pmatrix} \end{matrix}$$

$$= \begin{matrix} A' & B' & C' & D' \\ \begin{pmatrix} 2 & 6 & 8 & 4 \\ 2 & 5 & 9 & 6 \end{pmatrix} \end{matrix}$$

The transformed parallelogram A'B'C'D' and
the original parallelogram ABCD are shown
in Fig. 31.22. Note that A'B'C'D' is larger
than ABCD and hence the transformation
is not isometric.

Double Transformations

A double transformation is one transformation (e.g. a reflection) followed by a second
transformation (e.g. a translation). A double
transformation may be made into a single
operation by multiplying the two matrices,
which describe the transformations, together.
We pre-multiply the matrix describing the
first transformation by the matrix describing the second transformation.

Example 13

A rectangle ABCD has the coordinate matrix

$$\begin{matrix} A & B & C & D \\ \begin{pmatrix} 2 & 6 & 6 & 2 \\ 1 & 1 & 3 & 3 \end{pmatrix} \end{matrix}$$

It is given a double transformation under
$\begin{pmatrix} 1 & 3 \\ 0 & 1 \end{pmatrix}$ followed by $\begin{pmatrix} 1 & 0 \\ 4 & 1 \end{pmatrix}$. Find the
coordinate matrix for the image of ABCD.

The single matrix which will give the double
transformation is:

$$\mathbf{M} = \begin{pmatrix} 1 & 0 \\ 4 & 1 \end{pmatrix} \begin{pmatrix} 1 & 3 \\ 0 & 1 \end{pmatrix}$$

$$= \begin{pmatrix} 1 & 3 \\ 4 & 13 \end{pmatrix}$$

If A'B'C'D' is the image of ABCD, its coordinate matrix is found by pre-multiplying
the coordinate matrix for ABCD by **M**. Thus

$$\begin{matrix} & A & B & C & D \\ \begin{pmatrix} 1 & 3 \\ 4 & 13 \end{pmatrix} & \begin{pmatrix} 2 & 6 & 6 & 2 \\ 1 & 1 & 3 & 3 \end{pmatrix} \end{matrix}$$

$$= \begin{matrix} A' & B' & C' & D' \\ \begin{pmatrix} 5 & 9 & 15 & 11 \\ 21 & 37 & 63 & 47 \end{pmatrix} \end{matrix}$$

Glide Reflection

This transformation is produced by a translation followed by a reflection.

Example 14

The point P(2, 4) is given a glide reflection.
After translation under $\begin{pmatrix} 4 \\ 0 \end{pmatrix}$ it is reflected
in the *x*-axis. Determine the coordinates of
the image of P. The translation is

$$\begin{pmatrix} 2 \\ 4 \end{pmatrix} + \begin{pmatrix} 4 \\ 0 \end{pmatrix} = \begin{pmatrix} 6 \\ 4 \end{pmatrix}$$

The reflection is

$$\begin{pmatrix} 1 & 0 \\ 0 & -1 \end{pmatrix} \begin{pmatrix} 6 \\ 4 \end{pmatrix} = \begin{pmatrix} 6 \\ -4 \end{pmatrix}$$

As shown in Fig. 31.23, the coordinates
of the image of P are (6, −4).

Fig. 31.23

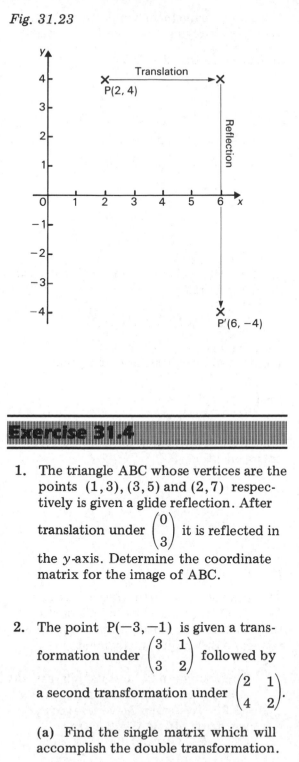

Exercise 31.4

1. The triangle ABC whose vertices are the points $(1,3)$, $(3,5)$ and $(2,7)$ respectively is given a glide reflection. After translation under $\begin{pmatrix} 0 \\ 3 \end{pmatrix}$ it is reflected in the y-axis. Determine the coordinate matrix for the image of ABC.

2. The point $P(-3,-1)$ is given a transformation under $\begin{pmatrix} 3 & 1 \\ 3 & 2 \end{pmatrix}$ followed by a second transformation under $\begin{pmatrix} 2 & 1 \\ 4 & 2 \end{pmatrix}$.

    (a) Find the single matrix which will accomplish the double transformation.

    (b) Determine the coordinates of the image of P under these transformations.

3. The triangle ABC with vertices $(10,10)$, $(20,15)$ and $(15,20)$ respectively is given a transformation under $\begin{pmatrix} 1 & 4 \\ 0 & 1 \end{pmatrix}$ followed by a reflection in the x-axis.

    (a) Determine the single matrix which will give the double transformation.

    (b) Find the coordinate matrix for the image of ABC under these transformations.

4. The point $P(2,6)$ is given an anticlockwise rotation of $45°$ about the origin followed by a translation under $\begin{pmatrix} -4 \\ -5 \end{pmatrix}$. Find by accurate drawing the coordinates of the image of P.

5. The square $A(2,1)$, $B(4,1)$, $C(4,3)$, $D(2,3)$ is given a rotation (anticlockwise) of $\frac{1}{4}$ turn about the origin followed by a translation under $\begin{pmatrix} -1 \\ -5 \end{pmatrix}$. Draw the transformed figure and state the coordinates of its vertices.

6. If **A** denotes the translation $\begin{pmatrix} 3 \\ 2 \end{pmatrix}$ and **B** denotes the translation $\begin{pmatrix} 1 \\ -1 \end{pmatrix}$, write down the coordinates of the points on to which $(2,1)$ is mapped under the following transformations:
    (a) **AB**   (b) **BA**   (c) $\mathbf{B}^2$
    (d) $\mathbf{A}^{-1}$ (e) $\mathbf{B}^{-1}$ (f) **ABA**
    (g) **BAB**

7. Let T be the transformation described by the matrix $\begin{pmatrix} 2 & 1 \\ 1 & 1 \end{pmatrix}$. After transformation under this matrix the image of $\triangle$ABC is $A'(8,5)$, $B'(12,7)$, $C'(11,7)$. Find the coordinates of the points A, B and C.

Enlargement

Enlargements are transformations which multiply all lengths by a scale factor. If the scale factor is greater than 1 then all lengths will be increased in size. If the scale factor is less than 1, i.e. a fraction, then all lengths will be decreased in size.

In Fig. 31.24, the points A(3, 2), B(8, 2), C(8, 3) and D(3, 3) have been plotted and joined to give the rectangle ABCD. In order to form the rectangle A'B'C'D', each of the points (x, y) has been mapped on to (x', y') by the mapping

$$(x, y) \rightarrow (2x, 2y)$$

Thus   A(3, 2) → A'(6, 4)

and   B(8, 2) → B'(16, 4)   and so on.

As can be seen from the diagram each length of A'B'C'D' is twice the corresponding length of ABCD, for instance, A'B' = 2AB. That is ABCD has been enlarged by a factor of 2. The same result could have been obtained by multiplying the coordinate matrix of ABCD by the scalar 2. Thus

$$
2\begin{matrix} A & B & C & D \\ \begin{pmatrix} 3 & 8 & 8 & 3 \\ 2 & 2 & 3 & 3 \end{pmatrix} \end{matrix} = \begin{matrix} A' & B' & C' & D' \\ \begin{pmatrix} 6 & 16 & 16 & 6 \\ 4 & 4 & 6 & 6 \end{pmatrix} \end{matrix}
$$

Fig. 31.24

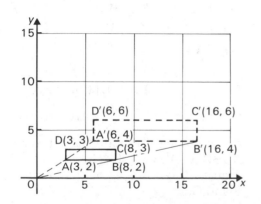

The same enlargement would be produced by pre-multiplying the coordinate matrix of ABCD by the 2×2 matrix $\begin{pmatrix} 2 & 0 \\ 0 & 2 \end{pmatrix}$.

Thus,

$$
\begin{pmatrix} 2 & 0 \\ 0 & 2 \end{pmatrix} \begin{matrix} A & B & C & D \\ \begin{pmatrix} 3 & 8 & 8 & 3 \\ 2 & 2 & 3 & 3 \end{pmatrix} \end{matrix}
$$

$$
= \begin{matrix} A' & B' & C' & D' \\ \begin{pmatrix} 6 & 16 & 16 & 6 \\ 4 & 4 & 6 & 6 \end{pmatrix} \end{matrix}
$$

If the centre of the enlargement is the origin then, in general,

$$(x, y) \rightarrow (kx, ky)$$

Which is equivalent to pre-multiplying $\begin{pmatrix} x \\ y \end{pmatrix}$ by $\begin{pmatrix} k & 0 \\ 0 & k \end{pmatrix}$.

k is called the **linear scale factor of the enlargement.**

Example 15

The triangle ABC with vertices A(2, 3), B(2, 1) and C(3, 2) is to be enlarged by a scale factor of 2, centre at the origin. Draw triangle ABC and its enlargement.

The triangle ABC and its enlargement A'B'C' have been constructed in Fig. 31.25. To obtain the vertices of A'B'C' we mark off OA' = 2 × OA, OB' = 2 × OB and OC' = 2 × OC.

We see that A'B'C' is similar to ABC and that corresponding lines are parallel.

Fig. 31.25

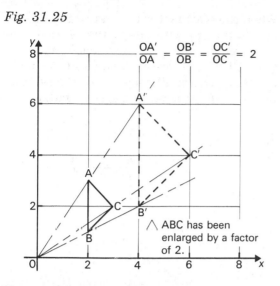

$$\frac{OA'}{OA} = \frac{OB'}{OB} = \frac{OC'}{OC} = 2$$

△ ABC has been enlarged by a factor of 2.

The centre of an enlargement need not be at the origin. In this case no simple matrix equivalent exists and enlargements can be found by drawing.

Example 16

The triangle ABC with vertices A(2, 1), B(4, 2) and C(3, 3) is transformed to A'(7, 6), B'(11, 8) and C'(9, 10) by means of an enlargement. Find the coordinates of the centre of the enlargement and the scale factor.

Fig. 31.26

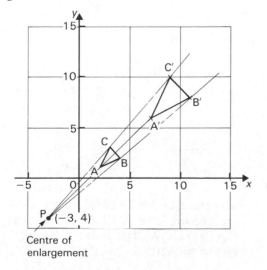

Centre of enlargement

The centre of the enlargement is found by joining AA', BB' and CC' and producing them to intersect at a single point P (Fig. 31.26). P is the centre of the enlargement and by scaling we find $\frac{PA'}{PA} = 2$. Thus the scale factor is 2.

The transformation is an enlargement, centre at the origin, and a translation.

The centre of enlargement can also be found by solving a matrix equation of the type:

$$\begin{pmatrix} k & 0 \\ 0 & k \end{pmatrix} \begin{pmatrix} x \\ y \end{pmatrix} + \begin{pmatrix} m \\ n \end{pmatrix} = \begin{pmatrix} x \\ y \end{pmatrix}$$

where k is the enlargement factor, x and y are the coordinates of the centre of rotation and $\begin{pmatrix} m \\ n \end{pmatrix}$ is a vector describing a translation.

Example 17

The rectangle A(2, 1), B(4, 1), C(4, 4), D(2, 4) is mapped on to A"B"C"D" by the enlargement

$$\begin{pmatrix} x \\ y \end{pmatrix} \rightarrow \begin{pmatrix} 4 & 0 \\ 0 & 4 \end{pmatrix} \begin{pmatrix} x \\ y \end{pmatrix} + \begin{pmatrix} 6 \\ 9 \end{pmatrix}$$

(a) Find the coordinates of the centre of the enlargement.

(b) On a diagram show the transformation and the centre of enlargement.

(a) To find the coordinates of the centre of enlargement we solve the matrix equation:

$$\begin{pmatrix} 4 & 0 \\ 0 & 4 \end{pmatrix} \begin{pmatrix} x \\ y \end{pmatrix} + \begin{pmatrix} 6 \\ 9 \end{pmatrix} = \begin{pmatrix} x \\ y \end{pmatrix}$$

$$\begin{pmatrix} 4x \\ 4y \end{pmatrix} + \begin{pmatrix} 6 \\ 9 \end{pmatrix} = \begin{pmatrix} x \\ y \end{pmatrix}$$

$$\begin{pmatrix} 4x + 6 \\ 4y + 9 \end{pmatrix} = \begin{pmatrix} x \\ y \end{pmatrix}$$

$4x + 6 = x \quad 3x = -6 \quad \text{and} \quad x = -2$

$4y + 9 = y \quad 3y = -9 \quad \text{and} \quad y = -3$

The coordinates of the centre of enlargement are therefore $(-2, -3)$.

(b) The enlargement gives the image:

$$\begin{matrix} & A & B & C & D \\ \begin{pmatrix} 4 & 0 \\ 0 & 4 \end{pmatrix} & \begin{pmatrix} 2 & 4 & 4 & 2 \\ 1 & 1 & 4 & 4 \end{pmatrix} \end{matrix}$$

$$= \begin{matrix} A' & B' & C' & D' \\ \begin{pmatrix} 8 & 16 & 16 & 8 \\ 4 & 4 & 16 & 16 \end{pmatrix} \end{matrix}$$

The translation gives the image:

$$\begin{matrix} A'' & B'' & C'' & D'' \\ \begin{pmatrix} 14 & 22 & 22 & 14 \\ 13 & 13 & 25 & 25 \end{pmatrix} \end{matrix}$$

The transformation and the centre of enlargement are shown in Fig. 31.27.

Fig. 31.27

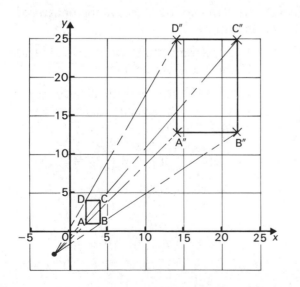

When the scale factor is negative the image lies on the opposite side of the centre of the enlargement and is turned upside down (Fig. 31.28). However, triangles OAB and OA′B′ are still similar and OA′ = kOA and OB′ = kOB, where k is the scale factor.

Fig. 31.28

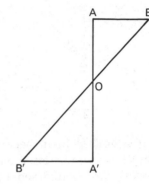

Example 18

(a) Triangle ABC has the following co-ordinate matrix:

$$\begin{matrix} A & B & C \\ \begin{pmatrix} 1 & 3 & 3 \\ 2 & 2 & 4 \end{pmatrix} \end{matrix}$$

Find the image of ABC after the enlargement $\begin{pmatrix} -3 & 0 \\ 0 & -3 \end{pmatrix}$.

$$\begin{matrix} & A & B & C \\ \begin{pmatrix} -3 & 0 \\ 0 & -3 \end{pmatrix} & \begin{pmatrix} 1 & 3 & 3 \\ 2 & 2 & 4 \end{pmatrix} \end{matrix}$$

$$= \begin{matrix} A' & B' & C' \\ \begin{pmatrix} -3 & -9 & -9 \\ -6 & -6 & -12 \end{pmatrix} \end{matrix}$$

The triangle ABC and its image A′B′C′ are shown in Fig. 31.29. Note that OA′ = 3 × OA, OB′ = 3 × OB and OC′ = 3 × OC.

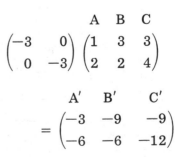

Fig. 31.29

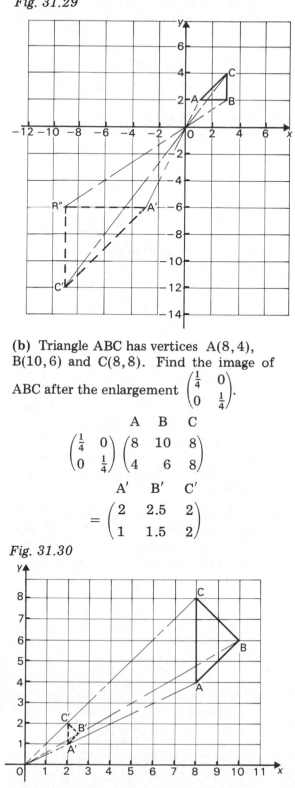

(b) Triangle ABC has vertices A(8, 4), B(10, 6) and C(8, 8). Find the image of ABC after the enlargement $\begin{pmatrix} \frac{1}{4} & 0 \\ 0 & \frac{1}{4} \end{pmatrix}$.

$$\begin{matrix} & A & B & C \\ \begin{pmatrix} \frac{1}{4} & 0 \\ 0 & \frac{1}{4} \end{pmatrix} & \begin{pmatrix} 8 & 10 & 8 \\ 4 & 6 & 8 \end{pmatrix} \end{matrix}$$

$$\begin{matrix} & A' & B' & C' \\ = & \begin{pmatrix} 2 & 2.5 & 2 \\ 1 & 1.5 & 2 \end{pmatrix} \end{matrix}$$

Fig. 31.30

The triangle ABC and its image A'B'C' are shown in Fig. 31.30. Note that A'B'C' is nearer the origin than is ABC and it is reduced in size. $OA' = \frac{1}{4} \times OA$, $OB' = \frac{1}{4} \times OB$ and $OC' = \frac{1}{4} \times OC$.

Areas of Enlargements

When an enlargement of $k:1$ is required the coordinate matrix of the figure is pre-multiplied by the 2×2 matrix $\begin{pmatrix} k & 0 \\ 0 & k \end{pmatrix}$. The area of the enlarged figure is then k^2 times the area of the original figure.

Thus $\dfrac{\text{Area of image}}{\text{Area of object}} = k^2$

Note that if $M = \begin{pmatrix} k & 0 \\ 0 & k \end{pmatrix}$ then $|M| = k^2$ and hence

$$\frac{\text{Area of image}}{\text{Area of object}} = |M|$$

Example 19

Triangle ABC has an area of 6 cm^2. It is enlarged by pre-multiplying its coordinate matrix by the matrix $M = \begin{pmatrix} 3 & 0 \\ 0 & 3 \end{pmatrix}$. What is the area of the image of ABC?

$$|M| = 3 \times 3$$
$$= 9$$

$$\frac{\text{Area of A'B'C'}}{\text{Area of ABC}} = 9$$

$$\text{Area of A'B'C'} = 6 \times 9 \text{ cm}^2$$
$$= 54 \text{ cm}^2$$

Exercise 31.5

1. In Fig. 31.31:

   (a) find the scale factor of the enlargement

   (b) state the coordinates of the centre of the enlargement.

Fig. 31.31

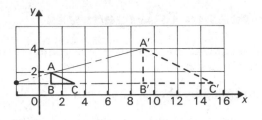

2. Copy the diagram shown in Fig. 31.32 and draw the straight lines AA', BB' and CC'. Hence find:

   (a) the scale of the enlargement

   (b) the coordinates of the centre of the enlargement.

Fig. 31.32

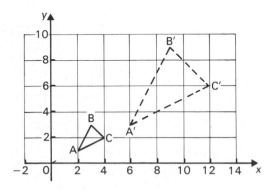

3. Plot the points A(2,0), B(2,2), C(4,3) and D(4,0) on graph paper. Join them in alphabetical order to form the trapezium ABCD. Draw an enlargement of ABCD with a scale factor of 2 and the origin as the centre of the enlargement. Label this enlargement A'B'C'D'. Write down the coordinates of A', B', C' and D' and state how many times bigger in area is A'B'C'D' than ABCD.

4. The rectangle A(3,2), B(8,2), C(8,3), D(3,3) is mapped on to A'(9,8), B'(19,8), C'(19,10), D'(9,10) by an enlargement.

   (a) What is the scale factor?

   (b) What are the coordinates of the centre of the enlargement?

5. The vertices of triangle ABC are A(2,3), B(4,1), C(4,4). Find the image of ABC under the transformations whose matrices are:

   (a) $\begin{pmatrix} 4 & 0 \\ 0 & 4 \end{pmatrix}$ (b) $\begin{pmatrix} \frac{1}{2} & 0 \\ 0 & \frac{1}{2} \end{pmatrix}$

   (c) $\begin{pmatrix} -2 & 0 \\ 0 & -2 \end{pmatrix}$

6. The quadrilateral ABCD is formed by joining the points A(1,1), B(1,4), C(7,7) and D(4,1). ABCD is enlarged by a scale factor of $\frac{1}{2}$ with P(−6,4) as the centre of the enlargement. Find the coordinate matrix for A'B'C'D' the image of ABCD under this enlargement.

7. In Fig. 31.33, CD is mapped on to AB by an enlargement. Find the scale factor of the enlargement and then find the lengths of AB and AC. If the area of OAB is 3.41 cm² what is the area of OCD?

Fig. 31.33

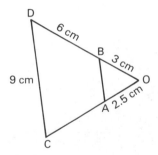

Miscellaneous Exercise 31

Section A

1. The point $A(4, -2)$ is reflected in the line $y = x$. Find the coordinates of A', the image of A.

2. P is the point $(4, 3)$. Plot this point on graph paper.

   (a) The image of P when reflected in the x-axis is the point A. Mark A on the graph paper and state its coordinates.

   (b) The image of P when reflected in the y-axis is the point B. Mark B on the graph paper and state its coordinates.

3. Write down the coordinates (p, q) of the image of the point $R(3, 1)$ under the translation $\begin{pmatrix} 3 \\ 2 \end{pmatrix}$.

4. The point $P(3, -2)$ is given a rotation of $90°$ anticlockwise about the point $(2, 1)$. Find the coordinates of P', the image of P, after this rotation.

5. A given translation maps $P(3, 5)$ on to $P'(7, 8)$. What is the image of the point $Q(-2, 4)$ under this translation?

6. The triangle ABC has vertices $A(0, 0)$, $B(2, 0)$ and $C(2, 1)$. Find the image of ABC under an enlargement whose centre is the origin and whose scale factor is 2.

7. A trapezium is formed by joining, in alphabetical order, the points $A(2, 1)$, $B(7, 1)$, $C(6, 3)$ and $D(3, 3)$. Draw the trapezium on graph paper and mark on it all the lines of symmetry.

Section B

1. The matrix of a transformation is $\begin{pmatrix} 1 & 0 \\ 0 & -1 \end{pmatrix}$. Find the image of $(-4, 3)$ under this transformation.

2. The matrix of a transformation is $\begin{pmatrix} 2 & 0 \\ 0 & 2 \end{pmatrix}$. Find the image of $(-2, 3)$ under this transformation.

3. Fig. 31.34 shows a triangle ABC. Under reflection in the line $x = 5$, the triangle is mapped on to the triangle $A'B'C'$. The triangle $A'B'C'$ when reflected in the line $y = 0$ is mapped on to triangle $A''B''C''$. Draw a diagram showing the triangles $A'B'C'$ and $A''B''C''$ and write down the coordinates of their vertices.

Fig. 31.34

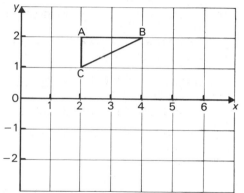

4. On graph paper draw the lines $y + x = 3$, $y = 2$, $y + x = 1$ and $y = 1$. Use a scale of $2\,\text{cm} = 1\,\text{unit}$ on both axes and cover the range -4 to 4 on both axes.

   (a) Write down the coordinates of the points of intersection of the straight lines.

   (b) Name the figure produced by these intersections.

   (c) Calculate the coordinates of the images of these points of intersection when transformed by the matrix $\begin{pmatrix} 2 & 0 \\ 0 & -2 \end{pmatrix}$.

(d) Plot these points on your graph.

(e) Name the figure produced when these images are joined.

(f) Calculate the ratio:

$$\frac{\text{Area of original shape}}{\text{Area of image}}$$

5. A trapezium is formed by joining, in alphabetical order, the four points A(0,0), B(2,0), C(1,1), D(0,1).

(a) Write down the coordinates of A, B, C and D as a 2×4 matrix.

(b) A transformation is represented by the matrix $\begin{pmatrix} 0 & -1 \\ -1 & 0 \end{pmatrix}$. Find the image of ABCD under this transformation and call it A'B'C'D'.

(c) Translate A'B'C'D' by the vector $\begin{pmatrix} 4 \\ 4 \end{pmatrix}$. Find the image and call it A"B"C"D".

6. A triangle ABC has vertices A(1,2), B(3,4) and C(5,1). Plot these points on graph paper using a scale of 1 cm = 1 unit on both axes. Cover the range -6 to 6 on the x-axis and -8 to 8 on the y-axis. Draw the image of ABC when it is

(a) translated by the vector $\begin{pmatrix} 0 \\ 3 \end{pmatrix}$, and label it T

(b) reflected in the y-axis, and label it Q

(c) rotated through $\frac{1}{4}$ turn clockwise about the origin, and label it P

(d) rotated $\frac{1}{4}$ turn clockwise about the origin followed by a reflection in the y-axis, and label it R.

7. (a) The triangle ABC has vertices A(2,0), B(2,2) and C(6,0). It is translated by the vector $\begin{pmatrix} 2 \\ 4 \end{pmatrix}$. Find the coordinates of the vertices of ABC after this translation.

(b) The square ABCD can be described by the coordinate matrix $\begin{pmatrix} 2 & 4 & 4 & 2 \\ 0 & 0 & 2 & 2 \end{pmatrix}$. Transform the square by enlarging it by a scale factor of 2 about the origin.

Multi-Choice Questions 31

1. Under the reflection $(2,-1) \rightarrow (6,-1)$ what is the mirror line?

   A $y = -1$ B $x = 4$

   C $x = -1$ D $y = 4$

2. Under the transformation $\begin{pmatrix} x \\ y \end{pmatrix} \rightarrow \begin{pmatrix} 2y + x \\ x \end{pmatrix}$ the point $(5,-1) \rightarrow P(r,s)$. What are the values of r and s.

   A $r = 3, s = -1$ B $r = 3, s = 5$

   C $r = -3, s = 5$ D $r = 9, s = 5$

3. Under the transformation $\begin{pmatrix} x \\ y \end{pmatrix} \rightarrow \begin{pmatrix} 2y + x \\ x \end{pmatrix}$ the point $P(x, y) \rightarrow P'(5,-1)$. What are the coordinates of P?

   A $(-1, 3)$ B $(3, -1)$

   C $(5, -3)$ D $(-3, 5)$

4. Under which of the following trans-
    formations is the point $(2, 3)$ the
    image of the point $(3, 2)$?

    A Translation by the vector $\begin{pmatrix} 1 \\ -1 \end{pmatrix}$

    B Reflection in the line $y = -x$

    C Reflection in the line $y = x$

    D An enlargement from the origin
       with a scale factor of $1\frac{1}{2}$

5. What transformation does the mapping
    $\begin{pmatrix} x \\ y \end{pmatrix} \rightarrow \begin{pmatrix} 3x \\ 3y \end{pmatrix}$ give?

    A translation   B enlargement

    C reflection   D rotation

6. A translation which maps the point
    $P(-1, -2)$ on to the point $P'(3, 4)$
    also maps the point $Q(a, b)$ on to
    $Q'(2, 5)$. The coordinates of Q are

    A $(2, 1)$ B $(-2, 1)$

    C $(-6, 10)$ D $(-2, -1)$

7. The transformation represented by the
    matrix $\begin{pmatrix} -1 & 0 \\ 0 & 1 \end{pmatrix}$ is

    A a reflection in the x-axis

    B a reflection in the y-axis

    C a reflection in the line $y = x$

    D a rotation of $90°$ clockwise about
       the origin

Symmetry, Patterns and Tessellation

Line Symmetry

In Chapter 31 we discovered that if a plane figure contains its own reflection in a line m, then the figure is symmetrical about m. The line m is called the mirror line or the **axis of symmetry**.

Fig. 32.1 shows a kite ABCD. If we fold the shape along the line AC one half of the shape will cover the other half exactly. The kite is therefore symmetrical about AC which is therefore the axis of symmetry. Note that triangle ADC is the image of ABC, AC being the mirror line. It follows that:

(1)   AC is the perpendicular bisector of BD, since the point D is a reflection of the point B.

(2)   AC bisects the angles DAB and DCB.

(3)   Triangles ABC and ADC are congruent.

Fig. 32.1

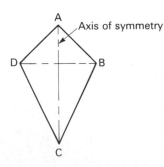

It is possible for a plane figure to have several lines of symmetry. For instance, the square shown in Fig. 32.2 has the four lines of symmetry shown.

Fig. 32.2

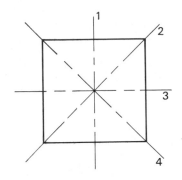

Some plane figures have no axes of symmetry. The parallelogram shown in Fig. 32.3 is an example.

Fig. 32.3

A line of symmetry may be horizontal, vertical or oblique.

Exercise 32.1

1. Each of the shapes shown in Fig. 32.4 has one line of symmetry. Copy the shapes on to squared paper and draw the lines of symmetry.

Fig. 32.4

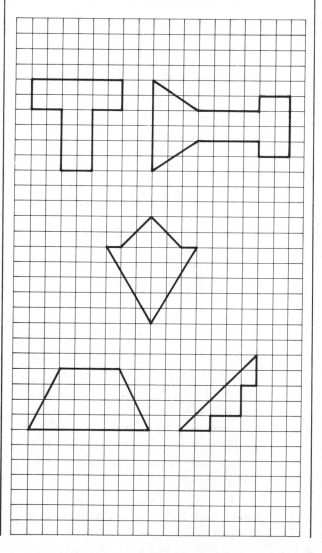

2. Fig. 32.5 shows a number of half shapes with the line of symmetry indicated in chain dot. On squared paper, draw the complete symmetrical shape.

Fig. 32.5

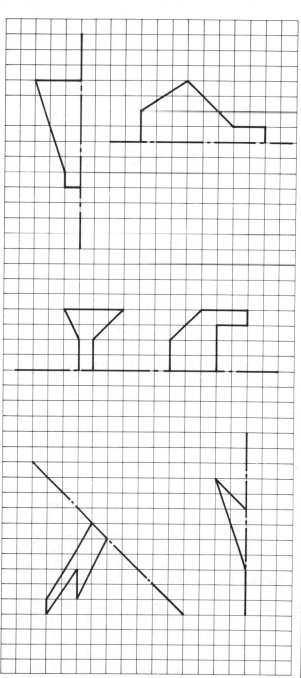

3. Fig. 32.6 shows a number of shapes which have two lines of symmetry. Copy these shapes on to squared paper and then draw, for each of them, the two lines of symmetry.

Fig. 32.6

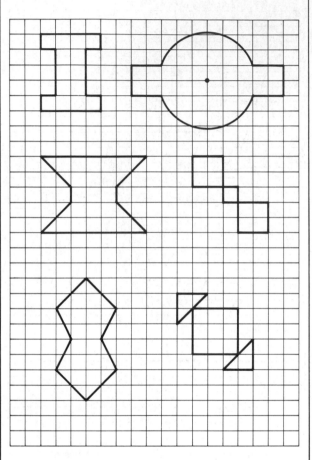

4. Fig. 32.7 shows a rhombus drawn inside a rectangle. Sketch the figure and show on it all the lines of symmetry.

Fig. 32.7

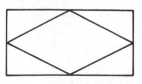

5. Draw a triangle

   (a) with one axis of symmetry

   (b) with three axes of symmetry.

6. Draw a regular pentagon (five-sided polygon) and mark on it all the axes of symmetry.

7. For each of the letters shown in Fig. 32.8 write down the number of axes of symmetry that each possesses.

Fig. 32.8

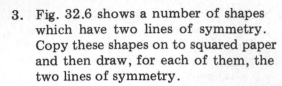

8. Plot the points given below on graph paper and then join them, in alphabetical order, to give plane figures. For each shape draw all the axes of symmetry.

   (a) A(2, 5), B(4, 12), C(6, 10)

   (b) A(4, 5), B(6, 5), C(6, 10), D(4, 10)

   (c) A(4, 5), B(6, 5), C(6, 10), D(4, 12)

   (d) A(4, 10), B(8, 10), C(10, 15), D(6, 15)

   (e) A(4, 3), B(6, 10), D(8, 3)

   (f) A(2, 2), B(6, 5), C(6, 15), D(2, 15)

   (g) A(0, 10), B(5, 5), C(10, 5), D(15, 10), E(15, 15), F(10, 20), G(5, 20), H(0, 15)

   (h) A(0, 5), B(5, 0), C(0, −5), D(−5, 0)

   (i) A(0, 2), B(2, 3), C(6, 3), D(6, 1), E(2, 1)

9. Plot each of the following points on graph paper and join them up, in alphabetical order: A(0, 0), B(0, 2), C(2, 4), D(2, 6), E(4, 6), F(4, 4), G(6, 4), H(6, 2) and I(4, 0).

   (a) Reflect this shape in the *x*-axis.

   (b) Reflect the original shape in the *y*-axis.

   (c) Complete the drawing so that a shape with two lines of symmetry is produced.

10. Draw an equilateral triangle. By drawing the axis of symmetry obtain an angle of 30°.

Rotational Symmetry

Fig. 32.9 shows a square ABCD, whose diagonals intersect at O. In the diagram the square has been rotated about O through 90°, 180°, 270° and 360° and it does not appear to have moved. Because there are four positions where it appears not to have moved, the square is said to have rotational symmetry of order 4. Of course if we label the corners as in the diagram the movement becomes apparent.

Fig. 32.9

Original position   Rotated through
and rotated   90°
through 360°

Rotated through   Rotated through
180°   270°

Fig. 32.10 shows a rectangle ABCD whose diagonals intersect at O. On rotating the rectangle about O we see that only when it is rotated through 180° and 360° does it appear not to have moved. Hence the rectangle has only rotational symmetry of order 2.

Fig. 32.10

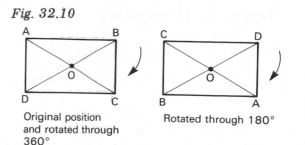

Original position   Rotated through 180°
and rotated through
360°

Every shape has rotational symmetry of 1 since rotation through 360° will bring it back to its original position.

Point Symmetry

A parallelogram has no axes of symmetry. However, if we draw the diagonals AC and BD to intersect at O (Fig. 32.11), this point is called the **point of symmetry**. This is because for any line passing through O, there are two points, one on each side of O which are equidistant from O. The parallelogram is said to have point symmetry.

When a plane figure has point symmetry it appears not to have moved after rotation through 180°.

Fig. 32.11

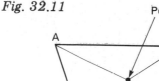

Symmetry of Quadrilaterals			
Figure	Number of axes of symmetry	Order of rotational symmetry	Point symmetry
Square	4	4	Yes
Rectangle	2	2	Yes
Parallelogram	0	2	Yes
Rhombus	2	2	Yes
Kite	1	1	No
Trapezium (isosceles)	1	1	No

Planes of Symmetry

If a solid figure such as a sphere is cut into two equal parts as shown in Fig. 32.12, the plane of the cut is called a **plane of symmetry**. The cuboid (Fig. 32.13) has been cut into two equal parts by the plane ABDC. Hence the plane ABDC is a plane of symmetry for the cuboid. Note that solid figures cannot be reflected in a line; they must be reflected in a plane mirror, hence the term plane symmetry.

Fig. 32.12

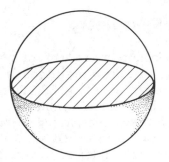

Fig. 32.13

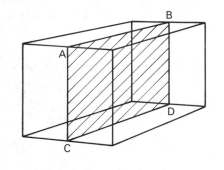

Exercise 32.2

For each of the shapes shown in Fig. 32.14, write down

(a) the number of axes of symmetry

(b) the order of rotational symmetry

(c) whether or not the shape has point symmetry.

Fig. 32.14

1. Rhombus

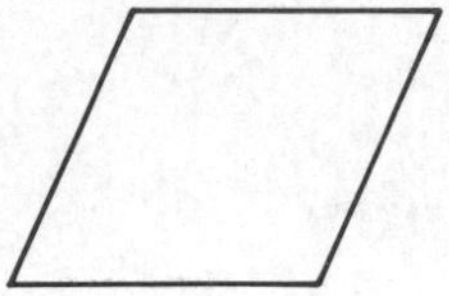

2. Regular pentagon

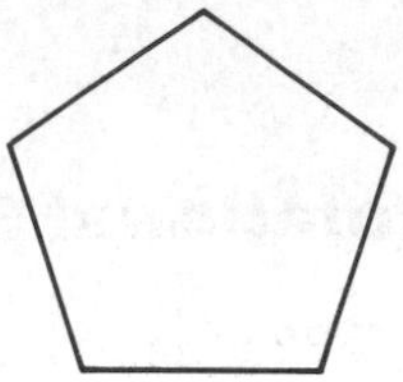

3. Ellipse

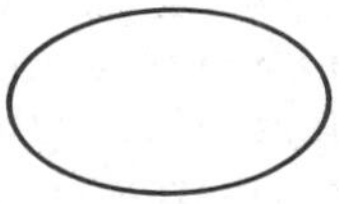

4. Regular octagon

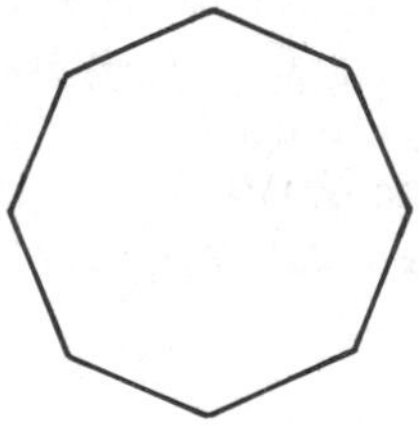

5. Star

6. Letter X

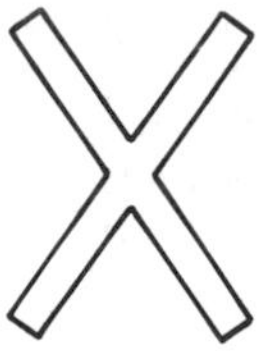

7. Letter E

8. Equilateral triangle

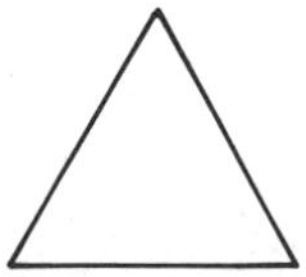

9. How many planes of symmetry has

(a) a sphere

(b) a cone

(c) a cube?

10. What is the order of rotational symmetry for the shape shown in Fig. 32.15?

Fig. 32.15

Patterns

Number sequences and transformations are often used to form patterns of various kinds. The most popular number sequences are:

(1) **Square numbers** whose sequence is $1, 4, 9, 16 \ldots$ i.e. $1^2, 2^2, 3^2, 4^2 \ldots$

(2) **Rectangular numbers** which can be represented as a pattern of dots in the form of a rectangle.

$$6 = 2 \times 3 \qquad 24 = 6 \times 4$$

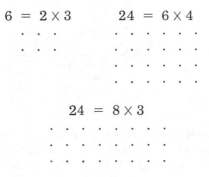

$$24 = 8 \times 3$$

Note that 1 is not regarded as being a rectangular number. The sequence of of rectangular numbers is $4, 6, 8, 9, 10, \ldots$

(3) **Triangular numbers** which can be represented as a pattern of dots in the shape of a triangle.

$$3 \qquad\qquad 6 \qquad\qquad 10$$

The sequence of triangular numbers $1, 3, 6, 10, 15, \ldots$

Note that:

$$3 = 2 + 1, \quad 6 = 3 + 2 + 1,$$
$$10 = 4 + 3 + 2 + 1,$$
$$15 = 5 + 4 + 3 + 2 + 1.$$

(4) **Pascal's Triangle**

The pattern is as follows:

$$
\begin{array}{ccccccccc}
 & & & & 1 & & & & \\
 & & & 1 & & 1 & & & \\
 & & 1 & & 2 & & 1 & & \\
 & 1 & & 3 & & 3 & & 1 & \\
1 & & 4 & & 6 & & 4 & & 1
\end{array} \quad \text{and so on.}
$$

The numbers in any line are formed from the line above. For instance in line 5, the pattern is $1, 1 + 3 = 4$, $3 + 3 = 6$, $3 + 1 = 4, 1$.

Example 1

Fig. 32.16 shows a pattern made with matches. Draw the next pattern in the series and complete the table below.

Number of triangles	1	4	9	?	?	?
Number of matches	3	9	18	?	?	?
Number of extra matches	—	6	9	?	?	?

Fig. 32.16

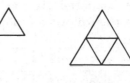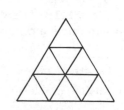

The next pattern is shown in Fig. 32.17.

Fig. 32.17

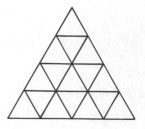

The number of triangles is a sequence of square numbers which is $1, 4, 9, 16, 25, 36, \ldots$

The number of matches is given by the sequence $3, 3 + 6 = 9, \ 9 + 9 = 18, \ 18 + 12 = 30, \ 30 + 15 = 45, \ 45 + 18 = 63, \ldots$

The number of extra matches is given by the sequence $6, 9, 12, 15, 18, \ldots$

The completed table is as follows:

Number of triangles	1	4	9	16	25	36
Number of matches	3	9	18	30	45	63
Number of extra matches	—	6	9	12	15	18

Example 2

A small firm specialising in tiling bathrooms offers patterns based upon those shown in Fig. 32.18. Draw the next pattern in the sequence and complete the following table:

Pattern number	1	2	3	4	5	6	7
Number of shaded tiles	4	8	16	?	?	?	?

Fig. 32.18

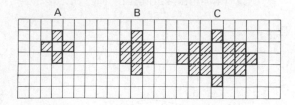

Counting the shaded tiles row by row: Diagram A consists of $1, 2, 1$; diagram B consists of $1, 3, 3, 1$; diagram C consists of $1, 4, 6, 4, 1$. Hence the series of patterns follows the pattern of Pascal's triangle. The next pattern in the series is $1, 5, 10, 10, 5, 1$ (see Fig. 32.19). The completed table is shown below:

Pattern number	1	2	3	4	5	6	7
Number of shaded tiles	4	8	16	32	64	128	256

Fig. 32.19

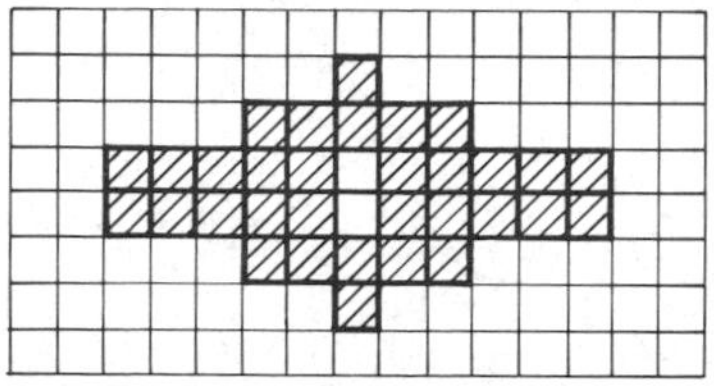

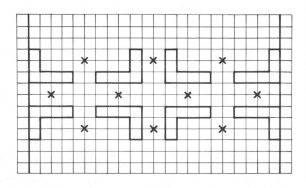

Wallpaper and carpet patterns are often composed of reflections, rotations and translations. Fig. 32.20 is a typical example.

Fig. 32.20

Tessellation

A tessellation is a pattern made by repeating a plane figure (or figures) to cover a plane completely without leaving any gaps. It may be considered to be a pattern which could never end.

A regular tessellation consists of regular polygons such that every vertex is identical. As shown in Fig. 32.21 there are three such tesselations.

Fig. 32.21

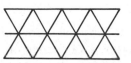

Equilateral triangles

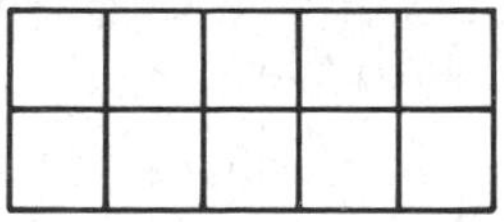

Squares

Regular hexagons

Fig. 32.22 is a tessellation of rectangles and Fig. 32.23 is a tessellation composed of regular octagons (eight-sided polygons) and squares.

Tessellations are used for tiling patterns in many shapes.

Fig. 32.22

Fig. 32.23

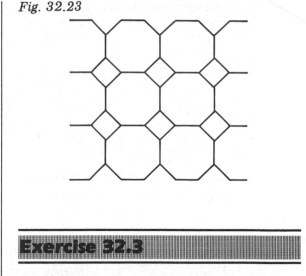

Exercise 32.3

1. Fig. 32.24 shows part of a tessellation based upon irregular octagons and squares. Using spotted or squared paper, complete the tessellation.

Fig. 32.24

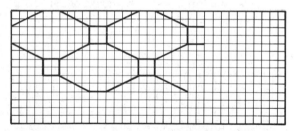

2. Patterns based upon rotations, reflections and translations are often used for carpet designs. Such a design is shown in Fig. 32.25. On spotted or squared paper complete the pattern.

Fig. 32.25

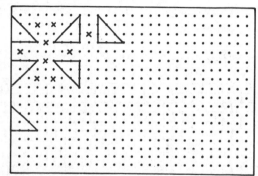

3. In the flower borders of a public park rose bushes are always planted in a special pattern. Fig. 32.26 shows three of these patterns.

  (a) Draw the patterns when there are
     (i) 4 white roses
     (ii) 6 white roses.

  (b) How many red roses would be needed if there were 20 white roses?

  (c) How many white roses would be needed if there were 50 red roses?

Fig. 32.26

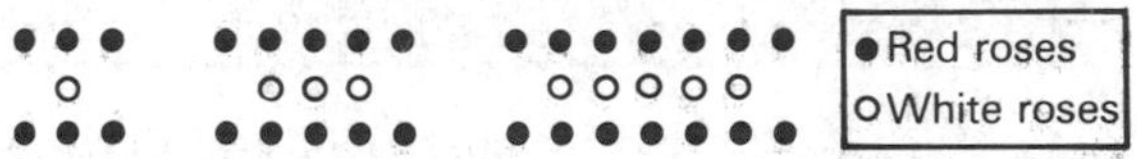

● ● ●   ● ● ● ● ●   ● ● ● ● ● ● ● | ● Red roses
  ○   ○ ○ ○   ○ ○ ○ ○ ○ | ○ White roses
● ● ●   ● ● ● ● ●   ● ● ● ● ● ● ●

4. The wallpaper pattern shown in Fig. 32.27 uses reflections, rotations and translations. Using squared paper, complete the pattern.

Fig. 32.27

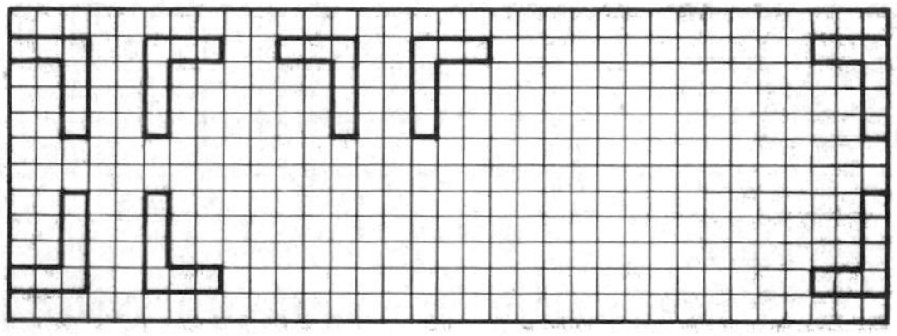

5. Fig. 32.28 shows three tiling patterns which form a mathematical series. Draw diagrams showing the next two patterns in the series.

Fig. 32.28

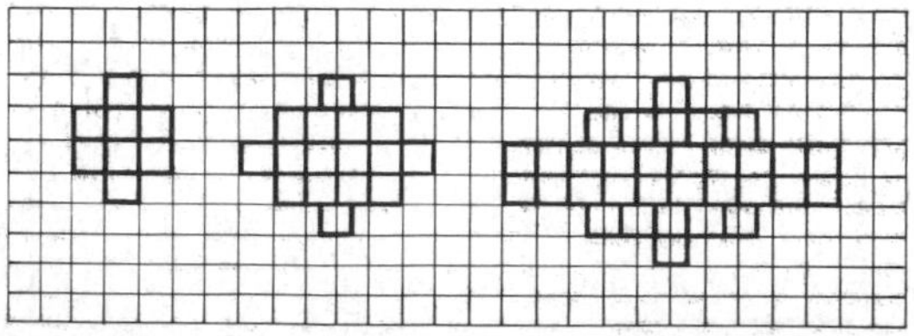

Copy and complete the following table:

Pattern number	1	2	3	4	5	6	7	8
Total number of tiles	8	16	32					

6. The diagram (Fig. 32.29) shows 2 shaded tiles surrounded by 10 white tiles.

  (a) How many white tiles are needed to surround a line of
     (i)   3 shaded tiles
     (ii) 5 shaded tiles
     (iii) 15 shaded tiles?

  (b) Establish a rule for the number of white tiles (W) needed to surround a row of shaded tiles of any size (S).

  (c) Use this rule to find the number of white tiles needed to surround a row of 80 shaded tiles.

  (d) How many shaded tiles are surrounded by 306 white tiles?

Fig. 32.29

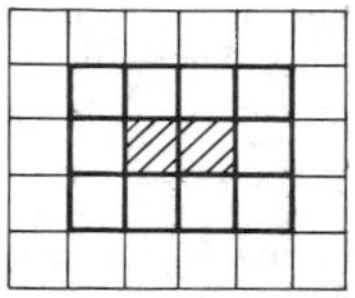

7. A gardener plants crocus bulbs in groups, the size of the group depending upon the area of his flower beds. The first three groups he uses consist of 3, 6 and 10 bulbs. If he continues with this pattern how many bulbs will there be in his next three groups?

8. Use the basic shape shown in Fig. 32.30 to draw two different tessellations. Your tessellations should consist of a minimum of eight basic shapes and they should clearly show the pattern of the tessellations.

Fig. 32.30

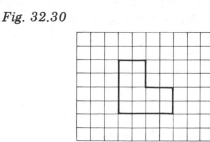

9. The tile shown in Fig. 32.31 is in the form of a regular pentagon. Tiles of this shape will not fit together without leaving gaps between them. By sketching a tessellation find the shape of the tiles needed to fill these gaps.

Fig. 32.31

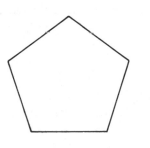

10. Fig. 32.32 shows the top left-hand quarter of a crossword puzzle skeleton. The pattern is to be symmetrical about both its vertical and horizontal centre lines. By using squared paper, complete the skeleton.

Fig. 32.32

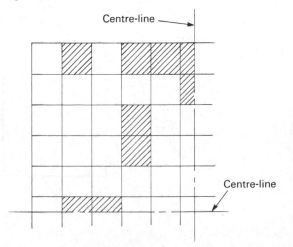

Centre-line

Centre-line

11. Fig. 32.33 shows a stack of tins of peas in a supermarket. Copy and complete the following table:

Rows	1 and 2	1, 2 and 3
Total number of tins	3	6
Rows	1, 2, 3 and 4	1, 2, 3, 4 and 5
Total number of tins		

Fig. 32.33

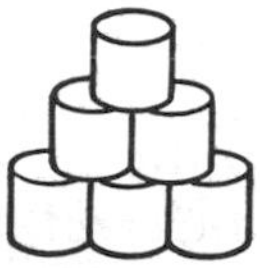

12. Fig. 32.34 shows a nest of snooker balls. How many balls are there in the first

   (a) 5 rows

   (b) 8 rows?

Fig. 32.34

Multi-Choice Questions 32

1. How many lines of symmetry can be drawn on the letter H shown in Fig. 32.35?

   A 0   B 1   C 2   D 3

Fig. 32.35

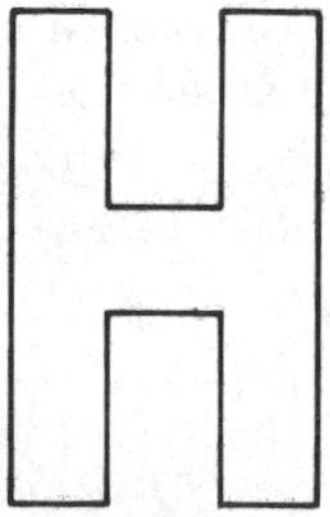

2. Which one of the following statements is not correct?

    I   Any square has 4 axes of symmetry.
    II Any rhombus has 2 axes of symmetry.
    III Any kite has 1 axis of symmetry.
    IV Any trapezium has 1 axis of symmetry.

    A I only   B II only
    C III only   D IV only

3. Which one of the letters O, N, T, V, has no axis of symmetry?

    A O   B N   C T   D V

4. Fig. 32.36 shows an isosceles triangle. How many different lines of symmetry can be drawn on it?

    A 0   B 1   C 2   D 4

Fig. 32.36

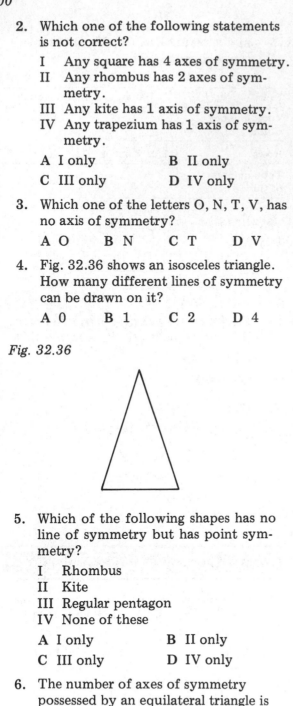

5. Which of the following shapes has no line of symmetry but has point symmetry?

    I   Rhombus
    II Kite
    III Regular pentagon
    IV None of these

    A I only   B II only
    C III only   D IV only

6. The number of axes of symmetry possessed by an equilateral triangle is

    A less than 2   B 2
    C 3   D 4

7. Fig. 32.37 shows a tessellation. How many different lines of symmetry can be drawn on it?

    A 0   B 1   C 2   D 4

Fig. 32.37

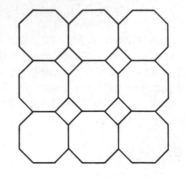

8. In Fig. 32.38, which of the shapes is **not** symmetrical about a horizontal axis?

Fig. 32.38

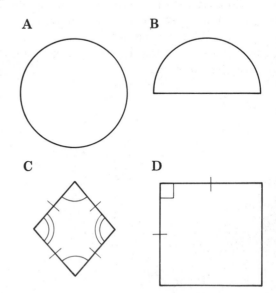

Angles and Straight Lines

Definitions

(1) A point has position but no size, although to show it on paper it must be given some size.

(2) If two points A and B are chosen (Fig. 33.1) then only one straight line can contain them.

(3) If the two points chosen are the end points of the line then AB is called a line segment (Fig. 33.2).

Fig. 33.1

A ⚬ ———————————— ⚬ B

Fig. 33.2

A ⚬———————————————————⚬ B

Angles

When two lines meet at a point they form an angle. The size of the angle depends only upon the amount by which one of the lines would need to be rotated to lie on the other. It does not depend upon the lengths of the lines forming the angle. In Fig. 33.3 the angle A is larger than the angle B despite the fact that the lengths of the arms are shorter.

Fig. 33.3

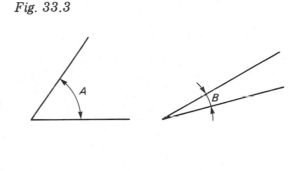

Angular Measurement

An angle may be looked upon as the amount of rotation or turning. In Fig. 33.4 the line OA has been turned about O until it takes up the position OB. The angle through which the line has turned is the amount of opening between the lines OA and OB.

Fig. 33.4

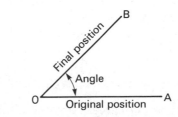

When writing angles we write, for example, seventy degrees as $70°$. The small $°$ at the right-hand corner of the figure replaces the word 'degrees'. Thus '$87°$' reads '87 degrees'.

If the line OA is rotated until it returns to its original position it will have described one revolution. Hence we can measure an angle as a fraction of a revolution. Fig. 33.5 shows a circle divided up into 36 equal parts. The first division is split up into 10 equal parts so that each small division is $\frac{1}{360}$ of a complete revolution. We call this division a **degree**.

$$1 \text{ degree} = \frac{1}{360} \text{ of a revolution}$$

$$360 \text{ degrees} = 1 \text{ revolution}$$

Fig. 33.5

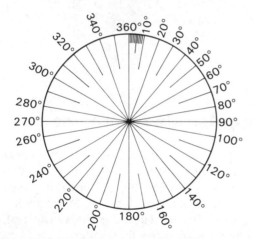

The right angle is $\frac{1}{4}$ of a revolution and hence it contains $\frac{1}{4}$ of $360° = 90°$. Two right angles contain $180°$ and three right angles contain $270°$.

Example 1

Find the angle in degrees corresponding to $\frac{1}{8}$ of a revolution.

$$1 \text{ revolution} = 360°$$
$$\tfrac{1}{8} \text{ revolution} = \tfrac{1}{8} \times 360°$$
$$= 45°$$

Example 2

Find the angle in degrees corresponding to 0.6 of a revolution.

$$1 \text{ revolution} = 360°$$
$$0.6 \text{ revolution} = 0.6 \times 360°$$
$$= 216°$$

Most angles are stated in degrees and decimals of a degree. A typical angle might be $36.7°$ or $206.4°$.

Types of Angle

An **acute angle** (Fig. 33.6) is less than $90°$.

Fig. 33.6

Acute angle

A **right angle** (Fig. 33.7) is equal to $90°$.

Fig. 33.7

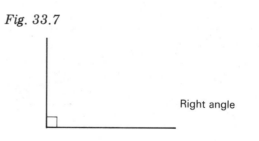

Right angle

A **reflex angle** (Fig. 33.8) is greater than $180°$.

Fig. 33.8

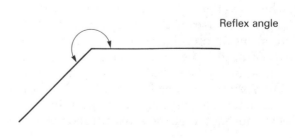

Reflex angle

An **obtuse angle** (Fig. 33.9) lies between 90° and 180°.

Fig. 33.9

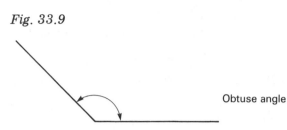

Obtuse angle

Complementary angles are angles whose sum is 90°.

Supplementary angles are angles whose sum is 180°.

Exercise 33.1

1. How many degrees are there in $1\frac{1}{2}$ right angles?

2. How many degrees are there in $\frac{3}{5}$ of a right angle?

3. How many degrees are there in $\frac{2}{3}$ of a right angle?

4. How many degrees are there in 0.7 of a right angle?

Find the angle in degrees corresponding to the following:

5. $\frac{1}{20}$ revolution

6. $\frac{3}{8}$ revolution

7. $\frac{4}{5}$ revolution

8. 0.8 revolution

9. 0.25 revolution

10. 0.38 revolution

11. Look at each of the angles shown in Fig. 33.10. State which are acute, which are obtuse and which are reflex.

Fig. 33.10

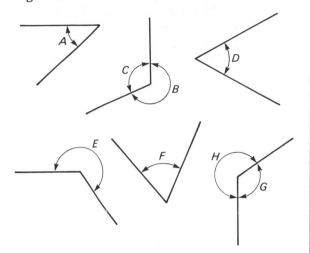

12. **(a)** Two angles are complementary. One of them is 63°. What is the size of the other.

    (b) Two angles are complementary. One is 23°. What is the size of the other?

    (c) Two angles are supplementary. One is 37°. What is the size of the other?

    (d) Angles A and B are supplementary. If $A = 108°$, what is the size of B?

Properties of Angles and Straight Lines

(1) The total angle on a straight line is 180° (Fig. 33.11). The angles A and B are called adjacent angles. They are also supplementary.

Fig. 33.11

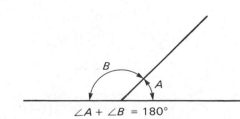

$\angle A + \angle B = 180°$

(2) When two straight lines intersect, the opposite angles are equal (Fig. 33.12). The angles A and C are called vertically opposite angles. Similarly the angles B and D are also vertically opposite angles.

Fig. 33.12

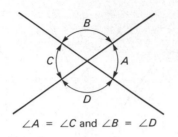

$$\angle A = \angle C \text{ and } \angle B = \angle D$$

Parallel Lines

Two straight lines in a plane that have no points in common, no matter how far they are produced, are called **parallel lines**.

When two parallel lines are cut by a transversal (Fig. 33.13):

(1) The corresponding angles are equal:
$a = l$; $b = m$; $c = p$; $d = q$.

(2) The alternate angles are equal: $d = m$; $c = l$.

(3) The interior angles are supplementary:
$d + l = 180°$; $c + m = 180°$.

Fig. 33.13

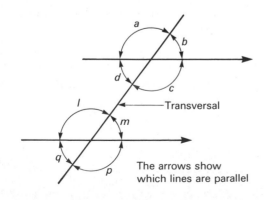

Transversal

The arrows show which lines are parallel

Conversely if two straight lines are cut by a transversal the lines are parallel if any **one** of the following is true:

(1) Two corresponding angles are equal.

(2) Two alternate angles are equal.

(3) Two interior angles are supplementary.

Example 3

Find the angle A shown in Fig. 33.14.

Fig. 33.14

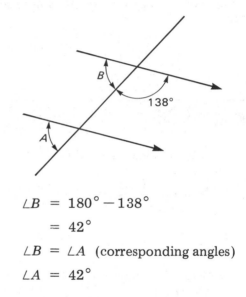

138°

$$\angle B = 180° - 138°$$
$$= 42°$$
$$\angle B = \angle A \quad \text{(corresponding angles)}$$
$$\angle A = 42°$$

Drawing Parallel Lines

Two set-squares or a set-square and a rule may be used to draw a line parallel to a given line. The procedure is as follows:

(1) Place the longest edge of a set-square along the given line (Fig. 33.15).

(2) Place the ruler or the second set-square along one of the other edges of the set-square as shown in the diagram.

(3) Slide the set-square along the rule until the required position is reached and then draw the parallel line.

Fig. 33.15

Fig. 33.15

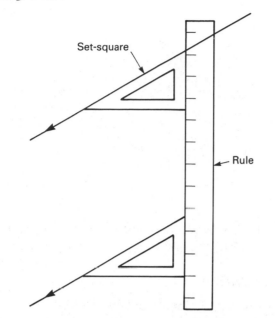

Dividing a Straight Line into a Number of Equal Parts

To divide a straight line AB into a number of equal parts.

Construction

Suppose that AB has to be divided into four equal parts. Draw AC at any angle to AB. Set off on AC, four equal parts AP, PQ, QR, RS of any convenient length. Join SB. Draw RV, QW and PX each parallel to SB. Then AX = XW = WV = VB (Fig. 33.16).

Fig. 33.16

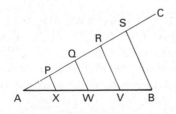

Perpendicular Lines

If two lines intersect (or meet) at right angles then the two lines are said to be **perpendicular** to each other.

Example 4

In Fig. 33.17 the line BF bisects $\angle$ABC. Find the value of the angle α.

Fig. 33.17

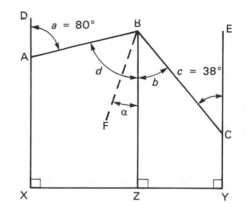

The lines AX, BZ and EY are all parallel because they are all perpendicular to XY.

$$c = b \quad \text{(alternate angles: BZ} \parallel \text{EY)}$$
$$b = 38° \quad \text{(since } c = 38°)$$
$$a = d \quad \text{(alternate angles: XD} \parallel \text{BZ)}$$
$$d = 80° \quad \text{(since } a = 80°)$$
$$\angle ABC = b + d$$
$$= 38° + 80°$$
$$= 118°$$
$$\angle FBC = 118° \div 2$$
$$= 59° \quad \text{(since BF bisects } \angle ABC)$$
$$b + \alpha = 59°$$
$$38° + \alpha = 59°$$
$$\alpha = 59° - 38°$$
$$= 21°$$

Constructing Perpendiculars

(1) To divide a line AB into two equal parts.

Construction

With A and B as centres and a radius greater than $\frac{1}{2}$AB, draw circular arcs which intersect at X and Y (Fig. 33.18). Join XY. The line XY divides AB into two equal parts and it is also perpendicular to AB.

Fig. 33.18

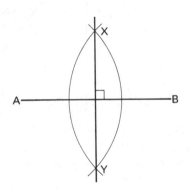

(2) To draw a perpendicular from a given point A on a straight line.

Construction

With centre A and any radius draw a circle to cut the straight line at points P and Q (Fig. 39.19). With centres P and Q and a radius greater than AP (or AQ) draw circular arcs to intersect at X and Y. Join XY. This line will pass through A and it is perpendicular to the given line.

Fig. 33.19

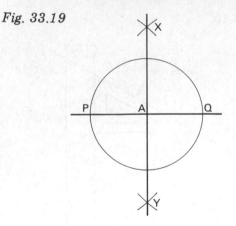

(3) To draw a perpendicular from a point A at the end of a line (Fig. 33.20).

Construction

From any point O outside the line and radius OA draw a circle to cut the line at B. Draw the diameter BC and join AC. AC is perpendicular to the straight line (because the angle in a semicircle is 90°).

Fig. 33.20

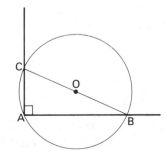

(4) To draw the perpendicular to a line AB from a given point P which is not on the line.

Construction

With P as centre draw a circular arc to cut AB at points C and D. With C and D as centres and a radius greater than $\frac{1}{2}$CD, draw circular arcs to intersect at E. Join PE. The line PE is the required perpendicular (Fig. 33.21).

Fig. 33.21

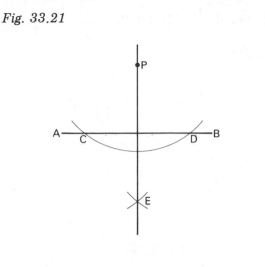

Finding a Mirror Line

In Chapter 31 we saw that the mirror line is the perpendicular bisector of the line joining the object and its image. We make use of this fact when finding the equation of a mirror line.

Example 5

The point $P(2, 1)$ is mapped on to $P'(5, 3)$ by a reflection. Draw a suitable construction to find the mirror line. Establish the equation of the mirror line.

    The construction is shown in Fig. 33.22 where the mirror line is the perpendicular bisector of the line PP'.

    To find the gradient of the mirror line we draw the right-angled triangle ABC in which $BC = 2$ units and $AB = 3$ units. Hence

$$\text{Gradient} = -\frac{3}{2}$$

$$= -1.5$$

    The intercept on the y-axis is seen to be 7.25 and therefore the equation of the mirror line is $y = -1.5x + 7.25$.

Fig. 33.22

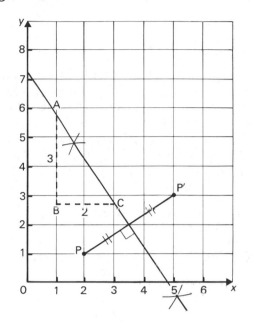

Constructing Angles

(1)   To construct an angle of $60°$.

Construction

Draw a line AB. With A as centre and any radius draw a circular arc to cut AB at D. With D as centre and the same radius draw a second arc to cut the first arc at C. Join AC. The angle CAD is then $60°$ (Fig. 33.23).

Fig. 33.23

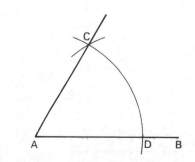

(2) To bisect a given angle ∠BAC.

Construction

With centre A and any radius draw an arc to cut AB at D and AC at E. With centres D and E and a radius greater than $\frac{1}{2}$DE draw arcs to intersect at F. Join AF, then AF bisects ∠BAC (Fig. 33.24). Note that by bisecting an angle of 60°, an angle of 30° is obtained. An angle of 45° is obtained by bisecting a right angle.

All other angles may be obtained by using a protractor.

Fig. 33.24

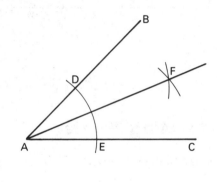

1. Find *x* in Fig. 33.25.

Fig. 33.25

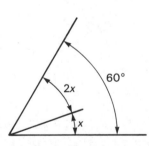

2. Find *A* in Fig. 33.26.

Fig. 33.26

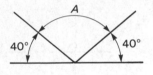

3. Find *x* in Fig. 33.27.

Fig. 33.27

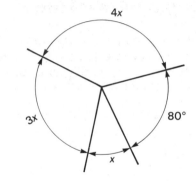

4. In Fig. 33.28 find *a*, *b*, *c* and *d*.

Fig. 33.28

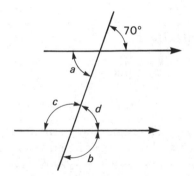

5. Find the angle x in Fig. 33.29.

Fig. 33.29

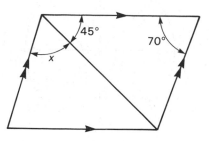

6. Find x in Fig. 33.30.

Fig. 33.30

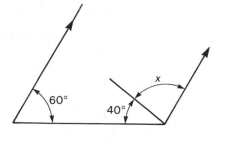

7. In Fig. 33.31, find A.

Fig. 33.31

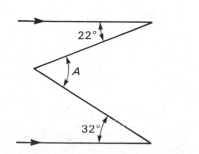

8. In Fig. 33.32, AB is parallel to ED. Find the angle x.

Fig. 33.32

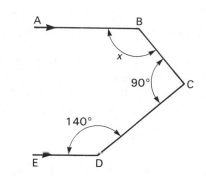

9. Find A in Fig. 33.33.

Fig. 33.33

10. In Fig. 33.34 the lines AB, CD and EF are parallel. Find the values of x and y.

Fig. 33.34

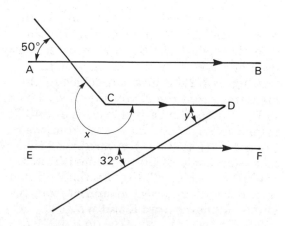

11. Draw the line AB = 8 cm. At A construct an angle of 60° and at B an angle of 45°. Hence complete the triangle ABC.

12. Construct an angle of 60°. Bisect this angle and so obtain an angle of 30°.

13. Draw a line AB = 8 cm. Construct the perpendicular bisector of AB to cut AB at E. Mark off EC = 3 cm and ED = 3 cm. Join A and C, C and B, B and D and D and A to form the quadrilateral ABCD.

14. Draw a line AB = 9 cm and divide it into 7 equal parts.

15. Draw the line XY = 7 cm. Mark off PX = 3 cm. Erect a perpendicular through P. Mark off PZ = 5 cm, Z being above XY. Hence complete the triangle XYZ.

16. The point A(2, 4) is mapped on to the point A'(6, 12) by a reflection. Draw an accurate construction to show the mirror line and hence find its equation.

17. The point P(1, 2) is mapped on to the point P'(6, 17) by a reflection. Find the equation of the mirror line.

Bearings

The four cardinal directions are north, south, east and west (Fig. 33.35). The directions NE, NW, SE and SW are frequently used and are as shown in Fig. 33.35. A bearing of N20°E means an angle of 20° measured from the N towards E as shown in Fig. 33.36. Similarly a bearing of S40°E means an angle 40° measured from the S towards E (Fig. 33.37). A bearing of N50°W means an angle of 50° measured from N towards W (Fig. 33.38). **Bearings quoted in this way are always measured from N and S and never from E and W.**

Fig. 33.35

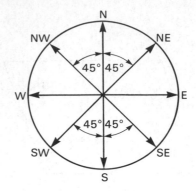

Fig. 33.36

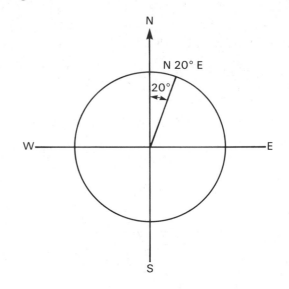

Fig. 33.37

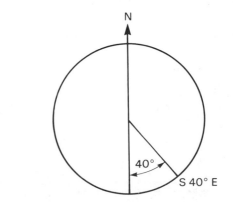

Fig. 33.38

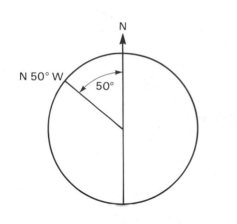

However, bearings are also measured from north in a clockwise direction, N being taken as 0°. Three figures are always stated. For example 005° is written instead of 5° and 035° instead of 35° and so on. East will be 090°, south 180° and west 270°. Some typical bearings are shown in Fig. 33.39.

Fig. 33.39

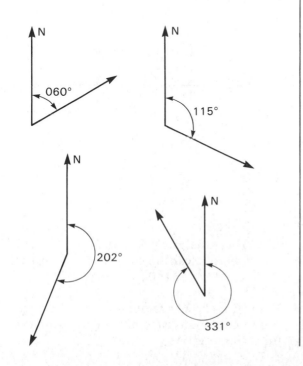

Example 6

(a) B is a point due east of a point A on the coast. C is another point on the coast and is 6 km due south of A. The distance BC is 7 km. Calculate the bearing of C from B.

Fig. 33.40

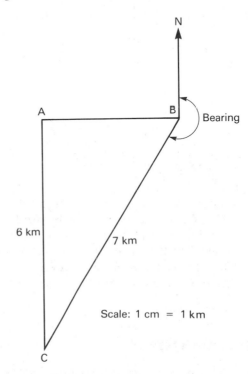

Scale: 1 cm = 1 km

This problem may be solved by making a scale drawing (as below) or by using trigonometry (see Chapter 38).

To make a scale drawing we first choose a suitable scale to represent the distances AB and AC (see Fig. 33.40). A scale of 1 cm = 1 km is suggested. The bearing of C from B is found by using a protractor and it is 211°.

(b) B is 5 km due north of P and C is 2 km due east of P. A ship started from C and steamed in a direction N30°E. Calculate the distance the ship had to go before it was due east of B. Find also the distance it is then from B (Fig. 33.41).

Fig. 33.41

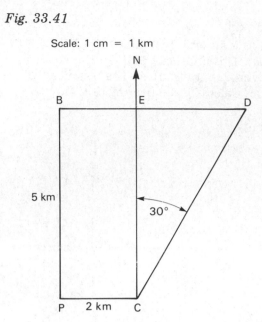

Scale: 1 cm = 1 km

Using a scale drawing we find that CD = 5.8 km and hence the ship has to sail 5.8 km to be due east of B. The length of BD represents the distance the ship is from B. By scaling the drawing the ship is found to be 4.9 km from B.

Back Bearings

In Fig. 33.42, the bearing of Q from P is θ. The bearing of P from Q is known as the **back bearing**.

Fig. 33.42

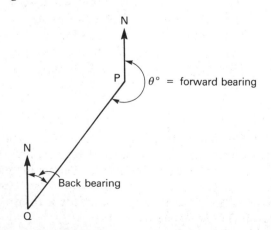

Example 7

(a) The bearing of a point B from a point A is 065°. What is the bearing of A from B?

> The situation is shown in Fig. 33.43 where it can be seen that the bearing of A from B is 245°.

Fig. 33.43

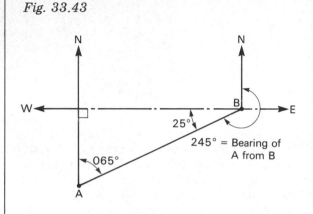

(b) The bearing of a point Q from a point P is 220°. What is the bearing of P from Q?

Fig. 33.44

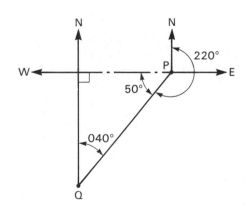

The situation is shown in Fig. 33.44 where it can be seen that the bearing of P from Q is 040°.

From the above examples we see that we can calculate the back bearing by using the following rules:

(1) If the forward bearing is less than 180° then to find the back bearing add 180° to the forward bearing. Thus in Example 6(a)

$$\text{Back bearing} = 065° + 180°$$
$$= 245°$$

(2) If the forward bearing is greater than 180° then to find the back bearing subtract 180° from the forward bearing. Thus in Example 6(b).

$$\text{Back bearing} = 220° - 180°$$
$$= 40°$$

Exercise 33.3

1. Write each of the following as three-digit bearings:
   - (a) N39°E
   - (b) N54°W
   - (c) S33°W
   - (d) S72°E

2. Write as three-digit bearings the following directions:
   - (a) south-west
   - (b) south-east
   - (c) north-west
   - (d) west
   - (e) south

3. A point Q lies on a bearing of N48°E from a point P. What is the bearing of P from Q?

4. If AN (Fig. 33.45) is due north, write down the bearing of A from B.

Fig. 33.45

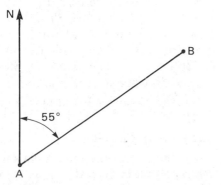

5. A ship is on a bearing of S65°E from a lighthouse. What is the bearing of the lighthouse from the ship?

6. P is a point due west of a harbour H and Q is a point which is 5 km due south of H. If the distance PH is 7 km, find by making a scale drawing the bearing of Q from P.

7. A ship is on a bearing of 068° from a lighthouse. What is the bearing of the lighthouse from the ship?

8. B and C (Fig. 33.46) are both 100 km from A. C is on a bearing of 225° from B. Find by scale drawing the bearing of A from B and the size of the angle ABC.

Fig. 33.46

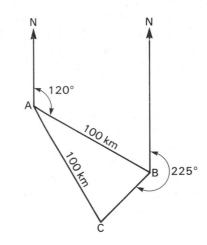

9. The bearing of Y from X is 320°. Find the bearing of X from Y.

10. Point C is 50 km from A on a bearing of 320°. Point B is 60 km from A on a bearing of 230°. Make a scale drawing showing the relative positions A, B and C using a scale of 1 cm to 10 km. Hence find the distance between the points B and C and state the bearings of B from C and C from B.

11. A ship sets out from a point A and sails due north to a point B, a distance of 120 km. It then sails due east to a point C. If the bearing of C from A is 037°, find:

    (a) the distance BC

    (b) the distance AC.

12. A boat leaves a harbour A on a course of S60°E and it sails 50 km in this direction until it reaches a point B. How far is B east of A? What distance south of A is B?

13. X is a point due west of a point P. Y is a point due south of P. If the distances PX and PY are 10 km and 15 km respectively, find the bearing of X from Y.

14. B is 10 km north of P and C is 5 km due west of P. A ship starts from C and sails in a direction of 330°. Find the distance the ship has to sail before it is due west of B and find also the distance it is then from B.

15. A fishing boat places a float on the sea at A, 50 metres due north of a buoy B. A second boat places a float at C, whose bearing from A is S30°E. A taut net connecting the floats at A and C is 80 metres long. Find the distance BC and the bearing of C from B.

Miscellaneous Exercise 33

Section A

1. The points A, B and C (Fig. 33.47) form an equilateral triangle. If the bearing of B from A is 050°, find by making a scale drawing:

    (a) the bearing of C from A

    (b) the bearing of B from C.

Fig. 33.47

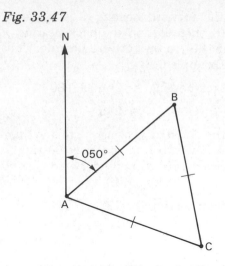

2. In Fig. 33.48, the straight lines ABC and EF are parallel. Find the values of the angles x and y.

Fig. 33.48

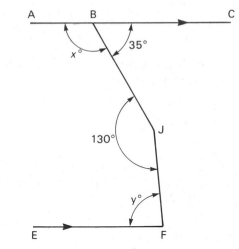

3. A and B are two points on a coast, B being 2000 m due east of A. A man sails from A on a bearing of 060° for 800 m to a buoy P, where he changes course to 070° and sails for another 1000 m to another buoy Q. Make a scale drawing and from it find:

    (a) the distances P and Q from the line AB

    (b) the distance QB

    (c) the course the man must set to sail directly from Q to B.

4. A point B is 10 km due north of a point A, and at mid-day a girl sets out to walk from A to B. After covering 5 km, she changes the direction of her walk to 036° and covers another 14 km to reach point C. By making a scale drawing find:

   (a) her distance from B when she was nearest to B

   (b) the time at which she was due east of B, assuming that she walked at an average speed of 4 km/h

   (c) the shortest distance from A to C.

5. Three towns A, B and C lie on a straight road running east from A. B is 6 km from A and C is 22 km from A. Another town D lies to the north of this road and it lies 10 km from both B and C. Make a scale drawing of this information and from it find the distance of D from A and the bearing of D from A. What is the bearing of A from D?

Section B

1. In Fig. 33.49, AB and CD are parallel lines and EF and EG are transversals. Calculate the sizes of the angles a, b, c and d.

Fig. 33.49

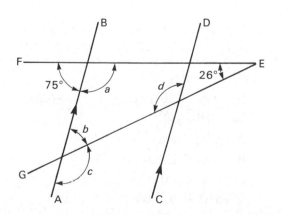

2. A ship leaves a port P and steams on a course of 045° for one hour at 18 km/h to reach a point A. At A the ship changes course to 130° and steams for a further $1\frac{1}{2}$ hours at the same speed to reach point B.

   (a) Using a scale of 1 cm to represent 3 km draw a diagram of the course of the ship.

   (b) Use your diagram to determine the bearing of B from P.

   (c) On reaching B the ship changes course and speed and steams for a further hour to a point C which is 21 km from P and on a bearing of 120° from P. Determine the speed and course set by the ship in going from B to C.

3. Fig. 33.50 shows three towns A, B and C.

   (a) Find the bearing of
      (i) B from A
      (ii) B from C.

   (b) Using a scale of 1 cm to 8 km make a scale drawing and from it find the distance of town C from town A.

Fig. 33.50

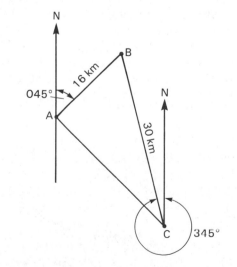

4. In Fig. 33.51, AD, BF and DH are straight lines and CE = CD. Calculate the sizes of the angles p, q, r, s, t and u.

Fig. 33.51

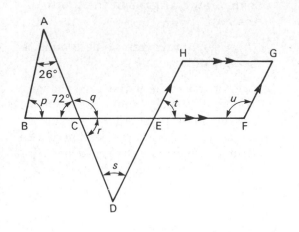

5. In Fig. 33.52, P and Q are two coast-guard stations 12 miles apart, Q being due east of P. Two markers X and Y are anchored in the sea on the straight line PQ so that PX = 2 miles and YQ = 4 miles. Using a scale of 1 cm = 1 mile show the positions of P, Q, X and Y on a scale drawing. Show the position of a third marker S, which is in the sea so that S is equidistant from X and Y and 8 miles from P. Use your diagram to find the distance and bearing of S from Q.

Fig. 33.52

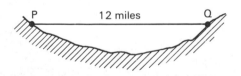

Multi-Choice Questions 33

1. If two parallel lines are cut by a third straight line, which of the following are true?

   (i)   Alternate angles are supplementary.

   (ii) Corresponding angles are equal.

   (iii) Co-interior angles are supplementary.

   (iv) Vertically opposite angles are complementary.

   A (i) and (ii) only

   B (ii) and (iii) only

   C (iii) and (iv) only

   D (i) and (iv) only

2. In Fig. 33.53, POQ and SOR are straight lines. The size of the angle TOR is

   A 35°   B 30°   C 25°   D 15°

Fig. 33.53

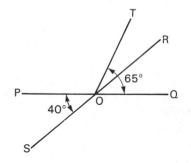

3. Fig. 33.54 shows the construction of the bisector of ∠XOY. Which of the following must be true if the construction is correct?

   (i)   OQ = OP   (ii) OQ = QR

   (iii) QR = PR

   A all of them

   B (i) and (ii) only

   C (i) and (iii) only

   D (ii) and (iii) only

Fig. 33.54

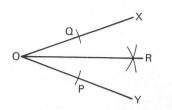

4. A bearing of S40°E can be expressed as a bearing of

   A 040° B 050° C 130° D 140°

5. In Fig. 33.55

   A $x = y$ B $x = 180° + y$

   C $x = y - 180°$ D $x + y = 180°$

Fig. 33.55

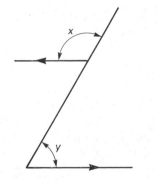

6. From a place 400 m north of X, a man walks eastwards to a place Y which is 800 m from X. What is the bearing of X from Y?

   A 270° B 240° C 210° D 180°

7. In the triangle PQR (Fig. 33.56), what is the bearing of Q from R?

   A N45°W   B S45°W

   C N20°W   D N45°E

Fig. 33.56

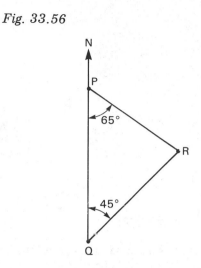

Use Fig. 33.57 to answer Questions 8 and 9.

8. The bearing of B from A is

   A 030° B 060° C 120° D 150°

9. The bearing of C from A is

   A 045° B 135° C 225° D 315°

Fig. 33.57

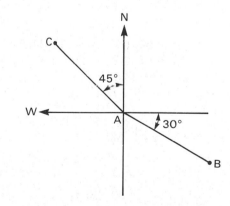

Triangles

Types of Triangle

(1) An **acute-angled** triangle has all its angles less than 90° (Fig. 34.1).

Fig. 34.1

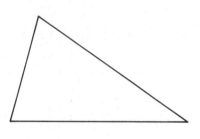

(2) A **right-angled** triangle has one of its angles equal to 90°. The side opposite to the right angle is the longest side and it is called the hypotenuse (Fig. 34.2).

Fig. 34.2

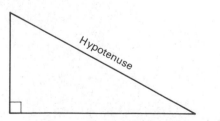

(3) An **obtuse-angled** triangle has one angle greater than 90° (Fig. 34.3).

Fig. 34.3

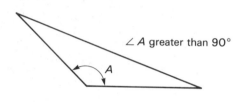

∠A greater than 90°

(4) A **scalene** triangle has all three sides of different length.

(5) An **isosceles** triangle has two sides and two angles equal. The equal angles lie opposite to the equal sides (Fig. 34.4).

Fig. 34.4

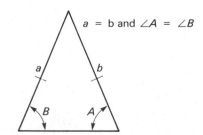

a = b and ∠A = ∠B

(6) An **equilateral** triangle has all its sides and angles equal. Each angle of the triangle is 60° (Fig. 34.5).

Fig. 34.5

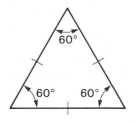

Angle Properties of Triangles

(1) The sum of the angles of a triangle are equal to 180° (Fig. 34.6).

Fig. 34.6

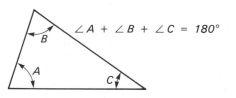

$$\angle A + \angle B + \angle C = 180°$$

(2) In every triangle the greatest angle is opposite to the longest side. The smallest angle is opposite to the shortest side. In every triangle the sum of the lengths of any two sides is always greater than the length of the third side (Fig. 34.7).

Fig. 34.7

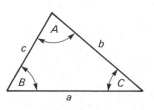

a is the longest side since it lies opposite to the greatest angle *A*. *c* is the shortest side since it lies opposite to the smallest angle *C*. *a* + *b* is greater than *c*, *a* + *c* is greater than *b* and *b* + *c* is greater than *a*.

(3) When the side of a triangle is produced the exterior angle so formed is equal to the sum of the opposite interior angles (Fig. 34.8).

Fig. 34.8

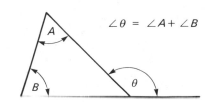

$$\angle \theta = \angle A + \angle B$$

Example 1

In Fig. 34.9, find the angles x and y.

Fig. 34.9

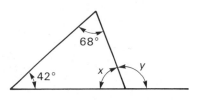

Since the three angles of a triangle add up to 180°.

$$x + 42° + 68° = 180°$$
$$x = 180° - 42° - 68°$$
$$= 70°$$

Hence the angle x is 70°.

The exterior angle of a triangle is equal to the sum of the opposite interior angles. Hence

$$y = 42° + 68°$$
$$= 110°$$

Therefore the angle y is 110°.

Exercise 34.1

Find the angles x and y shown in Fig. 34.10.

Fig. 34.10

1.

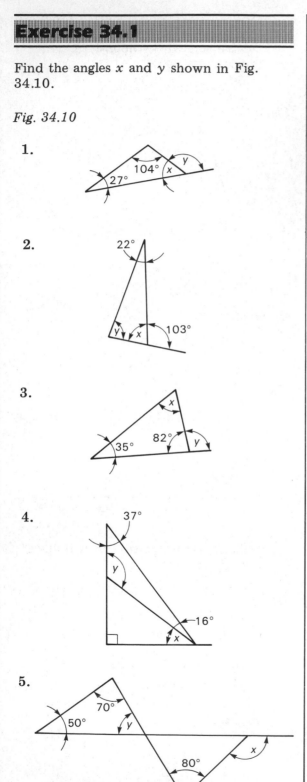

2.

3.

4.

5.

6.

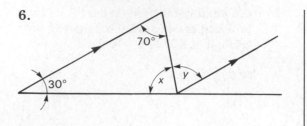

Standard Notation for a Triangle

Fig. 34.11 shows the **standard notation for a triangle**. The three vertices are marked A, B and C. The angles are called by the same letter as the vertices (see diagram). The side a lies opposite the angle A, b lies opposite the angle B and c lies opposite the angle C.

Fig. 34.11

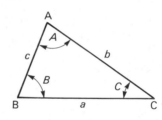

Pythagoras' Theorem

In any right-angled triangle the square on the hypotenuse is equal to the sum of the squares on the other two sides. In the diagram (Fig. 34.12),

$$AC^2 = AB^2 + BC^2$$

or $\qquad b^2 = a^2 + c^2$

Fig. 34.12

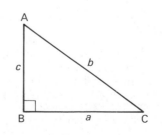

The hypotenuse is the longest side and it always lies opposite to the right angle. Thus in Fig. 34.12 the side *b* is the hypotenuse since it lies opposite to the right angle at B. It is worth remembering that triangles with sides of 3:4:5; 5:12:13; 7:24:25 are right-angled triangles.

Example 2

(a) In $\triangle ABC$, $\angle B = 90°$, $a = 4.2$ cm and $c = 3.7$ cm. Find b (Fig. 34.13).

Fig. 34.13

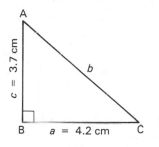

By Pythagoras' theorem,

$$b^2 = a^2 + c^2$$
$$b^2 = 4.2^2 + 3.7^2$$
$$= 17.64 + 13.69$$
$$= 31.33$$
$$b = \sqrt{31.33}$$
$$= 5.597 \text{ cm}$$

(b) In $\triangle ABC$, $\angle A = 90°$, $a = 6.4$ cm and $b = 5.2$ cm. Find c (Fig. 34.14).

Fig. 34.14

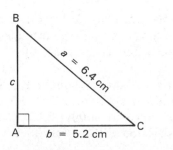

$$a^2 = b^2 + c^2$$
$$\text{or} \quad c^2 = a^2 - b^2$$
$$= 6.4^2 - 5.2^2$$
$$= 40.96 - 27.04$$
$$= 13.92$$
$$c = \sqrt{13.92}$$
$$= 3.731 \text{ cm}$$

Properties of the Isosceles Triangle

It will be remembered that an isosceles triangle (Fig. 34.15) has two sides and two angles equal. The equal angles lie opposite to the equal sides.

Fig. 34.15

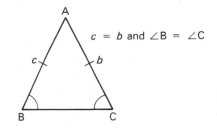

$c = b$ and $\angle B = \angle C$

Example 3

In Fig. 34.16, find the size of the angles A and C.

Fig. 34.16

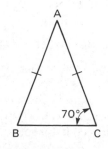

Since   AC = AB, △ABC is isosceles.

Hence

$$\angle B = \angle C$$
$$= 70°$$

Since the sum of the angles of a triangle is 180°,

$$\angle A = 180° - (70° + 70°)$$
$$= 180° - 140°$$
$$= 40°$$

In Fig. 34.17, ABC is isosceles and AD is perpendicular to the base BC. The two triangles ABD and ADC are congruent and hence ABC is symmetrical about AD which is the axis of symmetry. This means that:

(1)   AD bisects the unequal side (i.e. BD = DC).

(2)   AD bisects the apex angle BAC (i.e. $\angle BAD = \angle DAC$).

Fig. 34.17

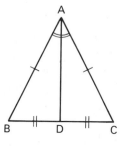

Example 4

An isosceles triangle has equal sides 6 cm long and a base 4 cm long.

(a) Find the altitude of the triangle.

(b) Calculate the area of the triangle.

   (a) The triangle is shown in Fig. 34.18. The altitude AD is perpendicular to the base and hence it bisects the base.

Fig. 34.18

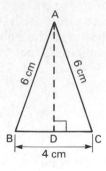

In triangle ABD, by Pythagoras' theorem,

$$AD^2 = AB^2 - BD^2$$
$$= 6^2 - 2^2$$
$$= 32$$
$$AD = \sqrt{32}$$
$$= 5.66$$

Hence the altitude of the triangle is 5.66 cm.

(b)   Area of triangle $= \frac{1}{2} \times$ base
$$\times \text{ altitude}$$
$$= \frac{1}{2} \times 4 \times 5.66$$
$$= 11.32 \text{ cm}^2$$

Properties of the Equilateral Triangle

An equilateral triangle has all of its sides equal in length and all of its angles equal in size. Each angle is therefore 60°.

In Fig. 34.19 △ABC is equilateral. It possesses three axes of symmetry AD, BE and CF. Hence each of the angles BAD, CAD, ACF, BCF, ABE and EBC equals 30°.

Fig. 34.19

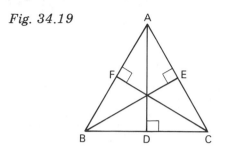

Example 5

Triangle ABC (Fig. 34.20) is equilateral with sides 8 cm long. Calculate BD, the vertical height of the triangle and work out its area.

Since BD is the perpendicular dropped from the vertex B,

$$AD = \tfrac{1}{2}AC$$
$$= \tfrac{1}{2} \times 8 \text{ cm}$$
$$- 4 \text{ cm}$$

In triangle ADB, AB is the hypotenuse and by Pythagoras' theorem,

$$BD^2 = AB^2 - AD^2$$
$$= 8^2 - 4^2$$
$$= 48$$
$$BD = \sqrt{48}$$
$$= 6.928 \text{ cm}$$
$$\text{Area of } \triangle ABC = \tfrac{1}{2} \times AC \times BD$$
$$= \tfrac{1}{2} \times 8 \times 6.928$$
$$= 27.71 \text{ cm}^2$$

Fig. 34.20

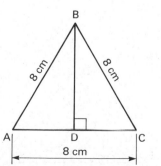

Exercise 34.2

1. Find the side a in Fig. 34.21.

Fig. 34.21

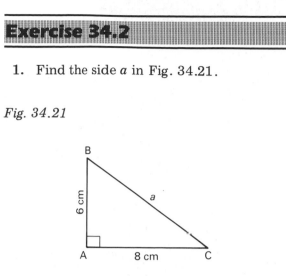

2. Find the side b in Fig. 34.22.

Fig. 34.22

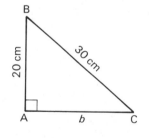

3. Find the side c in Fig. 34.23.

Fig. 34.23

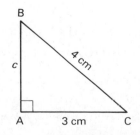

4. Find the sides marked x in Fig. 34.24.

Fig. 34.24

(a)

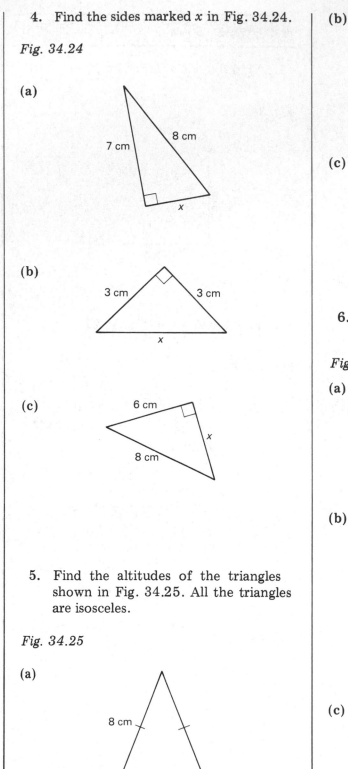

(b)

(c)

(b)

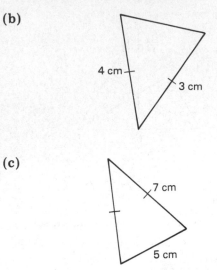

(c)

6. Find the angles marked θ for each of the isosceles triangles in Fig. 34.26.

Fig. 34.26

(a)

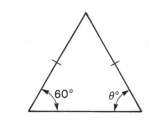

(b)

5. Find the altitudes of the triangles shown in Fig. 34.25. All the triangles are isosceles.

Fig. 34.25

(a)

(c)

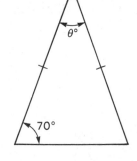

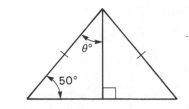

7. Find the angles marked x, y and z in Fig. 34.27.

Fig. 34.27

(a)

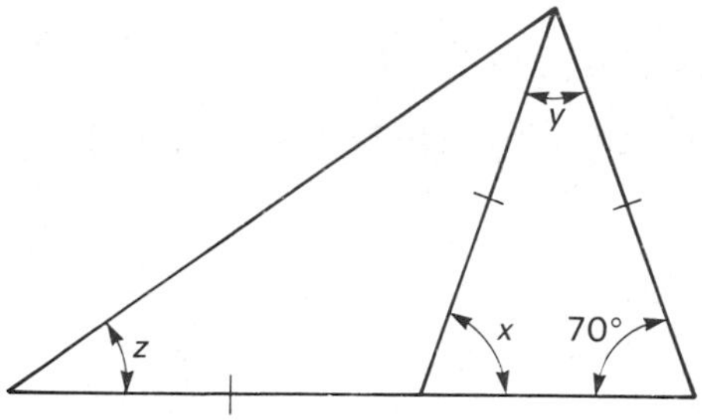

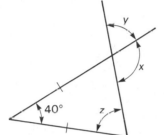

(b)

8. An equilateral triangle has sides 16 cm long. Calculate its vertical height and hence find its area.

9. What is the altitude of an equilateral triangle whose sides are 11 cm long?

10. Triangle ABC (Fig. 34.28) is isosceles with AB = BC. Also AD = BD. Calculate the size of the angles a, b, c, d and e.

Fig. 34.28

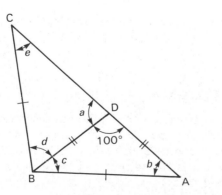

Construction of Triangles

(1) To construct a triangle given the lengths of each of the three sides.

Construction

Suppose $a = 6$ cm, $b = 3$ cm and $c = 4$ cm. Draw BC = 6 cm. With centre B and radius 4 cm draw a circular arc. With centre C and radius 3 cm draw a circular arc to cut the first arc at A. Join AB and AC. Then ABC is the required triangle (Fig. 34.29).

Fig. 34.29

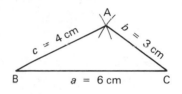

(2) To construct a triangle given two sides and the included angle between the two sides.

Construction

Suppose $b = 5$ cm and $c = 6$ cm and $\angle A = 60°$. Draw AB = 6 cm and draw AX such that $\angle BAX = 60°$. Along AX mark off AC = 5 cm. Then ABC is the required triangle (Fig. 34.30).

Fig. 34.30

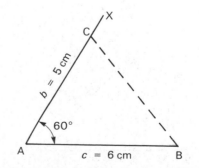

(3) To construct a triangle (or triangles) given the lengths of two of the sides and an angle which is not the included angle between the two given sides.

Construction

(a) Suppose $a = 5\text{ cm}$, $b = 6\text{ cm}$ and $\angle B = 60°$. Draw BC $= 5\text{ cm}$ and draw BX such that $\angle CBX = 60°$. With centre C and radius of 6 cm describe a circular arc to cut BX at A. Join CA then ABC is the required triangle ABC (Fig. 34.31).

Fig. 34.31

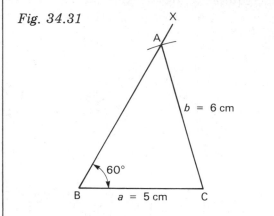

(b) Suppose that $a = 5\text{ cm}$, $b = 4.5\text{ cm}$ and $\angle B = 60°$. The construction is the same as before but the circular arc drawn with C as centre now cuts BX at two points A and A_1. This means that there are two triangles which meet the given conditions, i.e., $\triangle$s ABC and A_1BC (Fig. 34.32). For this reason this case is often called the **ambiguous case**.

Fig. 34.32

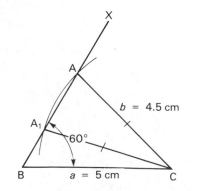

(4) To construct a triangle given two angles and the length of the side opposite to the third angle.

Construction

Suppose that in triangle ABC, $a = 4.7\text{ cm}$, $\angle B = 54°$ and $\angle C = 46°$.

Draw BC $= 4.7\text{ cm}$ (Fig. 34.33). At B, draw a line making an angle of 54° with BC.

Fig. 34.33

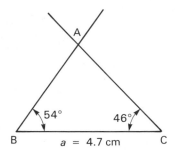

At C, and on the same side of BC as above, draw a line making an angle of 46° with BC. with BC.

The point of intersection of the two lines is A.

(5) To draw the circumscribed circle of a given triangle ABC.

Construction

Construct the perpendicular bisectors of the sides AB and AC so that they intersect at O. With centre O and radius AO draw a circle which is the required circumscribed circle (Fig. 34.34).

Fig. 34.34

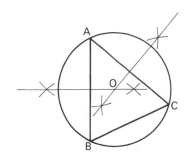

(6)   To draw the inscribed circle of a given triangle ABC.

Construction

Construct the internal bisectors of ∠B and ∠C to intersect at O. With centre O draw the inscribed circle of the triangle ABC (Fig. 34.35).

Fig. 34.35

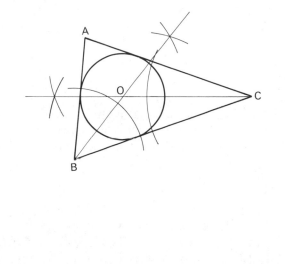

Construct each of the triangles ABC listed in the table below. a, b and c are stated in centimetres.

Triangle	∠A	∠B	∠C	a	b	c
1.	75°	34°		10		
2.	19°		105°			11
3.	116°		18°	8.5		
4.	50°			7		9
5.				5	8	7
6.	43°			7	9	
7.	62°				7	9
8.		84°		5	6	
9.				7	8	10
10.			85°		7	9
11.		60°	70°	8		
12.	110°	40°				7
13.	83°				5.3	4.4

14. Draw the inscribed circles for the triangles of questions 1, 5 and 9.

15. Draw the circumscribed circles for the triangles of questions 3, 4 and 12.

Congruent Triangles

Triangles which are exactly the same shape and size are said to be **congruent**.

Two triangles are congruent if the six elements of one triangle (i.e. three sides and three angles) are equal to the six elements of the second triangle. In Fig. 34.36,

AC = XZ   AB = XY   BC = YZ

∠B = ∠Y   ∠C = ∠Z   ∠A = ∠X

and the two triangles have a scale factor of 1 (see Chapter 31).

Fig. 34.36

Note that the angles which are equal lie opposite to the corresponding sides.

If two triangles are congruent they will also be equal in area. The notation used to express the fact that $\triangle ABC$ is congruent to $\triangle XYZ$ is $\triangle ABC \equiv \triangle XYZ$.

For two triangles to be congruent the six elements of one triangle (three sides and three angles) must be equal to the six elements of the second triangle. However, to prove that two triangles are congruent it is not necessary to prove all six equalities. Any of the following are sufficient to prove that two triangles are congruent:

(1)   One side and two angles in one triangle equal to one side and two similarly located angles in the second triangle (Fig. 34.37).

Fig. 34.37

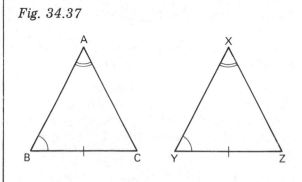

(2)   Two sides and the angle between them in one triangle equal to two sides and the angle between them in the second triangle (Fig. 34.38).

Fig. 34.38

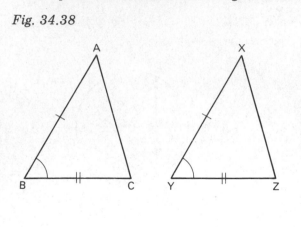

(3)   Three sides of one triangle equal to three sides of the other triangle (Fig. 34.39).

Fig. 34.39

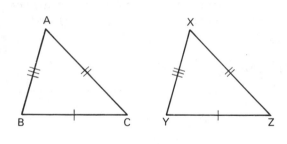

(4)   In right-angled triangles the hypotenuses are equal and one other side in each triangle is also equal (Fig. 34.40).

Fig. 34.40

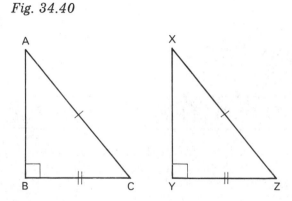

Note carefully that the following do **not** prove congruency (Fig. 34.41).

(i) Three angles in the one triangle equal to three angles in the second triangle.

(ii) Two sides and an angle which does not lie between these sides in the one triangle equal to two sides and a similarly located angle in the second triangle.

(iii) Two angles and a non-corresponding side. In each diagram below △ABC is not necessarily congruent to △XYZ.

Fig. 34.41

(i)

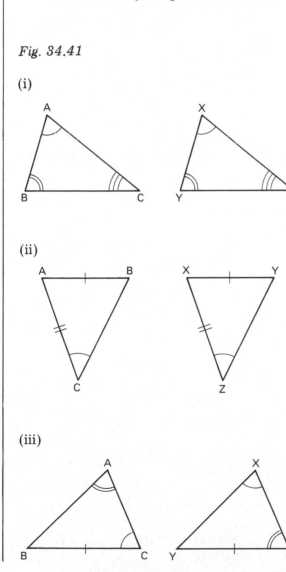

(ii)

(iii)

Example 6

(a) In Fig. 34.42, find the length of AB.

Fig. 34.42

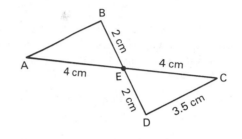

In triangles AEB and CED

$$BE = ED = 2\,cm \quad \text{(given)}$$

$$AE = EC = 4\,cm \quad \text{(given)}$$

∠AEB = ∠CED (vertically opposite angles)

We now have two sides and the angle between them in △AEB equal to two sides and the angle between them in △CED. Hence △AEB ≡ △CED.

∴ $$AB = CD$$

$$= 3.5\,cm$$

(b) The diagonals of the quadrilateral XYZW intersect at O. Given that OX = OW and OY = OZ show that XY = WZ.

Fig. 34.43

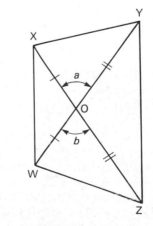

Referring to Fig. 34.43:

In △s XOY and WOZ

OX = OW   and   OY = OZ   (given)

$a = b$ (vertically opposite angles)

Hence the two sides and the included angle in △XOY equal two sides and the included angle in △WOZ. Hence △XOY ≡ △WOZ.

Therefore XY = WZ.

Exercise 34.4

1. State which of the pairs of triangles (Fig. 34.44) are definitely congruent.

Fig. 34.44

(a)

(b)

(c)

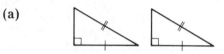

(d)

(e)

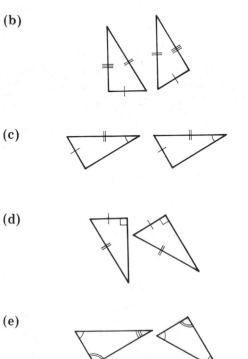

(f)

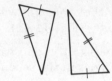

(g)

(h)

2. In Fig. 34.45, find the lengths CD and EC.

Fig. 34.45

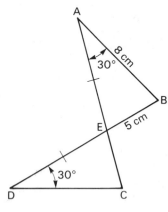

3. In Fig. 34.46, name all the triangles which are congruent. (G, H and J are the mid-points of DF, EF and DE respectively.)

Fig. 34.46

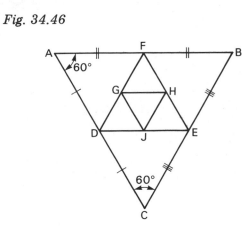

4. In Fig. 34.47, PQ = RS and PQ is
parallel to RS. If TS = 8 cm and
TR = 6 cm, find the lengths of TP
and TQ.

Fig. 34.47

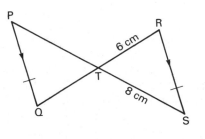

5. Figure 34.48 shows two right-angled
triangles ABC and XYZ. Find the length
of BC.

Fig. 34.48

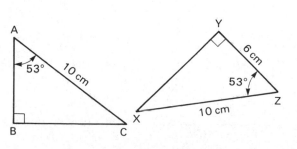

6. Figure 34.49 shows two isosceles
triangles LMP and RST. Find the
lengths of

(a) ST (b) LX (c) RY

Find the size of the angles
(d) MLP (e) SRY

Fig. 34.49

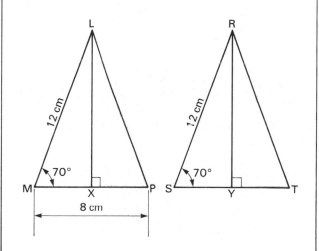

7. In Fig. 34.50, ABCD is a quadrilateral
with diagonals which meet at E. Find
the length of CD.

Fig. 34.50

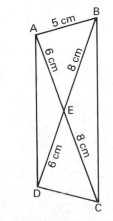

8. In Fig. 34.51, BC and AD are perpendicular to each other. Find the lengths
of CD and AB.

Fig. 34.51

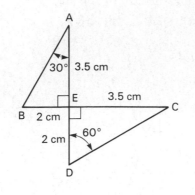

9. In Fig. 34.53, DE is parallel to BC, EF
is parallel to AB and AE = EC. Find

(a) the size of ∠ADE

(b) the size of ∠FEC

(c) the length of EF.

Fig. 34.52

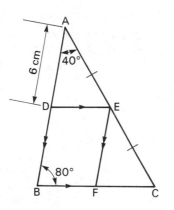

10. In Fig. 34.53, O is the centre of the
circle and PS and PT are tangents to
the circle. If PT = 8 cm and
∠TPS = 40°, find the length of PO
and the size of ∠TPO.

Fig. 34.53

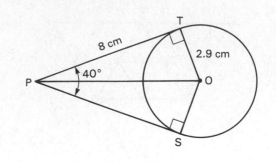

Similar Triangles

If two triangles have the same shape but one
is an enlargement of the other they are said
to be similar. The two triangles must be
equi-angular.

Thus in Fig. 34.54 if ∠A = ∠X, ∠B = ∠Y
and ∠C = ∠Z the triangles ABC and XYZ
are similar. In similar triangles the ratios
of corresponding sides are equal. Thus for
the triangles shown in Fig. 34.54

$$\frac{a}{x} = \frac{b}{y} = \frac{c}{z} = \frac{H}{h} = k$$

where k is the scale factor.

Fig. 34.54

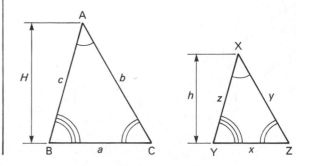

Note that by corresponding sides we mean the sides opposite to the equal angles. It helps in solving problems on similar triangles if we write the two triangles with the equal angles under each other. Thus in △s ABC and XYZ if $\angle A = \angle X$, $\angle B = \angle Y$ and $\angle C = \angle Z$

we write $\dfrac{ABC}{XYZ}$

The equations connecting the sides of the triangles are then easily obtained by writing any two letters in the first triangle over any two corresponding letters in the second triangle. Thus,

$$\frac{AB}{XY} = \frac{AC}{XZ} = \frac{BC}{YZ}$$

In Fig. 34.55 to prove △ABC is similar to △XYZ it is sufficient to prove any one of the following:

Fig. 34.55

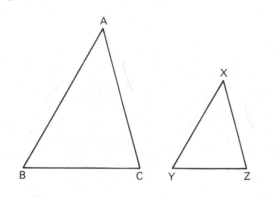

(1) Two angles in △ABC are equal to two angles in △XYZ. For instance, the triangles are similar if $\angle A = \angle X$ and $\angle B = \angle Y$, since it follows that $\angle C = \angle Z$.

(2) The three sides of △ABC are proportional to the corresponding sides of △XYZ. Thus △ABC is similar to △XYZ if,

$$\frac{AB}{XY} = \frac{AC}{XZ} = \frac{BC}{YZ}$$

(3) Two sides in △ABC are proportional to two sides in △XYZ and the angles included between these sides in each triangle are equal. Thus △ABC is similar to △XYZ if,

$$\frac{AB}{XY} = \frac{AC}{XZ} \quad \text{and} \quad \angle A = \angle X$$

Example 7

(a) In Fig. 34.56 find the dimension marked x.

Fig. 34.56

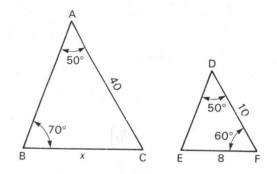

In △ABC,

$$\angle C = 180° - 50° - 70°$$
$$= 60°$$

In △DEF,

$$\angle E = 180° - 50° - 60°$$
$$= 70°$$

Therefore △ABC and △DEF are similar.

$$k = \frac{40}{10}$$
$$= 4$$

Since BC and EF are corresponding sides

$$x = 4 \times 8 \text{ cm}$$
$$= 32 \text{ cm}$$

(b) In Fig. 34.57 prove that △s PTS and PQR are similar and calculate the length of TS.

Fig. 34.57

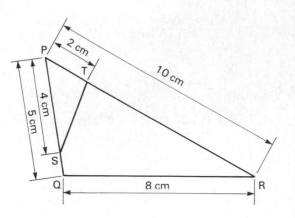

In △s PTS and PQR

$$\frac{PS}{PR} = \frac{4}{10}$$

$$= 0.4$$

$$\frac{PT}{PQ} = \frac{2}{5}$$

$$= 0.4$$

Therefore $\dfrac{PS}{PR} = \dfrac{PT}{PQ}$

Also ∠P is common to both triangles and it is the included angle between PS and PT in △PTS and PR and PQ in △PQR. Hence △s PTS and PQR are similar.

Writing $\dfrac{\triangle PTS}{\triangle PQR}$ we see that

$$\frac{TS}{QR} = \frac{PT}{PQ}$$

$$\frac{TS}{8} = \frac{2}{5}$$

$$TS = \frac{2 \times 8}{5}$$

$$= 3.2 \text{ cm}$$

Exercise 34.5

1. Figure 34.58 shows a number of triangles. Write down the letters representing the triangles which are similar. You should be able to find four sets of similar triangles.

Fig. 34.58

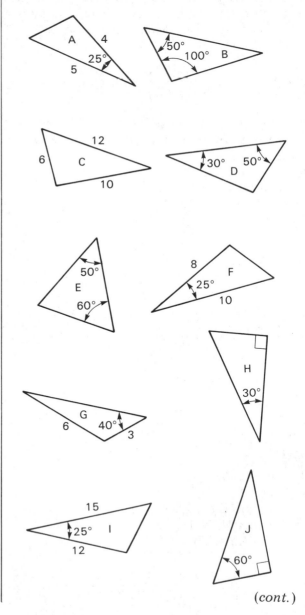

(cont.)

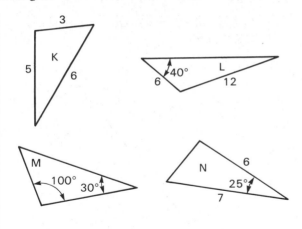

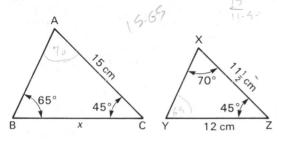

2. In Fig. 34.59, calculate the length marked x.

Fig. 34.59

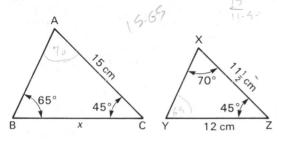

3. In Fig. 34.60, DE is parallel to BC, AD = 6 cm, BD = 3 cm and DE = 4 cm. Calculate:

(a) the length of BC

(b) the ratio $\dfrac{AE}{AC}$.

Fig. 34.60

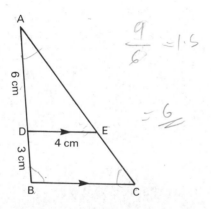

4. Figure 34.61 shows a triangle ABC. D is the mid-point of BC and E is the mid-point of AC. AD and BE intersect at G. The line through C parallel to AD intersects BE produced at H.

(a) Write down which angles of triangle BHC are equal to each of the angles of triangle BGD.

(b) State which angles of triangle CEH are equal to each of the angles of triangle AEG.

(c) Using the properties of similar triangles, find the ratio GD : HC.

Fig. 34.61

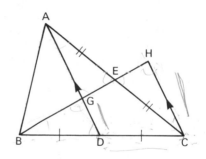

5. In Fig. 34.62, DE is parallel to the side BC of the triangle ABC. If AD = 3 cm, BD = 4 cm and DE = 6 cm, find the length of BC.

Fig. 34.62

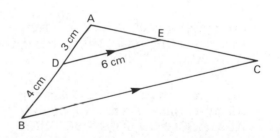

6. Figure 34.63 shows two triangles ABC and XYZ. Find the length of AB.

Fig. 34.63

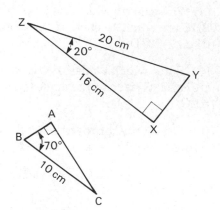

7. ABCD is an isosceles trapezium with AB parallel to CD (Fig. 34.64).

Fig. 34.64

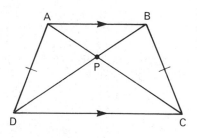

(a) State which of the following triangles is/are congruent to triangle ADP: ABC, ABP, PDC, ADC, BPC.

(b) State which of the following triangles is/are similar to triangle APB: ABC, ABD, ADC, PBC, PDC.

8. In Fig. 34.65, calculate the lengths of QR and ST.

Fig. 34.65

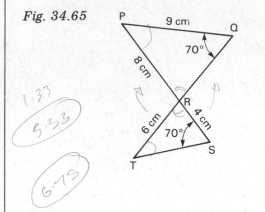

Areas of Similar Triangles

The ratio of the areas of similar triangles is equal to the square of the ratio of corresponding sides.

In Fig. 34.66, if triangles ABC and XYZ are similar, then

Fig. 34.66

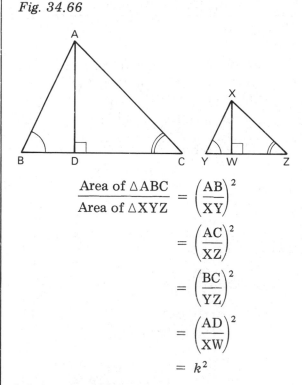

$$\frac{\text{Area of }\triangle ABC}{\text{Area of }\triangle XYZ} = \left(\frac{AB}{XY}\right)^2$$

$$= \left(\frac{AC}{XZ}\right)^2$$

$$= \left(\frac{BC}{YZ}\right)^2$$

$$= \left(\frac{AD}{XW}\right)^2$$

$$= k^2$$

k being the scale factor.

Example 8

Find the area of triangle XYZ given that the area of triangle ABC is $12\,\text{cm}^2$ (see Fig. 34.67).

Fig. 34.67

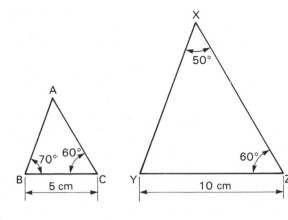

In triangle XYZ, $\angle Y = 70°$ and in triangle ABC, $\angle A = 50°$. Hence the two triangles are similar because they are equi-angular. BC and YZ correspond, therefore

$$k = \frac{\text{YZ}}{\text{BC}}$$

$$= \frac{10}{5}$$

$$= 2$$

$$\frac{\text{Area of } \triangle \text{XYZ}}{\text{Area of } \triangle \text{ABC}} = k^2$$

$$= 4$$

$$\frac{\text{Area of } \triangle \text{XYZ}}{12} = 4$$

$$\text{Area of } \triangle \text{XYZ} = 4 \times 12\,\text{cm}^2$$

$$= 48\,\text{cm}^2$$

Exercise 34.6

1. In Fig. 34.68, the triangles ABC and EFG are similar. If the area of $\triangle$ABC is $8\,\text{cm}^2$, calculate the area of $\triangle$EFG.

Fig. 34.68

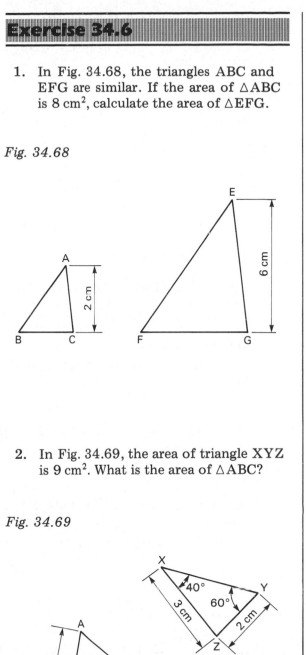

2. In Fig. 34.69, the area of triangle XYZ is $9\,\text{cm}^2$. What is the area of $\triangle$ABC?

Fig. 34.69

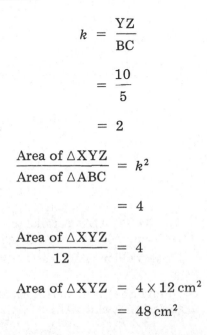

3. In the triangle LPR (Fig. 34.70), MN is parallel to PR. The area of triangle LMN = 24 cm² and the area of triangle LPR = 216 cm².

Fig. 34.70

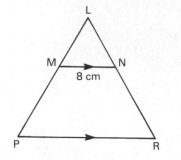

(a) Calculate the area of MNRP.

(b) Calculate the ratio $\dfrac{\text{area of } \triangle LMN}{\text{area of } \triangle LPR}$ as a fraction in its lowest terms.

(c) If MN = 8 cm, find the length of PR.

4. Fig. 34.71 shows two right-angled triangles ABC and XYZ. The area of triangle ABC is 137 cm².

(a) Calculate the area of triangle XYZ.

(b) Find the length of AC.

(c) What is the length of YZ?

Fig. 34.71

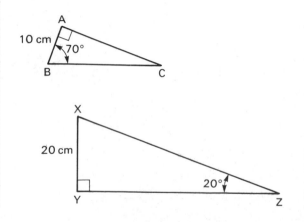

5. In Fig. 34.72, AB = 25 cm and ∠ABC = ∠BDC = 90°. The area of triangle ABC is 150 cm².

(a) Calculate the length of BC.

(b) Find the area of triangle BCD.

Fig. 34.72

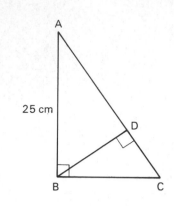

6. In Fig. 34.73, ABCD is a parallelogram. Calculate:

(a) the area of △ABC

(b) the area of △AED

(c) the length of AB

(d) the length of EC.

Fig. 34.73

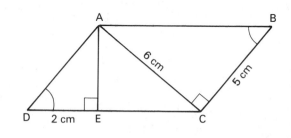

7. In Fig. 34.74, AB is parallel to CD, and AB = 5 cm, AX = 3 cm and CX = 4 cm.

(a) Use similar triangles to find the length of CD.

(b) State the ratio of the areas of triangles AXB and XDC.

Fig. 34.74

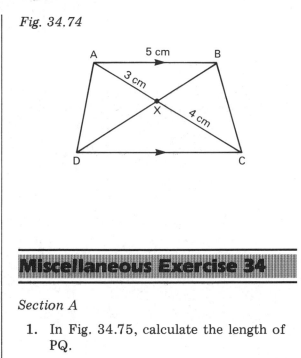

Miscellaneous Exercise 34

Section A

1. In Fig. 34.75, calculate the length of PQ.

Fig. 34.75

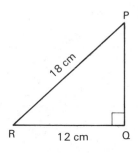

2. In Fig. 34.76, calculate the length of AB.

Fig. 34.76

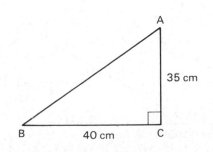

3. In Fig. 34.77, find:

   (a) the size of the angle marked *x*

   (b) the size of the angle marked *y*

   (c) the size of the angle ABC.

   (d) What kind of a triangle is ABD?

Fig. 34.77

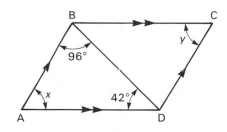

4. A piece of wire 12 cm long can be bent in different ways to form different triangles. The sides of the triangles must always be a whole number of centimetres. Write down the lengths of the three sides if the triangle is

   (a) equilateral

   (b) isosceles

   (c) right-angled.

5. In Fig. 34.78, find the size of the angles marked *x*, *y* and *z*. The line AC bisects the angle BAD.

Fig. 34.78

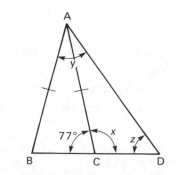

6. In Fig. 34.79, find the size of the angles marked p, q, r and s.

Fig. 34.79

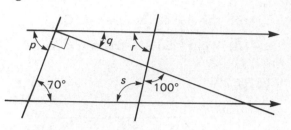

7. Triangle ABC has $a = b = 12$ cm and $c = 8$ cm. Sketch the triangle and then work out its vertical height.

Section B

1. In Fig. 34.80, the triangles ABC and WXY are isosceles. AB = AC = WX = XY and $\angle$BAC + $\angle$WXY = 180°. AD is perpendicular to BC and XZ is perpendicular to WY. If $\angle$BAC = $2x°$, write down, in terms of x, expressions for

(a) $\angle$WXY

(b) $\angle$XYW.

Hence prove that triangles ABD and WXY are congruent.

Fig. 34.80

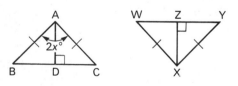

2. In Fig. 34.81, $\triangle$DEF is an enlargement of $\triangle$ABC.

(a) What is the scale factor?

(b) If AC is 3.5 cm, what is the length of DF?

(c) The area of $\triangle$DEF is 63 cm². What is the area of $\triangle$ABC?

Fig. 34.81

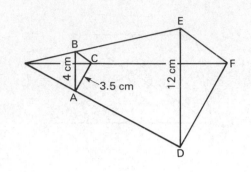

3. $\triangle$ABC is isosceles (Fig. 34.82). Calculate:

(a) $x°$ (b) $y°$

Fig. 34.82

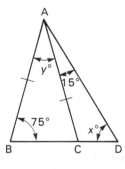

4. The area of triangle ABC (Fig. 34.83) is 24 cm². If AB = 8 cm,

(a) calculate the length of BC

(b) find the length of AC.

Fig. 34.83

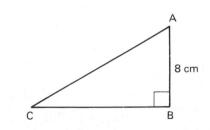

5. In Fig. 34.84, find the size of ∠PQS.

Fig. 34.84

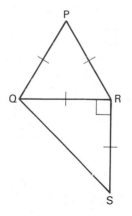

6. In Fig. 34.85, DE is parallel to BC. Find the length of BC. If the area of △ADE is 2 cm², calculate the area of the quadrilateral BDEC.

Fig. 34.85

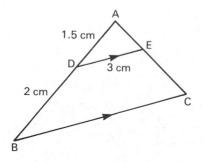

7. In Fig. 34.86, AB = AC = BD and ∠BAD = 70°. Find ∠CAD.

Fig. 34.86

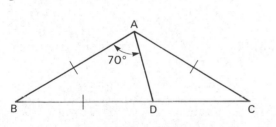

8. In Fig. 34.87, find

    (a) AD   **(b)** CD

    (c) area of △ABC.

Fig. 34.87

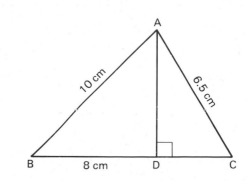

9. Construct a triangle PQR in which PQ = 7 cm, QR = 8.6 cm and ∠PQR = 60°. Ruler and compasses only are to be used.

10. In Fig. 34.88, find the size of the angles *x*, *y* and *z*.

Fig. 34.88

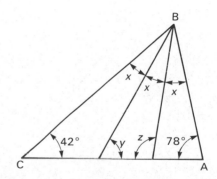

Multi-Choice Questions 34

1. In △PQR (Fig. 34.89), ST is parallel to PQ, ∠RQP = 60°. The size of ∠RST is

   A 50°   B 60°

   C 70°   D 110°

Fig. 34.89

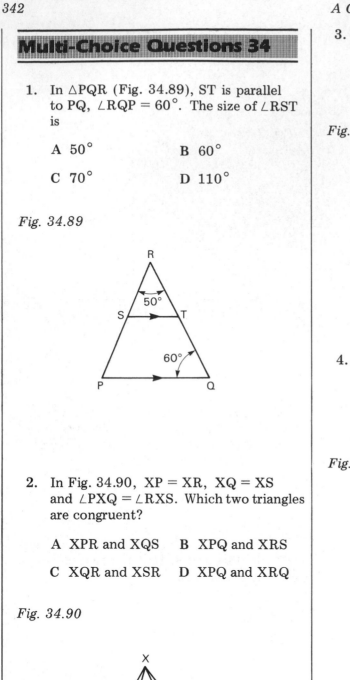

2. In Fig. 34.90, XP = XR, XQ = XS and ∠PXQ = ∠RXS. Which two triangles are congruent?

   A XPR and XQS   B XPQ and XRS

   C XQR and XSR   D XPQ and XRQ

Fig. 34.90

3. In Fig. 34.91, if KM = 10 cm, XQ is parallel to ML and KP = QR = RS = SL, the length of XM is

   A 5 cm   B 6 cm   C 7 cm   D 8 cm

Fig. 34.91

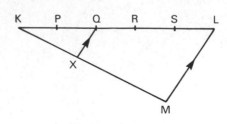

4. Given the triangles shown in Fig. 34.92, the value of the ratio $\dfrac{\text{area of } \triangle PQR}{\text{area of } \triangle XYZ}$ is

   A $\frac{1}{9}$ B $\frac{1}{3}$ C 3   D 9

Fig. 34.92

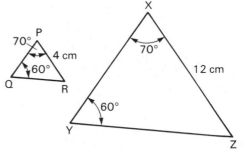

5. In Fig. 34.93, what is the length of FK?

   A 19 cm   B 15 cm

   C 13 cm   D 12 cm

Fig. 34.93

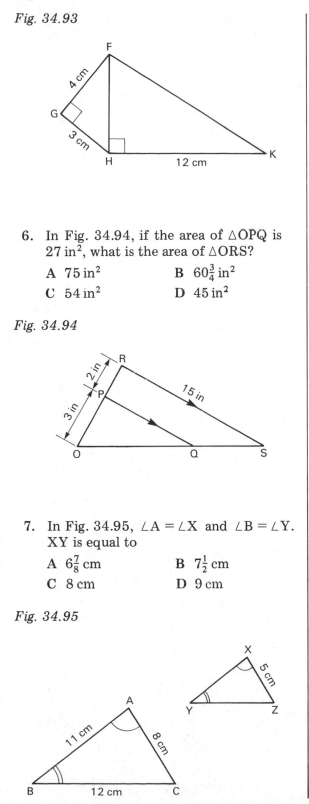

6. In Fig. 34.94, if the area of △OPQ is 27 in², what is the area of △ORS?

    A 75 in²   B $60\frac{3}{4}$ in²

    C 54 in²   D 45 in²

Fig. 34.94

7. In Fig. 34.95, ∠A = ∠X and ∠B = ∠Y. XY is equal to

    A $6\frac{7}{8}$ cm   B $7\frac{1}{2}$ cm

    C 8 cm   D 9 cm

Fig. 34.95

8. In Fig. 34.96, if the area of △XYZ is 5 cm², then the area of △ABC is

    A impossible to find from the given information

    B 20 cm²

    C 40 cm²

    D 80 cm²

Fig. 34.96

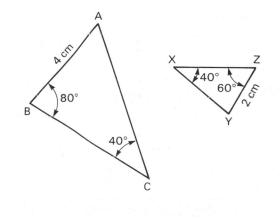

9. In Fig. 34.97, $\dfrac{AX}{XB} = \dfrac{2}{1}$. The area of △ABC is 72 cm². What is the area of the quadrilateral XYCB?

    A 24 cm²   B 32 cm²

    C 36 cm²   D 40 cm²

Fig. 34.97

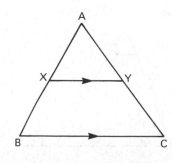

10. In Fig. 34.98, AB = AC. Which of the following is/are **not** necessarily true?

(i) ∠BAD = ∠DAC
(ii) ∠ABD = ∠ACD
(iii) BD = DC
(iv) AD = BD

A (i) and (iv) only
B (ii) and (iii) only
C (iii) only
D (iv) only

Fig. 34.98

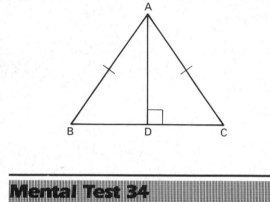

Mental Test 34

Try to answer the following questions without writing anything down except the answer.

1. Two angles of a triangle are 35° and 65°. What is the size of the third angle?

2. An isosceles triangle has an apex angle of 40°. Work out the size of the two equal angles.

3. In Fig. 34.99, find the size of the angle marked x. What is the size of the angle marked y?

Fig. 34.99

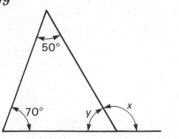

4. Using Fig. 34.100, find the missing angle for each of the triangles listed in the table below.

Triangle	∠A	∠B	x
(a)	20°	40°	?
(b)	100°	20°	?
(c)	40°	?	100°
(d)	?	80°	150°

Fig. 34.100

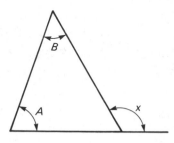

5. In Fig. 34.101, △ABC is isosceles. Write down the missing angles for each of the triangles listed below.

Triangle	∠A	∠B	∠C	x
(a)	70°	?	?	?
(b)	?	?	60°	?
(c)	?	30°	?	?
(d)	?	?	?	100°
(e)	?	?	80°	?
(f)	?	?	?	120°
(g)	45°	?	?	?

Fig. 34.101

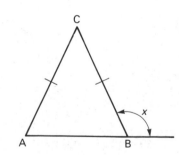

6. In Fig. 34.102, find the size of the angle marked *a*.

Fig. 34.102

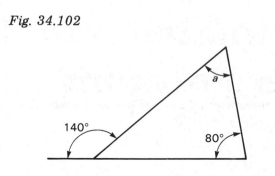

7. Two sides of a right-angled triangle are 5 and 12 cm long respectively. Write down the length of the third side which is the hypotenuse.

35

Quadrilaterals and Other Polygons

Introduction

A quadrilateral is any plane figure bounded by four straight lines.

If we draw a straight line from one corner to the opposite corner of a quadrilateral (Fig. 35.1) the quadrilateral is divided into two triangles. We have seen in Chapter 34 that the sum of the angles of a triangle is 180°. Hence **the sum of the angles of a quadrilateral is 360°**.

Fig. 35.1

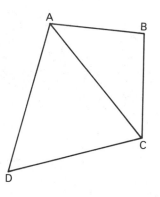

Example 1

Figure 35.2 shows a quadrilateral. Find the size of the angle marked x.

Fig. 35.2

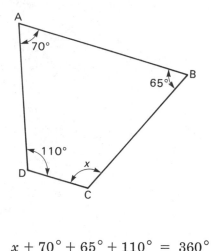

$$x + 70° + 65° + 110° = 360°$$
$$x + 245° = 360°$$
$$x = 115°$$

The Parallelogram

The parallelogram has both pairs of opposite sides parallel (Fig. 35.3(a)).

As shown in Chapter 32, a parallelogram has point symmetry. That is, if we rotate the parallelogram about O (Fig. 35.3(b)) through 180° it does not appear to have moved. However, we see that the vertices A, B, C and D become C′, D′, A′ and B′ respectively.

Fig. 35.3

(a)

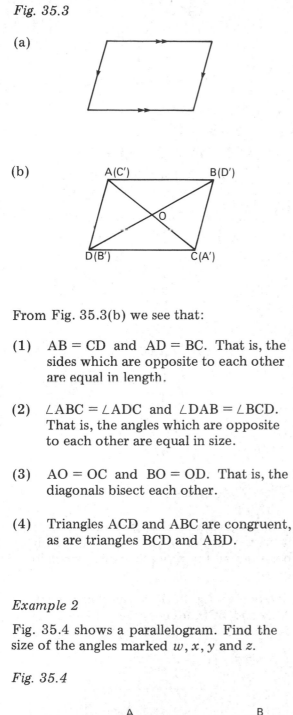

(b)

From Fig. 35.3(b) we see that:

(1) AB = CD and AD = BC. That is, the sides which are opposite to each other are equal in length.

(2) $\angle$ABC = $\angle$ADC and $\angle$DAB = $\angle$BCD. That is, the angles which are opposite to each other are equal in size.

(3) AO = OC and BO = OD. That is, the diagonals bisect each other.

(4) Triangles ACD and ABC are congruent, as are triangles BCD and ABD.

Example 2

Fig. 35.4 shows a parallelogram. Find the size of the angles marked w, x, y and z.

Fig. 35.4

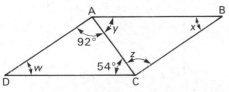

In $\triangle$ACD,

$$w + 92° + 54° = 180° \text{ (sum of the angles of a triangle)}$$

$$w + 146° = 180°$$

$$w = 180° - 146°$$

$$= 34°$$

In the parallelogram ABCD,

$$x = w$$

$$= 34° \quad \text{(opposite angles of a parallelogram are equal)}$$

$$y = 54° \quad \text{(AB\,\|\,CD, alternate angles)}$$

$$z + 54° = 92° + y \quad \text{(opposite angles of a parallelogram are equal)}$$

$$z = 92° + 54° - 54°$$

$$= 92°$$

Alternatively

$$z = 92° \text{ (BC\,\|\,AD, alternate angles)}$$

The Rhombus

The **rhombus** is a parallelogram with all its sides equal in length. Hence it possesses all the properties of a parallelogram. Whereas the parallelogram has no axes of symmetry, the rhombus is symmetrical about each of its diagonals. This leads to two further properties (Fig. 35.5):

(1) Triangles ABD and BCD are congruent. Hence $\angle$ABD = $\angle$DBC and $\angle$ADB = $\angle$BDC. That is, the diagonal bisects the angles through which it passes. Likewise the diagonal AC also bisects the angles through which it passes.

(2) From Fig. 35.5 it can be seen that the triangles AOB, BOC, COD and AOD are all congruent. Hence the four angles at O are all equal to 90°. This means that the diagonals of a rhombus bisect at right angles.

Fig. 35.5

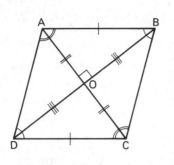

Example 3

ABCD (Fig. 35.6) is a rhombus whose sides are 12 cm long. The diagonal BD is 8.2 cm long. If $\angle ABD = 70°$, find the size of angles BDC, ADC and BAD and calculate the length of the diagonal AC.

Fig. 35.6

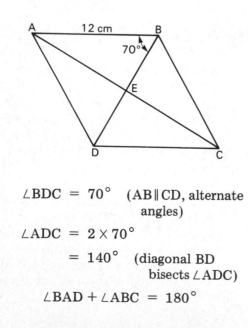

$$\angle BDC = 70° \quad (AB \parallel CD, \text{alternate angles})$$

$$\angle ADC = 2 \times 70°$$
$$\quad = 140° \quad (\text{diagonal BD bisects } \angle ADC)$$

$$\angle BAD + \angle ABC = 180°$$

$$\angle ABC = 2 \times \angle ABD$$
$$\quad = 2 \times 70°$$
$$\quad = 140°$$
$$\angle BAD = 180° - 140°$$
$$\quad = 40°$$

In $\triangle ABE$,

$$\angle AEB = 90° \quad \text{and} \quad BE = 4.1 \text{ cm}$$
$$\text{(diagonals of a rhombus bisect at right angles)}$$

Also AB = 12 cm (given). Using Pythagoras' theorem,

$$AE^2 = AB^2 - BE^2$$
$$\quad = 12^2 - 4.1^2$$
$$\quad = 144 - 16.81$$
$$\quad = 127.19$$
$$AE = \sqrt{127.19}$$
$$\quad \approx 11.3 \text{ cm}$$
$$AC = 2 \times AE$$
$$\quad \approx 2 \times 11.3$$
$$\quad = 22.6 \text{ cm}$$

The Rectangle

A **rectangle** is a parallelogram with each of its vertex angles equal to 90°. Hence it possesses all the properties of a parallelogram.

Looking at Fig. 35.7 we see that triangles ACD and BCD are congruent. Hence the diagonals AC and BD are equal in length.

Fig. 35.7

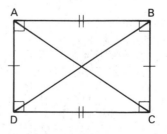

Example 4

A rectangle has sides 8 cm and 12 cm in length.

Calculate the length of its diagonals.

Fig. 35.8

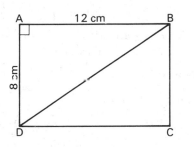

In △ABD (Fig. 35.8), by Pythagoras' theorem,

$$BD^2 = AB^2 + AD^2$$
$$= 12^2 + 8^2$$
$$= 144 + 64$$
$$= 208$$
$$BD = \sqrt{208}$$
$$\approx 14.4$$

Hence the length of the diagonals is 14.4 cm.

The Square

A **square** is a rectangle with all its sides equal in length. It has all the properties of a parallelogram, rhombus and rectangle. A square is symmetrical about the four axes shown in Fig. 35.9.

Fig. 35.9

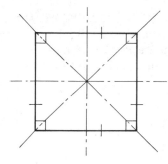

Example 5

A square has diagonals 8 cm long. Find the length of the sides of the square.

Fig. 35.10

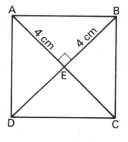

Since the diagonals bisect at right angles (Fig. 35.10), in △AEB, by Pythagoras' theorem,

$$AB^2 = AE^2 + BE^2$$
$$= 4^2 + 4^2$$
$$= 16 + 16$$
$$= 32$$
$$AB = \sqrt{32}$$
$$\approx 5.7$$

Hence the sides of the square are 5.7 cm long.

The Trapezium

A **trapezium** is a quadrilateral with one pair of sides parallel (Fig. 35.11). Generally speaking a trapezium has no axes of symmetry nor does it possess point symmetry.

Fig. 35.11

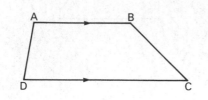

An **isosceles trapezium** has its two non-parallel sides equal in length (Fig. 35.12) and it has the one axis of symmetry shown in the diagram.

Fig. 35.12

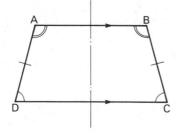

In the isosceles trapezium shown in Fig. 35.12,

$\angle BCD = \angle ADC$ and $\angle DAB = \angle CBA$.

Also $\angle CDA + \angle DAB = 180°$

and $\angle DCB + \angle ABC = 180°$.

Example 6

(a) ABCD (Fig. 35.13) is a trapezium with AB parallel to CD. Given the angles shown in the diagram, calculate the angle marked q.

Fig. 35.13

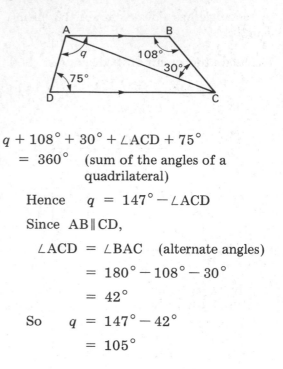

$q + 108° + 30° + \angle ACD + 75°$
$\quad = 360°$ (sum of the angles of a quadrilateral)

Hence $q = 147° - \angle ACD$

Since AB ∥ CD,

$\quad \angle ACD = \angle BAC$ (alternate angles)

$\qquad\qquad = 180° - 108° - 30°$

$\qquad\qquad = 42°$

So $q = 147° - 42°$

$\qquad = 105°$

(b) In Fig. 35.14, ABCD is an isosceles trapezium with AD = BC.

Find the size of the angles marked x and y.

Fig. 35.14

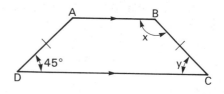

Since ABCD is an isosceles trapezium,

$\qquad\qquad y = 45°$

Also, $x + y = 180°$

$\qquad\quad x + 45° = 180°$

$\qquad\qquad x = 135°$

The Kite

The kite is a quadrilateral having two pairs of adjacent sides equal in length (Fig. 35.15). It is symmetrical about the line BD, and hence this line bisects the angles ADC and ABC. The angles DAB and DCB are also equal and the diagonals AC and BD intersect at right angles.

Fig. 35.15

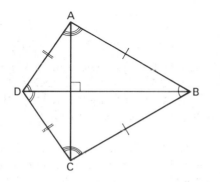

Since AD = DC, △ADC is isosceles and therefore

$$\angle ACD = \angle DAC$$
$$= 15°$$
$$\angle ADC = 180° - (15° + 15°)$$
$$= 180° - 30°$$
$$= 150°$$

Since $\angle CBD = 24°$, $\angle ABD = 24°$

$$b = 180° - (90° + 24°)$$
$$= 180° - 114°$$
$$= 66°$$
$$\angle BAD = 66° + 15°$$
$$= 81°$$

Exercise 35.1

1. Fig. 35.17 shows a kite. Find the angles marked x, y and z.

Fig. 35.17

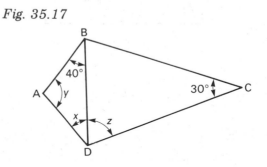

2. Calculate the angle x in Fig. 35.18.

Fig. 35.18

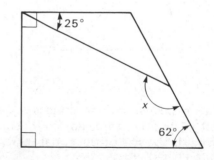

Example 7

Fig. 35.16 shows a kite ABCD. Given that $\angle DAC = 15°$ and $\angle CBD = 24°$, find the size of the angles ACD, ADC and BAD.

Fig. 35.16

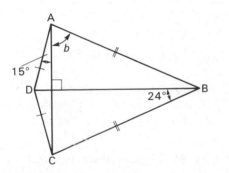

3. Find the angle x in Fig. 35.19.

Fig. 35.19

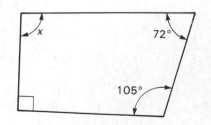

4. In Fig. 35.20, ABCD is a parallelogram. Calculate the angles x and y.

Fig. 35.20

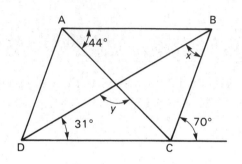

5. In Fig. 35.21, find p.

Fig. 35.21

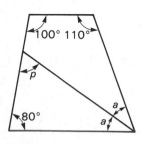

6. In a quadrilateral one angle is equal to $60°$. The other three angles are equal. What is the size of the equal angles?

7. Fig. 35.22 shows a rhombus. Are △s ABE and DEC congruent? Does ∠DAC equal ∠DCA? Is the angle DAB bisected by AC?

Fig. 35.22

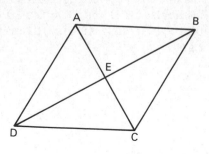

8. ABCD is a quadrilateral in which ∠A = $86°$, ∠C = $110°$ and ∠D = $40°$. The angle ABC is bisected to cut the side AD at E. Find ∠AEB.

9. In the quadrilateral ABCD, ∠DAB = $60°$ and the other three angles are equal. The line CE is drawn parallel to BA to meet AD at E. Calculate the angles ABC and ECD.

10. PQRS is a square. T is a point on the diagonal PR such that PT = PQ. The line through T, perpendicular to PR, cuts QR at X. Prove that QX = XT = TR.

11. The diagonals of a rhombus are 12 cm and 9 cm long. Calculate the length of the sides of the rhombus.

12. The diagonals of a quadrilateral XYZW intersect at O. Given that OX = OW and OY = OZ prove that

(a) XY = ZW

(b) YZ is parallel to XW.

13. ABCD is a parallelogram. Parallel lines BE and DF meet the diagonal AC at E and F respectively. Prove that

(a) AE = FC

(b) BEDF is a parallelogram.

14. Fig. 35.23 shows a parallelogram ABCD. Find:

    (a) the size of the angle marked x

    (b) the size of the angle marked y

    (c) the size of $\angle$ACB.

    (d) What type of triangle is ACD?

Fig. 35.23

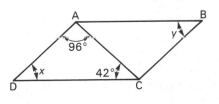

15. In a rhombus ABCD (Fig. 35.24), the diagonal AC = 24 cm and the diagonal BD = 10 cm. Find AO, BO and AB.

Fig. 35.24

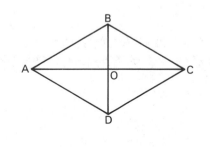

Other Polygons

A **polygon** is any plane figure bounded only by straight lines. Thus a triangle is a polygon with three sides, and a quadrilateral is a polygon with four sides.

A **convex polygon** (Fig. 35.25) has no interior angle greater than 180°.

Fig. 35.25

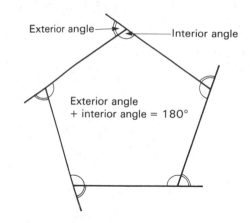

A **re-entrant polygon** (Fig. 35.26) has at least one angle greater than 180°.

Fig. 35.26

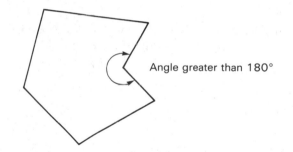

A **regular polygon** has all its sides and angles equal.

Sum of the Interior Angles of a Polygon

To find the sum of the interior angles of a polygon, choose any point such as O (Fig. 35.27) which lies within the polygon. Next draw lines from O to the vertices A, B, C ... to form triangles. The number of triangles is the same as the number of sides of the polygon: if the polygon has n sides then there are n triangles. Since the sum of the angles of each triangle is 180°, the sum of the angles of the triangles equals 180n degrees or 2n right angles. The sum of the angles round O equals 360° or 4 right angles. Hence **the sum of the interior angles of a polygon is 2n − 4 right angles.**

Fig. 35.27

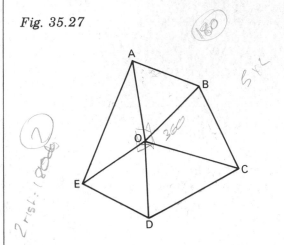

It makes no difference whether the polygon is convex or re-entrant, the sum of the interior angles is (2n − 4) right angles.

Names of Polygons

Name	Number of sides
Pentagon	5
Hexagon	6
Heptagon	7
Octagon	8
Nonagon	9
Decagon	10
Hendecagon	11
Dodecagon	12

Example 8

(a) Find the sum of the interior angles of a heptagon (seven-sided polygon).

Here we have the number of sides

$$= n = 7$$

Sum of the
interior angles $= (2n - 4)$ right angles

$= (2 \times 7 - 4)$ right angles

$= 10$ right angles

$= 10 \times 90$ degrees

$= 900$ degrees

(b) Find the size of each interior angle of a regular pentagon.

Since a pentagon has 5 sides,

$$n = 5.$$

Sum of the
interior angles $= (2n - 4)$ right angles

$= (2 \times 5 - 4)$ right angles

$= 6$ right angles

$= 540$ degrees

Size of each interior angle $= 540° \div 5$

$= 108°$

Sum of the Exterior Angles of a Convex Polygon

Fig. 35.28

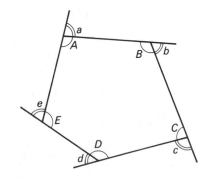

In Fig. 35.28, the interior angles of the polygon are A, B, C, D and E. The exterior angles are a, b, c, d and e. Now

$$A + a = 180°$$
$$= 2 \text{ right angles}$$
$$B + b = 180°$$
$$= 2 \text{ right angles}$$

and so on. Therefore:

Sum of the interior angles + Sum of the exterior angles = $2n$ right angles

Sum of the interior angles = $(2n - 4)$ right angles

Sum of the exterior angles = $[2n - (2n - 4)]$ right angles

$$= 4 \text{ right angles}$$
$$= 360°$$

No matter how many sides the convex polygon has, the sum of its exterior angles is 360°.

Example 9

(a) Each interior angle of a regular polygon is 150°. How many sides has it?

Each exterior angle $= 180° - 150°$
$$= 30°$$

Since the sum of the exterior angles is 360°,

Number of sides $= 360 \div 30$
$$= 12$$

(b) A regular polygon has 15 sides. What is the size of each interior angle?

Each exterior angle $= 360° \div 15$
$$= 24°$$

Each interior angle $= 180° - 24°$
$$= 156°$$

(c) In a regular polygon, each interior angle is greater by 140° than each exterior angle. How many sides has the polygon?

Fig. 35.29

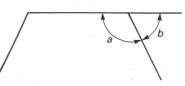

In Fig. 35.29 let a be the interior angle and b the exterior angle of a polygon having n sides.

Then $\quad a - b = 140°$ $\quad$ (1)

Also $\quad a + b = 180°$ $\quad$ (2)

Adding equations (1) and (2) gives

$$2a = 320°$$
$$a = 160°$$
$$b = 20°$$
$$n = 360 \div 20$$
$$= 18$$

Hence the polygon has 18 sides.

Symmetry of Regular Polygons

Regular polygons have the number of axes of symmetry equal to the number of sides of the polygon. Thus a pentagon (Fig. 35.30) has five axes of symmetry and an octagon (Fig. 35.31) has eight axes of symmetry.

Fig. 35.30

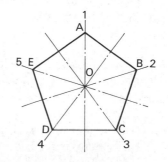

In Fig. 35.30, each of the triangles AOB, BOC, COD, DOE and EOA are congruent. The angle $AOB = 360° \div 5 = 72°$. Therefore $\angle ABO + \angle OBC = 180° - 72° = 108°$. Therefore each interior angle of a pentagon is $108°$. This method gives us a third way of finding the size of the interior angle of a regular polygon.

Fig. 35.31

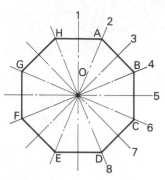

In Fig. 35.31 we note that there are 8 congruent triangles. Therefore the size of the interior angle of a regular octagon is $180° - (360° \div 8) = 180° - 45° = 135°$.

Exercise 35.2

1. Find the sum of the interior angles of a convex polygon with

    (a) 5   (b) 8

    (c) 10   (d) 12 sides

2. If the polygons in Question 1 are all regular, find the size of the interior angle of each.

3. A hexagon has interior angles of $100°$, $110°$, $120°$ and $128°$. If the remaining two angles are equal, what is their size?

4. Each interior angle of a regular polygon is $150°$. How many sides has it?

5. ABCDE is a regular pentagon and ABX is an equilateral triangle drawn outside the pentagon. Calculate $\angle AEX$.

6. In a regular polygon each interior angle is greater by $150°$ than each exterior angle. Calculate the number of sides of the polygon.

7. A polygon has n sides. Two of its angles are right angles and each of the remaining angles is $144°$. Calculate n.

8. In a pentagon ABCDE, $\angle A = 120°$, $\angle B = 138°$ and $\angle D = \angle E$. The sides AB, DC when produced meet at right-angles. Calculate $\angle C$ and $\angle E$.

9. In a regular pentagon ABCDE, the lines AD and BE intersect at P. Calculate the angles $\angle BAD$ and $\angle APE$.

10. Calculate the exterior angle of a regular polygon in which the interior angle is four times the exterior angle. Hence find the number of sides in the polygon.

11. Each exterior angle of a regular polygon of n sides exceeds by $6°$ each exterior angle of a regular polygon of $2n$ sides. Find an equation for n and solve it.

12. Calculate the number of sides of a regular polygon in which the exterior angle is one-fifth of the interior angle.

Drawing Quadrilaterals and Other Polygons

These may be drawn by using a rule, compasses, set-squares and a protractor.

Example 10

A field is in the shape of a quadrilateral ABCD with $AB = 90\,m$, $BC = 80\,m$ and $CD = 40\,m$. $\angle ABC = 80°$ and $\angle BCD = 130°$. Using a scale of 1 cm to represent $10\,m$, make an accurate scale drawing of the field.

(a) Use your drawing to find the length of the diagonal BD.

(b) A pylon is at the intersection of the bisector of ∠BCD and the perpendicular bisector of the side AB. Using ruler and compasses only, indicate the position of the pylon.

(c) Using your drawing, find the distance of the pylon from C.

Fig. 35.32

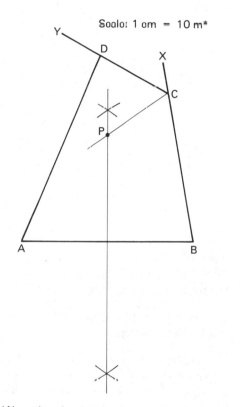

Scale: 1 cm = 10 m*

Note that the drawing is shown half size here. All lengths are therefore half those that would be obtained from the actual drawing.

To construct the quadrilateral (Fig. 35.32):

(1) Draw AB 9 cm long to represent 90 m.

(2) Draw BX inclined at 80° to AB and mark off BC = 8 cm.

(3) Draw CY inclined at 130° to BC and mark off CD = 4 cm.

(4) Join A and D. ABCD is then the required quadrilateral.

(a) The diagonal BD is 11 cm long representing a distance of 110 m.

(b) Using the standard construction ∠BCD is bisected and the perpendicular bisector of AB is drawn. The point P shows the position of the pylon.

(c) The distance of P from C is 3.9 cm representing 39 m.

Example 11

(a) The circumscribed circle of a hexagon has a radius of 2 cm. Construct the hexagon.

Fig. 35.33

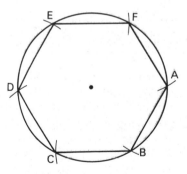

The construction is as follows:

(1) Draw the circumscribing circle (Fig. 35.33) having a radius of 2 cm.

(2) Starting from anywhere on the circumference of the circle and with the same radius, mark off the points A, B, C, D, E and F.

(3) Join A to B, B to C, C to D, D to E, E to F and F to A. Then ABCDEF is the required hexagon.

(b) The circumscribed circle of a regular pentagon has a radius of 2 cm. Construct the pentagon.

Fig. 35.34

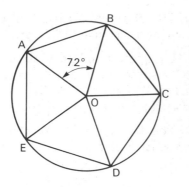

In the circumscribing circle, centre O (Fig. 35.34), the five congruent triangles AOB, BOC, COD, DOE and EOA are drawn. Each triangle has the angle at O equal to $360° − 5° = 72°$. The required pentagon is ABCDEF.

Exercise 35.3

1. In Fig. 35.35, the trapezium ABCD represents a plot of land. Using a scale of 1 cm to represent 10 m, make a scale drawing of this plot. A farmer placed a scarecrow at the intersection of the diagonals AC and BD.

   (a) Mark S, the point at which the scarecrow was placed.

   (b) Measure the distance of the scarecrow from A.

Fig. 35.35

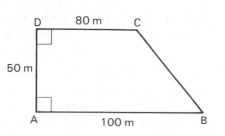

2. Fig. 35.36 shows the plan of a school playing field. Using a scale of 1 cm to represent 20 m, make a scale drawing of the field.

   (a) A post is placed at the intersection of the bisector of ∠ABC and the perpendicular bisector of AB. Show the position of the post and mark it P.

   (b) Using your scale drawing find the distance of B to P.

Fig. 35.36

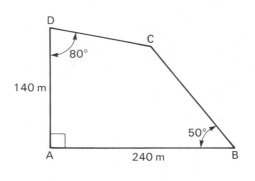

3. Fig. 35.37 represents the plan of a field. Using a scale of 2 cm to 40 m, make an accurate scale drawing of the field.

   (a) Draw the diagonals AC and BD and state their lengths.

   (b) At the intersection of these diagonals a telephone pole is to be sited. Write down the distance between the telephone pole and the corner A of the field.

Fig. 35.37

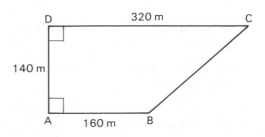

4. Using a scale of 1 cm to represent 10 m, make an accurate scale drawing of the plot of land shown in Fig. 35.38.

   (a) Using your drawing find the length of the side AD and the diagonal BD.

   (b) Calculate the area of the plot.

Fig. 35.38

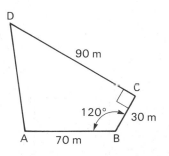

5. Make an accurate drawing of a regular hexagon with a distance across corners of 4 cm.

6. (a) Draw a regular octagon (eight-sided figure) whose distance across flats is 5 cm.

   (b) Work out the area of this octagon.

7. A heptagon is a seven-sided polygon. Draw a circle with a diameter of 6 cm and in it inscribe a regular heptagon. What is the size of the interior angles of this figure?

Miscellaneous Exercise 35

Section A

1. In Fig. 35.39, ABCD is a quadrilateral. Find the angles marked x and y.

Fig. 35.39

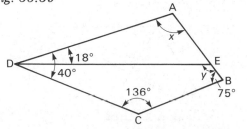

2. Fig. 35.40 shows a parallelogram. Write down the sizes of the angles marked *a* and *b*.

Fig. 35.40

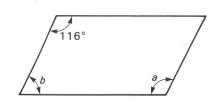

3. For the rhombus ABCD (Fig. 35.41), find the sizes of the angles DAC, BCD, ABC and DOC.

Fig. 35.41

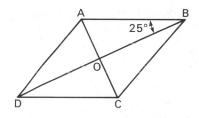

4. In a rhombus ABCD, the diagonal BD is 20 cm long and the diagonal AC is 12 cm long. Calculate the length of the sides of the rhombus.

5. A square has sides which are 8 cm long. Calculate the length of its diagonals.

6. The rectangle ABCD (Fig. 35.42) has its diagonal AC − 15 cm and its side AB = 11 cm. Calculate the length of the side BC.

Fig. 35.42

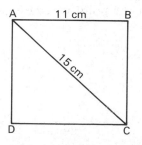

7. Fig. 35.4? shows an iron rectangular gate made from thin tubing. What length of tubing is needed to make it?

Fig. 35.43

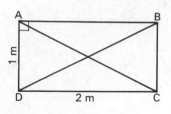

8. In Fig. 35.44, ABCD is a rhombus. BD = 10 cm, AC = 6 cm and XE is a straight line parallel to BC.

   (a) What is the size of the angle AXD?

   (b) What is the length of AX?

   (c) What is the name of the figure EXCB?

Fig. 35.44

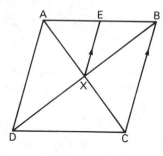

Section B

1. Fig. 35.45 shows two sides of a quadrilateral ABCD which is symmetrical about the point O. Draw the figure.

Fig. 35.45

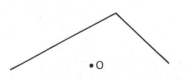

2. ABCDEF is a regular hexagon and ABXY is a square in the same plane. Calculate the size of the obtuse angle CXY (Fig. 35.46).

Fig. 35.46

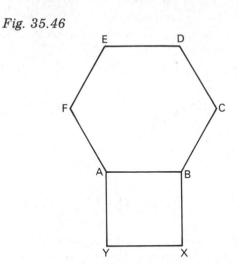

3. AB and BC are two adjacent sides of a regular twelve-sided polygon. The perpendicular from C meets AB produced at D. Calculate the size of BCD.

4. In this question use rule, compasses and set-squares only and show all construction lines clearly. Construct XYZ with YZ = 6 cm, XY = 5 cm and XYZ = 60°. Construct the perpendicular from X to ZY to meet ZY at W. Find, by measurement, the length of XW. Construct the point T such that XYTW is a parallelogram and find, by measurement, the length of XT.

5. Calculate the size of each interior angle of a regular sixteen-sided polygon.

6. Fig. 35.47 shows a rhombus. If x = 29°, find y.

Fig. 35.47

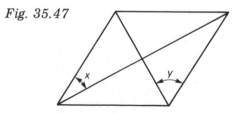

7. Using a scale of 2 cm to 10 m, make a scale drawing of a field ABCD from the following information: AB = 60 m, BC = 120 m, ∠B = 60°, ∠D = 90° and AD = DC. Measure and write down the length of the diagonal BD.

Multi-Choice Questions 35

1. In Fig. 35.48, x is equal to

   A $a + b + c$

   B $360° - (a + b + c)$

   C $a + b + c + 180°$

   D $360° - a + b + c$

Fig. 35.48

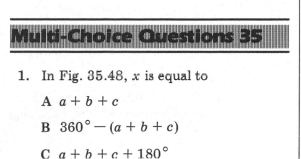

2. In Fig. 35.49, y is equal to
   A 100°   B 80°   C 70°   D 40°

Fig. 35.49

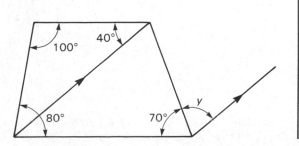

3. A regular polygon has each interior angle equal to 108°. How many sides has it?

   A 4   B 5   C 6   D 7

4. Fig. 35.50 shows a quadrilateral. Find the size of the angle marked x.

   A 120°   B 110°   C 70°   D 60°

Fig. 35.50

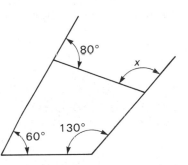

5. A regular polygon has each exterior angle equal to 40°. How many sides has it?

   A 7   B 8   C 9   D 10

6. A regular polygon has each interior angle greater by 60° than each exterior angle. How many sides has it?

   A 5   B 6   C 7   D 8

7. In Fig. 35.51, ABCD is a trapezium. Hence

   A ∠ADB = 40°   B ∠ADB = 70°
   C ∠ADC = 90°   D ∠ADC = 120°

Fig. 35.51

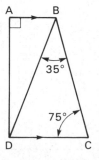

8. For a regular eight-sided polygon, what is the ratio of the exterior angle to the interior angle?

    A 1:1   **B** 1:3   **C** 1:4   **D** 1:5

9. Fig. 35.52 shows the unequal diagonals of a quadrilateral. If these diagonals do not cut at right angles, then the quadrilateral is

    A a rhombus   **B** a rectangle
    C a parallelogram   **D** a square

Fig. 35.52

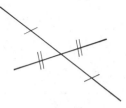

10. A quadrilateral has diagonals of the same length and they bisect at right angles. What is the name of the quadrilateral?

    A a parallelogram   **B** a rhombus
    C a square   **D** a rectangle

Mental Test 35

Try to answer the following questions without writing anything down except the answer.

1. Fig. 35.53 shows a quadrilateral. Find the size of the angle marked x.

Fig. 35.53

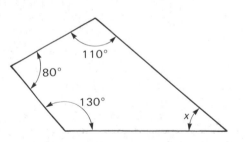

2. Fig. 35.54 shows a parallelogram. What is the size of the angle marked x?

Fig. 35.54

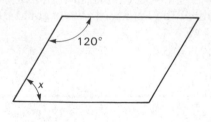

3. In the rhombus (Fig. 35.55) what is the size of the angle marked x and the angle marked y.

Fig. 35.55

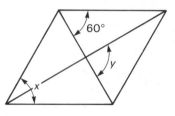

4. A rectangle ABCD (Fig. 35.56) has AB = 8 cm and AD = 6 cm. What is the length of the diagonal AC?

Fig. 35.56

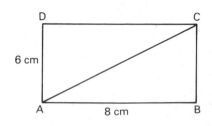

5. How many axes of symmetry has a regular seven-sided polygon?

6. In the kite ABCD (Fig. 35.57) find the size of the angles ACD and ACB.

Fig. 35.57

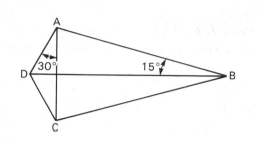

7. The exterior angle of a regular polygon is 40°.

   (a) How many sides has it?

   (b) How many axes of symmetry has it?

   (c) What is the size of its interior angles?

8. A regular polygon has interior angles of 135°. How many sides has it?

9. A regular polygon has 10 sides. What is the size of its interior angles?

10. How many axes of symmetry does an octagon possess?

The Circle

Introduction

A circle is a curve consisting of all the points at a constant distance, the **radius**, from a fixed point, the **centre**.

Figure 36.1 shows the main parts of a circle.

Fig. 36.1

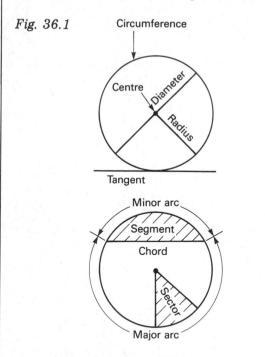

Angles in Circles

(1) The angle which an arc of a circle subtends at the centre is twice the angle at which the arc subtends at the circumference.

Thus in Fig. 36.2, ∠AOB = 2 × ∠APB.

Fig. 36.2

(2) If a triangle is drawn in a circle with a diameter as the base of the triangle, then the angle opposite that diameter is a right angle.

Because the diameter AB (Fig. 36.3) forms the angle (180°) subtended by the semicircle ABD, the angle subtended by this arc at the circumference is 90°.

Fig. 36.3

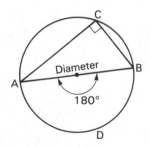

Example 1

In Fig. 36.4, O is the centre of the circle. If $\angle AOB = 60°$, find $\angle ACB$.

Fig. 36.4

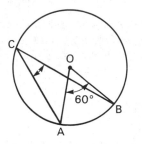

Since $\angle AOB$ is the angle subtended by the arc AB at the centre O and $\angle ACB$ is the angle subtended by AB at the circumference,

$$\angle AOB = 2 \times \angle ACB$$

Since $\quad \angle AOB = 60°$

$$\angle ACB = 30°$$

Example 2

In Fig. 36.5, find the length of the diameter BC.

Fig. 36.5

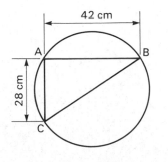

Since BC is a diameter, the angle A is the angle in a semicircle and hence it is a right angle.

In $\triangle ABC$, by Pythagoras' theorem,

$$BC^2 = 28^2 + 42^2$$
$$= 784 + 1764$$
$$= 2548$$
$$BC = \sqrt{2548}$$
$$= 50.48$$

Hence the length of the diameter BC is 50.48 cm.

The chord AB (Fig. 36.6) divides the circle into two arcs. ABP is called the **major arc** and ABQ the **minor** arc. The regions ABP and ABQ are called **segments**.

Fig. 36.6

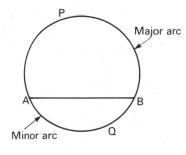

The angles ARB and ASB (Fig. 36.7) are called angles in the segment APB. The angle ATB is called an angle in the segment ABQ.

Fig. 36.7

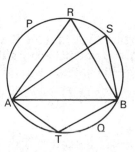

Angles in the same segment of a circle are equal. Thus in Fig. 36.8, ∠APB = ∠AQB since they are angles in the same segment ABQP.

Fig. 36.8

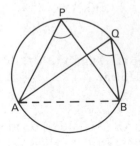

Example 3

In Fig. 36.9, find the size of the angles marked *x* and *y*.

Fig. 36.9

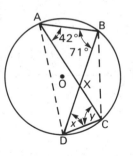

First draw the chords AD and BC.

The angles BAX and XDC are angles in the same segment because they stand on the same chord BC. Hence

$$∠XDC = ∠BAX$$

$$= 42°$$

i.e. $$x = 42°$$

The angles ABX and XCD are angles in the same segment because they stand on the same chord AD. Hence

$$∠XCD = ∠ABX$$

$$= 71°$$

i.e. $$y = 71°$$

Cyclic Quadrilaterals

The opposite angles of any quadrilateral inscribed in a circle are supplementary (i.e. their sum is 180°). A quadrilateral inscribed in a circle is called a **cyclic quadrilateral**.

Thus the cyclic quadrilateral ABCD (Fig. 36.10) has ∠A + ∠C = 180° and ∠B + ∠D = 180°. Also ∠CDX = ∠B and ∠BCY = ∠A etc.

Fig. 36.10

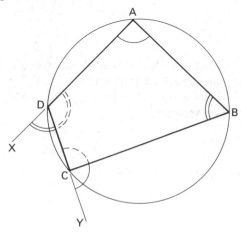

Example 4

ABCD (Fig. 36.11) is a cyclic quadrilateral with ∠A = 100° and ∠B and ∠D equal. Find the angles C and D.

Fig. 36.11

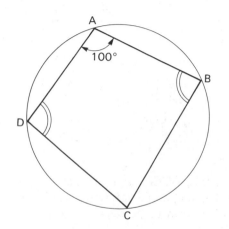

Since ABCD is a cyclic quadrilateral

$$\angle A + \angle C = 180°$$
$$\angle C = 180° - \angle A$$
$$= 180° - 100°$$
$$= 80°$$

Also

$$\angle B + \angle D = 180°$$
$$\angle B = \angle D$$

Therefore $\angle D = 90°$

Exercise 36.1

1. In Fig. 36.12, if $\angle AOB = 76°$, find $\angle ACB$.

Fig. 36.12

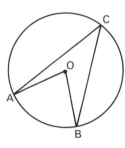

2. In Fig. 36.13, ABC is an equilateral triangle inscribed in a circle whose centre is O. Find $\angle BOC$.

Fig. 36.13

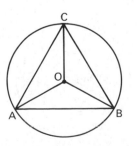

3. ABC is an isosceles triangle inscribed in a circle whose centre is O (Fig. 36.14). If $\angle AOB = 116°$, find $\angle ABC$.

Fig. 36.14

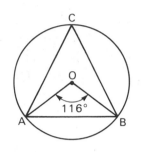

4. In Fig. 36.15, determine the angle x.

Fig. 36.15

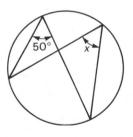

5. Find the angles C and D of the cyclic quadrilateral ABCD shown in Fig. 36.16.

Fig. 36.16

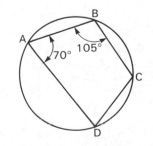

6. In Fig. 36.17, find the angles x and y.

Fig. 36.17

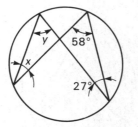

7. Figure 36.18 shows a cyclic quadrilateral. Find the angles marked x and y.

Fig. 36.18

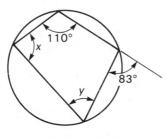

8. In Fig. 36.19, AB is a diameter. Find AC.

Fig. 36.19

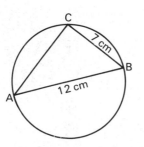

9. In Fig. 36.20, find each of the angles marked a and b.

Fig. 36.20

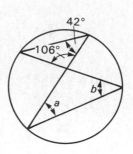

10. In Fig. 36.21, determine the angles x and y, AB being a diameter.

Fig. 36.21

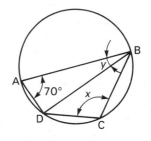

11. In Fig. 36.22, calculate the diameter of the circle.

Fig. 36.22

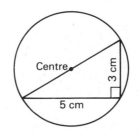

Chords

A **chord** is a straight line which joins two points on the circumference of a circle. A **diameter** is a chord drawn through the centre of the circle (Fig. 36.1).

(1) If a diameter of a circle is at right-angles to a chord then it divides the chord into two equal parts (Fig. 36.23). The converse is also true.

Fig. 36.23

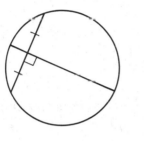

(2) Chords which are equal in length are equidistant from the centre of the circle. Thus, in Fig. 36.24 if the chords AB and CD are equal in length, then the distances OX and OY are also equal. The converse is also true, i.e., chords which are equidistant from the centre of the circle are equal in length.

Fig. 36.24

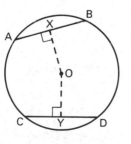

(3) If two chords intersect inside or outside a circle the product of the segments of one chord is equal to the product of the segments of the other chord. Thus in Fig. 36.25,

AE × EB = CE × ED. The converse is also true, i.e., if AE × EB = CE × ED then the points ABCD are concyclic (i.e., lie on the circumference of a circle).

Fig. 36.25

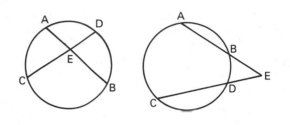

Example 5

Figure 36.26 shows a segment of a circle. Find the diameter of the circle.

Fig. 36.26

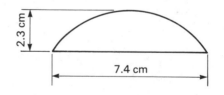

In Fig. 36.27 draw the diameter CD at right-angles to the chord. The chord AB is bisected and AE = EB = 3.7 cm.

Fig. 36.27

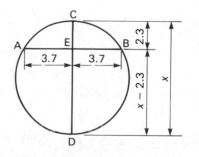

By the theorem of intersecting chords,

$$AE \times EB = CE \times ED$$

Since $CE = 2.3\,cm$, $ED = x - 2.3\,cm$, where x is the length of the diameter CD.

$$3.7 \times 3.7 = 2.3 \times (x - 2.3)$$
$$13.69 = 2.3x - 2.3 \times 2.3$$
$$13.69 = 2.3x - 5.29$$
$$2.3x = 13.69 + 5.29$$
$$= 18.98$$
$$x = \frac{18.98}{2.3}$$
$$= 8.25$$

Hence the diameter of the circle is 8.25 cm.

Example 6

A chord AB (Fig. 36.28) is drawn in a circle of 5 cm radius. If it is 4 cm from the centre of the circle, find its length.

Fig. 36.28

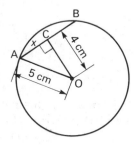

In $\triangle OAC$, by Pythagoras' theorem,

$$x^2 = 5^2 - 4^2$$
$$= 25 - 16$$
$$= 9$$
$$x = \sqrt{9}$$
$$= 3$$

Hence the chord is $2 \times 3\,cm = 6\,cm$.

Exercise 36.2

1. In Fig. 36.29 find the distance x.

Fig. 36.29

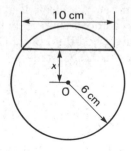

2. In Fig. 36.30, find the length of the chord AB.

Fig. 36.30

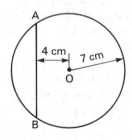

3. Figure 36.31 shows an equilateral triangle inscribed in a circle. Calculate the diameter of the circle.

Fig. 36.31

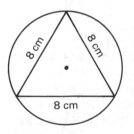

4. In Fig. 36.32, calculate the diameter of the circle.

Fig. 36.32

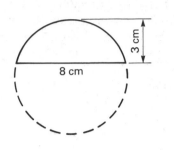

5. Find the length *y* in Fig. 36.33.

Fig. 36.33

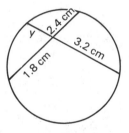

6. Find the length *x* in Fig. 36.34.

Fig. 36.34

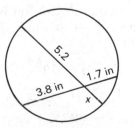

7. In Fig. 36.35, calculate the length of the chord *x*.

Fig. 36.35

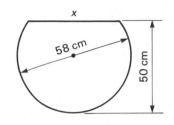

8. In Fig. 36.36, calculate the height *y* of the segment.

Fig. 36.36

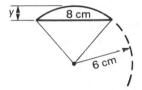

Tangent Properties of a Circle

A **tangent** is a line which touches a circle at one point. This point is called the **point of tangency** (Fig. 36.37).

Fig. 36.37

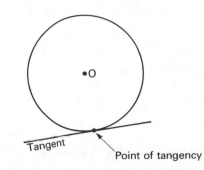

(1) A tangent to a circle lies at right angles to a radius drawn to the point of tangency.

Thus in Fig. 36.38, ∠OTS is a right angle.

Fig. 36.38

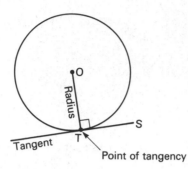

Example 7

In Fig. 36.39, TS is a tangent, T being the point of tangency and O the centre of the circle. If the radius of the circle is 5 cm and OS = 7 cm, find the distance TS.

Fig. 36.39

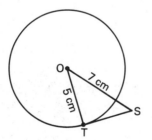

    Since OT is a radius drawn to the point of tangency, ∠OTS = 90°.

By Pythagoras' theorem

$$TS^2 = OS^2 - OT^2$$
$$= 7^2 - 5^2$$
$$= 49 - 25$$
$$= 24$$
$$TS = \sqrt{24}$$
$$= 4.90$$

Hence the distance TS is 4.90 cm.

(2) If, from a point outside a circle, tangents are drawn to the circle, then the two tangents are equal in length.

Thus in Fig. 36.40, PX = PY. The quadrilateral OXPY is a kite since OX = OY (radii) and it is symmetrical about the line OP. Hence ∠XPY is bisected and ∠KXP = ∠KYP.

Fig. 36.40

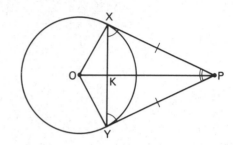

Example 8

In Fig. 36.41, calculate the size of the angles marked *a* and *b*.

Fig. 36.41

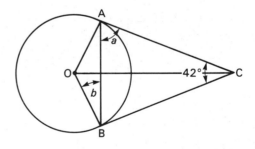

The two tangents AC and BC meet at C. Hence AC = BC and ABC is isosceles. Therefore

$$a = \tfrac{1}{2} \times (180° - 42°)$$
$$= \tfrac{1}{2} \times 138°$$
$$= 69°$$

Since OBC is the angle between a tangent and a radius, it equals 90°. Therefore

$$b = \angle OBC - \angle ABC$$

$$= 90° - 69°$$

$$= 21°$$

(3) If two circles touch internally or externally then the line which passes through their centres also passes through the point of tangency (Fig. 36.42).

Fig. 36.42

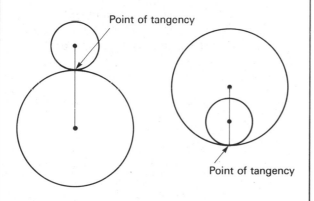

Point of tangency

Point of tangency

Example 9

Three circles are arranged as shown in Fig. 36.43. Find the distance h.

Fig. 36.43

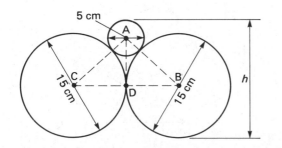

5 cm

15 cm

15 cm

h

Because the circles are tangential to each other

$$AC = 7.5 + 2.5$$

$$= 10 \text{ cm}$$

$$AB = 7.5 + 2.5$$

$$= 10 \text{ cm}$$

$$BC = 7.5 + 7.5$$

$$= 15 \text{ cm}$$

Therefore triangle ABC is isosceles, and hence

$$CD = \tfrac{1}{2} \text{ of } 15$$

$$= 7.5 \text{ cm}$$

In $\triangle ACD$, using Pythagoras' theorem,

$$AD^2 = AC^2 - CD^2$$

$$= 10^2 - 7.5^2$$

$$= 100 - 56.25$$

$$= 43.75$$

$$AD = \sqrt{43.75}$$

$$= 6.61$$

$$h = 7.5 + 6.61 + 2.5$$

$$= 16.61 \text{ cm}$$

Constructing Tangents

(1) To construct a common tangent to two given circles.

Construction

The two given circles have centres X and Y and radii x and y respectively (Fig. 36.44). With centre X draw a circle whose radius is $(x - y)$. With diameter XY and centre the mid-point of XY draw an arc to cut the previously drawn circle at M. Join XM and produce to P at the circumference of the circle. Draw YQ parallel to XP, Q being at the circumference of the circle. Join PQ which is the required tangent.

Fig. 36.44

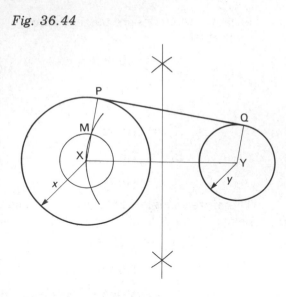

(2)   To construct a pair of tangents from an external point to a given circle (Fig. 36.45).

Construction

It is required to draw a pair of tangents from the point P to the circle centre O. Join OP. With OP as diameter and centre the mid-point of OP draw an arc to cut the given circle at points A and B. Join PA and PB which are the required pair of tangents.

Fig. 36.45

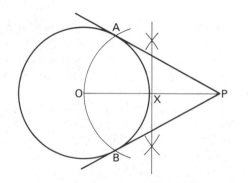

Exercise 36.3

1. In Fig. 36.46, calculate the distance OP given that OA is a radius and PA and PB are tangents meeting at P.

Fig. 36.46

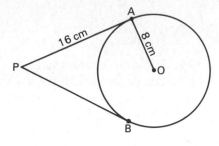

2. In Fig. 36.47, OA is a radius and AP and BP are tangents meeting at P. Find the size of the angles marked a and b.

Fig. 36.47

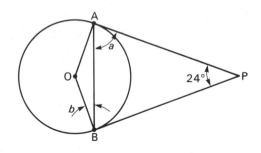

3. In Fig. 36.48, OA is the radius of the circle and CA and CB are tangents meeting at C. Find the length of CB.

Fig. 36.48

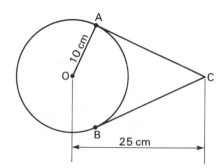

4. In Fig. 36.49, AB and AC are tangents to the circle whose centre is O. AOD is a straight line. Find the size of the angles AOB, OBD and CBD.

Fig. 36.49

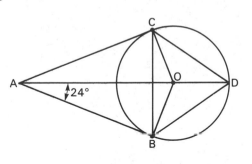

5. In Fig. 36.50, C is the centre of the circle and AP is a tangent to the circle. Calculate the radius of the circle.

Fig. 36.50

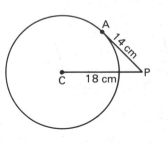

6. Figure 36.51 shows two circles which are just touching. Find *x* by using Pythagoras' theorem.

Fig. 36.51

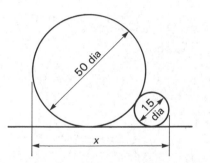

7. In Fig. 36.52 apply Pythagoras' theorem and hence find *h*.

Fig. 36.52

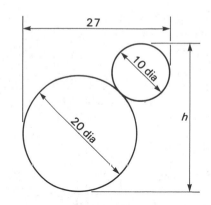

8. In Fig. 36.53, find *h*.

Fig. 36.53

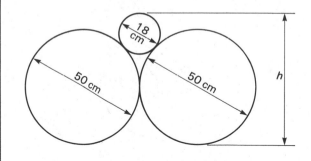

9. Draw two circles whose centres are 8 cm apart and whose diameters are 6 cm and 8 cm. Draw the common tangent to these circles.

10. Draw a circle, centre O, whose radius is 4 cm. Mark off any point P so that OP = 8 cm. Construct a pair of tangents from P to the circle.

Miscellaneous Exercise 36

Section A

1. In Fig. 36.54, O is the centre of the circle. Find:

    (a) ∠OAB   **(b)** ∠ACB

Fig. 36.54

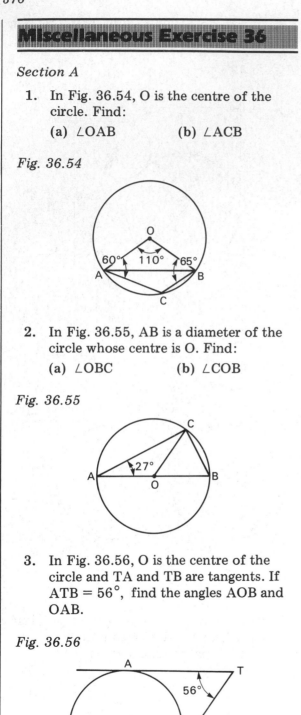

2. In Fig. 36.55, AB is a diameter of the circle whose centre is O. Find:

    (a) ∠OBC   **(b)** ∠COB

Fig. 36.55

3. In Fig. 36.56, O is the centre of the circle and TA and TB are tangents. If ATB = 56°, find the angles AOB and OAB.

Fig. 36.56

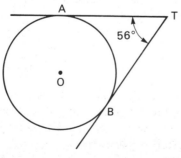

4. Copy Fig. 36.57 and construct the tangents PA and PB. From your construction measure the length of AB.

Fig. 36.57

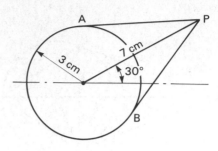

5. In Fig. 36.58, ABC is a straight line and so is CDE. If DE is a diameter of the circle, find ∠BAE and ∠DAE.

Fig. 36.58

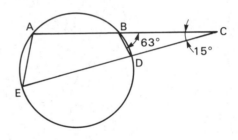

6. In Fig. 36.59, AB, BC, CD and AD are tangents to the circle touching it at P, Q, R and S respectively. If BC is parallel to AD, calculate each of the angles of the quadrilateral PQRS.

Fig. 36.59

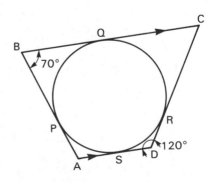

7. In Fig. 36.60, O is the centre of the circle. If BAC = 42°, calculate ∠ABC and ∠AOC.

Fig. 36.60

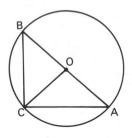

Section B

1. In Fig. 36.61, BD is a diameter of the circle and AEF is a straight line which is parallel to BD. If ∠BAC = 45° and ∠CAE = 65°,

   (a) calculate ∠BDC

   (b) prove that BC = CD

   (c) calculate ∠DEF.

Fig. 36.61

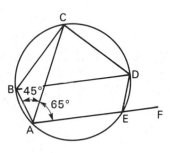

2. In Fig. 36.62, calculate

   (a) OA

   (b) area of △AOB

   (c) area of sector AOBC given that OP = 4.2 cm, AP = 5 cm and O is the centre of the circle.

Fig. 36.62

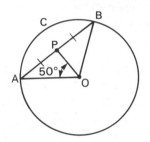

3. In Fig. 36.63, O is the centre of the circle whose radius is 13 cm. If the chord AB is 24 cm long, calculate the shortest distance of the chord from O.

Fig. 36.63

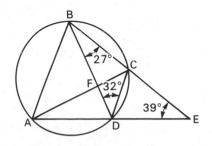

4. In Fig. 36.64,

   (a) find ∠ADB

   (b) name two pairs of similar triangles.

Fig. 36.64

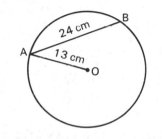

5. Two chords PQ and RS are parallel to each other and they lie on opposite sides of the centre of the circle. If PQ = 10 cm and RS = 8 cm and the circle is 12 cm in radius, find the distance between the chords.

6. Three points X, Y and Z are marked on the circumference of a circle, so that YZ is a diameter. If YZ = 62 mm and XZ = 41 mm find XY.

7. Two chords AB and CD intersect at E. If CE = 3 cm, ED = 2 cm and AE = 4 cm find BE.

8. In Fig. 36.65, prove that AE.CD = CE.AB.

Fig. 36.65

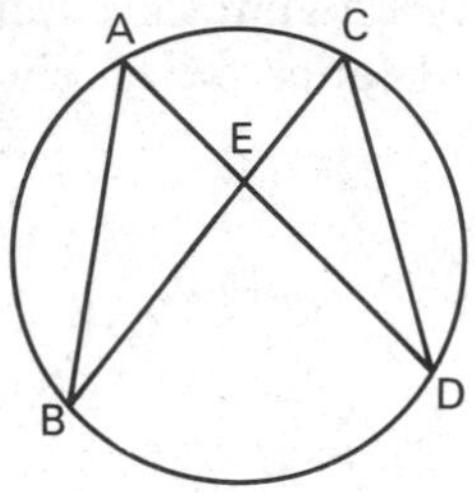

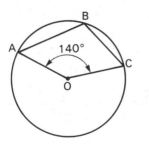

Multi-Choice Questions 36

1. In Fig. 36.66, O is the centre of the circle. What is the size of ∠ABC?
   A 40°   B 70°   C 110°   D 140°

Fig. 36.66

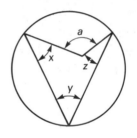

2. In Fig. 36.67, O is the centre of the circle and QT and PT are tangents to the circle at P and Q respectively. What is the size of ∠QTP?
   A 126°   B 63°   C 54°   D 27°

Fig. 36.67

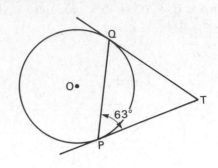

3. In Fig. 36.68, a is equal to
   A $x + y + z$
   B $360° - (x + y + z)$
   C $180° + x + y + z$
   D $360° - x + y + z$

Fig. 36.68

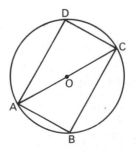

4. In Fig. 36.69, O is the centre of the circle and AB = CD. It is necessarily true that ABCD is a
   A parallelogram   B square
   C rhombus   D rectangle

Fig. 36.69

5. In Fig. 36.70, AB is the diameter of the circle and ∠CBX = ∠ABX. Hence ∠CAX is equal to

    A 15°   B 30°   C 45°   D 60°

Fig. 36.70

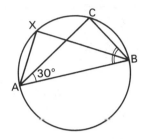

6. From the information given in Fig. 36.71, the obtuse angle MOL is equal to

    A 360° − 2a − 2b

    B 2a + 2b

    C 270° − 2a − b

    D 180° − 2a

Fig. 36.71

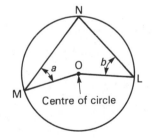

7. In Fig. 36.72, O is the centre of the circle and XY is a diameter. What is the size of ∠PQY?

    A 5°   B 15°   C 20°   D 25°

Fig. 36.72

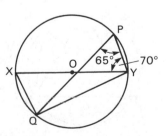

8. In Fig. 36.73, the size of ∠LYM is

    A 80°   B 100°   C 160°   D 200°

Fig. 36.73

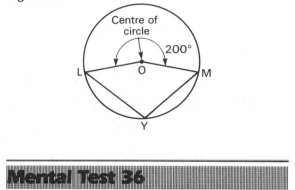

Mental Test 36

Try to answer the following questions without writing anything down except the answer.

1. The circle in Fig. 36.74 has a centre O and COD is a straight line.

    (a) What is the size of the obtuse angle AOB?

    (b) What is the size of ∠DAC?

    (c) Which angle is the same size as ∠ADC?

Fig. 36.74

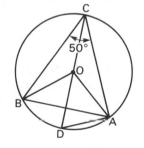

2. In Fig. 36.75, O is the centre of the circle. Calculate the size of the angle *x*.

Fig. 36.75

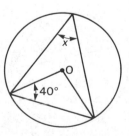

3. In Fig. 36.76, O is the centre of the circle. What is the size of the angle *x*?

Fig. 36.76

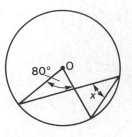

4. In Fig. 36.77, O is the centre of the circle and ∠PSR = 50°. Find the size of ∠OPR.

Fig. 36.77

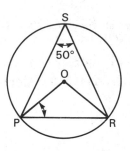

5. In Fig. 36.78, O is the centre of the circle and ROP is a diameter. Copy the diagram and mark on your copy the sizes of the angles

   (a) PQR   **(b)** PRQ

   (c) PSQ   **(d)** SQR

Fig. 36.78

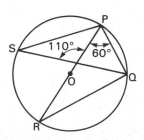

6. In Fig. 36.79, find the size of the angle marked *x*.

Fig. 36.79

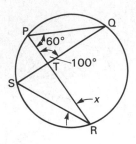

7. In Fig. 36.80, what is the size of the angles

   (a) OAB   **(b)** CAO?

Fig. 36.80

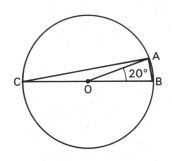

8. In Fig. 36.81, ABC is an equilateral triangle. What is the size of ∠CYB?

Fig. 36.81

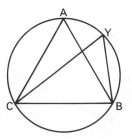

9. In Fig. 36.82, O is the centre of the circle. What is its radius?

Fig. 36.82

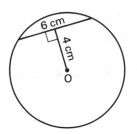

10. In Fig. 36.83, find the length of CE.

Fig. 36.83

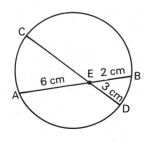

Loci

Introduction

A locus is a set of points traced out by a point which moves according to some law.

Example 1

Find the locus of a point P which is always 3 cm from a fixed point A.

As shown in Fig. 37.1, the locus is a circle with a radius of 3 cm.

Fig. 37.1

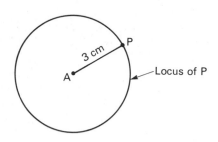

It often helps if we mark off a few points according to the given law. By doing this we may gain some idea of what the locus is. Sometimes three or four points are sufficient but sometimes ten or more points are required before the locus can be recognised.

Example 2

Given a straight line AB of length 6 cm, find the locus of a point P so that $\angle APB$ is always a right angle.

By drawing a number of points P_1, P_2, ..., etc. so that $\angle AP_1B$, $\angle AP_2B$ ... etc. (Fig. 37.2) are all right angles it appears that the locus is a circle with AB as a diameter. We now try to prove that this is so.

Fig. 37.2

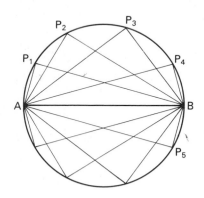

Since the angle in a semicircle is a right angle, all angles subtended by the diameter AB at the circumference of the circle will be right angles. Hence the locus of P is a circle with AB as diameter.

Standard Loci

The following standard loci should be remembered.

(1) The locus of a point equidistant from two given points A and B is the perpendicular bisector of AB (Fig. 37.3).

Fig. 37.3

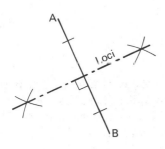

(2) The locus of a point equidistant from the arms of an angle is the bisector of the angle (Fig. 37.4).

Fig. 37.4

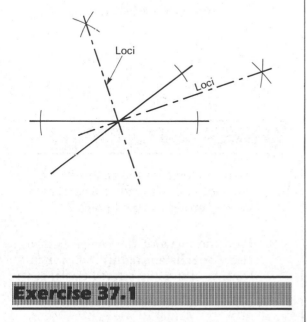

Exercise 37.1

1. Find the locus of a point P which is always 5 cm from a fixed straight line of infinite length.

2. Find the locus of a point P which moves so that it is always 3 cm from a given straight line AB which is 8 cm long.

3. XYZ is a triangle whose base XY is fixed. If $XY = 5 \text{ cm}$ and the area of $\triangle XYZ = 10 \text{ cm}^2$ find the locus of Z.

4. Given a square of 10 cm side, find all the points which are 8 cm from two of the vertices. How many points are there?

5. Find all the points which are 5 cm from each of two intersecting straight lines inclined at an angle of 45°. How many points are there?

6. Find the locus of the centre P of a circle of constant radius 3 cm which passes through a fixed point A.

7. XY is a fixed line of given length. State the locus of a point P which moves so that $\angle XPY$ is constant and sketch the locus.

8. Draw the locus of a point P which is equidistant from two parallel straight lines AB and CD which are 6 cm apart.

Intersecting Loci

Frequently two pieces of information are given about the position of a point. Each piece of information should then be dealt with separately, since any attempt to comply with the two conditions at the same time will lead to a trial and error method which is not acceptable. Each piece of information will partially locate the point and the intersection of the two loci will determine the required position of the point.

Example 3

A point P lies 3 cm from a given straight line and it is also equidistant from two fixed points not on the line and not perpendicular to it. Find the two possible positions of P.

Fig. 37.5

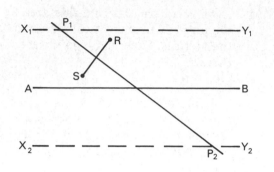

Condition 1

The point P lies 3 cm from the given straight line (AB in Fig. 37.5). To meet this condition draw two straight lines X_2Y_2 and X_1Y_1 parallel to AB and on either side of it.

Condition 2

P is equidistant from the two fixed points (R and S in Fig. 37.5). To meet this condition we draw the perpendicular bisector of RS (since $\triangle$RSP must be isosceles).

The intersections of the two loci give the required position of the point P. These are the points P_1 and P_2 in Fig. 37.5.

Three-Dimensional Loci

(1)   The locus of a point P which moves so that the length PA is constant (Fig. 37.6) is a sphere centre A.

Fig. 37.6

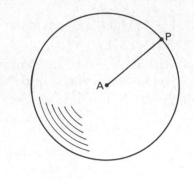

(2)   The locus of a point P which moves so that it is a fixed distance from a line AB is a cylinder whose centre-line is AB (Fig. 37.7).

Fig. 37.7

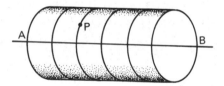

Exercise 37.2

1.   Find the locus of the centre P of a circle of constant radius 3 cm which passes through a fixed point A.

2.   Find the locus of the centre Q of a circle of constant radius 2 cm which touches externally a fixed circle, centre B and radius 4 cm.

3.   Find the locus of the centre of a variable circle which passes through two fixed points A and B.

4. AB is a fixed line of length 8 cm and P is a variable point. The distance of P from the middle point of AB is 5 cm and the distance of P from AB is 4 cm. Construct a point P so that both of these conditions are satisfied. State the number of possible positions of P.

5. X is a point inside a circle, centre C, and Q is the mid-point of a chord which passes through X. Determine the locus of Q as the chord varies. If CX is 5 cm and the radius of the circle is 8 cm construct the locus of Q accurately and hence construct a chord which passes through X and has its mid-point 3 cm from C.

6. XY is a fixed line of given length. State the locus of a point P which moves so that the size of ∠XPY is constant and sketch the locus.

7. AB is a fixed line of 4 cm and R is a point such that the area of △ABR is 5 cm². S is the mid-point of AR. State the locus of R and the locus of S.

8. T is a fixed point outside a fixed circle whose centre is O. A variable line through T meets the circle at X and Y. Show that the locus of the mid-point of XY is an arc of the circle with OT as diameter.

9. Chords of a circle, centre C, are drawn through a fixed point A within the circle. Show that the mid-points of all these chords lie on a circle and state the position of the centre of the circle.

10. Draw a circle centre O and radius 4 cm. Construct the locus of the mid-points of all chords of this circle which are 6.5 cm long.

11. AB is a fixed vertical line of length 6 cm. Sketch the locus of a point P so that PAB is constant at 40°.

12. The plane ABCD is a rectangle in which AB = 5 cm and BC = 3 cm. A point P moves so that its perpendicular distance from ABCD is always 4 cm. Sketch the locus of P.

Trigonometry

The Notation for a Right-Angled Triangle

The sides of a right-angled triangle are given special names. In Fig. 38.1 the side AB lies opposite the right angle and it is called the **hypotenuse.** The side BC lies opposite to the angle A and it is called the side opposite to A. The side AC is called the side adjacent to A.

Fig. 38.1

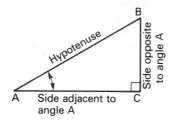

When we consider the angle B (Fig. 38.2) the side AB is still the hypotenuse but AC is now the side opposite to B and BC is the side adjacent to B.

Fig. 38.2

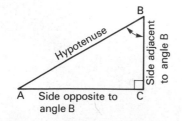

The Trigonometrical Ratios

Consider any angle θ which is bounded by the lines OA and OB as shown in Fig. 38.3. Take any point P on the boundary line OB. From P draw line PM perpendicular to OA to meet it at the point M. Then,

the ratio $\dfrac{MP}{OP}$ is called the **sine** of $\angle AOB$

the ratio $\dfrac{OM}{OP}$ is called the **cosine** of $\angle AOB$

and

the ratio $\dfrac{MP}{OM}$ is called the **tangent** of $\angle AOB$.

Fig. 38.3

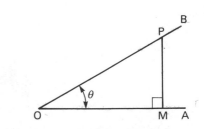

The Sine of an Angle

The abbreviation sin is usually used for sine.
In any right-angled triangle (Fig. 38.4):

$$\text{The sine of angle} = \frac{\text{Side opposite the angle}}{\text{Hypotenuse}}$$

$$\sin A = \frac{BC}{AC}$$

$$\sin C = \frac{AB}{AC}$$

Fig. 38.4

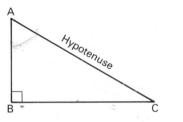

The sine of an angle may be found by using
the table of sines of angles or by using a
scientific calculator.

Example 1

(a) Find the length of AB in Fig. 38.5.

Fig. 38.5

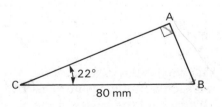

AB is the side opposite ∠ACB. BC is
the hypotenuse since it is opposite to
the right angle.

Therefore

$$\frac{AB}{BC} = \sin 22°$$

$$AB = BC \times \sin 22°$$

$$= 80 \times \sin 22°$$

$$= 29.97\,mm$$

A scientific calculator may be used for
this calculation as follows:

Input	Display
22	22.
sin	0.3746 ...
×	0.3746 ...
80	80.
=	29.968 ...

(b) Find the length of AB in Fig. 38.6.

Fig. 38.6

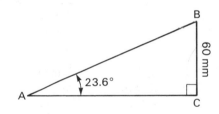

BC is the side opposite to ∠BAC and
AB is the hypotenuse.

Therefore

$$\frac{BC}{AB} = \sin 23.6°$$

$$AB = \frac{BC}{\sin 23.6°}$$

$$= \frac{60}{\sin 23.6°}$$

$$= 149.9\,mm$$

Input	Display
23.6	23.6
sin	0.4003 ...
1/x	2.4978 ...
×	2.4978 ...
60	60.
=	149.869 ...

(c) Find the angles CAB and ABC in △ABC which is shown in Fig. 38.7.

Fig. 38.7

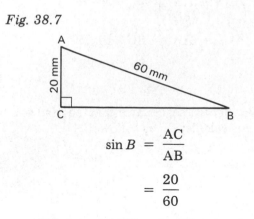

$$\sin B = \frac{AC}{AB}$$

$$= \frac{20}{60}$$

Using a scientific calculator (see below):

$$\angle B = 19.47°$$

$$\angle A = 90° - 19.47°$$

$$= 70.53°$$

Input	Display
20	20.
÷	20.
60	60.
=	0.3333 ...
INV sin	19.4712 ...
+/−	−19.4712 ...
+	−19.4712 ...
90	90.
=	70.5287 ...

Exercise 38.1

1. Find the lengths of the sides marked *x* in Fig. 38.8 the triangles being right-angled.

Fig. 38.8

(a)   **(b)**

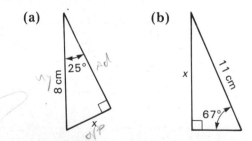

(c)

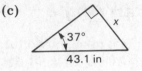

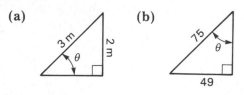

2. Find the angles marked θ in Fig. 38.9 the triangles being right-angled.

Fig. 38.9

(a)   **(b)**

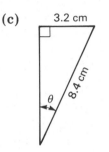

(c)

3. In △ABC, $C = 90°$, $B = 23.3°$ and $AC = 11.2$ cm. Find AB.

4. In △ABC, $B = 90°$, $A = 67.5°$ and $AC = 0.86$ m. Find BC.

5. An equilateral triangle has an altitude of 18.7 cm. Find the length of the equal sides.

6. Find the altitude of an isosceles triangle whose vertex angle is 38° and whose equal sides are 7.9 m long.

7. The equal sides of an isosceles triangle are each 27 cm long and the altitude is 19 cm. Find the angles of the triangle.

8. In △ABC, $A = 90°$, $C = 54°$ and $a = 6.4$ cm. Find the lengths of the sides *b* and *c*.

9. Fig. 38.10 shows part of a framework. Find the lengths of the members AD, AB, BC and CD.

Fig. 38.10

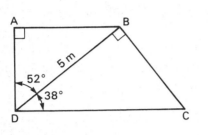

10. Fig. 38.11 shows the plan of a plot of land. Find the length of the diagonal AC and the size of the angle ACB.

Fig. 38.11

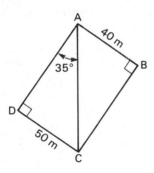

The Cosine of an Angle

In any right-angled triangle (Fig. 38.12):

$$\text{The cosine of an angle} = \frac{\text{Side adjacent to the angle}}{\text{Hypotenuse}}$$

$$\cos A = \frac{AB}{AC}$$

$$\cos C = \frac{BC}{AC}$$

Fig. 39.12

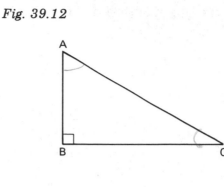

Example 2

(a) Find the length of the side BC in Fig. 38.13.

Fig. 38.13

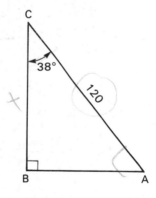

BC is the side adjacent to ∠BCA and AC is the hypotenuse.

Therefore

$$\frac{BC}{AC} = \cos 38°$$

$$BC = AC \times \cos 38°$$

$$= 120 \times \cos 38°$$

$$= 94.56 \text{ mm}$$

(b) Find the length of the side AC in Fig. 38.14.

Fig. 38.14

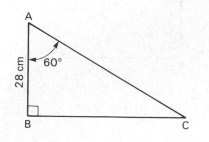

AB is the side adjacent to ∠BAC and AC is the hypotenuse.

Therefore

$$\frac{AB}{AC} = \cos 60°$$

$$AC = \frac{AB}{\cos 60°}$$

$$= \frac{28}{\cos 60°}$$

$$= 56 \text{ cm}$$

(c) Find the angle θ shown in Fig. 38.15.

Fig. 38.15

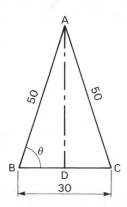

Since △ABC is isosceles the perpendicular AD bisects the base BC and hence BD = 15 mm.

$$\cos \theta = \frac{BD}{AB}$$

$$= \frac{15}{50}$$

$$\theta = 72.54°$$

Exercise 38.2

1. Find the lengths of the sides marked x in Fig. 38.16, the triangles being right-angled.

Fig. 38.16

(a)

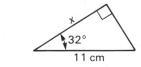

(b)

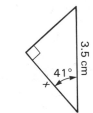

(c)

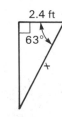

2. Find the angles marked θ in Fig. 38.17, the triangles being right-angled.

Fig. 38.17

(a)

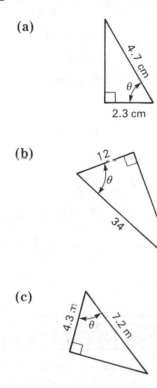

2.3 cm

(b)

12

θ

34

(c)

4.3 m   7.2 m

θ

3. An isosceles triangle has a base of 3.4 cm and the equal sides are each 4.2 cm long. Find the angles of the triangle and also its altitude.

4. In $\triangle ABC$, $C = 90°$, $B = 33.5°$ and $BC = 2.4$ cm. Find AB.

5. In $\triangle ABC$, $B = 90°$, $A = 62.8°$ and $AC = 4.3$ cm. Find AB.

6. In Fig. 38.18, calculate $\angle BAC$ and the length BC.

Fig. 38.18

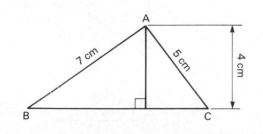

7 cm   5 cm   4 cm

B   C

7. In Fig. 38.19, calculate BD, AD, AC and BC.

Fig. 38.19

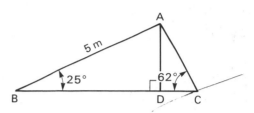

A

5 m

25°   62°

B   D   C

8. Fig. 38.20 shows a trapezium with CD parallel to AB. Calculate the lengths of the sides AB, BC and AD.

Fig. 38.20

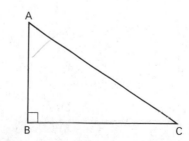

D   8 cm   C

55°

A   B

The Tangent of an Angle

In any right-angled triangle (Fig. 38.21):

The tangent of an angle $=$ $\dfrac{\text{Side opposite to the angle}}{\text{Side adjacent to the angle}}$

$$\tan A = \frac{BC}{AB}$$

$$\tan C = \frac{AB}{BC}$$

Fig. 38.21

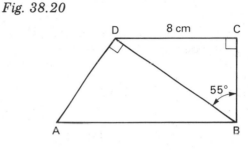

A

B   C

Example 3

(a) Find the length of the side AB in Fig. 38.22.

Fig. 38.22

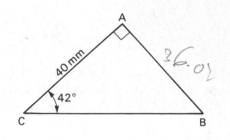

AB is the side opposite ∠ACB and AC is the side adjacent to ∠ACB.

Hence $\dfrac{AB}{AC} = \tan 42°$

$AB = AC \times \tan 42°$

$= 40 \times \tan 42°$

$= 36.02 \text{ mm}$

(b) Find the length of the side BC in Fig. 38.23.

Fig. 38.23

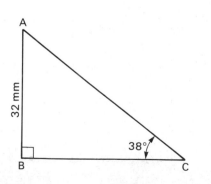

There are two ways of doing this problem.

(i) $\dfrac{AB}{BC} = \tan 38°$ or $BC = \dfrac{AB}{\tan 38°}$

Therefore $BC = \dfrac{32}{\tan 38°}$

$= 40.96 \text{ mm}$

(ii) Since $C = 38°$,

$A = 90° - 38°$

$= 52°$

Now

$\dfrac{BC}{AB} = \tan A$ or $BC = AB \times \tan A$

$BC = 32 \times \tan 52°$

$= 40.96 \text{ mm}$

Exercise 38.3

1. Find the lengths of the sides marked y in Fig. 38.24, the triangles being right-angled.

Fig. 38.24

(a)

(b)

(c)

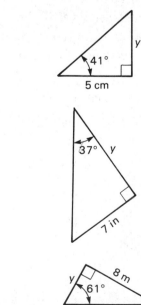

2. Find the angles marked α in Fig. 38.25, the triangles being right-angled.

Fig. 38.25

(a)

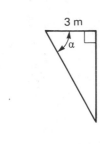

3 m

5 m

(b)

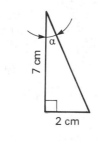

7 cm

2 cm

(c)

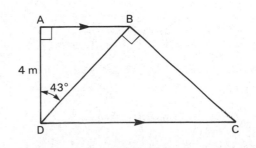

23

55

α

3. An isosceles triangle has a base 10 cm long and the two equal angles are each 57°. Calculate the altitude of the triangle.

4. In $\triangle ABC$, $B = 90°$, $C = 49°$ and $AB = 3.2$ cm. Find BC.

5. In $\triangle ABC$, $A = 12.38°$, $B = 90°$ and $BC = 7.31$ cm. Find AB.

6. Calculate the distance x in Fig. 38.26.

Fig. 38.26

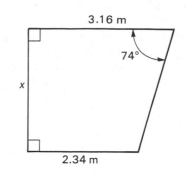

3.16 m

74°

x

2.34 m

7. Calculate the distance d in Fig. 38.27.

Fig. 38.27

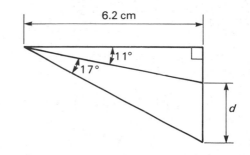

6.2 cm

11°

17°

d

8. Fig. 38.28 shows a framework in the form of a trapezium with AB parallel to CD. Find the lengths of the members AB, BD and BC.

Fig. 38.28

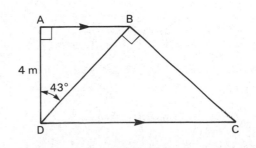

A   B

4 m

43°

D   C

Trigonometrical Ratios for 30°, 60° and 45°

Ratios for 30° and 60°

Fig. 38.29 shows an equilateral triangle ABC with each of the sides equal to 2 units. From C draw the perpendicular CD which bisects the base AB and also bisects $\angle ACB$.

Fig. 38.29

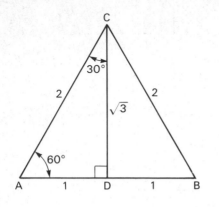

In $\triangle ACD$,

$$CD^2 = AC^2 - AD^2$$

$$= 2^2 - 1^2$$

$$= 3$$

Therefore $\quad CD = \sqrt{3}$

Since all the angles of $\triangle ABC$ are 60° and $\angle ACD = 30°$,

$$\sin 60° = \frac{\sqrt{3}}{2}$$

$$\tan 60° = \frac{\sqrt{3}}{1}$$

$$= \sqrt{3}$$

$$\cos 60° = \frac{1}{2}$$

$$\sin 30° = \frac{1}{2}$$

$$\tan 30° = \frac{1}{\sqrt{3}}$$

$$= \frac{\sqrt{3}}{3}$$

$$\cos 30° = \frac{\sqrt{3}}{2}$$

Ratios for 45°

Fig. 38.30 shows a right-angled isosceles triangle ABC with the equal sides each 1 unit in length. The equal angles are each 45°. Now,

$$AC^2 = AB^2 + BC^2$$

$$= 1^2 + 1^2$$

$$= 2$$

Therefore $\quad AC = \sqrt{2}$

$$\sin 45° = \frac{1}{\sqrt{2}}$$

$$= \frac{\sqrt{2}}{2}$$

$$\cos 45° = \frac{1}{\sqrt{2}}$$

$$= \frac{\sqrt{2}}{2}$$

$$\tan 45° = \frac{1}{1}$$

$$= 1$$

Fig. 38.30

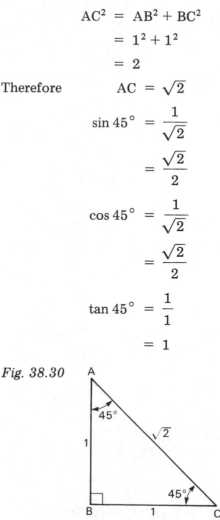

Complementary Angles

Complementary angles are angles whose sum is 90°.

Consider the triangle ABC shown in Fig. 38.31.

Fig. 38.31

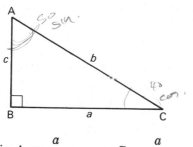

$$\sin A = \frac{a}{b} \qquad \cos C = \frac{a}{b}$$

Hence
$$\sin A = \cos C$$
$$0.766 = \cos(90° - A)$$

Similarly,
$$\cos A = \sin(90° - A)$$
$$0.643$$

Therefore, **the sine of an angle is equal to the cosine of its complementary angle and vice versa.**

$$\sin 26° = \cos 64°$$
$$= 0.4384$$
$$\cos 70° = \sin 20°$$
$$= 0.3420$$

The Squares of the Trigonometrical Ratios

The square of $\sin A$ is usually written as $\sin^2 A$. Thus,

$$\sin^2 A = (\sin A)^2$$

and similarly for the remaining trigonometrical ratios. That is,

$$\cos^2 A = (\cos A)^2$$
and $$\tan^2 A = (\tan A)^2$$

Example 4

(a) Find the value of $\cos^2 37°$.
$$\cos 37° = 0.7986$$
$$\cos^2 37° = (0.7986)^2$$
$$= 0.6378$$

(b) Find the value of $\tan^2 60° + \sin^2 60°$.
$$\tan 60° = \sqrt{3}$$
$$\therefore \qquad \tan^2 60° = 3$$
$$\sin 60° = \frac{\sqrt{3}}{2}$$
$$\therefore \qquad \sin^2 60° = \frac{3}{4}$$
$$\tan^2 60° + \sin^2 60° = 3 + \frac{3}{4}$$
$$= 3\tfrac{3}{4}$$

It is sometimes useful to remember that
$$\sin^2 A + \cos^2 A = 1$$
which may easily be proved by considering a right-angled triangle (Fig. 38.32).

Fig. 38.32

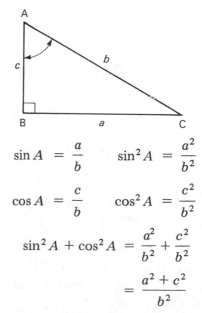

$$\sin A = \frac{a}{b} \qquad \sin^2 A = \frac{a^2}{b^2}$$

$$\cos A = \frac{c}{b} \qquad \cos^2 A = \frac{c^2}{b^2}$$

$$\sin^2 A + \cos^2 A = \frac{a^2}{b^2} + \frac{c^2}{b^2}$$
$$= \frac{a^2 + c^2}{b^2}$$

But, by Pythagoras' theorem:

$$a^2 + c^2 = b^2$$

$$\therefore \qquad \sin^2 A + \cos^2 A = \frac{b^2}{b^2}$$

$$= 1$$

Example 5

The angle A is acute and
$5 \sin^2 A - 2 = \cos^2 A$.
Find the angle A.

Since $\sin^2 A + \cos^2 A = 1$

$$\cos^2 A = 1 - \sin^2 A$$

$$\therefore \qquad 5 \sin^2 A - 2 = 1 - \sin^2 A$$

$$6 \sin^2 A - 3 = 0$$

$$6 \sin^2 A = 3$$

$$\sin^2 A = 0.5$$

$$\sin A = \pm\sqrt{0.5}$$

$$= \pm 0.7071$$

Since A is acute $\sin A$ must be positive (see Chapter 40).

Hence $\sin A = 0.7071$

$$A = 45°$$

Exercise 38.4

1. Show that $\cos 60° + \cos 30° = \dfrac{1 + \sqrt{3}}{2}$.

2. Show that $\sin 60° + \cos 30° = \sqrt{3}$.

3. Show that

    $\cos 45° + \sin 60° + \sin 30°$

    $= \dfrac{\sqrt{2} + \sqrt{3} + 1}{2}$.

4. Given that $\sin 48° = 0.7431$ find the value of $\cos 42°$, without using tables or a calculator.

5. If $\cos 63° = 0.4540$, what is the value of $\sin 27°$?

6. (a) Find the value of $\cos^2 30°$.
    (b) Find the value of $\tan^2 30°$.
    (c) Find the value of $\sin^2 60°$.

7. Evaluate:
    (a) $\cos^2 41°$ (b) $\sin^2 27°$
    (c) $\tan^2 58°$

8. If the angle A is acute and
    $2 \sin^2 A - \dfrac{1}{3} = \cos^2 A$ find A.

9. If the angle θ is acute and
    $\cos^2 \theta + \dfrac{1}{8} = \sin^2 \theta$ find θ.

It is important that the trigonometrical ratios be remembered. One way of doing this is to use the word SOHCAHTOA meaning 'Sine Opposite Hypotenuse, Cosine Adjacent Hypotenuse, Tangent Opposite Adjacent.'

Angle of Elevation

If you look upwards at an object, say the top of a tree, the angle formed between the horizontal and your line of sight is called the **angle of elevation** (Fig. 38.33). It is the angle through which the line of your sight has been elevated.

Fig. 38.33

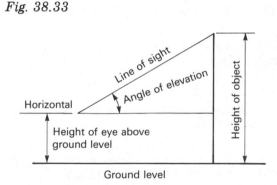

Example 6

To find the height of a tower a surveyor sets up his theodolite 100 m from the base of the tower. He finds that the angle of elevation to the top of the tower is 30°. If the instrument is 1.5 m from the ground, what is the height of the tower?

Fig. 38.34

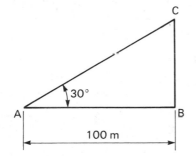

In Fig. 38.34,

$$\frac{BC}{AB} = \tan 30°$$

$$BC = AB \times \tan 30°$$

$$= 100 \times 0.5774$$

$$= 57.74$$

Hence, height of tower

$$= 57.74 + 1.5$$

$$= 59.24 \text{ m}$$

Example 7

To find the height of a pylon, a surveyor sets up his theodolite some distance from the pylon and finds the angle of elevation to the top of the pylon to be 30°. He then moves 60 m nearer to the pylon and finds the angle of elevation to be 42°. Find the height of the pylon, assuming the ground to be horizontal and that the instrument is 1.5 m above ground level.

This problem is much more difficult than Example 6 because we have to use two triangles and some algebra.

Fig. 38.35

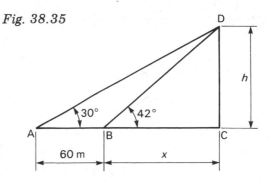

Looking at Fig. 38.35, let $BC = x$ and $DC = h$. Our problem is to find h.

In $\triangle ACD$,

$$\frac{DC}{AC} = \tan 30°$$

$$DC = AC \times \tan 30°$$

$$h = 0.5774(x + 60) \qquad (1)$$

In $\triangle BDC$

$$\frac{DC}{BC} = \tan 42°$$

$$DC = BC \times \tan 42°$$

$$h = 0.9004x \qquad (2)$$

$$x = \frac{h}{0.9004}$$

$$= 1.1106h$$

Substituting for x in equation (1) we have,

$$h = 0.5774(1.1106h + 60)$$

$$h = 0.6413h + 34.64$$

$$h - 0.6413h = 34.64$$

$$0.3587h = 34.64$$

$$h = \frac{34.64}{0.3587}$$

$$= 96.57 \text{ m}$$

Hence

Height of the pylon $= 96.57 + 1.5$

$$= 98.07 \text{ m}$$

Altitude of the Sun

The altitude of the sun is simply the angle of elevation of the sun (Fig. 38.36).

Fig. 38.36

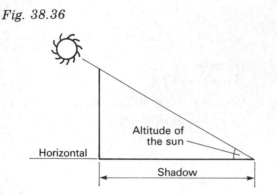

Example 8

A flagpole is 15 m high. What length of shadow will it cast when the altitude of the sun is 57°?

Fig. 38.37

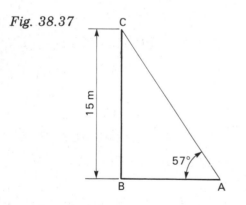

Looking at Fig. 38.37 we have

$$\angle ACB = 90° - 57°$$

$$= 33°$$

$$\frac{AB}{BC} = \tan \angle ACB$$

$$AB = BC \times \tan \angle ACB$$

$$= 15 \times \tan 33°$$

$$= 9.741$$

Hence the flagpole will cast a shadow 9.741 m long.

✳ Angle of Depression

If you look downwards at an object, the angle formed between the horizontal and your line of sight is called the **angle of depression** (Fig. 38.38). It is therefore the angle through which the line of sight is depressed from the horizontal. Note carefully that both the angle of elevation and the angle of depression are measured from the horizontal.

Fig. 38.38

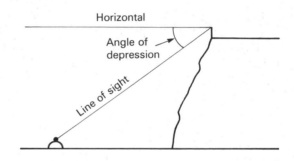

Example 9

A person standing on top of a cliff 50 m high is in line with two buoys whose angles of depression are 18° and 20°. Calculate the distance between the buoys.

The problem is illustrated in Fig. 38.39 where the buoys are C and D and the observer is A.

Fig. 38.39

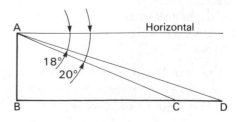

In △BAC,

$$BAC = 90° - 20°$$
$$= 70°$$

$$\frac{BC}{AB} = \tan \angle BAC$$

$$BC = AB \times \tan \angle BAC$$
$$= 50 \times \tan 70°$$
$$= 137.4 \, m$$

In △ABD,

$$\angle BAD = 90° - 18°$$
$$= 72°$$

$$\frac{BD}{AB} = \tan \angle BAD$$

$$BD = AB \times \tan \angle BAD$$
$$= 50 \times \tan 72°$$
$$- 153.9 \, m$$

The distance between the buoys

$$= BD - BC$$
$$= 153.9 - 137.4$$
$$= 16.5 \, m$$

Exercise 38.5

1. From a point, the angle of elevation of a tower is 30°. If the tower is 20 m distant from the point, what is the height of the tower?

2. A man 1.8 m tall observes the angle of elevation of a tree to be 26°. If he is standing 16 m from the tree, find the height of the tree.

3. A man 1.5 m tall is 15 m away from a tower 20 m high. What is the angle of elevation of the top of the tower from his eyes?

4. A man, lying down on top of a cliff 40 m high, observes the angle of depression of a buoy to be 20°. If he is in line with the buoy, calculate the distance between the buoy and the foot of the cliff (which may be assumed to be vertical).

5. A tree is 20 m tall. When the altitude of the sun is 62°, what length of shadow will it cast?

6. A flagpole casts a shadow 18 m long when the altitude of the sun is 54°. What is the height of the flagpole?

7. A man standing on top of a mountain 1200 m high observes the angle of depression of a steeple to be 43°. How far is the steeple from the mountain?

8. In Fig. 38.40, a vertical cliff is 40 m high and is observed from a boat which is 50 m from the foot of the cliff. Calculate the angle of elevation of the top of the cliff from the boat.

Fig. 38.40

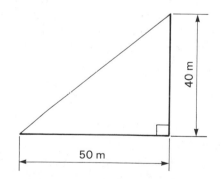

9. To find the height of a tower a surveyor stands some distance away from its base and he observes the angle of elevation to the top of the tower to be 45°. He then moves 80 m nearer to the tower and he then finds the angle of elevation to be 60°. Find the height of the tower.

10. A tower is known to be 60 m high. A man using a theodolite stands some distance away from the tower and measures its angle of elevation as 38°. How far away from the tower is he if the theodolite stands 1.5 m above the ground? If the man now moves 80 m further away from the tower what is now the angle of elevation of the tower?

11. A man standing 20 m away from a tower observes the angles of elevation to the top and bottom of a flagstaff standing on the tower as 62° and 60° respectively. Calculate the height of the flagstaff.

12. A surveyor stands 100 m from the base of a tower on which an aerial stands. He measures the angles of elevation to the top and bottom of the aerial as 58° and 56°. Find the height of the aerial.

13. Figure 38.41 shows a tower TR and an observer at O. From O which is 100 m from the base R of tower TR the angle of elevation of the top of the tower is found to be 35°.

Fig. 38.41

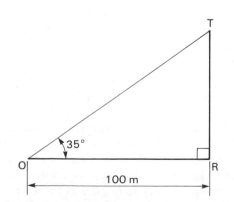

(a) Calculate the height of the tower TR, to 3 significant figures.

(b) The observer walks forward towards the tower until the angle of elevation of the top T is 45°. How far does he walk forward?

(c) Find the distance of the observer from the base of the tower when the angle of elevation of the top is 65°. Give your answer to 3 significant figures.

14. A man standing on top of a cliff 80 m high is in line with two buoys whose angles of depression are 17° and 21°. Calculate the distance between the buoys.

15. Figure 38.42 represents two plotting stations, A and B, 4000 m apart. T is a stationary target in the same vertical plane as A and B. It is recorded that, when the distance of the target from station A is 10 000 m, the angle of elevation is 29°. Calculate:

(a) the vertical height of the target, TX

(b) the distance AX

(c) the distance BX

(d) the angle of elevation of the target, T, from B

(e) the distance TB.

Fig. 38.42

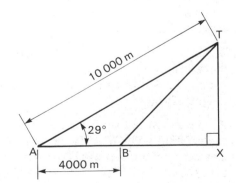

Calculating Bearings

Bearings were discussed in Chapter 33 where scale drawings were used to solve problems with them. In this section we show how problems connected with bearings may be solved by using trigonometry.

Example 10

(a) B is a point due east of a point A on the coast. C is another point on the coast and is 6 km due south of A. The distance BC is 7 km. Calculate the bearing of C from B.

Fig. 38.43

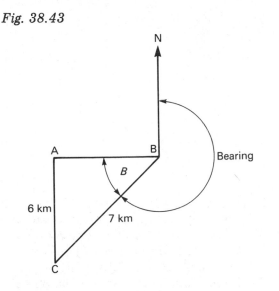

Referring to Fig. 38.43,

$$\sin B = \frac{AC}{BC}$$

$$= \frac{6}{7}$$

$$B = 59°$$

The bearing of C from B

$$= 270° - 59°$$

$$= 211°.$$

(b) B is 5 km due north of P and C is 2 km due east of P. A ship started from C and steamed in a direction 030°. Calculate the distance the ship had to go before it was due east of B. Find also the distance it was then from B.

Fig. 38.44

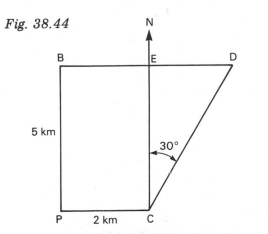

Referring to Fig. 38.44, the ship will be due east of B when it has sailed the distance CD. The bearing 030° makes the angle ∠ECD equal to 30°. In △CED, EC = 5 km and ∠ECD = 30°. Hence,

$$\frac{EC}{CD} = \cos 30°$$

$$CD = \frac{EC}{\cos 30°}$$

$$= \frac{5}{\cos 30°}$$

$$= 5.77 \text{ km}$$

Hence the ship had to sail 5.77 km to become due east of B.

$$\frac{ED}{CD} = \sin 30°$$

$$ED = CD \times \sin 30°$$

$$= 5.77 \times \sin 30°$$

$$= 2.885 \text{ km}$$

$$BD = BE + ED$$

$$= 2 + 2.885$$

$$= 4.885 \text{ km}$$

Hence the ship will be 4.885 km due east of B.

Exercise 38.6

1. A ship sets out from a point A and sails due north to a point B, a distance of 120 km. It then sails due east to a point C. If the bearing of C from A is 037.67° (Fig. 38.45) calculate:

    (a) the distance BC

    (b) the distance AC.

Fig. 38.45

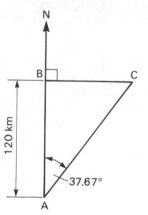

2. A boat leaves a harbour A on a course of 150° and it sails 50 km in this direction until it reaches a point B (Fig. 38.46). How far is B east of A? What distance south of A is B?

Fig. 38.46

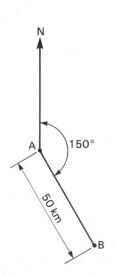

3. X is a point due west of a point P. Y is a point due south of P. If the distances PX and PY are 10 km and 15 km respectively, calculate the bearing of X from Y (Fig. 38.47).

Fig. 38.47

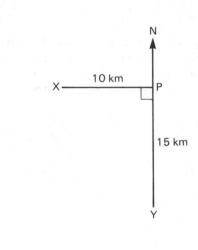

4. B is 10 km north of P and C is 5 km due west of P. A ship starts from C and sails in a direction of 330°. Calculate the distance the ship has to sail before it is due west of B and find also the distance it is then from B (Fig. 38.48).

Fig. 38.48

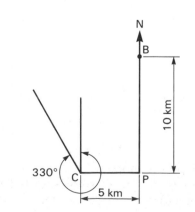

5. A fishing boat places a float on the sea at A, 50 m due north of a buoy B. A second boat places a float at C, whose bearing from A is 130°. A taut net connecting the floats at A and C is 80 m long. Calculate the distance BC and the bearing of C from B (Fig. 38.49).

Fig. 38.49

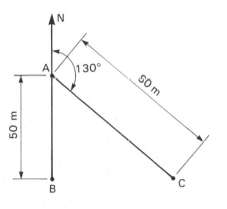

6. An aircraft starts to fly from Λ to B a distance of 140 km, B being due north of A. The aircraft flies on course 18°E of N for a distance of 80 km. Calculate how far the aircraft is then from the line AB and in what direction it should then fly to reach B.

7. X and Y are two lighthouses, Y being 20 km due east of X. From a ship due south of X, the bearing of Y was 055°. Find:

(a) the distance of the ship from Y

(b) the distance of the ship from X.

8. An aircraft flies 50 km from an aerodrome A on a bearing of 065° and then flies 80 km on a bearing of 040°. Find the distance of the aircraft from A and also its bearing from A.

9. A boat sails 10 km from a harbour H on a bearing of S30°E. It then sails 15 km on a bearing of N20°E. How far is the boat from H? What is its bearing from H?

10. Three towns A, B and C lie on a straight road running east from A. B is 6 km from A and C is 22 km from A. Another town D lies to the north of this road and it lies 10 km from both B and C. Calculate the distance of D from A and the bearing of D from A.

Miscellaneous Exercise 38

Section A

1. ABCD (Fig. 38.50) is a trapezium with AB parallel to CD. Calculate:

(a) AD   (b) BC

(c) the area of ABCD.

Fig. 38.50

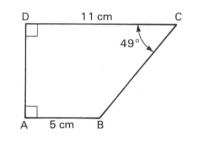

2. Fig. 38.51 shows a parallelogram ABCD with DN perpendicular to AB. Calculate:

(a) the length AN

(b) the area of △ABD

(c) the area of the parallelogram ABCD.

Fig. 38.51

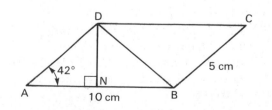

3. In Fig. 38.52, calculate:

    (a) WX   **(b)** ∠YZV

Fig. 38.52

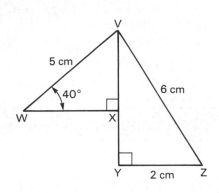

4. In Fig. 38.53:

    (a) calculate the length x

    (b) write down the value of $\tan y$

    (c) calculate the length p.

Fig. 38.53

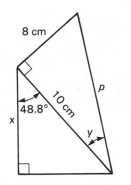

5. In Fig. 38.54, AB = 6 cm and BC = 8 cm. The perpendicular distance from A to BC is 5 cm. Calculate the perpendicular distance from C to AB.

Fig. 38.54

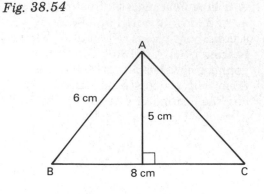

6. A boat starts from a harbour A and sails 4 km due south to a point B. It then changes direction and sails on a bearing of 050° to a point C. If C is due east of A, calculate the distance BC, in kilometres, correct to 2 decimal places.

7. In Fig. 38.55, AC = 80 in, AD = 18 in, BD = 18 in and BE = 30 in. Calculate:

    (a) the size of ∠CAD

    (b) the length DE.

Fig. 38.55

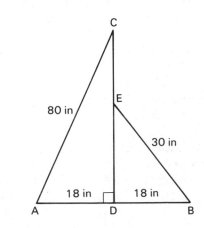

8. In △ABC (Fig. 38.56) BC = 5 cm.

   (a) Given that sin ∠ADB = 0.7, write down the value of sin ∠BDC.

   (b) Given also that tan A = 0.5, calculate the length of AB.

Fig. 38.56

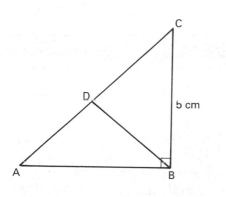

Section B

1. Calculate from Fig. 38.57:

   (a) the length of CD

   (b) the value of the ratio

   Area of triangle ACB
   ─────────────────────
   Area of triangle BCD

Fig. 38.57

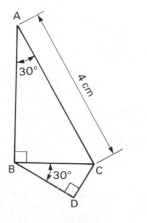

2. In Fig. 38.58, the lengths of the line segments AB and BC are 10 cm and 5 cm respectively and angle ABC is a right-angle. Calculate:

   (a) the perpendicular distance from B to the line *q*

   (b) the perpendicular distance from C to *q*.

   (Give your answers correct to one decimal place.)

Fig. 38.58

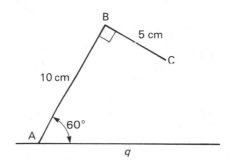

3. In Fig. 38.59, DB is perpendicular to the line ABC. Calculate the length of DE.

Fig. 38.59

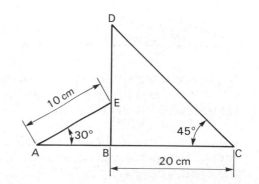

4. In Fig. 38.60, angle ABC = 50°, angle AED = 20°, BC = AE = 10 cm. Angle ACB = angle ADE = 90°. Calculate:

    (a) the length of AC

    (b) the length of AD.

Fig. 38.60

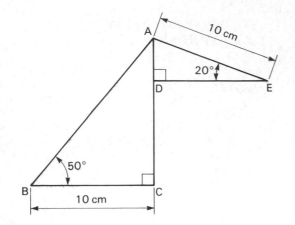

5. Calculate the length of AB in Fig. 38.61.

Fig. 38.61

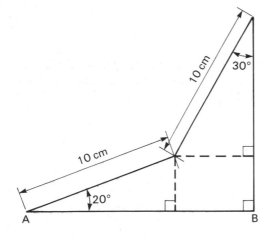

6. In the acute-angled △PQR, PQ = 6 cm, QR = 4 cm and the area of the triangle is 10.8 cm². Calculate ∠PQR.

7. Given that $\sin^2 x = 0.2652$, find the value of the acute angle x, correct to the nearest degree.

8. A ship sets out from a point A and sails due North to a point B a distance of 120 km. It then sails due East to a point C. If the bearing of C from A is 037° find:

    (a) the distance BC

    (b) the distance AC.

Multi-Choice Questions 38

1. If $\sin A = \dfrac{3}{5}$ then $\cos A$ equals

    A $\dfrac{2}{5}$ **B** $\dfrac{3}{4}$ **C** $\dfrac{4}{5}$ **D** $\dfrac{5}{3}$

2. From the top of a tower the angle of depression of a boat is 30°. If the tower is 20 metres high, how far is the boat from the foot of the tower?

    A 40 m   **B** $10\sqrt{3}$ m

    C $20\sqrt{2}$ m   **D** $20\sqrt{3}$ m

3. From a place 400 metres north of X, a man walks eastwards to a place Y which is 800 m from X. What is the bearing of X from Y?

    A 270° **B** 240° **C** 210° **D** 180°

4. The expression for the length AC in Fig. 38.62 is

    A 40 sin 50°   **B** 40 cos 50°

    C $\dfrac{40}{\sin 50°}$ **D** $\dfrac{40}{\cos 50°}$

Fig. 38.62

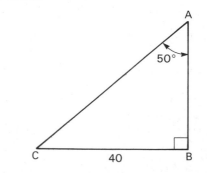

5. In Fig. 38.63, an expression for the side AC is

    A $\dfrac{9\sqrt{3}}{2} + 3$ B $3 + \dfrac{9}{2}$

    C $\dfrac{6\sqrt{3}}{2} + \dfrac{9}{2}$ D $\dfrac{6\sqrt{3}}{2} + \dfrac{9\sqrt{3}}{2}$

Fig. 38.63

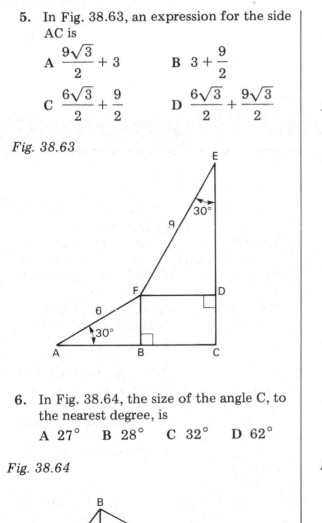

6. In Fig. 38.64, the size of the angle C, to the nearest degree, is

    A 27°   B 28°   C 32°   D 62°

Fig. 38.64

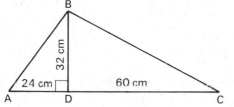

7. From the information given in Fig. 38.65, what is the value of $\cos R$?

    A $\dfrac{5}{8}$ B $\dfrac{3}{5}$ C $\dfrac{4}{\sqrt{41}}$ D $\dfrac{5}{\sqrt{41}}$

Fig. 38.65

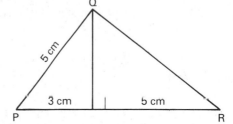

8. In Fig. 38.66, $\sin \angle MNL$ equals

    A 0.342   B 0.313
    C 0.425   D 0.612

Fig. 38.66

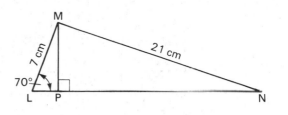

Solid Trigonometry

The Plane

A plane is a surface such as the top of a table or the cover of a book.

The Angle between a Line and a Plane

In Fig. 39.1 the line PA intersects the plane WXYZ at A. To find the angle between PA and the plane draw PL which is perpendicular to the plane and join AL. The angle between PA and the plane is ∠PAL.

Fig. 39.1

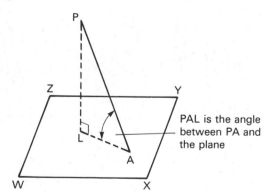

PAL is the angle between PA and the plane

The Angle between Two Planes

Two planes which are not parallel intersect in a straight line. Examples of this are the floor and a wall of a room and two walls of a room. To find the angle between two planes draw a line in each plane which is perpendicular to the common line of intersection. The angle between the two lines is the same as the angle between the two planes.

Three planes usually intersect at a point as, for instance, two walls and the floor of a room.

Problems with solid figures are solved by choosing suitable right-angled triangles in different planes. It is essential to make a clear three-dimensional drawing in order to find these triangles. The examples which follow show the methods that should be adopted.

Example 1

Figure 39.2 shows a cuboid. Calculate the length of the diagonal AG.

Fig. 39.2

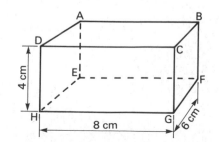

Figure 39.3 shows that in order to find AG we must use the right-angled triangle AGE. EG is the diagonal of the base rectangle.

Fig. 39.3

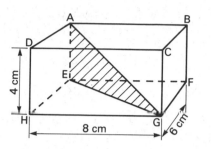

In $\triangle EFG$, $EF = 8\,cm$, $GF = 6\,cm$ and $\angle EFG = 90°$.

Using Pythagoras' theorem,

$$EG^2 = EF^2 + GF^2$$
$$= 8^2 + 6^2$$
$$= 64 + 36$$
$$= 100$$
$$EG = \sqrt{100}$$
$$= 10\,cm$$

In $\triangle AGE$, $AE = 4\,cm$, $EG = 10\,cm$ and $\angle AEG = 90°$.

Using Pythagoras' theorem,

$$AG^2 = AE^2 + EG^2$$
$$= 4^2 + 10^2$$
$$= 16 + 100$$
$$= 116$$
$$AG = \sqrt{116}$$
$$= 10.77\,cm$$

Example 2

Figure 39.4 shows a solid triangular cross-section with the face YDC inclined as shown. Find the angle that this sloping face YDC makes with the base.

Fig. 39.4

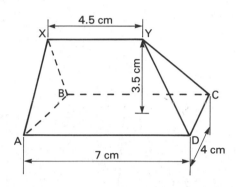

As shown in Fig. 39.5, if we use $\triangle YEF$ then the required angle is YFE. In $\triangle YEF$,

$$EF = 7 - 4.5$$
$$= 2.5\,cm$$
$$YE = 3.5\,cm$$
$$\tan \angle YFE = \frac{YE}{EF}$$
$$= \frac{3.5}{2.5}$$
$$\therefore \quad \angle YFE = 54.5°$$

Fig. 39.5

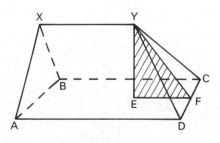

Hence the angle that the sloping face YDC makes with the base is $54.5°$.

Example 3

Figure 39.6 shows a pyramid with a square base. The base has sides 6 cm long and the edges of the pyramid, VA, VB, VC and VD are each 10 cm long. Find the altitude of the pyramid.

Fig. 39.6

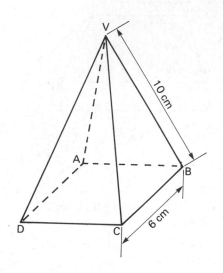

The right-angled triangle VBE (Fig. 39.7) allows the altitude VE to be found, but first we must find BE from the right-angled triangle BEF.

Fig. 39.7

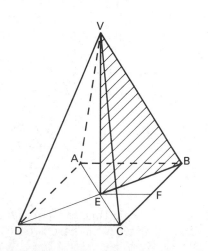

In $\triangle$BEF, BF = EF = 3 cm and $\angle$BFE = 90°.

Using Pythagoras' theorem,

$$BE^2 = BF^2 + EF^2$$
$$= 3^2 + 3^2$$
$$= 9 + 9$$
$$= 18$$
$$BE = \sqrt{18}$$
$$= 4.243 \text{ cm}$$

In $\triangle$VBE, BE = 4.243 cm, VB = 10 cm and $\angle$VEB = 90°.

Using Pythagoras' theorem,

$$VE^2 = VB^2 - BE^2$$
$$= 10^2 - 18$$
$$= 100 - 18$$
$$= 82$$
$$VE = \sqrt{82}$$
$$= 9.055 \text{ cm}$$

Hence the altitude of the pyramid is 9.055 cm.

Exercise 39.1

1. Figure 39.8 shows a cuboid.

   (a) Sketch the rectangle EFGH.

   (b) Calculate the diagonal FH of rectangle EFGH.

   (c) Sketch the rectangle FHDB adding known dimensions.

   (d) Calculate the diagonal BH of rectangle FHDB.

Fig. 39.8

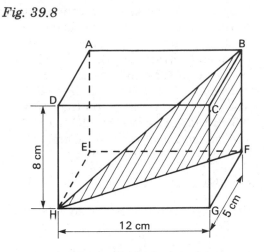

2. Figure 39.9 shows a pyramid on a square base of side 8 cm. The altitude of the pyramid is 12 cm.

Fig. 39.9

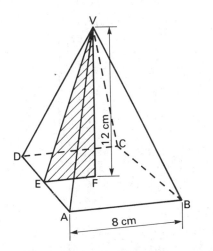

(a) Calculate EF.

(b) Draw the triangle VEF adding known dimensions.

(c) Find the angle VEF.

(d) Calculate the slant height VE.

(e) Calculate the area of △VAD.

(f) Calculate the complete surface area of the pyramid.

3. Figure 39.10 shows a pyramid on a rectangular base. Calculate the length VA.

Fig. 39.10

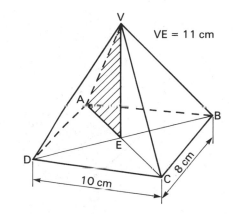

VE = 11 cm

4. Figure 39.11 shows a pyramid on a square base with VA = VB = VC = VD = 5 cm. Calculate the altitude of the pyramid.

Fig. 39.11

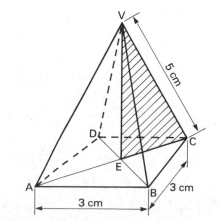

5. Figure 39.12 shows a wooden wedge. Find the length EA and the angle q. The end faces ADF and EBC are isosceles triangles.

Fig. 39.12

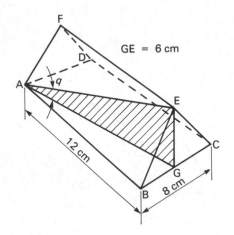

7. Figure 39.14 represents a wooden block in the shape of a triangular prism in which the edges AD, BE and CF are equal and vertical and the base DEF is horizontal. AB = BC = DE = EF = 14 cm and $\angle ABC = \angle DEF = 40°$. The points G and H are on the edges AB and BC respectively. BG = BH = 4 cm and DG = FH = 20 cm. Calculate:

(a) the length of BE

(b) the angle between FH and the base DEF

(c) the angle between the plane GHFD and the base DEF

(d) the distance between the mid-points of GH and DF.

Fig. 39.14

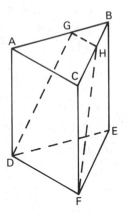

6. The base of the triangular prism shown in Fig. 39.13 is a rectangle 8 cm long and 6 cm wide. The vertical faces ABC and PQR are equilateral triangles of side 6 cm. Calculate:

(a) the angle between the diagonals PB and PC

(b) the angle between the plane PBC and the base

(c) the angle between the diagonal PC and the base.

Fig. 39.13

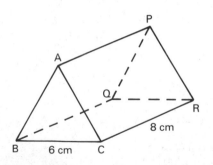

8. Figure 39.15 shows a shed with a slanting roof ABCD. The rectangular base ABEF rests on level ground and the shed has three vertical sides. Calculate:

(a) the angle of inclination of the roof to the ground

(b) the volume of the shed in cubic metres.

Fig. 39.15

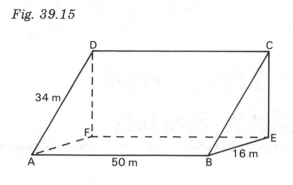

9. The base of a pyramid consists of a regular hexagon ABCDEF of side 4 cm. The vertex of the pyramid is V and VA = VB = VC = VD = VE = VF = 7 cm. Sketch a general view of the solid. Indicate on your diagram the angles p and q described below and calculate the size of these angles:

   (a) the angle p between VA and the base

   (b) the angle q between the face VCD and the base.

10. In Fig. 39.16, ABCD represents part of a hillside. A line of greatest slope AB is inclined at 36° to the horizontal AE and runs due North from A. The line AC bears 050° (N 50° E) and C is 2500 m East of B. The lines BE and CF are vertical. Calculate:

    (a) the height of C above A

    (b) the angle between AB and AC.

Fig. 39.16

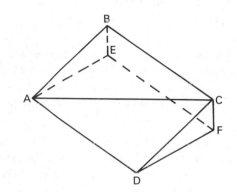

40

The Sine and Cosine Rules

Trigonometrical Ratios between 0° and 180°

In Chapter 38 the definition for the sine, cosine and tangent of an angle between $0°$ and $90°$ were given. In this chapter we show how to deal with angles between $0°$ and $180°$.

In Fig. 40.1, the axes XOX′ and YOY′, have been drawn at right-angles to each other to form the four quadrants. In each of these four quadrants we make use of the sign convention used when drawing graphs.

Fig. 40.1

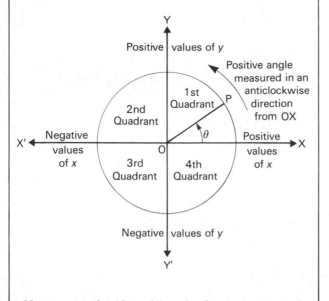

Now an angle, if positive, is always measured in an anticlockwise direction from OX and an angle is formed by rotating a line (such as OP) in an anticlockwise direction. It is convenient to make the length of OP equal to 1 unit. Referring to Fig. 40.2, we see:

In the first quadrant,

$$\sin \theta_1 = \frac{P_1 M_1}{OP_1}$$

$$= P_1 M_1$$

$$= y \text{ coordinate of } P_1$$

$$\cos \theta_1 = \frac{OM_1}{OP_1}$$

$$= OM_1$$

$$= x \text{ coordinate of } P_1$$

Fig. 40.2

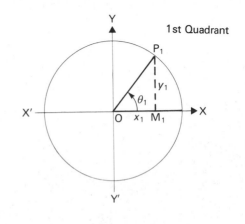

Hence in the first quadrant both the sine and cosine of an angle are positive.

In the second quadrant (Fig. 40.3),

$$\sin \theta_2 = \frac{P_2 M_2}{OP_2}$$

$$= P_2 M_2$$

$$= y \text{ coordinate of } P_2$$

The y coordinate of P_2 is positive and hence in the second quadrant the sine of an angle is positive.

Fig. 40.3

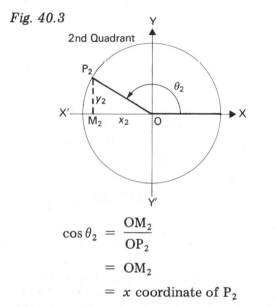

$$\cos \theta_2 = \frac{OM_2}{OP_2}$$

$$= OM_2$$

$$= x \text{ coordinate of } P_2$$

The x coordinate of P_2 is negative and hence in the second quadrant the cosine of an angle is negative.

The trigonometrical tables usually give values of the trigonometrical ratios for angles between $0°$ and $90°$. In order to use these tables for angles greater than $90°$ we make use of the triangle $OP_2 M_2$, where we see that

$$P_2 M_2 = OP_2 \sin (180° - \theta_2)$$

$$= \sin (180° - \theta_2)$$

But $\qquad P_2 M_2 = \sin \theta_2$

$\therefore \qquad \sin \theta_2 = \sin (180° - \theta_2)$

Also $\qquad OM_2 = -OP_2 \cos (180° - \theta_2)$

$$= -\cos (180° - \theta_2)$$

$\therefore \qquad \cos \theta_2 = -\cos (180° - \theta_2)$

Example 1

Find the values of $\sin 158°$ and $\cos 158°$.

Referring to Fig. 40.4,

$$\sin 158° = \frac{MP}{OP}$$

$$= \sin \angle POM$$

$$= \sin (180° - 158°)$$

$$= \sin 22°$$

$$= 0.3746$$

$$\cos 158° = \frac{OM}{OP}$$

$$= -\cos \angle POM$$

$$= -\cos (180° - 158°)$$

$$= -\cos 22°$$

$$= -0.9272$$

Fig. 40.4

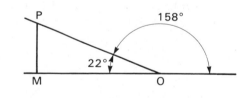

However, it is recommended that a scientific calculator be used in which case

$$\sin 158° = 0.374\,606\,5$$

i.e. 0.3746 correct to 4 s.f.

$$\cos 158° = -0.927\,183\,8$$

i.e. -0.9272 correct to 4 s.f.

Example 2

Find all the angles between $0°$ and $180°$:

(a) whose sine is 0.4676

(b) whose cosine is -0.3572.

(a) The angles whose sines are 0.4676 occur in the first and second quadrants.

In the first quadrant:
$$\sin \theta = 0.4676$$
$$\theta = 27.88°$$

In the second quadrant:
$$\theta = 180° - 27.88°$$
$$= 152.12° \text{ (see Fig. 40.5)}$$

Fig. 40.5

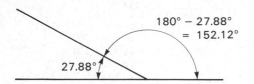

Note that when a calculator is used only the angle in the first quadrant will be displayed.

(b) The angle whose cosine is -0.3572 occurs in the second quadrant (i.e., the angle is between 90° and 180°).

If
$$\cos \theta = 0.3572$$
$$\theta = 69.07°$$

In the second quadrant:
$$\theta = 180° - 69.07°$$
$$\theta = 110.93° \text{ (see Fig. 40.6)}$$

Fig. 40.6

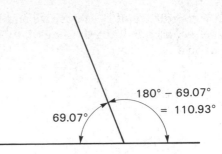

In this case the calculator will give the angle in the second quadrant.

Fig. 40.7 shows a sine curve which is obtained by first drawing a circle of unit radius with radial lines inclined at 30°, 60°, 90°, etc. From the curve we see that in the first quadrant, values of $\sin \theta$ range from 0 to 1, whilst in the second quadrant they range from 1 to 0. In the third and fourth quadrants the values of $\sin \theta$ are negative and range from 0 to -1 and -1 to 0 respectively. A sine

Fig. 40.7

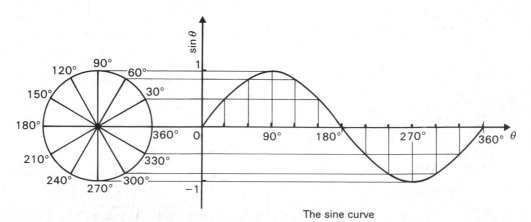

The sine curve

Fig. 40.8

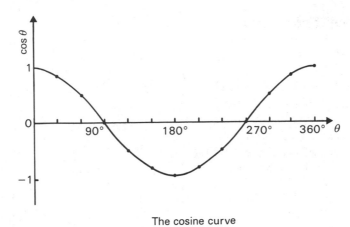

The cosine curve

curve is often said to be a sine wave. Fig. 40.8 shows a cosine curve which may be obtained by a construction similar to that used for the sine curve. We see from the curve that in the first quadrant the values of $\cos \theta$ are positive and range from 1 to 0. In the second and third quadrants the values of $\cos \theta$ are negative and range from 0 to -1 and -1 to 0 respectively. In the fourth quadrant the values of $\cos \theta$ are positive and range from 0 to 1.

	Trigonometrical ratio	Angle in	
		1st quadrant	2nd quadrant
9.	$\sin \theta = 0.5163$		
10.	$\sin \theta = 0.2167$		
11.	$\sin \theta = -0.4069$		
12.	$\cos \theta = 0.8817$		
13.	$\cos \theta = -0.7613$		
14.	$\cos \theta = -0.0812$		

Exercise 40.1

Copy and complete the following tables.

	θ	$\sin \theta$	$\cos \theta$
1.	108°		
2.	163°		
3.	95°		
4.	115°		
5.	134°		
6.	168°		
7.	146°		
8.	175°		

The Standard Notation for a Triangle

In $\triangle ABC$ (Fig. 40.9) the angles are denoted by the capital letters as shown in the diagram. The side a lies opposite the angle A, the side b opposite the angle B and the side c opposite the angle C. This is the standard notation for a triangle.

Fig. 40.9

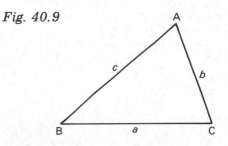

The Solution of Triangles

We now deal with triangles which are not right-angled. Every triangle consists of six elements — three sides and three angles. If we are given three suitable elements we can find the other three elements by using either the sine rule or the cosine rule. When we have found the values of the three missing elements we are said to have solved the triangle.

The Sine Rule

The sine rule may be used when we are given:

(1)   One side and any two angles.

(2)   Two sides and an angle opposite to one of the sides.

Using the notation of Fig. 40.9

$$\frac{a}{\sin A} = \frac{b}{\sin B}$$

$$= \frac{c}{\sin C}$$

Example 3

Solve $\triangle ABC$ given that $A = 42°$, $C = 72°$ and $b = 61.8\ mm$.

The triangle should be drawn for reference as shown in Fig. 40.10 but there is no need to draw it to scale.

Fig. 40.10

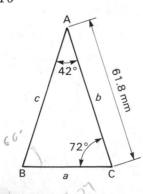

Since

$$A + B + C = 180°$$

$$B = 180° - 42° - 72°$$

$$= 66°$$

The sine rule states:

$$\frac{a}{\sin A} = \frac{b}{\sin B}$$

$$a = \frac{b \sin A}{\sin B}$$

$$= \frac{61.8 \times \sin 42°}{\sin 66°}$$

$$= 45.27\ mm$$

A scientific calculator may be used for this calculation. The program for a Casio fx-31 calculator is given below, but the method will be similar for any other type of calculator.

Input	Display	
66	66.	B
sin	0.9135 ...	$\sin B$
M+	0.9135 ...	
42	42.	A
sin	0.6691 ...	$\sin A$
×	0.6691 ...	
61.8	61.8	b
=	41.3522 ...	$b \sin A$
÷	41.3522 ...	
MR	0.9135 ...	$\sin B$
=	45.2656 ...	a

Also,

$$\frac{c}{\sin C} = \frac{b}{\sin B}$$

$$c = \frac{b \sin C}{\sin B}$$

$$= \frac{61.8 \times \sin 72°}{\sin 66°}$$

$$= 64.34\ mm$$

$$\frac{a}{\sin A} = \frac{b}{\sin B} = \frac{c}{\sin c}$$

The complete solution is

$$B = 66°$$

$$a = 45.27 \text{ mm}$$

$$c = 64.34 \text{ mm}$$

A rough check on sine rule calculations may be made by remembering that in any triangle the longest side lies opposite to the largest angle and the shortest side lies opposite to the smallest angle. Thus in the previous example:

$$\text{Smallest angle} = 42°$$

$$= A$$

$$\text{Shortest side} = a$$

$$= 45.27 \text{ mm}$$

$$\text{Largest angle} = 72°$$

$$= C$$

$$\text{Longest side} = c$$

$$= 64.34 \text{ mm}$$

Use of the Sine Rule to find the Diameter of the Circumscribing Circle of a Triangle

Using the notation of Fig. 40.11

$$\frac{a}{\sin A} = \frac{b}{\sin B}$$

$$= \frac{c}{\sin C}$$

$$= D$$

Fig. 40.11

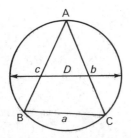

Example 4

In △ABC, $B = 41°$, $b = 112.5$ mm and $a = 87.63$ mm. Find the diameter of the circumscribing circle.

Referring to Fig. 40.12.

$$D = \frac{b}{\sin B}$$

$$= \frac{112.5}{\sin 41°}$$

$$= 171.5 \text{ mm}$$

Fig. 40.12

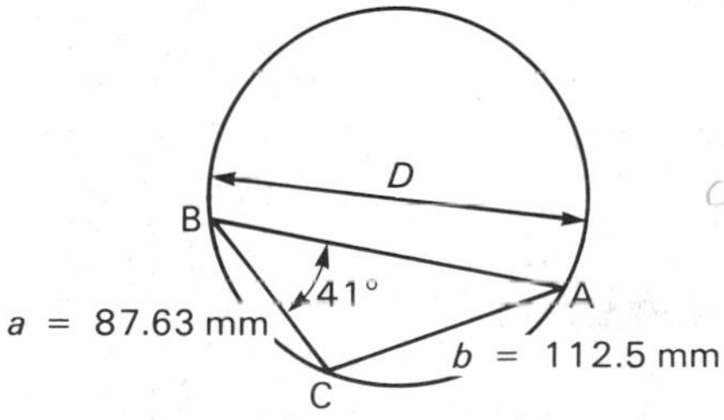

$$\frac{a}{\sin A} = \frac{b}{\sin B}$$

$$a = \frac{\sin A \times b}{\sin B}$$

$$a \times \sin B$$

B $a = 87.63$ mm   41° A $b = 112.5$ mm C D

$$\frac{b}{\sin B} \qquad \frac{112.5}{\sin 41}$$

Exercise 40.2

Solve the following triangles ABC by using the sine rule, given:

1. $A = 75°$, $B = 34°$, $a = 10.2$ cm
2. $C = 61°$, $B = 71°$, $b = 91$ mm
3. $A = 19°$, $C = 105°$, $c = 11.1$ m
4. $A = 116°$, $C = 18°$, $a = 17$ cm
5. $A = 36°$, $B = 77°$, $b = 2.5$ m
6. $A = 49.18°$, $B = 67.28°$, $c = 11.22$ mm
7. $A = 17.25°$, $C = 27.12°$, $b = 22.15$ m
8. $A = 77.05°$, $C = 21.05°$, $a = 9.793$ m
9. $B = 115.07°$, $C = 11.28°$, $c = 516.2$ mm

10. $a = 7$ m, $c = 11$ m, $C = 22.12°$

11. $b = 15.13$ cm, $c = 11.62$ cm,
$B = 85.28°$

12. $a = 23$ cm, $c = 18.2$ cm, $A = 49.32°$

13. $a = 9.217$ cm, $b = 7.152$ cm,
$A = 105.07°$

Find the diameter of the circumscribing circle of the following triangles ABC given:

14. $A = 75°$, $B = 48°$, $a = 21$ cm

15. $C = 100°$, $B = 50°$, $b = 90$ mm

16. $A = 20°$, $C = 102°$, $c = 11$ m

17. $A = 70°$, $C = 35°$, $a = 8.5$ cm

18. $a = 16$ cm, $b = 14$ cm, $B = 40°$

The Cosine Rule

This rule is used when we are given:

(1) Two sides of a triangle and the angle between them.

(2) Three sides of a triangle.

In all other cases the sine rule is used. The cosine rule states:

$$a^2 = b^2 + c^2 - 2bc \cos A$$
$$b^2 = a^2 + c^2 - 2ac \cos B$$
$$c^2 = a^2 + b^2 - 2ab \cos C$$

When using the cosine rule remember that if an angle is greater than 90° its cosine is negative.

Example 5

(a) Solve $\triangle ABC$ if $a = 70$ mm, $b = 40$ mm and $C = 64°$.

Referring to Fig. 40.13, to find the side c we use:

$$c^2 = a^2 + b^2 - 2ab \cos C$$
$$= 70^2 + 40^2 - 2 \times 70 \times 40 \times \cos 64°$$
$$= 4044$$
$$c = \sqrt{4044}$$
$$= 63.59 \text{ mm}$$

Fig. 40.13

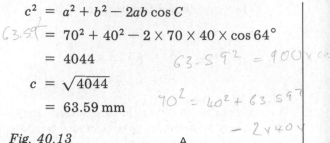

We now use the sine rule to find the angle A:

$$\frac{a}{\sin A} = \frac{c}{\sin C}$$

$$\sin A = \frac{a \sin C}{c}$$

$$= \frac{70 \times \sin 64°}{63.59}$$

$$A = 81.7°$$

$$B = 180° - 81.7° - 64°$$

$$= 34.3°$$

The complete solution is

$$A = 81.7°$$
$$B = 34.3°$$

and $\quad c = 63.59$ mm

(b) Find the side b in $\triangle ABC$ if $a = 160$ mm, $c = 200$ mm and $B = 124.25°$.

Fig. 40.14

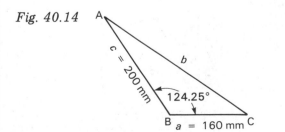

Referring to Fig. 40.14, to find the side *b* we use:

$$b^2 = a^2 + c^2 - 2ac \cos B$$

$$= 160^2 + 200^2 - 2 \times 160$$

$$\times 200 \times \cos 124.25°$$

Now $\cos 124.25°$

$$= -\cos(180° - 124.25°)$$

$$= -\cos 55.75°$$

$$= -0.5628$$

$$\therefore \quad b^2 = 160^2 + 200^2 - 2 \times 160$$

$$\times 200 \times (-0.5628)$$

$$= 160^2 + 200^2 + 2 \times 160$$

$$\times 200 \times 0.5628$$

$$b = 318.8 \, \text{mm}$$

This calculation may be done on a scientific calculator, as follows:

Input	Display	
124.25	124.25	*B*
cos	−0.5628 . . .	cos *B*
×	−0.5628 . . .	
2	2.	
×	−1.1256 . . .	
160	160.	*a*
×	−180.09 . . .	
200	200.	*c*
=	−36 019.5 . . .	2*ac* cos *B*
+/−	36 019.5 . . .	
M+	36 019.5 . . .	
160	160.	*a*
×	160.	
M+	25 600.	*a*^{2}
200	200.	*c*
×	200.	
M+	40 000	*c*^{2}
MR	101 619.51	*b*^{2}
$\sqrt{}$	318.77 . . .	*b*

Exercise 40.3

Solve the following triangles ABC using the cosine rule, given:

1. $a = 9 \, \text{cm}, \quad b = 11 \, \text{cm}, \quad C = 60°$

2. $b = 10 \, \text{cm}, \quad c = 14 \, \text{cm}, \quad A = 56°$

3. $a = 8.16 \, \text{m}, \quad c = 7.14 \, \text{m}, \quad B = 37.3°$

4. $a = 5 \, \text{m}, \quad b = 8 \, \text{m}, \quad c = 7 \, \text{m}$

5. $a = 312 \, \text{mm}, \quad b = 527.3 \, \text{mm}, \quad c = 700 \, \text{mm}$

6. $a = 7.912 \, \text{cm}, \quad b = 4.318 \, \text{cm}, \quad c = 11.08 \, \text{cm}$

7. $a = 12 \, \text{cm}, \quad b = 9 \, \text{cm}, \quad C = 118°$

8. $b = 8 \, \text{cm}, \quad c = 12 \, \text{cm}, \quad A = 132°$

Area of a Triangle

In Chapter 21, two formulae for the area of a triangle were used. They are:

(1) Given the base and altitude of the triangle (Fig. 40.15)

$$\text{Area} = \tfrac{1}{2} \times \text{base} \times \text{altitude}$$

or $\quad A = \tfrac{1}{2}bh$

Fig. 40.15

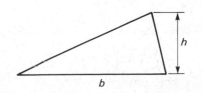

(2) Given the three sides of the triangle (Fig. 40.16)

$$A = \sqrt{s(s-a)(s-b)(s-c)}$$

where $\quad s = \frac{1}{2}(a+b+c)$

Fig. 40.16

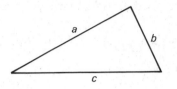

However, a third formula can be used to find the area of a triangle when we are given two sides of the triangle and the angle included between these sides. Referring to Fig. 40.17,

$$\text{Area} = \frac{1}{2}ab \sin C$$

$$\text{Area} = \frac{1}{2}ac \sin B$$

$$\text{Area} = \frac{1}{2}bc \sin A$$

Fig. 40.17

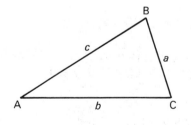

Example 6

Find the area of the triangle shown in Fig. 40.18.

$$\text{Area} = \frac{1}{2}ac \sin B$$

$$= \frac{1}{2} \times 4 \times 3 \times \sin 30°$$

$$= 3 \text{ cm}^2$$

Fig. 40.18

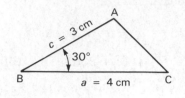

Example 7

Find the area of the obtuse-angled triangle shown in Fig. 40.19.

Fig. 40.19

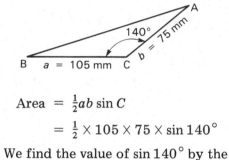

$$\text{Area} = \frac{1}{2}ab \sin C$$

$$= \frac{1}{2} \times 105 \times 75 \times \sin 140°$$

We find the value of $\sin 140°$ by the method used earlier in this chapter or by using a calculator.

$$\sin 140° = \sin(180° - 140°)$$

$$= \sin 40°$$

$$= 0.6428$$

$$\text{Area} = \frac{1}{2} \times 105 \times 75 \times 0.6428$$

$$= 2531 \text{ mm}^2$$

Exercise 40.4

In this exercise any of the three formulae given above may be needed.

1. Obtain the area of a triangle whose sides are 39.3 cm and 41.5 cm if the angle between them is 41.5°.

2. Find the area of the template shown in Fig. 40.20.

Fig. 40.20

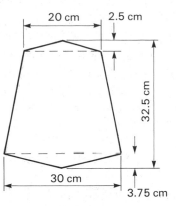

3. Calculate the area of a triangle ABC if
   (a) $a = 4$ cm, $b = 5$ cm and $C = 49°$.
   (b) $a = 3$ m, $c = 6$ m and $B = 63.73°$.

4. Find the areas of the quadrilaterals shown in Fig. 40.21.

Fig. 40.21

(a)

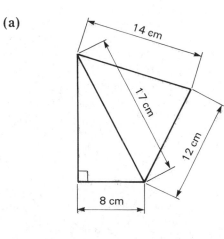

(b)

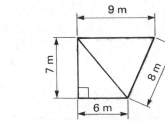

(c)

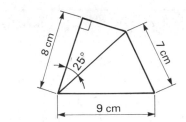

5. Find the areas of the following triangles
   (a) $a = 5$ cm, $b = 7$ cm and $C = 105°$
   (b) $b = 7.3$ cm, $c = 12.2$ cm and $A = 135°$
   (c) $a = 9.6$ cm, $c - 11.2$ cm and $B = 163°$.

6. Find the area of the parallelogram shown in Fig. 40.22.

Fig. 40.22

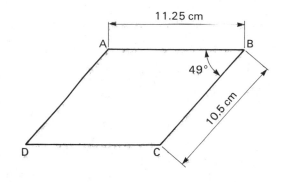

7. Find the area of the trapezium shown in Fig. 40.23.

Fig. 40.23

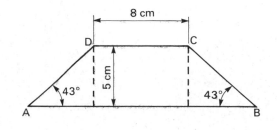

8. Find the area of a regular octagon which has sides 2 cm long.

Miscellaneous Exercise 40

1. In triangle ABC, $a = 4$ cm, $b = 8$ cm and $c = 7$ cm. Find the angle A.

2. Find the area of an equilateral triangle of side 4 cm. Give the answer correct to three significant figures.

3. In a triangle ABC, $AB = 25$ m, $B = 50°$ and $C = 30°$. Calculate the length of AC.

4. In the acute angled $\triangle PQR$, $PQ = 6$ cm, $QR = 4$ cm and the area of the triangle is 10.8 cm^2. Calculate $\angle PQR$.

5. In triangle ABC, $AB = 7$ cm, $BC = 8$ cm and $\angle ACB = 53.22°$. Calculate $\angle BAC$, given that it is acute.

6. ABC is a triangle with $AB = 10$ cm, $BC = 12$ cm and angle $ABC = 35°$. Calculate:

    (a) the length of the perpendicular from A to BC

    (b) the area of the triangle ABC.

7. A and B are points on a coast, B being 2000 m due east of A. A man sails from A on a bearing of $060°$ (N 60° E) for 800 m to a buoy P, where he changes course to $070°$ (N 70° E) and sails for 1000 m to another buoy Q. Calculate:

    (a) the distances of P and Q from the line AB

    (b) the distance QB

    (c) the course he must set to sail directly from Q to B.

Vectors

Vector Quantities

Scalar quantities are fully described by size or magnitude. Some examples of scalar quantities are time (35 seconds), temperature (8 degrees Celsius) and mass (25 kilograms).

Vector quantities need both direction and magnitude to describe them fully. Some examples of vector quantities are:

(1) A displacement from one point to another, e.g. 15 m due east.

(2) A velocity in a given direction, e.g. 30 km/h due north.

(3) A force of 15 newtons acting vertically downwards.

Representing Vectors

Vectors can be represented by an accurately scaled line.

Example 1

A woman walked 8 km due east. Draw the vector.

We first choose a suitable scale to represent the magnitude of the vector. In Fig. 41.1 a scale of 1 cm = 2 km has been chosen. We then draw a horizontal line 4 cm long. An arrow is placed on the line to show that the woman walked in an easterly direction (and not a westerly direction).

Fig. 41.1

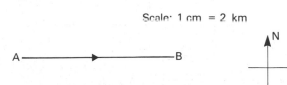

When we want to refer to a vector several times we need a shorthand way of doing this. The usual way is to name the end points of the vector. The vector in Example 1 starts at A and ends at B. If we now want to refer to the vector we write $\overrightarrow{AB}$ which means 'the vector from A to B'.

Sometimes vectors are written in heavy type. Thus if a line AB represents a vector, then the vector may also be written as **AB** or as **a**.

Exercise 41.1

Draw the following vectors using a convenient scale for each. Label each vector correctly.

1. $\overrightarrow{AB}$ 5 m due north

2. $\overrightarrow{XY}$ 8 km due south

3. $\overrightarrow{PQ}$ 7 km due west

4. $\overrightarrow{MN}$ 10 m due east

5. $\overrightarrow{RS}$ 6 m south-east

Fig. 41.2 shows the vector $\overrightarrow{AB}$ which is inclined at 30° to the horizontal. Its magnitude is 6 N and its direction is 30° to the horizontal. The arrowhead shows that the vector is acting away from A and the way in which the arrowhead points gives the sense of the vector.

Fig. 41.2

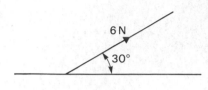

Instead of stating the magnitude of the vector and the angle denoting its direction vectors are often stated as being so many units horizontally and so many units vertically. Thus in Fig. 41.3, the vector **a** is given as 4 units along and 3 units upwards. This way of stating a vector allows us to find the magnitude of the vector and its direction.

Fig. 41.3

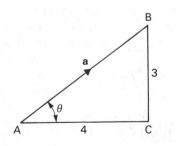

Using Pythagoras' theorem, the magnitude of the vector is

$$|\mathbf{a}| = \sqrt{3^2 + 4^2}$$

$$= 5 \text{ units}$$

To state its direction we find the size of the angle θ:

$$\tan \theta = \tfrac{3}{4}$$

$$\theta = 36.9°$$

In Fig. 41.3, $\overrightarrow{AC}$ is said to be the horizontal component of **a** and $\overrightarrow{CB}$ is called the vertical component of **a**.

When vectors are stated in component form they can be written as column matrices. Thus

$$\mathbf{a} = \begin{pmatrix} 4 \\ 3 \end{pmatrix}$$

If **i** and **j** are unit vectors (i.e. vectors having a magnitude of 1 unit) **i** being taken horizontally and **j** being taken vertically, then

$$\overrightarrow{AC} = 4\mathbf{i} \quad \text{and} \quad \overrightarrow{BC} = 3\mathbf{j}$$

The vector **a** is then described as

$$\mathbf{a} = 4\mathbf{i} + 3\mathbf{j}$$

The magnitude of a vector is sometimes called its **modulus.** If the vector is **v** then its modulus is written as $|\mathbf{v}|$.

Example 2

(a) Calculate the magnitude and direction of the vector $\mathbf{v} = 3\mathbf{i} + 5\mathbf{j}$.

From Fig. 41.4,

$$|\mathbf{v}| = \sqrt{3^2 + 5^2}$$

$$= \sqrt{34}$$

$$= 5.83$$

$$\tan \theta = \tfrac{5}{3}$$

$$\theta = 59.0°$$

Fig. 41.4

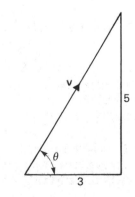

(b) Calculate the magnitude and direction of the vector $\mathbf{a} = \begin{pmatrix} 7 \\ 4 \end{pmatrix}$.

From Fig. 41.5,

$$|\mathbf{a}| = \sqrt{7^2 + 4^2}$$
$$= \sqrt{65}$$
$$= 8.06$$
$$\tan \theta = \tfrac{4}{7}$$
$$\theta = 29.7°$$

Fig. 11.5

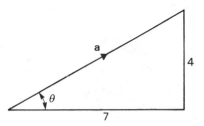

(c) Calculate the magnitude and direction of the vector $\mathbf{b} = \begin{pmatrix} 4 \\ -2 \end{pmatrix}$.

From Fig. 41.6,

Fig. 41.6

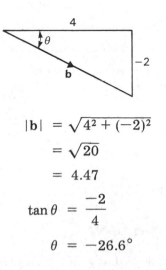

$$|\mathbf{b}| = \sqrt{4^2 + (-2)^2}$$
$$= \sqrt{20}$$
$$= 4.47$$
$$\tan \theta = \frac{-2}{4}$$
$$\theta = -26.6°$$

It pays always to sketch the vector; otherwise you may make a mistake in the direction of the vector, especially if using a calculator.

1. Draw the following vectors using a convenient scale for each. Mark each of the angles clearly.

   (a) 6 newtons acting at $45°$ to the horizontal

   (b) 7 m/s acting at $75°$ to the horizontal

   (c) 20 km at an angle of $-30°$ to the horizontal

   (d) 4 m/s² at an angle of $-90°$ to the horizontal.

2. For each of the vectors shown in Fig. 41.7, find the horizontal and vertical components and write each of them in the equivalent component form (i.e. as column matrices).

   Fig. 41.7

   (a) (b) (c) (d) (e)

   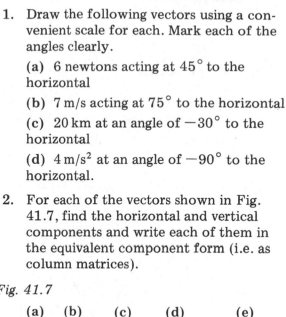

3. Using squared paper draw lines to represent each of the following:

   (a) $\begin{pmatrix} 3 \\ 3 \end{pmatrix}$ (b) $\begin{pmatrix} 4 \\ -5 \end{pmatrix}$

   (c) $4\mathbf{i} + 3\mathbf{j}$ (d) $5\mathbf{i} - 3\mathbf{j}$

   (e) $-4\mathbf{i} - 3\mathbf{j}$ (f) $\begin{pmatrix} 6 \\ -4 \end{pmatrix}$

   (g) $\begin{pmatrix} -4 \\ -5 \end{pmatrix}$ (h) $\begin{pmatrix} -5 \\ 0 \end{pmatrix}$

   (i) $\begin{pmatrix} 0 \\ 4 \end{pmatrix}$

   (i) By measurement find the magnitude and direction of each of these vectors.

(ii) Check your answers to part (i) by calculating the magnitude and direction of the vectors.

Equal Vectors

Two vectors are equal if they have the same magnitude and the same direction. Hence in Fig. 41.8, the vector $\overrightarrow{AB}$ is equal to the vector $\overrightarrow{CD}$, because the length of $\overrightarrow{AB}$ is equal to that of $\overrightarrow{CD}$ and $\overrightarrow{AB}$ is parallel to $\overrightarrow{CD}$.

Fig. 41.8

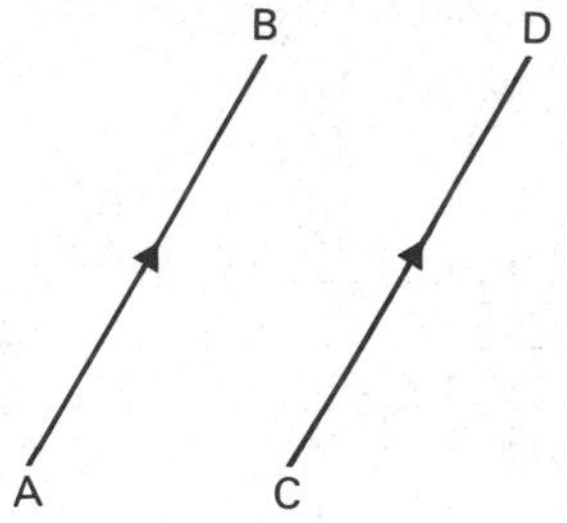

Inverse Vectors

In Fig. 41.9, the vector $\overrightarrow{BA}$ has the same magnitude as $\overrightarrow{AB}$ but its direction is reversed. Hence $\overrightarrow{BA} = -\overrightarrow{AB}$. $\overrightarrow{BA}$ is said to be the inverse of $\overrightarrow{AB}$.

Fig. 41.9

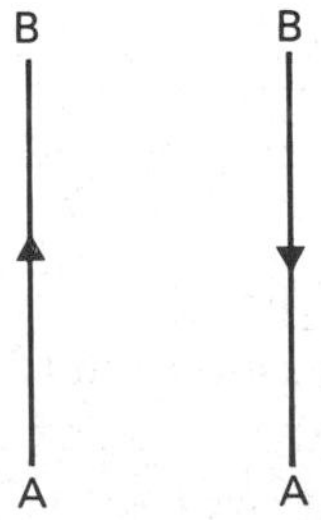

Resultant Vectors

In Fig. 41.10, A, B and C are three points marked out in a field. A man walks from A to B (i.e. he describes $\overrightarrow{AB}$) and then walks from B to C (i.e. he describes $\overrightarrow{BC}$). Instead, the man could have walked from A to C, thus describing the vector $\overrightarrow{AC}$.

Fig. 41.10

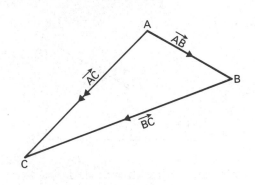

Now going from A to C direct has the same result as going from A to C via B. We therefore call $\overrightarrow{AC}$ the resultant of the vectors $\overrightarrow{AB}$ and $\overrightarrow{BC}$. Note carefully that the arrows on the vectors $\overrightarrow{AB}$ and $\overrightarrow{BC}$ follow nose to tail. The resultant $\overrightarrow{AC}$ is often marked with a double arrow.

Triangle Law

The resultant of any two vectors is equal to the length and direction of the line needed to complete the triangle. This is called the triangle law.

Example 3

Two vectors act as shown in Fig. 41.11.
Find the resultant of these two vectors.

Fig. 41.11

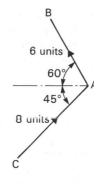

The vector triangle is drawn in Fig.
41.12. To find the resultant vector we
scale CB which is found to be 11.2 units.

Fig. 41.12

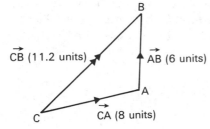

We say that the resultant vector $\overrightarrow{CB}$ is

the sum of the two vectors $\overrightarrow{CA}$ and $\overrightarrow{AB}$,
and we write

$$\overrightarrow{CB} = \overrightarrow{CA} + \overrightarrow{AB}$$

When the vectors are given in component
form the components of the resultant are
found by adding together separately the
horizontal and the vertical components of
the vectors to be added.

Example 4

Find the resultant of the vectors $\begin{pmatrix} 6 \\ 2 \end{pmatrix}$ and
$\begin{pmatrix} 7 \\ 8 \end{pmatrix}$.

As shown in Fig. 41.13, the components
of the resultant $\overrightarrow{AC}$ are found by adding
the two column matrices together. Thus

$$\overrightarrow{AC} = \begin{pmatrix} 6 \\ 2 \end{pmatrix} + \begin{pmatrix} 7 \\ 8 \end{pmatrix}$$

$$= \begin{pmatrix} 13 \\ 10 \end{pmatrix}$$

$$|\overrightarrow{AC}| = \sqrt{13^2 + 10^2}$$

$$= \sqrt{269}$$

$$= 16.4$$

$$\tan \theta = \frac{10}{13}$$

$$\theta = 37.6°$$

Fig. 41.13

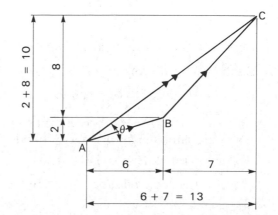

Vectors stated in the **i, j** form are treated in exactly the same way.

Example 5

Find the resultant **r** of the vectors **a** = 3i + 5j and **b** = 4i + 6j.

$$\mathbf{r} = \mathbf{a} + \mathbf{b}$$

$$= (3i + 5j) + (4i + 6j)$$

$$= 7i + 11j$$

$$|\mathbf{r}| = \sqrt{7^2 + 11^2}$$

$$= \sqrt{170}$$

$$= 13.0$$

$$\tan \theta = \frac{11}{7}$$

$$\theta = 57.5°$$

Example 6

ABCD is a parallelogram (Fig. 41.14). $\overrightarrow{AB} = \mathbf{x}$ and $\overrightarrow{BC} = \mathbf{y}$.

Fig. 41.14

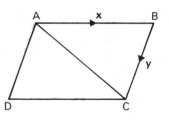

(a) Express $\overrightarrow{DC}$ in terms of **x**.

(b) Express $\overrightarrow{AD}$ in terms of **y**.

(c) Express $\overrightarrow{AC}$ in terms of **x** and **y**.

   (a) $\overrightarrow{DC}$ is equal to $\overrightarrow{AB}$ because they have the same magnitude and AB and DC are parallel. Hence $\overrightarrow{DC} = \mathbf{x}$.

   (b) For the same reasons, $\overrightarrow{AD} = \overrightarrow{BC}$ and hence $\overrightarrow{AD} = \mathbf{y}$.

(c) AC is the resultant of AB and BC because the arrows on AB and AC follow nose to tail. Hence

$$\overrightarrow{AC} = \overrightarrow{AB} + \overrightarrow{BC}$$

$$= \mathbf{x} + \mathbf{y}$$

More than two vectors may be added together. The resultant of the four vectors shown in Fig. 41.15 is found by drawing the vector diagram as shown. Thus

$$\mathbf{a} + \mathbf{b} + \mathbf{c} + \mathbf{d} = \mathbf{r}$$

Fig. 41.15

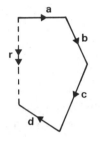

The addition of vectors is commutative. That is, it does not matter in which order the vectors are added. In Fig. 41.16 in (a) **a** has been added to **b** whilst in (b) **b** has been added to **a**. In each case the resultant is the same.

Fig. 41.16

(a)

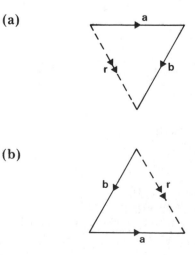

(b)

Subtraction of Vectors

To subtract a vector we add its inverse.
Hence

$$\overrightarrow{AC} - \overrightarrow{BC} = \overrightarrow{AC} + \overrightarrow{CB}$$
$$= \overrightarrow{AB}$$

as shown in Fig. 41.17.

Fig. 41.17

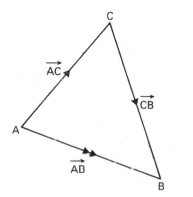

Example 7

If $\mathbf{a} = \begin{pmatrix} 1 \\ 4 \end{pmatrix}$ and $\mathbf{b} = \begin{pmatrix} 5 \\ -1 \end{pmatrix}$ find $\mathbf{a} - \mathbf{b}$.

The two vectors $\mathbf{a}$ and $\mathbf{b}$ are shown in Fig. 41.18.

Fig. 41.18

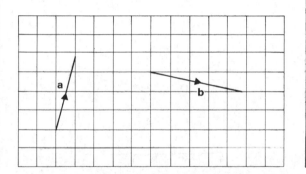

To get $-\mathbf{b}$ we reverse the direction of $\mathbf{b}$. We then perform the addition

$$\mathbf{r} = \mathbf{a} + (-\mathbf{b})$$

The graphical solution is given in Fig. 41.19 where it will be seen that

$$|\mathbf{r}| = 6.4 \text{ units} \quad \text{with} \quad \theta = 51.3°$$

Fig. 41.19

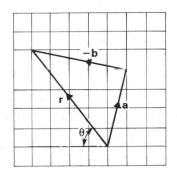

Using column matrices,

$$\mathbf{r} = \mathbf{a} - \mathbf{b}$$
$$= \begin{pmatrix} 1 \\ 4 \end{pmatrix} - \begin{pmatrix} 5 \\ -1 \end{pmatrix}$$
$$= \begin{pmatrix} -4 \\ 5 \end{pmatrix}$$
$$|\mathbf{r}| = \sqrt{(-4)^2 + 5^2}$$
$$= \sqrt{41}$$
$$= 6.4$$
$$\tan \theta = \frac{5}{-4}$$
$$\theta = -51.3°$$

Multiplying by a Scalar

If a vector $\overrightarrow{AB}$ is 5 m due North, then the vector $3\overrightarrow{AB}$ is 15 m due North. In general, if k is any number, the vector $k\overrightarrow{AB}$ is a vector in the same direction as $\overrightarrow{AB}$ whose length is k times the length of $\overrightarrow{AB}$.

Example 8

C is the mid-point of AD (Fig. 41.20) and AC = **a**. Express DA in terms of **a**.

Fig. 41.20

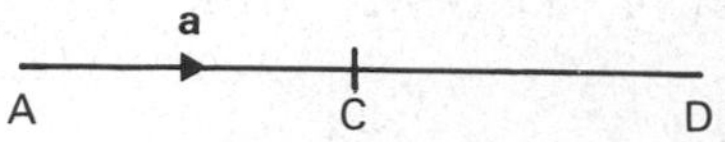

$$\overrightarrow{AD} = \overrightarrow{AC} + \overrightarrow{CD}$$
$$= \mathbf{a} + \mathbf{a}$$
$$= 2\mathbf{a}$$

Since $\overrightarrow{DA}$ is the inverse of $\overrightarrow{AD}$,
$$\overrightarrow{DA} = -2\mathbf{a}$$

Distributive Law for Vectors

From our previous work
$$3\mathbf{a} = \mathbf{a} + \mathbf{a} + \mathbf{a}$$
$$\therefore \ 3(\mathbf{a}+\mathbf{b}) = (\mathbf{a}+\mathbf{b}) + (\mathbf{a}+\mathbf{b}) + (\mathbf{a}+\mathbf{b})$$
$$= \mathbf{a}+\mathbf{b}+\mathbf{a}+\mathbf{b}+\mathbf{a}+\mathbf{b}$$
$$= (\mathbf{a}+\mathbf{a}+\mathbf{a}) + (\mathbf{b}+\mathbf{b}+\mathbf{b})$$
$$= 3\mathbf{a} + 3\mathbf{b}$$

Similarly it can be shown that, if k is a scalar,
$$k(\mathbf{a}+\mathbf{b}) = k\mathbf{a} + k\mathbf{b}$$

This is the distributive law for vectors.

Example 9

In Fig. 41.21, $\overrightarrow{PX} = \mathbf{a}$ and $\overrightarrow{XQ} = \mathbf{b}$. $\overrightarrow{YP} = 2\mathbf{a}$ and $\overrightarrow{QZ} = 2\mathbf{b}$. Find $\overrightarrow{YZ}$ in terms of **a** and **b** and show that PQ and YZ are parallel.

Fig. 41.21

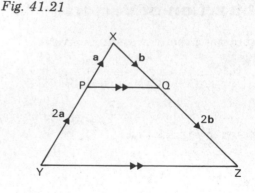

Referring to Fig. 41.21, we see that
$$\overrightarrow{PQ} = \mathbf{a} + \mathbf{b}$$
$$\overrightarrow{YZ} = \overrightarrow{YX} + \overrightarrow{XZ}$$
$$= (2\mathbf{a} + \mathbf{a}) + (2\mathbf{b} + \mathbf{b})$$
$$= 3\mathbf{a} + 3\mathbf{b}$$
$$= 3(\mathbf{a} + \mathbf{b})$$

Hence $\overrightarrow{YZ}$ is 3 times the magnitude of $\overrightarrow{PQ}$. Since $\overrightarrow{PQ} = (\mathbf{a} + \mathbf{b})$ and $\overrightarrow{YZ} = 3(\mathbf{a} + \mathbf{b})$, $\overrightarrow{PQ}$ and $\overrightarrow{YZ}$ must be in the same direction; i.e. they are parallel to each other.

Example 10

In Fig. 41.22, M is the mid-point of RS. PQ = **a**, QR = **b** and MR = **c**. Express each of the following in terms of **a**, **b** and **c**.

(a) $\overrightarrow{SR}$ (b) $\overrightarrow{PR}$ (c) $\overrightarrow{QM}$ (d) $\overrightarrow{PM}$

Fig. 41.22

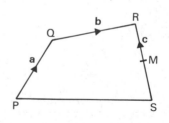

(a) In Fig. 41.23(a),

$$\overrightarrow{SR} = 2\overrightarrow{MR}$$
$$= 2\mathbf{c}$$

(b) In $\triangle PQR$ (Fig. 41.23(b)), $\overrightarrow{PR}$ is the resultant of $\overrightarrow{PQ}$ and $\overrightarrow{QR}$. Hence

$$\overrightarrow{PR} = \overrightarrow{PQ} + \overrightarrow{QR}$$
$$= \mathbf{a} + \mathbf{b}$$

(c) In $\triangle QRM$ (Fig. 41.23(c)), $\overrightarrow{QR}$ is the resultant of $\overrightarrow{QM}$ and $\overrightarrow{MR}$. Hence

$$\overrightarrow{QR} = \overrightarrow{QM} + \overrightarrow{MR}$$
$$\overrightarrow{QM} = \overrightarrow{QR} - \overrightarrow{MR}$$
$$= \mathbf{b} - \mathbf{c}$$

(d) In $\triangle PRM$ (Fig. 41.23(d)), $\overrightarrow{PR}$ is the resultant of $\overrightarrow{PM}$ and $\overrightarrow{MR}$. Hence

$$\overrightarrow{PR} = \overrightarrow{PM} + \overrightarrow{MR}$$
$$\overrightarrow{PM} = \overrightarrow{PR} - \overrightarrow{MR}$$
$$= \mathbf{a} + \mathbf{b} - \mathbf{c}$$

Fig. 41.23

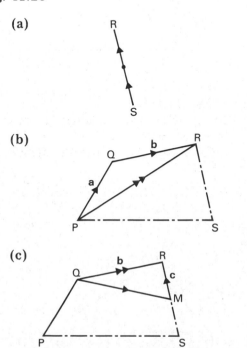

(d)

Exercise 41.3

1. Find the resultant (in magnitude and direction) of the following pairs of vectors (Fig. 41.24).

Fig. 41.24

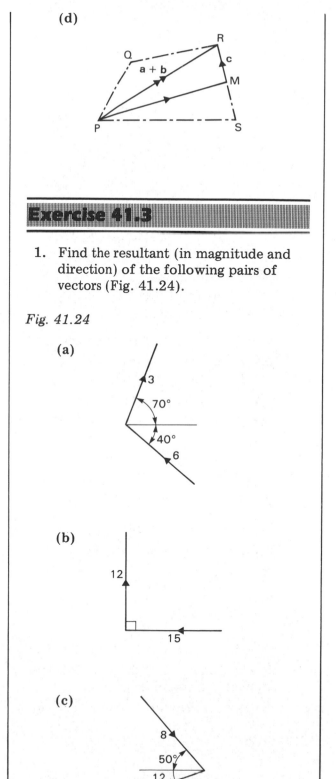

(cont.)

(d)

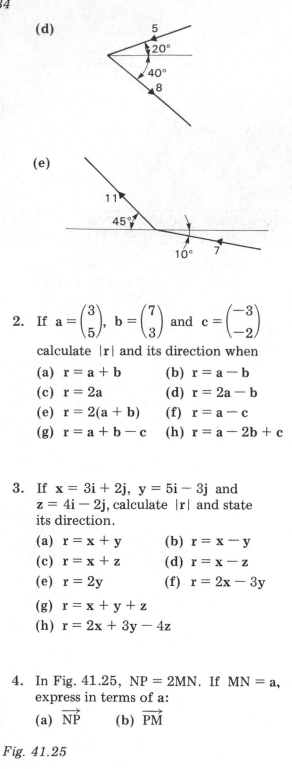

(e)

2. If $a = \begin{pmatrix} 3 \\ 5 \end{pmatrix}$, $b = \begin{pmatrix} 7 \\ 3 \end{pmatrix}$ and $c = \begin{pmatrix} -3 \\ -2 \end{pmatrix}$ calculate $|r|$ and its direction when

    (a) $r = a + b$ (b) $r = a - b$

    (c) $r = 2a$ (d) $r = 2a - b$

    (e) $r = 2(a + b)$ (f) $r = a - c$

    (g) $r = a + b - c$ (h) $r = a - 2b + c$

3. If $x = 3i + 2j$, $y = 5i - 3j$ and $z = 4i - 2j$, calculate $|r|$ and state its direction.

    (a) $r = x + y$ (b) $r = x - y$

    (c) $r = x + z$ (d) $r = x - z$

    (e) $r = 2y$ (f) $r = 2x - 3y$

    (g) $r = x + y + z$

    (h) $r = 2x + 3y - 4z$

4. In Fig. 41.25, $NP = 2MN$. If $MN = a$, express in terms of a:

    (a) $\overrightarrow{NP}$ (b) $\overrightarrow{PM}$

Fig. 41.25

5. In Fig. 41.26, ABCD is a parallelogram. $AB = a$ and $BC = b$. Express in terms of a and b:

    (a) $\overrightarrow{AD}$ (b) $\overrightarrow{CD}$ (c) $\overrightarrow{AC}$

Fig. 41.26

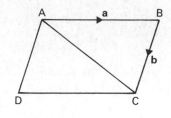

6. In Fig. 41.27, XYZ is an isosceles triangle. W is the mid-point of XZ. If $\overrightarrow{XY} = a$ and $\overrightarrow{YZ} = b$, express in terms of a and b:

    (a) $\overrightarrow{XZ}$ (b) $\overrightarrow{ZW}$ (c) $\overrightarrow{YW}$

Fig. 41.27

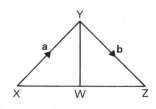

7. In Fig. 41.28, ABC is a triangle in which $AC = a$, $BC = b$ and $AD = c$. D is the mid-point of AB. Express in terms of a, b and c:

    (a) $\overrightarrow{DB}$ (b) $\overrightarrow{BA}$ (c) $\overrightarrow{CD}$

Fig. 41.28

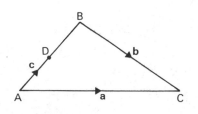

8. In Fig. 41.29, D and E are the mid-
    points of AB and AC respectively.
    AD = a and AE = b. Express in
    terms of **a** and **b**:

    (a) $\overrightarrow{ED}$ (b) $\overrightarrow{DE}$ (c) $\overrightarrow{BD}$

    (d) $\overrightarrow{AB}$ (e) $\overrightarrow{CE}$ (f) $\overrightarrow{AC}$

Fig. 41.29

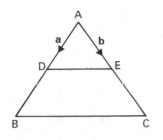

9. In Fig. 41.30, AB is parallel to DC and
    DC = 2AB. If $\overrightarrow{DA}$ = a and $\overrightarrow{AB}$ = b,
    express in terms of **a** and **b**:

    (a) $\overrightarrow{DB}$ (b) $\overrightarrow{DC}$ (c) $\overrightarrow{BC}$

Fig. 41.30

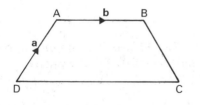

10. In Fig. 41.31, ST = 2TP. If $\overrightarrow{TP}$ = a,
    $\overrightarrow{PQ}$ = b and $\overrightarrow{QR}$ = c, express in terms
    of **a**, **b** and **c**:

    (a) $\overrightarrow{RT}$ (b) $\overrightarrow{ST}$ (c) $\overrightarrow{SR}$

Fig. 41.31

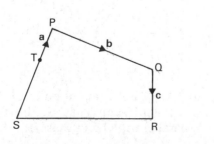

11. In Fig. 41.32, ABCDEF is a regular
    hexagon. If AF = a, AB = b and
    BC = c, express each of the following
    in terms of **a**, **b** and **c**:

    (a) $\overrightarrow{DC}$ (b) $\overrightarrow{DE}$ (c) $\overrightarrow{FE}$

    (d) $\overrightarrow{FC}$ (e) $\overrightarrow{AE}$ (f) $\overrightarrow{AD}$

Fig. 41.32

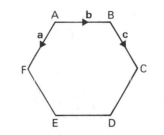

Application of Vectors to Geometry

It will be recalled that the vectors a and ka,
where k is a scalar, are parallel, and also that
$k(a + b) = k a + k b$.

These two facts are useful in solving certain
geometric problems.

Example 11

ABCD is a parallelogram. DN and BM are
perpendicular to the diagonal AC. Show that
BM = DN.

Fig. 41.33

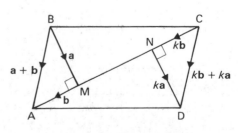

BM and DN are parallel since they both
lie at 90° to AC. Let $\overrightarrow{BM}$ = a and
$\overrightarrow{ND}$ = ka (Fig. 41.33).

Also, AM and NC lie in the same straight line.

Let $\overrightarrow{MA} = \mathbf{b}$ and $\overrightarrow{CN} = k\mathbf{b}$.

$$\overrightarrow{BA} = \overrightarrow{BM} + \overrightarrow{MA}$$

$$= \mathbf{a} + \mathbf{b}$$

$$\overrightarrow{CD} = \overrightarrow{CN} + \overrightarrow{ND}$$

$$= k\mathbf{b} + k\mathbf{a}$$

$$= k(\mathbf{a} + \mathbf{b})$$

Since $\overrightarrow{BA} = \overrightarrow{CD}$,

$$\mathbf{a} + \mathbf{b} = k(\mathbf{a} + \mathbf{b})$$

$$\therefore \qquad k = 1$$

$$\therefore \qquad \mathbf{a} = k\mathbf{a}$$

Hence   BM = DN

Exercise 41.4

The following questions should be solved using vectors.

1. In $\triangle$PQR (Fig. 41.34), the line ST is drawn parallel to QR so that PS = 3SQ. Prove that PT = 3TR.

Fig. 41.34

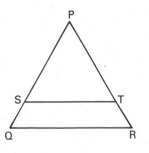

2. In the parallelogram ABCD, the point P is taken on the side AB such that AP = 2PB. The lines BD and PC intersect at O. Show that OC = 3PO.

3. In $\triangle$ABC, the point P is the mid-point of AB. A line through P parallel to BC meets AC at Q. Prove that $PQ = \frac{1}{2}BC$.

4. In Fig. 41.35, ABCD is a parallelogram. Q is the mid-point of CD. Show that 2AQ = AD + AC = 2AD + AB.

Fig. 41.35

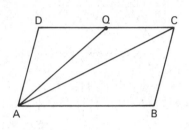

5. AB is parallel to DC and AB = DC. Prove that BC is parallel to AD and that BC = AD.

6. In Fig. 41.36, D is the mid-point of BC and $AG = \frac{2}{3}AD$. Prove that 2BD + AB = 3BG.

Fig. 41.36

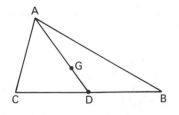

7. In $\triangle$ABC, P is the mid-point of AB and Q is the mid-point of AC. Show that BC is parallel to PQ and that BC = 2PQ.

8. Two parallelograms ABCD and ABEF are shown in Fig. 41.37. Prove that DCEF is a parallelogram.

Fig. 41.37

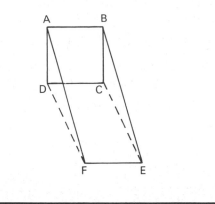

Miscellaneous Exercise 41

1. Find the modulus of each of the following vectors and state its direction:

(a) $\begin{pmatrix} 3 \\ 1 \end{pmatrix}$ (b) $\begin{pmatrix} 5 \\ -7 \end{pmatrix}$ (c) $\begin{pmatrix} -3 \\ -8 \end{pmatrix}$

2. If $U = \begin{pmatrix} 1 \\ 4 \end{pmatrix}$ and $V = \begin{pmatrix} 3 \\ 2 \end{pmatrix}$ find $R = U + V$. State the magnitude of R and its direction.

3. If $P = \begin{pmatrix} 2 \\ 6 \end{pmatrix}$ and $Q = \begin{pmatrix} 4 \\ 2 \end{pmatrix}$ find $R = P - Q$. State the modulus of R and give its direction.

4. $R = M + N$. If $R = 7i + 6j$ and $N = 2i + 4j$, find M. Calculate the magnitude of M and determine its direction.

5. In Fig. 41.38, $\overrightarrow{BQ} = 2x$, $\overrightarrow{QA} = x$, $\overrightarrow{RC} = y$ and $\overrightarrow{AR} = 2y$. BC is produced to a point P such that $\overrightarrow{CP} = k\overrightarrow{BC}$.

(a) Find in terms of x and y (i) $\overrightarrow{QR}$

(ii) $\overrightarrow{BR}$.

(b) Find in terms of x, y and k (i) $\overrightarrow{CP}$

(ii) $\overrightarrow{RP}$.

Fig. 41.38

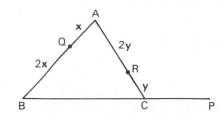

6. Fig. 41.39 shows the triangle ABC, P, Q and R being the mid-points of its sides. $\overrightarrow{PB} = 3a$ and $\overrightarrow{BQ} = 3b$. Express each of the following in terms of **a** and **b**:

(a) $\overrightarrow{AP}$ (b) $\overrightarrow{BC}$ (c) $\overrightarrow{PQ}$

(d) $\overrightarrow{AC}$ (e) $\overrightarrow{RC}$

Fig. 41.39

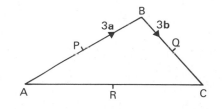

7. (a) Express each of the vectors $\overrightarrow{AB}$, $\overrightarrow{BC}$ and $\overrightarrow{CD}$ (Fig. 41.40) in the i, j form and hence calculate the magnitude of each and state its direction.

(b) Express the resultant of $\overrightarrow{AB}$, $\overrightarrow{BC}$ and $\overrightarrow{CD}$ in the i, j form and hence calculate the modulus of the resultant, stating also its direction.

Fig. 41.40

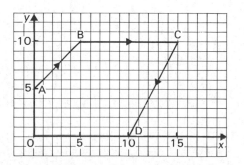

8. P is the point $(3, 5)$ and $\overrightarrow{PQ} = \begin{pmatrix} 3 \\ 5 \end{pmatrix}$ and

$\overrightarrow{QR}$ is $\begin{pmatrix} 4 \\ 1 \end{pmatrix}$.

(a) On graph paper plot the points P, Q and R using a scale of 1 cm to 1 unit on both axes and a range of 0 to 9 on the x-axis and 0 to 8 on the y-axis.

(b) Write down the coordinates of the points Q and R.

(c) Mark on your diagram a point S such that PQRS is a parallelogram.

(d) What are the coordinates of the point S?

(e) Write $\overrightarrow{RS}$ as a column vector.

(f) If **a** is a vector such that $2\mathbf{a} = \overrightarrow{PR}$, write down **a** as a column vector.

42 Statistics

Introduction

Statistics is the name given to the science of collecting and analysing facts. Originally, statistics was only used to gain information about the state. Hence the name statistics. Nowadays however, in almost all business reports, newspapers and government publications use is made of statistical methods.

Statistical methods range from the use of tables and diagrams to the use of statistical averages, the object being to assist the reader to understand what conclusions can be drawn from the quantities under discussion.

Raw Data

Raw data is collected information which is not organised numerically, that is, it is not arranged in any sort of order.

Consider the marks of 50 students obtained in a test:

```
4 3 5 4 3 5 5 4 3 6 5 4 5 3
4 4 5 5 7 4 3 4 3 4 5 4 3 6
1 3 6 3 2 6 6 3 5 2 7 5 7 1
7 6 5 8 6 4 3 5
```

This is an example of raw data and we see that it is not organised into any sort of order.

Frequency Distributions

One way of organising raw data into order is to arrange it in the form of a frequency distribution. The number of students obtaining 3 marks is found, the number obtaining 4 marks is found, and so on. A tally chart is the best way of doing this.

On examining the raw data we see that the smallest mark is 1 and the greatest is 8. The marks from 1 to 8 inclusive are written in column 1 of the tally chart. We now take each figure in the raw data, just as it comes, and for each figure we place a tally mark opposite the appropriate mark.

The fifth tally mark for each number is usually made in an oblique direction thereby tying the tally marks into bundles of five.

When the tally marks are complete they are counted and the numerical value recorded in the column headed 'frequency'. Hence the frequency is the number of times each mark occurs. From the tally chart below it will be seen that the mark 1 occurs twice (a frequency of 2), the mark 5 occurs twelve times (a frequency of 12) and so on.

TABLE 1

Mark	Tally	Frequency
1	I I	2
2	I I	2
3	⊦⊦⊦⊦ ⊦⊦⊦⊦ I	11
4	⊦⊦⊦⊦ ⊦⊦⊦⊦ I	11
5	⊦⊦⊦⊦ ⊦⊦⊦⊦ I I	12
6	⊦⊦⊦⊦ I I	7
7	I I I I	4
8	I	1
Total		50

Grouped Distributions

When dealing with a large amount of numerical data it is useful to group the numbers into classes or categories. We can then find out the number of items belonging to each class thus obtaining a class frequency.

Example 1

The numbers shown below are the times in seconds (to the nearest second) for 40 children to complete a length of a swimming pool. The swimmers were divided into heats as the pool had eight lanes.

Heat	Lane number							
	1	2	3	4	5	6	7	8
I	40	49	43	35	42	43	46	36
II	42	36	37	44	39	41	31	45
III	38	48	44	51	38	53	35	32
IV	30	43	41	52	46	43	50	40
V	39	41	48	47	32	52	47	42

Display this data in the form of a grouped frequency table using intervals 30-34, 35-39, etc.

In order to obtain the grouped frequency distribution we use a tally chart as shown in Table 2.

TABLE 2

Class	Tally	Frequency
30-34	I I I I	4
35-39	⊥⊦⊦⊤ I I I I	9
40-44	⊥⊦⊦⊤ ⊥⊦⊦⊤ I I I I	14
45-49	⊥⊦⊦⊤ I I I	8
50-54	⊥⊦⊦⊤	5
	Total	40

The main advantage of grouping is that it produces a clear overall picture of the distribution. However, too many groups will destroy the pattern of the distribution whilst too few will destroy much of the detail which was present in the raw data. Depending upon the volume of raw data the number of classes is usually between 5 and 20.

Class Intervals

In Table 2, the first class is 30-34. These figures give the class interval. For the second class, the class interval is 35-39. The end numbers 35 and 39 are called the **class limits** for the second class, 35 being the lower class limit and 39 the upper class limit.

Class Boundaries

In Table 2, the times have been recorded to the nearest second. The class interval 35-39 theoretically includes all the times between 34.5 seconds and 39.5 seconds. These numbers are called the **lower and upper class boundaries** respectively for the second class. For any distribution, the class boundaries may be found by adding the upper class limit of one class to the lower class limit of the next class and dividing this sum by two.

Example 2

The figures below show part of a frequency distribution. State the lower and upper class boundaries for the second class.

LIFETIME OF ELECTRIC BULBS

Lifetime (hours)	Frequency
400-449	22
450-499	38
500-549	62

For the second class:

$$\text{Lower class boundary} = \frac{449 + 450}{2}$$

$$= 449.5 \text{ hours}$$

$$\text{Upper class boundary} = \frac{499 + 500}{2}$$

$$= 499.5 \text{ hours}$$

Discrete and Continuous Variables

A variable which can take any value between two given values is called a **continuous variable**. Thus the height of an individual which can be 158 cm, 164.2 cm or 177.832 cm, depending upon the accuracy of measurement, is a continuous variable.

A variable which can only have certain values is called a **discrete variable**. Thus the number of children in a family can only take whole number values such as $0, 1, 2, 3$, etc. It cannot be $2\frac{1}{2}$, $3\frac{1}{4}$, etc., and it is therefore a discrete variable.

Note that the values of a discrete variable need not be whole numbers. The size of shoes is a discrete variable but can be $4\frac{1}{2}$, 5, $5\frac{1}{2}$, 6, etc.

The Histogram

The **histogram** is a diagram which is used to represent a frequency distribution. It consists of a set of rectangles whose areas represent the frequencies of the various classes. If all the classes have the same width then all the rectangles will be the same width and the frequencies are then represented by the heights of the rectangles. Figure 42.1 shows the histogram for the frequency distribution of Table 1.

Fig. 42.1

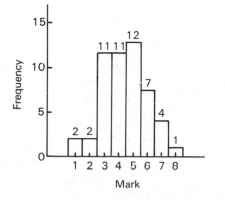

Histogram for a Grouped Distribution

A histogram for a grouped distribution may be drawn by using the mid-points of the class intervals as the centres of the rectangles. The histogram for the distribution of Table 2 is shown in Fig. 42.2. Note that the extremes of the base of each rectangle represent the lower and upper class boundaries.

Fig. 42.2

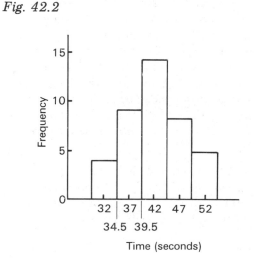

Discrete Distributions

The histogram shown in Fig. 42.2 represents a distribution in which the variable is continuous. The data in Example 3 is discrete and we shall see how a discrete distribution is represented.

Example 3

Five coins were tossed 100 times and after each toss the number of heads was recorded. The table below gives the number of tosses during which $0, 1, 2, 3, 4$ and 5 heads were obtained. Represent this data in a suitable diagram.

Number of heads	Number of tosses (frequency)
0	4
1	15
2	34
3	29
4	16
5	2
Total	100

Since the data is discrete (there cannot be 2.3 or 3.6 heads) Fig. 42.3 seems the most natural diagram to use. This diagram is in the form of a vertical bar chart in which the bars have zero width. Figure 42.4 shows the same data represented as a histogram. Note that the area under the diagram gives the total frequency of 100 which is as it should be.

Discrete data is often represented as a histogram as was done in Fig. 42.4, despite the fact that in doing this we are treating the data as though it was continuous.

Fig. 42.3

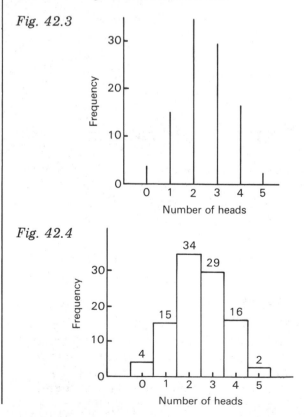

Fig. 42.4

Frequency Polygons

The frequency polygon provides a second way of representing a frequency distribution. It is drawn by connecting the mid-points of the tops of the rectangles in the histogram by straight lines.

Example 4

Draw a frequency polygon to represent the information given below:

Age of employee	Frequency
15–19	5
20–24	23
25–29	58
30–34	104
35–39	141
40–44	98
45–49	43
50–54	19
55–59	6

The frequency polygon is drawn in Fig. 42.5. It is customary to add the extensions PQ and RS to the next lower and next higher class mid-points as shown in the diagram. When this is done the area of the polygon is equal to the area of the histogram.

Fig. 42.5

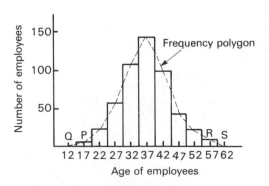

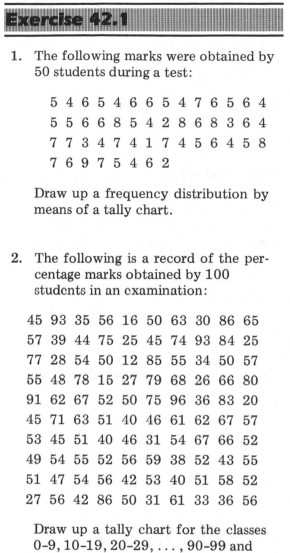

Exercise 42.1

1. The following marks were obtained by 50 students during a test:

    5 4 6 5 4 6 6 5 4 7 6 5 6 4

    5 5 6 6 8 5 4 2 8 6 8 3 6 4

    7 7 3 4 7 4 1 7 4 5 6 4 5 8

    7 6 9 7 5 4 6 2

    Draw up a frequency distribution by means of a tally chart.

2. The following is a record of the percentage marks obtained by 100 students in an examination:

    45 93 35 56 16 50 63 30 86 65

    57 39 44 75 25 45 74 93 84 25

    77 28 54 50 12 85 55 34 50 57

    55 48 78 15 27 79 68 26 66 80

    91 62 67 52 50 75 96 36 83 20

    45 71 63 51 40 46 61 62 67 57

    53 45 51 40 46 31 54 67 66 52

    49 54 55 52 56 59 38 52 43 55

    51 47 54 56 42 53 40 51 58 52

    27 56 42 86 50 31 61 33 36 56

    Draw up a tally chart for the classes 0–9, 10–19, 20–29, ... , 90–99 and hence form a frequency distribution.

3. Draw a histogram and a frequency polygon of the following data which relates to the earnings of full-time girl employees in 1986.

Wage £	Frequency
41–50	2
51–60	5
61–70	8
71–80	6
81–90	2

4. Draw a histogram and a frequency polygon for the data shown below.

Mass (kg)	Frequency
0–4	50
5–9	64
10–14	43
15–19	26
20–25	17

5. The data below gives the diameter of machined parts:

Diameter (mm)	Frequency
14.96–14.98	3
14.99–15.01	8
15.02–15.04	12

    Write down:

    (a) the upper and lower class boundaries for the second class

    (b) the class width of the classes shown in the table

    (c) the class interval for the first class.

6. Classify each of the following as continuous or discrete variables:

    (a) The diameters of ball bearings.

    (b) The number of shirts sold per day.

    (c) The mass of packets of chemical.

    (d) The number of bunches of daffodils packed by a grower.

    (e) The daily temperature.

    (f) The lifetime of electric light bulbs.

    (g) The number of telephone calls made per day by a person.

7. An industrial organisation gives an aptitude test to all applicants for employment. The results of 150 people taking the test were:

Score (out of 10)	Frequency
1	6
2	12
3	15
4	21
5	35
6	24
7	20
8	10
9	6
10	1

Draw a histogram of this information.

8. The lengths of 100 pieces of wood were measured with the following results:

Length (cm)	Frequency
29.5	2
29.6	4
29.7	11
29.8	18
29.9	31
30.0	22
30.1	8
30.2	3
30.3	1

Draw a histogram and a frequency polygon of this information.

Statistical Averages

We have seen that a mass of raw data does not mean very much until it is arranged into a frequency distribution or until it is represented as a histogram.

A second way of making the data more understandable is to try to find a single value which will represent all the values in a distribution. This single representative value is called an average.

In statistics several kinds of average are used. The more important are:

(a) the arithmetic mean, often referred to as the mean

(b) the median

(c) the mode.

The Arithmetic Mean

This is found by adding up all the values in a set and dividing this sum by the number of values making up the set. That is,

$$\text{Arithmetic mean} = \frac{\text{Sum of all the values}}{\text{The number of values}}$$

Example 5

The heights of 5 men were measured as follows: 177.8, 175.3, 174.8, 179.1, 176.5 cm. Calculate the mean height of the 5 men.

Mean =
$$\frac{177.8 + 175.3 + 174.8 + 179.1 + 176.5}{5}$$

$$= \frac{883.5}{5}$$

$$= 176.7 \text{ cm}$$

Note that the unit of the mean is the same as the unit used for each of the quantities in the set.

The Mean of a Frequency Distribution

When finding the mean of a frequency distribution we must take into account the frequencies as well as the measured observations.

Example 6

Five packets of chemical have a mass of 20.01 grams, 3 have a mass of 19.98 grams and 2 have a mass of 20.03 grams. What is the mean mass of the packets?

Mass of 5 packets @ 20.01 grams = 100.05
Mass of 3 packets @ 19.98 grams = 59.94
Mass of 2 packets @ 20.03 grams = 40.06
Total mass of 10 packets = 200.05

$$\text{Mean mass} = \frac{\text{Total mass}}{10}$$

$$= \frac{200.05}{10}$$

$$= 20.005 \text{ grams}$$

This example gives the clue whereby we may find the mean of a frequency distribution.

Example 7

Each of 200 similar engine components are measured correct to the nearest millimetre and recorded as follows:

Length (mm)	Frequency
198	8
199	30
200	132
201	24
202	6

Calculate the mean length of the 200 components.

$$\text{Mean length} = [(198 \times 8) + (199 \times 30)$$
$$+ (200 \times 132) + (201 \times 24)$$
$$+ (202 \times 6)] \div 200$$
$$= 199.95 \text{ mm}$$

The calculation is often set out in tabular form as shown below. This method reduces the risk of making errors when performing the calculation.

Length (mm)	Frequency	Length × frequency
198	8	1 584
199	30	5 970
200	132	26 400
201	24	4 824
202	6	1 212
Total	200	39 990

$$\text{Mean} = \frac{\text{Total of (length} \times \text{frequency)}}{\text{Total frequency}}$$

$$= \frac{39\,990}{200}$$

$$= 199.95 \text{ mm}$$

The Median

If a set of values is arranged in ascending (or descending) order of size the median is the value which lies half-way along the series. Thus the median of $3, 4, 4, 5, 6, 8, 8, 9, 10$ is 6 because there are four numbers below this value and four numbers above it.

When there are an even number of values in the set the median is found by taking the mean of the middle two values. Thus the median of $3, 3, 5, 7, 9, 10, 13, 15$ is $\frac{7 + 9}{2} = 8$.

The median of a discrete frequency distribution may be found by setting out the scores in numerical order and finding the middle value.

Example 8

The table below shows the distribution of numbers obtained when a die is thrown 30 times. Determine the median.

Number obtained	1	2	3	4	5	6
Frequency		2	7	5	7	3	6

The total frequency is 30. Hence the median must lie between the 15th and 16th items in the distribution. We now look at the values of the 15th and 16th items. These are both 4 and hence the median is 4.

It is unnecessary to write down all the values in numerical order to find the median but if we do this we obtain:

$$1, 1, 2, 2, 2, 2, 2, 2, 2, 3, 3, 3, 3, 3,$$
$$4, 4, 4, 4, 4, 4, 4, 5, 5, 5, 6, 6, 6, 6,$$
$$6, 6$$

Looking at this set of values we see that the two middle values are 4 and hence the median is 4.

The Mode

The mode of a set of values is the value which occurs most frequently. That is, it is the most common value. Thus the mode of $2, 3, 3, 4, 4, 4, 5, 6, 6, 7$ is 4 because this number occurs three times, which is more than any of the other numbers in the set.

Sometimes, in a set of numbers, no mode exists, as for the set $2, 4, 7, 8, 9, 11$ in which each number occurs once. It is possible for there to be more than one mode. The set $2, 3, 3, 3, 4, 4, 5, 6, 6, 6, 7, 8$ has two modes, 3 and 6, because each of these numbers occurs three times, which is more than any of the other numbers.

A set of values which has two modes is called **bimodal**. If the set has only one mode it is said to be **unimodal** but if there are more than two modes the set is called **multimodal**.

The Mode of a Frequency Distribution

To find the mode of a frequency distribution we draw a histogram. By drawing the diagonals as shown in Fig. 42.6 the mode is found.

Example 9

The heights of a group of boys are measured to the nearest centimetre with the following results:

Height (cm)	157	158	159	160	161	162
Frequency	20	36	44	46	39	30

Height (cm)	163	164	165	166	167
Frequency	22	17	10	4	2

Find the mode of the distribution.

By constructing the histogram (Fig. 42.6) the mode is found to be 159.7. It is worthwhile noting that the modal class is 159.5 to 160.5 cm.

Fig. 42.6

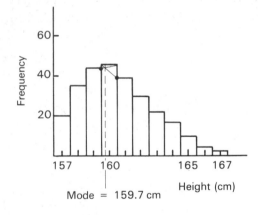

Mode = 159.7 cm

Which Average to Use

The arithmetic mean is the most familiar kind of average and it is extensively used in business and other statistical work, such as with sales data, income and expenditure, operation costs, rates of pay, etc. It is easy to understand but in some circumstances it can be definitely misleading. For instance, if the hourly wages of five employees in an office are £1.52, £1.64, £1.88, £4.60 and £1.76. The mean wage is £2.28, but this is affected by the extreme value of £4.60 and hence the value of the mean gives a false impression of the wages paid in the office under discussion.

The median is not affected by extreme values, and it will give a better indication of the wages paid in the office discussed above (note that the median wage is £1.76 per hour). For distributions which are sharply peaked or very skew the median is usually the best average to use.

The mode is used when the commonest value is required. For instance, a manufacturer is not particularly interested in the mean length of men's legs because it might not represent a stock size in trousers. It may, in fact, be some point between stock sizes and in such cases the mode is probably the best average to use. However, which average is used will depend upon particular circumstances.

Exercise 42.2

1. Find the mean of £23, £27, £30, £28 and £32.

2. The heights of some men are as follows: 172, 170, 181, 175, 179 and 173 cm. Calculate the mean height of the men.

3. Five people earn £84 per week, 3 earn £76 per week and 2 earn £88 per week. What is the mean wage for the 10 people?

4. Calculate the mean length from the following table:

Length (mm)	Frequency
198	1
199	4
200	17
201	2
202	1

5. Calculate the mean height of 50 people from the table below:

Height (cm)	Frequency
160	1
161	5
162	10
163	16
164	10
165	6
166	2

6. Find the median of the numbers 5, 3, 8, 6, 4, 2, 8.

7. Find the median of the numbers 2, 4, 6, 5, 3, 1, 8, 9.

8. The marks of a student in five examinations were: 54, 63, 49, 78 and 57. What is his median mark?

9. Find the mode of the following set of numbers 3, 5, 2, 7, 5, 8, 5, 2, 7.

10. Find the mode of 38.7, 29.6, 32.1, 35.8, 43.2.

11. Find the modes of 8, 4, 9, 3, 5, 3, 8, 5, 3, 8, 9, 5, 6, 7.

12. The marks of 100 students were as follows:

Mark	Frequency
1	2
2	8
3	20
4	32
5	18
6	9
7	6
8	3
9	2

Obtain the median mark.

13. Find the mode of the distribution of Question 4.

14. Find the mode of the distribution of Question 12.

Cumulative Frequency Distribution

A cumulative frequency curve is an alternative to a histogram for presenting a frequency distribution. The way in which it is obtained is shown in Example 10.

Example 10

Obtain a cumulative frequency distribution for the data below.

Height (cm)	Frequency
150–154	8
155–159	16
160–164	43
165–169	29
170–174	4

The class boundaries are 149.5, 154.5, 159.5, 164.5, 169.5 and 174.5 cm.

In drawing up a cumulative frequency distribution the lower boundary limit for each class is used as shown in Table 3.

The distribution may be represented by a cumulative frequency polygon (Fig. 42.7) or by a cumulative frequency curve (Fig. 42.8). A cumulative frequency curve is known as an ogive, after the architectural term used for a shape of this kind.

TABLE 3

Height (cm)	Cumulative frequency
Less than 149.5	0
Less than 154.5	8
Less than 159.5	8 + 16 = 24
Less than 164.5	24 + 43 = 67
Less than 169.5	67 + 29 = 96
Less than 174.5	96 + 4 = 100

Fig. 42.7

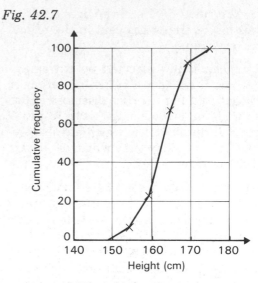

Cumulative frequency polygon

Fig. 42.8

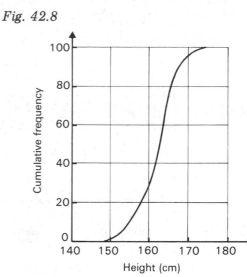

Cumulative frequency curve or ogive

The Median of a Grouped Distribution

When the information is grouped into class intervals it is possible, by looking at the distribution, to determine in which class the median lies. To determine its value more exactly we draw an ogive. The median is then the value corresponding to half the total frequency.

Example 11

The table below shows the annual rents of people living in council houses in 1972.

Rent (£ per annum)	Under 40	40–	80–	120–	160–
Percentage of households	2	15	25	27	18

Rent (£ per annum)	200–	240–280	280 and over
Percentage of households	8	3	2

Draw an ogive and from it estimate the median rent.

Annual rent £	Cumulative %
Less than 40	2
Less than 80	17
Less than 120	42
Less than 160	69
Less than 200	87
Less than 240	95
Less than 280	98

Fig. 42.9

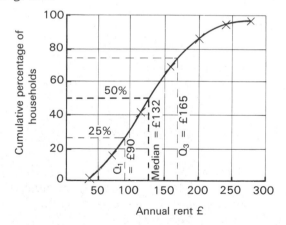

The ogive is shown in Fig. 42.9, and the median rent is found to be £132.

Quartiles

We have seen that the median divides a set of values into two equal parts. Using a similar method we can divide a set of values into four equal parts. The values which so divide the set are called the quartiles. They are usually denoted by the symbols Q_1, Q_2 and Q_3, Q_1 being the lower quartile and Q_3 the upper quartile, Q_2 being equal to the median value.

Example 12

Use the ogive of Fig. 42.9 to find the values of the lower and upper quartiles.

The lower quartile is the value of the annual rent corresponding to 25% of the households. From the ogive we see that $Q_1 = £90$ per annum. The upper quartile is the value of the annual rent corresponding to 75% of the households. From the ogive we see that $Q_3 = £165$ per annum.

Measures of Dispersion

A statistical average gives us some idea about the magnitude of the quantities in a distribution, but it tells us nothing about the spread of the distribution (Fig. 42.10). Two of the measures of dispersion which are used are:

(1)   the range

(2)   the semi-interquartile range.

Fig. 42.10

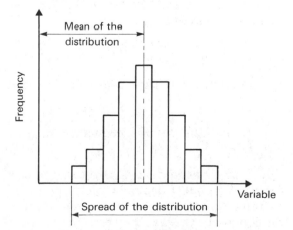

The Range

The range is the difference between the largest and smallest values in a distribution. That is,

Range = Largest value − Smallest value

Example 13

The weekly wages of five office workers are: £142.50, £155, £210, £171 and £200. Find the range of wages.

Lowest wage = £142.50

Highest wage = £210

Range of wages = £210 − £142.50

= £67.50

The Semi-Interquartile Range

This is found from the formula:

Semi-interquartile range = $\frac{1}{2}$(upper quartile

− lower quartile)

= $\frac{1}{2}(Q_3 - Q_1)$

Example 14

An examination of the wages paid by a certain company showed that the upper quartile was £256 per week whilst the lower quartile was £192 per week. Find the semi-interquartile range.

We are given that Q_1 = £192 and Q_3 = £256.

Semi-interquartile range = $\frac{1}{2}(256 - 192)$

= 32

Hence the semi-interquartile range is £32.

The range gives some idea of the spread of a distribution but it depends solely upon the extreme values of the data. It gives no information about the way in which the data is

dispersed and hence it is seldom used as a measure of dispersion for a frequency distribution. However, when dealing with small samples like that in Example 13, the range is a very effective measure of dispersion.

The semi-interquartile range has the advantage that it ignores the extreme values of a distribution. However, this measure of dispersion covers that half of the distribution which is centred only around the median and therefore it does not show the dispersion of the distribution as a whole. A small value of the semi-interquartile range shows that the data has only a small amount of spread between the quartiles. The semi-interquartile range is used extensively in business and education.

Exercise 42.3

1. The table below gives the lifetimes of TV tubes. Draw a cumulative frequency curve for this information.

Lifetime (hours)	Frequency
300–399	7
400–499	23
500–599	30
600–699	28
700–799	18
800–899	4

2. Draw an ogive for the following data which relates to the weekly earnings of women aged 18 and over in 1986.

Weekly earnings £	Frequency
Under £50	3
50–	19
75–	28
100–	22
125–	13
150–	6
175–	5
200–225	4

3. The table below gives the number of bags of agricultural foodstuffs produced by a firm in 1976.

Number produced	Frequency
100-	6
150-	15
200-	29
250-300	2

Draw a cumulative frequency curve.

4. The table below gives the weights of 250 boys, each weight being recorded to the nearest 100 grams.

Weight (kg)	Number
44.0-47.9	3
48.0-51.9	17
52.0-55.9	50
56.0-57.9	45
58.0-59.9	46
60.0-63.9	57
64.0-67.9	23
68.0-71.9	9

Draw an ogive.

5. The data below gives the dead-weight tonnage of U.K. tankers in 1973. Find, by drawing a cumulative frequency curve, the median tonnage.

Tonnage (tonnes)	Number of tankers
Under 10 000	123
10 000 and under 20 000	98
20 000 and under 50 000	138
50 000 and under 100 000	53
100 000 and under 250 000	59

6. The table below gives the distribution of the maximum loads supported by a certain make of cable.

Max. load (kN)	Frequency
19.2-19.5	4
19.6-19.9	12
20.0-20.3	18
20.4-20.7	3

Draw a cumulative frequency curve and hence find the median load.

7. The table shows the distribution of the scores obtained by 200 people in a particular test.

Score	No. of pupils
26-30	3
31-35	6
36-40	10
41-45	21
46-50	38
51-55	55
56-60	32
61-65	19
66-70	9
71-75	5
76-80	2

(a) Draw a cumulative frequency curve. Use your curve to estimate:

(b) the median score

(c) the lower quartile

(d) the upper quartile.

8. The wages of five office juniors are: £59, £55.60, £64.80, £98.40 and £74.80 per week. Find the range of wages.

9. The largest of 50 measurements is 29.88 mm. If the range is 0.12 mm, find the smallest measurement.

10. The histogram (Fig. 42.11) shows the amounts of pocket money received by a group of 12 year old children.

Fig. 42.11

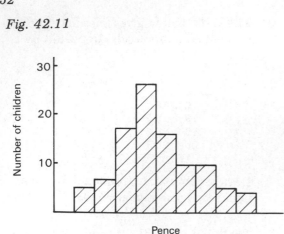

(a) Use the histogram to complete the table for the distribution:

Pocket money (p)	1-9	10-19	20-29	30-39	40-49
Number of children	0	5	7		

Pocket money (p)	50-59	60-69	70-79	80-89	90-99
Number of children					

(b) Complete the cumulative frequency table and use it to draw a cumulative frequency diagram.

Pocket money (p) Less than	10	20	30	40	50	60	70
Number of children	0	5	12	29			

Pocket money (p) Less than	80	90	100
Number of children			

(c) From your cumulative frequency diagram estimate the median.

11. Determine the range of the following observations: 2, 5, 12, 9, 1, 7, 10, 4, 3, 11.

12. After an investigation it was found that the lower quartile was 135 mm whilst the upper quartile was 147 mm. What is the semi-interquartile range?

13. Draw an ogive for the following frequency distribution and find the lower and upper quartiles. Hence find the value of the semi-interquartile range.

Family income £	Frequency
500–	2
1000–	12
1500–	37
2000–	78
2500–	121
3000–	60
3500–	15

14. The table below shows the age distribution of the UK at the Census in a certain year.

(a) Copy and complete the cumulative frequency table below.

Age group	Population (millions)	Cumulative frequency
Under 10	9	
10–20	8	
20–30	7	
30–40	7	
40–50	7	
50–60	7	
60–70	5	
70–80	3	
80–100	1	

(b) Draw a cumulative frequency graph.

(c) From the graph determine the values of the lower quartile, the median and the upper quartile.

(d) Write down the value of the semi-interquartile range.

Miscellaneous Exercise 42

Section A

1. A survey was made one evening of the ages of 30 members of a youth club, with the following results:

14	16	16	15	15	14
14	17	18	15	14	16
16	16	14	14	15	17
15	14	13	14	14	18
16	15	14	13	17	17

Tabulate the above results in a suitable frequency distribution table and draw a histogram to illustrate your results.

2. In the game of bridge, one method of evaluating the strength of a hand is to count 4 points for each ace, 3 points for a king, 2 points for a queen, and 1 point for a jack. A player keeps a record of the points he received over 50 deals. They are as follows:

16	1	17	8	9	11	19	13	10	18
21	2	13	10	14	15	10	4	11	20
12	10	28	9	11	12	9	11	9	5
6	16	8	24	7	9	8	7	6	0
3	7	12	10	10	13	5	15	14	8

Compile a grouped frequency distribution for the classes 0–3, 4–7, 8–11, etc. and hence draw a histogram to depict this distribution.

3. The histogram (Fig. 42.12) illustrates the frequency distribution of the heights (in centimetres) of a group of 54 boys aged 14.

Fig. 42.12

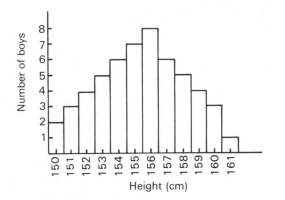

Copy and complete the following table:

Height (cm)	150	151	152	153	154	155
Frequency						

Height (cm)	156	157	158	159	160	161
Frequency						

4. The table below shows the number of calls per day received by a fire station over a given year. Determine the median of this distribution.

No. of calls per day	0	1	2	3	4	5	6 and over
No. of days	139	102	57	30	19	12	6

5. The temperature at noon in a certain city was measured every day for the month of June. The distribution was as follows:

Temperature (°C)	16	17	18	19	20	21	22
Number of days	1	4	6	8	9	1	1

Calculate the mean temperature to the nearest degree Celsius.

6. The following figures are the maximum temperatures recorded in degrees Celsius on 25 consecutive days at a local ice rink.

2	5	3	2	4
5	3	7	6	7
6	7	4	2	5
7	6	3	7	3
3	4	7	5	5

(a) Complete the following frequency table:

Maximum temperature (°C)	2	3	4	5	6	7
Frequency (number of days)						

(b) Draw a histogram to illustrate these statistics.

(c) (i) What is the modal temperature?
(ii) Why does the modal temperature give a misleading impression of the average temperature?

(d) Find the median temperature by inspection.

7. The numbers of peas per pod were counted for a number of pods and the results are tabulated below. Determine the mode.

No. of peas per pod	5	6	7	8	9	10	11
No. of pods		12	25	18	16	10	6	2

Section B

1. In an experiment in a biology laboratory, the weights of 100 leaves were found. The distribution of weights is given in the following table:

Weight (in grams)	Frequency
0–4	1
5–9	2
10–14	15
15–19	21
20–24	26
25–29	28
30–34	4
35–39	2
40–44	1

(a) Draw up a table to show the cumulative frequencies.

(b) Draw a cumulative frequency graph (ogive) and from your graph determine the upper and lower quartiles and the semi-interquartile range.

2. The following table shows the number of runs scored by a sample of cricketers, in a limited over competition.

Number of runs scored	Frequency
18	5
19	10
20	12
21	28
22	36
23	41
24	30
25	25
26	10
27	3

(a) How many cricketers are there in the sample?

(b) What is the modal score?

(c) What is the median score?

(d) What is the mean score of the cricketers correct to one decimal place?

3. The bar chart (Fig. 42.13) shows the number of meals served in a small canteen during one week. What was the average number of meals served per day?

Fig. 42.13

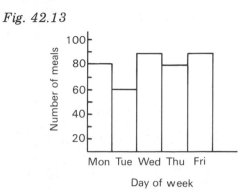

4. The following table shows the heights of 80 boys leaving the 5th form of a school in the summer term of 1976, measured to the nearest centimetre.

Height (cm)	Frequency
153–157	4
158–162	11
163–167	20
168–172	24
173–177	17
178–182	4

(a) Draw a histogram to represent this data and state the modal class.

(b) Calculate the mean height of the boys, giving your answer to the nearest centimetre.

5. A batch of 100 samples of crockery showed the following frequency distribution for flaws.

No. of flaws	Frequency
0–2	10
3–5	22
6–8	30
9–11	21
12–14	12
15–17	5

(a) Calculate the mean of this grouped data by using mid-point values 1, 4, 7, etc.

(b) Draw a cumulative frequency curve for the data and use the diagram to find the median. Use a scale of 1 cm to 10 flaws on the frequency axis.

6. Number of hours of part-time work in 15 weeks: 5, 8, 8, 11, 11, 11, 12, 13, 14, 15, 15, 16, 18, 20, 21. State:
   (a) the median
   (b) the upper quartile
   (c) the lower quartile
   (d) the full range
   (e) the semi-interquartile range.

7.

Length (cm)	30–	32–	34–	etc.
Frequency	4	8	14	etc.

Write down the limits of the second class if the lengths were measured
   (a) to the nearest centimetre
   (b) to the nearest millimetre.

Multi-Choice Questions 42

The table below gives the ages of a group of pupils in a school who were born in March.

Age in years	13	14	15	16	17
Number of pupils	1	4	2	2	1

Use the table to answer Questions 1 and 2.

1. What is the mean age of the pupils?
   A 14 years   B 14.8 years
   C 15 years   D 15.5 years

2. What is the mode?
   A 13 years   B 14 years
   C 15 years   D 17 years

3. The figures below give a record of the number of goals scored by a football team in 13 successive matches. What is the mode of this set of goals?

   2, 3, 2, 0, 2, 1, 0, 4, 2, 5, 2, 6, 0

   A 6   B 5   C 4   D 2

4. The table below shows the frequency distribution for the radii of ball bearings. What is the class width?

Radius (mm)	14–15	16–17	18–19
Frequency	8	11	14

    A 1 mm   B 2 mm

    C 3 mm   D 4 mm

5. What is the modal score for the following distribution?

Score	0 1 2 3 4 5 6 7 8 9 10
Frequency	0 3 2 6 4 8 2 3 2 5 6

    A 10   B 8   C 5   D 4

6. Which of the following is a discrete variable?

    A The number of goals scored in a hockey match

    B The radius of a ball bearing

    C The temperature of an oven

    D The volume of gas in a container

7. The arithmetic mean of ten numbers is 36. If one of the numbers is 18, what is the mean of the other nine?

    A 18   B 27   C 38   D 54

8. The median of the numbers $5, 3, 8, 6, 4, 2, 8$ is

    A 8   B 6   C 5   D 4

9. The mean of the numbers $12, 24$ and y is the same as the mean of the numbers $9, 12, 18$ and 21. Find the value of y.

    A 9   B 15   C 18   D 24

10. The marks of ten students in a test are $8, 4, 5, 10, 9, 8, 6, 5, 8, 3$. What is the modal mark?

    A 10   B $8\frac{1}{2}$ C 8   D 6.9

43 Probability

Simple Probability

If a fair coin is tossed the outcome is equally likely to be heads or tails. We say that the events of tossing heads or tossing tails are **equi-probable events**.

Similarly if we throw an unbiased die (plural: dice) the events of throwing a 1, or a 2, or a 3, or a 4, or a 5, or a 6 are equi-probable events.

The probability of an event occurring is defined as

$$P(E) = \frac{\text{Total number of favourable outcomes}}{\text{Total number of possible outcomes}}$$

Example 1

(a) A fair coin is tossed once. What is the probability that it will come down heads?

Number of favourable outcomes = 1
Total number of possible outcomes = 2

$$P(\text{head}) = \frac{1}{2}$$

(b) What is the probability of scoring a 5 in the single roll of a fair die?

Number of favourable outcomes = 1
Total number of possible outcomes = 6

$$P(\text{five}) = \frac{1}{6}$$

(c) What is the probability of drawing an ace from a deck of 52 playing cards when the deck is cut once?

Number of favourable outcomes = 4
(since there are four aces in the deck)
Total number of possible outcomes = 52

$$P(\text{ace}) = \frac{4}{52}$$

$$= \frac{1}{13}$$

Probability Scale

When an event is absolutely certain to happen we say that the probability of it happening is 1. Thus the probability that one day each of us will die is 1. When an event can never happen we say the probability of it happening is 0. Thus the probability that any one of us can jump a height of 5 metres unaided is 0.

All probabilities must therefore have a value between 0 and 1. They can be expressed as either a fraction or a decimal. Thus

$$P(\text{head}) = \frac{1}{2}$$

$$= 0.5$$

$$P(\text{ace}) = \frac{1}{13}$$

$$= 0.077$$

Probabilities can be expressed on a probability scale (Fig. 43.1).

Fig. 43.1

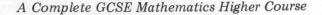

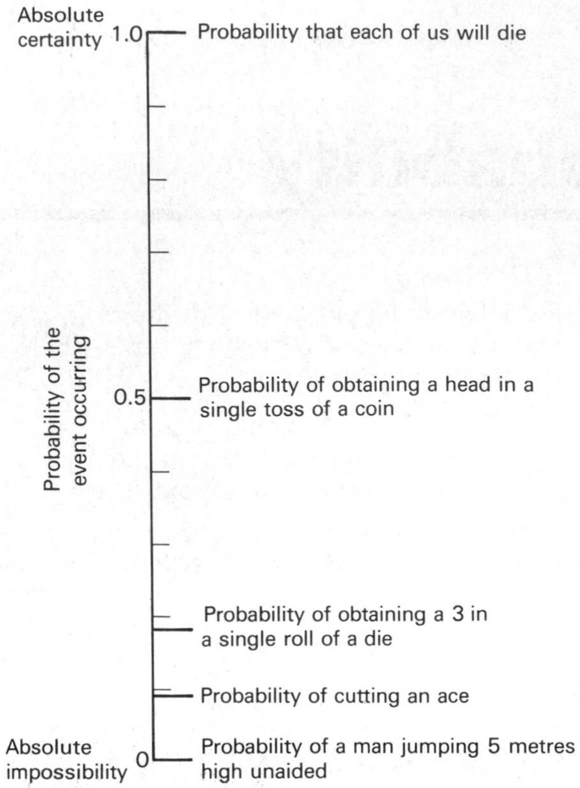

Example 2

A bag contains 5 blue balls, 3 red balls and 2 black balls. A ball is drawn at random from the bag. Calculate the probability that it will be

(a) blue   **(b)** red

(c) not black

$$\text{(a)} \quad P(\text{blue}) = \frac{5}{10} = 0.5$$

$$\text{(b)} \quad P(\text{red}) = \frac{3}{10} = 0.3$$

$$\text{(c)} \quad P(\text{black}) = \frac{2}{10} = 0.2$$

$$P(\text{not black}) = 1 - 0.2 = 0.8$$

Meaning of Probability

When we say that the probability of an event happening is $\frac{1}{3}$ we do **not** mean that if we repeat the experiment three times the event will happen once. Even if we repeat the experiment 30 times it is very unlikely that the event will happen exactly 10 times.

Probability tells us what to expect in the long run. If the experiment is repeated 300 times we would expect the event to happen **about** 100 times. However we would not be worried if it happened only on 94 occasions or if it happened on 106 of them.

In cases where the probability has to be determined by experiment, the probability of an event happening is calculated from the formula:

$P(E) =$

Number of trials with favourable outcomes

Total number of trials

Probabilities determined by experiment are often called **empirical probabilities**.

Total Probability

If we toss a fair coin it will come down either heads or tails. That is:

$$P(\text{heads}) = \frac{1}{2} \quad \text{and} \quad P(\text{tails}) = \frac{1}{2}$$

The total probability covering all possible outcomes is $\frac{1}{2} + \frac{1}{2} = 1$. Another way of saying this is

$P(\text{favourable outcomes})$

$+ P(\text{unfavourable outcomes}) = 1$

Example 3

50 invoices are checked for errors. If 3 of them are found to contain errors, determine the probability that an invoice, chosen at random from these 50, will contain errors.

Treating the event of finding an error as a favourable outcome:

Number of favourable outcomes = 3
Total number of trials = 50

$$P(\text{finding an error}) = \frac{3}{50}$$

$$= 0.06$$

Relative Frequency and Probability

The relative frequency of a class in a frequency distribution is found by dividing the class frequency by the total frequency. That is,

$$\text{Relative frequency} = \frac{\text{Class frequency}}{\text{Total frequency}}$$

Example 4

A loaded die was thrown 200 times with the following results:

Score	1	2	3	4	5	6
Frequency	28	36	32	30	34	40

Calculate the relative frequencies.

The relative frequencies are found by dividing each of the class frequencies by 200. (Note that the sum of the relative frequencies must equal 1.)

Score	1	2	3	4	5	6
Relative frequency	0.14	0.18	0.16	0.15	0.17	0.20

In the case of the loaded die (Example 4), the probabilities of obtaining each of the scores on the die can only be determined by experiment. We can estimate the probabilities by assuming that they are the same as the relative frequencies. Certainly, if the number of observations is very large the relative frequencies will give results which are very near the truth.

Consider the following as an example. If 1000 tosses of a coin result in 542 heads, the relative frequency of heads is $542/1000 = 0.542$. If in another 1000 tosses the number of heads is 492, the total number of heads in 2000 tosses is $542 + 492 = 1034$. The relative frequency is then $1034/2000 = 0.517$. By continuing in this way we should get closer and closer to 0.5, which is the calculated probability of throwing a head in a single toss of the coin.

Exercise 43.1

1. A die is rolled. Calculate the probability that it will give
   (a) a five
   (b) a score less than 3
   (c) an even number.

2. A card is drawn at random from a pack of 52 playing cards. Calculate the probability that it will be
   (a) the Jack of Hearts
   (b) a king
   (c) an ace, king, queen, or jack
   (d) the King of Hearts or the Ace of Spades.

3. A letter is chosen from the word TERRIFIC. Determine the probability that it will be
   (a) an R   (b) a vowel
   (c) a consonant.

4. A bag contains three red balls, five blue balls and two green balls. A ball is chosen at random from the bag. Calculate the probability that it will be

   (a) green   (b) blue   (c) not red

5. Two dice are thrown and their scores are added together. Find the probability that the total will be

   (a) 5   (b) less than 5

   (c) more than 5

6. Determine the probability for each of the following situations:

   (a) A sample of 9000 industrial workers were questioned regarding industrial injuries. 600 reported sustaining such injuries during a 12-month period. What is the probability of a worker sustaining an industrial injury?

   (b) A wholesaler of electrical goods finds that of 150 deliveries from a certain firm 10 are late. Calculate the probability of the next delivery being late.

   (c) A new component is fitted to an engine. 20 engines, fitted with the new part are tested and 2 fail to function correctly. Calculate the probability that an engine fitted with the new component will not function correctly.

7. The table below shows the family income of 500 families.

Income range (£)	Number of families
Less than 3000	70
3000 and less than 4000	120
4000 and less than 5000	180
5000 and less than 6000	80
More than 6000	50

   If a family is chosen at random from these 500 families, calculate the probability that its income will be

   (a) less than £3000

   (b) £4000 but less than £5000

   (c) more than £6000.

8. Over a period of 100 days the following numbers of absentees were recorded:

Number of absentees	0	1	2	3
Number of days	14	28	28	18

   (a) What is the probability that on any one day there will be no absentees?

   (b) On any one day what is the probability that there will be more than 3 absentees?

   (c) On any one day what is the probability that there will be less than 3 absentees?

Independent Events

An independent event is one which has no effect on subsequent events. If a die is rolled three times what happens on the first roll does not affect what happens on the second or third rolls. Hence the three rolls of the die are independent events. Similarly the events of tossing a coin and then cutting a deck of playing cards are independent events because the way in which the coin lands has no effect on the cut.

The probability that two events E_1 and E_2 both happen and happen in that order is

$$P(E_1 \text{ then } E_2) = P(E_1) \times P(E_2)$$

where $P(E_1)$ = probability of first event happening and $P(E_2)$ = probability of second event happening.

In general,

$$P(E_1 \text{ then } E_2 \text{ then } E_3 \ldots \text{ then } E_n)$$
$$= P(E_1) \times P(E_2) \times P(E_3) \times \ldots P(E_n)$$

The order in which the events occur must be taken into account even if no order is specified. Thus

$$P(E_1 \text{ and } E_2 \text{ in any order})$$
$$= P(E_1 \text{ then } E_2) + P(E_2 \text{ then } E_1)$$

Example 5

A die is thrown twice. What is the probability of obtaining

(a) a 3 on the first throw and a 4 on the second throw?

(b) a 3 and a 4 in any order?

$$\text{(a)} \quad P(3 \text{ on first throw}) = \frac{1}{6}$$

$$P(4 \text{ on second throw}) = \frac{1}{6}$$

$$P(3 \text{ then } 4) = \frac{1}{6} \times \frac{1}{6}$$

$$= \frac{1}{36}$$

$$\text{(b)} \quad P(3 \text{ and } 4 \text{ in any order})$$

$$= \frac{1}{6} \times \frac{1}{6} + \frac{1}{6} \times \frac{1}{6}$$

$$= \frac{1}{18}$$

Example 6

A coin is tossed five times. What is the probability of each toss resulting in a head?

The five tosses of the coin are independent events since what happens on the first toss in no way affects subsequent tosses. Similarly what happens on the second toss in no way affects the third and subsequent tosses and so on.

$$P(5 \text{ heads}) = \frac{1}{2} \times \frac{1}{2} \times \frac{1}{2} \times \frac{1}{2} \times \frac{1}{2}$$

$$= \frac{1}{32}$$

A Venn diagram (Fig. 43.2) may be used to illustrate the probability of two events E_1 and E_2 occurring. The shaded area gives the required probability.

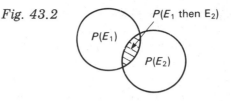

Fig. 43.2

Dependent Events

Consider a bag containing 3 red balls and 2 blue balls. A ball is drawn at random from the bag and **not** replaced. The probability that it is red is 3/5. Now, assuming we drew a red ball, let us choose a second ball. The probability that this is also red is 2/4. Hence the probability of drawing two red balls is $3/5 \times 2/4 = 3/10$. The events of drawing one red ball followed by drawing a second red ball are **dependent events** because the probability of the second event depends upon what happened in the first event.

Example 7

A bag contains 5 white balls, 3 black balls and 2 green balls. A ball is chosen at random from the bag and not replaced. In three draws find the probability of obtaining white, black and green in that order.

Let $P(E_1)$ = the probability of drawing a white ball on the first draw = $\dfrac{5}{10}$

$P(E_2)$ = the probability of drawing a black ball on the second draw = $\dfrac{3}{9}$

$P(E_3)$ = the probability of drawing a green ball on the third draw = $\dfrac{2}{8}$

The probability of drawing white, then black, then green

$$= \frac{5}{10} \times \frac{3}{9} \times \frac{2}{8}$$

$$= \frac{1}{24}$$

Mutually Exclusive Events

If two events could not happen at the same time they are said to be **mutually exclusive**. For instance, suppose we want to know the probability of a 3 or a 4 occurring in the single roll of a die. In a single roll either a 3 can occur or a 4 can occur. It is not possible for a 3 and a 4 to occur together. Hence the events of throwing a 3 or a 4 in a single roll of a die are mutually exclusive. Similarly, it is impossible to cut a jack and a king in a single cut of a pack of playing cards. Hence these two events are mutually exclusive.

If $E_1, E_2, \ldots, E_n$ are mutually exclusive events then the probability of one of the events occurring is

$$P(E_1 \text{ or } E_2 \text{ or } \ldots \text{ or } E_n)$$
$$= P(E_1) + P(E_2) + \ldots + P(E_n)$$

Example 8

A die with faces numbered 1 to 6 is rolled once. What is the probability of obtaining either a 3 or a 4?

$$P(3) = P(E_1)$$
$$= \frac{1}{6}$$
$$P(4) = P(E_2)$$
$$= \frac{1}{6}$$
$$P(3 \text{ or } 4) = P(E_1 + E_2)$$
$$= \frac{1}{6} + \frac{1}{6}$$
$$= \frac{1}{3}$$

Example 9

A pack of playing cards is cut once. Find the probability that the card which is cut will be the Ace of Spades, a king, or the Queen of Hearts.

$$P(\text{Ace of Spades}) = P(E_1)$$
$$= \frac{1}{52}$$
$$P(\text{king}) = P(E_2)$$
$$= \frac{4}{52}$$
$$P(\text{Queen of Hearts}) = P(E_3)$$
$$= \frac{1}{52}$$

$$P(\text{Ace of Spades, a king} \atop \text{or the Queen of Hearts}) = P(E_1 + E_2 + E_3)$$
$$= \frac{1}{52} + \frac{4}{52} + \frac{1}{52}$$
$$= \frac{6}{52}$$
$$= \frac{3}{26}$$

Non-Mutually Exclusive Events

If a pack of playing cards is cut once, the events of drawing a jack and drawing a Diamond are not mutually exclusive events because the Jack of Diamonds can be cut. If $P(E_1)$ is the probability that an event E_1 will occur and if $P(E_2)$ is the probability that an event E_2 will occur then the probability that either E_1 or E_2 or both E_1 and E_2 will occur is

$$P(\text{Jack of Diamonds}) = P(E_1) + P(E_2) - P(E_1 E_2)$$

Example 10

If E_1 is the event of drawing a jack and E_2 is the event of drawing a Diamond find the probability of drawing a jack or a Diamond or both in a single cut of a pack of playing cards.

P(jack or diamond)
$$= P(E_1) + P(E_2) - P(E_1E_2)$$
$$= \frac{4}{52} + \frac{13}{52} - \frac{4}{52} \times \frac{13}{52}$$
$$= \frac{4}{52} + \frac{13}{52} - \frac{1}{52}$$
$$= \frac{16}{52}$$
$$= \frac{4}{13}$$

In the case of three mutually exclusive events:

$$P = P(E_1) + P(E_2) + P(E_3) - P(E_1E_2)$$
$$- P(E_1E_3) - P(E_2E_3) + P(E_1E_2E_3)$$

Note that:

$$P = 1 - P(\overline{E}_1\overline{E}_2\overline{E}_3)$$

where $P(\overline{E}_1) =$ the probability of E_1 not occurring $= 1 - P(E_1)$ and similarly for the other events.

The difference between mutually exclusive events and non-mutually exclusive events may be illustrated by drawing Venn diagrams as shown in Fig. 42.3.

Fig. 42.3

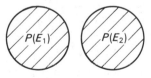

Mutually exclusive events
Shaded area gives
$P(E_1 + E_2) = P(E_1) + P(E_2)$

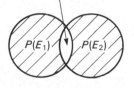

Non-mutually exclusive events
Shaded area gives
$P(E_1 + E_2) = P(E_1) + P(E_2) - P(E_1E_2)$

Example 11

Three people A, B and C work independently at solving a crossword puzzle. The probability that A will solve the problem is 2/3, the probability that B will solve it is 3/4 and the probability that C will solve it is 4/5. Determine the probability that the puzzle will be solved.

We have $P(E_1) = \dfrac{2}{3}$, $P(E_2) = \dfrac{3}{4}$ and $P(E_3) = \dfrac{4}{5}$.

Hence

$$P(\overline{E}_1) = 1 - P(E_1)$$
$$= 1 - \frac{2}{3}$$
$$= \frac{1}{3}$$
$$P(\overline{E}_2) = 1 - P(E_2)$$
$$= 1 - \frac{3}{4}$$
$$= \frac{1}{4}$$
$$P(\overline{E}_3) = 1 - P(E_3)$$
$$= 1 - \frac{4}{5}$$
$$= \frac{1}{5}$$
$$P = 1 - P(\overline{E}_1\overline{E}_2\overline{E}_3)$$
$$= 1 - \frac{1}{3} \times \frac{1}{4} \times \frac{1}{5}$$
$$= 1 - \frac{1}{60}$$
$$= \frac{59}{60}$$

Probability Tree

Suppose that we toss a coin three times. What are the various possibilities and what are their respective probabilities? One way of finding out is to draw a probability tree.

On the first toss the coin can show either a head or a tail. The probability of a head is $\frac{1}{2}$ and the probability of a tail is also $\frac{1}{2}$. Showing possible heads by a full line and possible tails by a dashed line we can draw Fig. 43.4(a).

Fig. 43.4

(a)

(b)

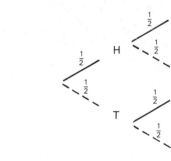

(c)

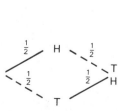

(d)

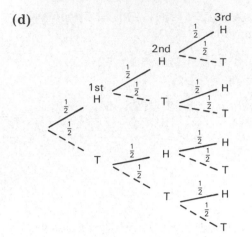

On the second toss, for each of the branches in Fig. 43.4(a) we may obtain either a head or a tail. Hence from each of the branches in diagram (a) we draw two more branches as shown in diagram (b). Diagram (b) tells us that the probability of a head occurring in both tosses is

$$P(\text{HH}) = \frac{1}{2} \times \frac{1}{2}$$

$$= \frac{1}{4}$$

One head may be obtained in two ways, i.e. a head followed by a tail or a tail followed by a head as shown in Fig. 43.4(c). Hence

$$P(\text{one head}) = \frac{1}{2} \times \frac{1}{2} + \frac{1}{2} \times \frac{1}{2}$$

$$= \frac{1}{4} + \frac{1}{4}$$

$$= \frac{1}{2}$$

Carrying on the same way the tree diagram is completed for the three tosses of the coin as shown in Fig. 43.4(d).

Example 12

Using the tree diagram (Fig. 43.4):

(a) write down all the possibilities that can occur when the coin is tossed three times

(b) calculate the probability of three heads occurring

(c) calculate the probability that only one head will appear in the three tosses

(d) calculate the probability of three tails occurring.

(a) The possibilities are as shown in Fig. 43.5.

Fig. 43.5

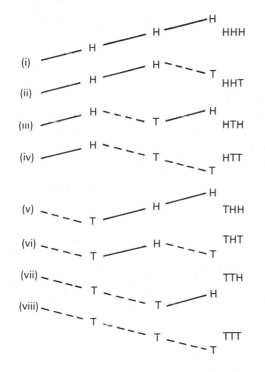

(b) Using branch (i) of Fig. 43.5 gives us

$$P(\text{HHH}) = P(\text{three heads})$$
$$= \frac{1}{2} \times \frac{1}{2} \times \frac{1}{2}$$
$$= \frac{1}{8}$$

(c) Using branches (iv), (vi) and (vii) of Fig. 43.5 gives us

$$P(\text{one head}) = \frac{1}{2} \times \frac{1}{2} \times \frac{1}{2} + \frac{1}{2} \times \frac{1}{2}$$
$$\times \frac{1}{2} + \frac{1}{2} \times \frac{1}{2} \times \frac{1}{2}$$
$$= \frac{3}{8}$$

(d) Using branch (viii) of Fig. 43.5 gives us

$$P(\text{TTT}) = P(\text{three tails})$$
$$= \frac{1}{2} \times \frac{1}{2} \times \frac{1}{2}$$
$$= \frac{1}{8}$$

Example 13

A box contains 4 black and 6 red balls. A ball is drawn from the box and it is not replaced. A second ball is then drawn. Find the probabilities of:

(a) red then black being drawn

(b) black then red being drawn

(c) red then red being drawn

(d) black then black being drawn.

The probability tree and the probabilities are shown in Fig. 43.6.

Fig. 43.6

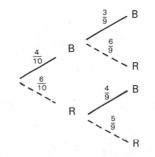

(a) $P(RB) = \dfrac{6}{10} \times \dfrac{4}{9}$

$= \dfrac{4}{15}$

(b) $P(BR) = \dfrac{4}{10} \times \dfrac{6}{9}$

$= \dfrac{4}{15}$

(c) $P(RR) = \dfrac{6}{10} \times \dfrac{5}{9}$

$= \dfrac{1}{3}$

(d) $P(BB) = \dfrac{4}{10} \times \dfrac{3}{9}$

$= \dfrac{2}{15}$

Repeated Trials

If in a single trial, p is the probability of success and q is the probability of failure, then $p + q = 1$, since the trial must result in either success or failure.

If there are n trials then the successive terms of $(q + p)^n$ give the probabilities of $0, 1, 2, \ldots, n$ successes.

Pascal's triangle, as shown below, may be used to expand expressions of the type $(q + p)^n$.

Each number in the triangle is obtained from the line immediately above by adding together the two numbers which lie on either side of it. Thus in the line $n = 6$, the number 20 is obtained by adding together the numbers 10 and 10 in the line $n = 5$. Also in the line $n = 8$, the number 28 is obtained by adding the numbers 21 and 7 in the line $n = 7$.

The expansion of $(q + p)^n$ is easily obtained if we remember that the powers of q decrease from q^n (which is always the first term) and that powers of p increase to p^n (which is always the last term). Thus

$$(q + p)^4 = q^4 + 4q^3p + 6q^2p^2 + 4qp^3 + p^4$$

$$(q + p)^6 = q^6 + 6q^5p + 15q^4p^2 + 20q^3p^3$$
$$+ 15q^2p^4 + 6qp^5 + p^6$$

Example 14

A die is rolled four times. Calculate the probabilities of obtaining no sixes, 1 six, 2 sixes, 3 sixes and 4 sixes in the four rolls.

Let the event of rolling a six in a single roll be regarded as a success. Then

$$p = \frac{1}{6}$$

and $$q = 1 - \frac{1}{6}$$

$$= \frac{5}{6}.$$

Value of n		Numerical coefficients of $(q + p)^n$								
						1				
1					1		1			
2				1		2		1		
3			1		3		3		1	
4		1		4		6		4		1
5	1		5		10		10		5		1
6	1		6		15		20		15		6		1
7	1		7		21		35		35		21		7		1
8	1		8		28		56		70		56		28		8		1

The expansion of $(q + p)^4$ is

$$(q + p)^4 = q^4 + 4q^3p + 6q^2p^2$$
$$+ 4qp^3 + p^4$$

Number of sixes obtained	Probability
0	$q^4 = \left(\dfrac{5}{6}\right)^4$ $= \dfrac{625}{1296}$
1	$4q^3p = 4 \times \left(\dfrac{5}{6}\right)^3 \times \dfrac{1}{6}$ $= \dfrac{500}{1296}$
2	$6q^2p^2 = 6 \times \left(\dfrac{5}{6}\right)^2 \times \left(\dfrac{1}{6}\right)^2$ $= \dfrac{150}{1296}$
3	$4qp^3 = 4 \times \dfrac{5}{6} \times \left(\dfrac{1}{6}\right)^3$ $= \dfrac{20}{1296}$
4	$p^4 = \left(\dfrac{1}{6}\right)^4$ $= \dfrac{1}{1296}$

Note that the total probability covering all possible events is

$$\frac{625}{1296} + \frac{500}{1296} + \frac{150}{1296} + \frac{20}{1296} + \frac{1}{1296} = 1$$

Exercise 43.2

1. A card is cut from a pack of playing cards. Determine the probability that it will be a jack or the Queen of Diamonds.

2. A coin is tossed and a die is rolled. Calculate the probabilities of:
   (a) a head and a six
   (b) a tail and an odd number.

3. A box contains 4 blue counters and 6 red ones. A counter is drawn from the box and then replaced. A second counter is then drawn. Find the probabilities that:
   (a) both counters will be blue
   (b) both counters will be red
   (c) one counter will be blue and the other red.

4. Out of 20 components 3 are defective. Two components are chosen at random from these 20 components. Determine the probability that they will both be defective.

5. A loaded die shows scores with the following probabilities:

Score	1	2	3	4	5	6
Probability	0.15	0.17	0.18	0.13	0.16	0.21

   (a) If I throw it once what is the probability of a score less than 3?
   (b) If I throw it twice what is the probability of (i) a 4 followed by a 6, (ii) a 4 and a 6 in any order?

6. From a shuffled pack of cards, two cards are dealt. Find the probability of
   (a) the first card being a king
   (b) the second card being a king if it is known that the first card was a king.

7. A, B and C are points on a toy train system. The probability of going straight on at each point is $\frac{2}{3}$ (see Fig. 43.7). Find the probability that

Fig. 43.7

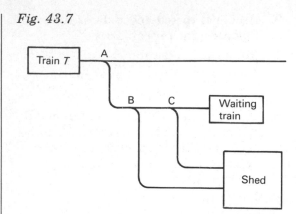

(a) the train *T* hits the waiting train

(b) the train *T* goes into the shed.

8. *Event*: spin the pointer (Fig. 43.8).

(a) Find the probability that the pointer stops in B section.

(b) Find the probability that the pointer stops in R or G sections.

(c) The results of two successive spins are noted. Find:

(i) the probability that the first spin is in R and the second spin is in G

(ii) the probability that the pointer stops in either R then G or G then R.

Fig. 43.8

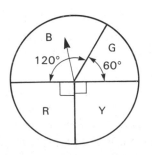

9. (a) If you cast an unbiased die what is the probability of getting a three?

(b) If you cast two such dice what is the probability of throwing (i) two threes, (ii) a three and a four?

(c) Write down all the various ways in which a total of seven can be obtained with two dice. Use this information to calculate the probability of throwing a total of seven with two dice.

10. Two ordinary dice, one coloured red and the other blue, are thrown at the same time.

(a) What is the probability that the number on the red die will be 4?

(b) What is the probability that the number on the blue die will be even?

(c) Copy and complete the table below which shows the total scores on the two dice.

Number on red die

		1	2	3	4	5	6
	1	2	3	4	5	6	7
	2	3					8
Number on blue die	3	4					9
	4	5					10
	5	6					11
	6	7	8	9	10	11	12

(d) What is the probability of getting a total score of (i) 5, (ii) at least 9?

11. The letters of the word STATISTICS are written, one letter on each of ten cards. The cards are placed face downwards on a table and shuffled, and one card is then turned face uppermost. What is the probability that it is

    (a) a letter A

    (b) a letter S

    (c) not a vowel (i.e. not A or I)?

    The experiment is repeated but two cards are turned over together. What is the probability that

    (d) both cards will be T's

    (e) one, at least, is a vowel?

12. A bag contains 7 red counters and 5 white counters. The counters are taken out in succession and not replaced.

    (a) Copy and complete the tree diagram (Fig. 43.9) by writing in the correct fractions for boxes A, B, C and D.

    (b) What is the probability of two red counters being taken out?

Fig. 43.9

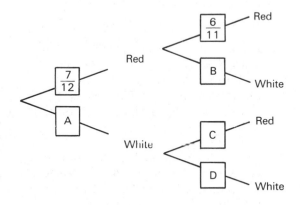

13. A box contains 3 red and 4 black balls. Draw a probability tree to show the probabilities of drawing one ball, then a second and finally a third, without replacement. From the tree answer the following questions:

    (a) What is the probability of red, black, red?

    (b) What are the chances of drawing red, red, black?

    (c) What is the probability of drawing black, red, black?

    (d) What is the probability of drawing only one black ball?

    (e) What is the chance of drawing three black balls?

14. Fig. 43.10 shows a circle divided into five equal sectors which are numbered 1–5. A pointer attached to the centre of the circle is free to spin. A trial consists of spinning the pointer twice. The result of a trial will be $(2, 3)$ if the pointer stops at 2 on the first spin and on 3 on the second spin. The total for the trial is found by adding the two numbers together. Thus for the trial described the total is $2 + 3 = 5$.

    (a) Show all the possible results of the trials.

    (b) What is the probability of obtaining each of the following totals: $1, 2, 3, 6, 10$.

Fig. 43.10

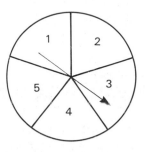

    (c) How many possible different results of trials would there be if the circle was divided into
    (i) two equal sectors numbered 1 and 2
    (ii) three equal sectors numbered 1, 2 and 3
    (iii) four equal sectors numbered 1, 2, 3 and 4?

15. Using Pascal's triangle obtain the expansion of the following:

    (a) $(q + p)^5$ (b) $(q + p)^7$

    (c) $(q + p)^9$

16. A coin is tossed four times. Calculate the probabilities of obtaining the following:

    (a) no heads   (b) one head

    (c) two heads   (d) three heads

    (e) four heads

17. A pack of cards is cut twice. Calculate the probability that

    (a) two aces will be obtained

    (b) two court cards will be obtained. (A court card is a king, queen or jack.)

18. 5% of the fuses produced by a firm are known to be defective. A sample of four fuses is taken at random from a large batch of fuses. Calculate the probability that the sample will contain

    (a) no defective fuses

    (b) two defective fuses

    (c) more than 1 defective fuse.

    (Hint: Take $p = 5/100$ and $n = 4$ and use the method of repeated trials.)

Multi-Choice Questions 43

1. A box contains 5 red balls and three green balls. A ball is picked at random from the box. What is the probability that it is green?

   A 0.25   B 0.375

   C 0.6   D 0.625

2. If two dice are thrown together, what is the probability of getting a total of exactly 5?

   A $\frac{5}{36}$ B $\frac{1}{18}$ C $\frac{1}{9}$ D $\frac{1}{6}$

3. If a fair coin is tossed and then a card dealt from a pack of 52 playing cards, what is the probability of a tail and a queen resulting?

   A $\frac{2}{53}$ B $\frac{1}{26}$ C $\frac{1}{2}$ D $\frac{15}{26}$

4. A coin is tossed twice and a head results each time. If the coin is tossed a third time, what is the probability that a head will again result?

   A 1   B $\frac{1}{2}$ C $\frac{1}{4}$ D $\frac{1}{8}$

5. A bag contains 5 red balls and 3 white balls. A ball is taken at random twice from the bag and replaced each time. What is the probability that each ball is white?

   A $\frac{3}{28}$ B $\frac{9}{64}$ C $\frac{5}{28}$ D $\frac{3}{4}$

6. Which of the following cannot be the probability of an event occurring?

   A $\frac{1}{2}$ B 0.7   C 1   D 2

7. A bag contains 30 balls of which 6 are red, 8 are blue and 9 are black. Each time a ball is drawn from the bag its colour is recorded and the ball is then replaced. The probability that the fourth ball drawn is black is

   A $\frac{1}{15}$ B $\frac{1}{5}$ C $\frac{3}{10}$ D $\frac{7}{15}$

8. A bag contains 5 red balls and 7 blue balls. Another bag contains 6 red balls and 9 blue balls. If a ball is drawn at random from each bag, what is the probability that each ball will be red?

   A $\frac{1}{6}$ B $\frac{11}{27}$ C $\frac{11}{16}$ D $\frac{49}{60}$

9. The following are the probabilities of certain events happening. There must be an error in one of the probabilities. Which?

   A −0.5   B 0   C 0.6   D 1

10. If p is a probability one of the following is necessarily true. Which?

    A p is negative

    B p is positive

    C p is greater than 1

    D p lies between 0.5 and 1

44 Coursework

Coursework tasks which are separately assessed at the examination centre may be set by the Examiner or the choice may be left to the teaching staff and the individual candidate. The content may or may not be directly related to the main core topics and the details connected with this aspect of the examination must be read through carefully in the chosen syllabus.

You will be asked to work on the various coursework assignments during the two years (or one year) before the final examinations. Coursework may include mathematical investigations as well as practical work, historical research, problem solving or perhaps a real life application of mathematics based on statistical surveys carried out locally.

A written report on each coursework study will be required to explain how the various results were obtained and in some cases the candidate may be asked for a verbal explanation to clarify some particular aspect of the work. The presentation should include all the relevant calculations and observations with diagrams, tables, graphs and constructions where necessary. The final package should contain sufficient explanation and detail for any outside reader to understand the development and conclusion. The percentage of marks allocated to coursework varies from one Examining Group to another, and an assessment will be made according to instructions received by the centre. However, the following points will certainly receive consideration:

(1) The candidate's comprehension of the task
(2) The clarity of the plan of attack
(3) Recognition and reaction to results at various stages
(4) Mathematical accuracy where appropriate
(5) Employment of equipment, graphics and suitable materials
(6) Final evaluation and presentation

An investigation begins with some facts and ideas concerning a given situation. The way in which the clues are applied is then left to the individual who will decide which particular direction the development should take. Two students may well follow quite different lines of enquiry and could arrive at different — but equally valid — conclusions for a given set of information. Even if no final solution is obtained, the reasoning processes involved are always valuable and should receive recognition in the final presentation. Topics requiring such an approach are included under the title 'Investigations'.

The syllabus may require some form of practical work. This could involve the construction of models or a sequence of practical tasks not normally associated with everyday mathematics lessons. Coursework may include investigations of real life situations; statistical surveys could be carried out, or you may be asked to undertake some historical research involving mathematical ideas and concepts. This element of the work is covered in the 'Further Studies' section.

Mathematical problem solving requiring a logical approach or even some trial-and-error tactics in search of a finite solution may also be included in certain coursework assignments. Examples of a variety of mathematical puzzles are offered in the 'Problems' part of this chapter.

At various times during the two-year (or one-year) GCSE preparation period you may like to attempt a selection of these different tasks. The investigations are varied and your choice will depend on the particular Examining Group's instructions that you are following. They should provide solid practice and may even stimulate a range of interesting ideas for inclusion in your own coursework.

Investigations

The first two are presented as examples and a logical sequence in pursuit of conclusions is outlined in some detail. Suggestions are made and questions are posed. In addition you may find other useful avenues of enquiry that are not mentioned in the text.

Investigation 1

If you know the game of chess, you will understand the move made by the knight. This diagram demonstrates the move:

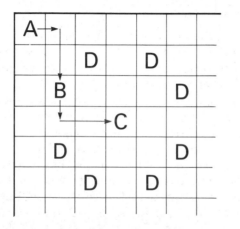

From square A, one square to the right and two more squares downward to B is an example of a knight's move. From B, one square down and then two squares to the right is another knight's move. The move is always a square in one direction followed by two squares at right angles to that direction. From square C, a knight could move to any of the squares marked D, or back to B.

Investigate the knight's move in a grid of nine squares and in other grids of various sizes.

(a) In this situation, the knight has already moved from A to B:

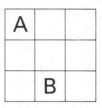

Without moving directly back to A, what is the minimum number of knight's moves needed for a return to square A?

If you are not allowed to follow any move with its inverse, (i.e. A to B cannot be followed by B to A), is it possible to start at A and make an infinite number of knight's moves without returning to A again?

Using the 3×3 grid, is it possible to start at any square and successively land on every square using the knight's move each time?

Solve this puzzle based on the knight's move:

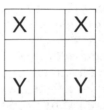

Using the knight's move, interchange the two letter Xs with the two letter Ys. A letter may not end a move sitting in the same square as another letter. Number the squares and record all your moves.

Repeat the operation and see whether you come up with the same solution.

(b) Investigate the knight's move inside these grids:

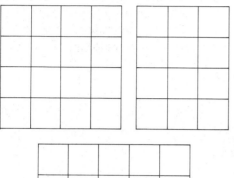

In the following notes, the first shape is used to illustrate how the investigation can proceed.

A	H		J
	K	D	G
E	B	I	
L		F	C

Starting at A, knight's moves have been carried out alphabetically up to the letter L. From A to B is a knight's move; B to C is a knight's move and so on. There is no square available for a knight's move to be made from square L without using the two squares labelled K and I, which have already appeared in the moves. There are four squares vacant.

Is it possible to use all the squares without visiting a square twice? If not, what is the smallest number of squares you can leave vacant?

If you start on different squares, are the results different? Investigate the 4×3 and the 5×4 grids in the same way.

Investigation 2

The diagram shows a rectangle split into squares of different sizes. Two of the squares are shown with sides measuring 9 units and 16 units in length. Make a rough drawing to calculate the size of the other squares and the length and breadth of the rectangle.

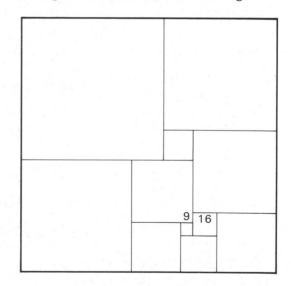

Attempt this further puzzle before proceeding with the investigation.

A rectangle measuring 112 by 75 units is made up of squares of different sizes. A square with a side of 11 units has a common vertex with a square of side 3 units. A square of side 14 units then appears adjacent to those two squares, sharing two of its vertices with them. A square of side 5 units shares one of the other vertices of the square of side 14 units. Another square of side 19 units is then adjacent to those two squares. The isolated side of the 19 square becomes part of the larger side of the rectangle. The remaining squares have sides of 24, 20, 9, 33, 42, 31, 39 and 36. Construct this rectangle of squares.

You may now investigate further rectangles constructed from squares.

It is not very likely that you will be able to produce a rectangle of squares which are all of different sizes, but can you construct one which contains two squares of equal area?

This attempt to produce a rectangle with a set of different squares failed. There is a gap measuring 10 × 2 on one side of the rectangle.

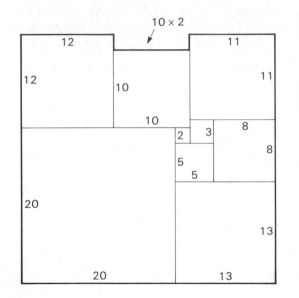

Some attempts by a group of students resulted in an extra area attached to one side of the rectangle.

Investigate the situation in a similar way.

What is the nearest approximation to a rectangle you can construct using squares whose sides are all different?

The remainder of the investigations do not include such a comprehensive range of suggestions. Some assistance is offered, but the depth and success of each enquiry will depend upon the extent to which the given facts are questioned and examined.

Investigation 3

This investigation concerning road junctions is in three parts.

(a) The towns of Alpha, Bravo, Charlie and Delta are situated such that no three lie in a straight line. The Highways Department is short of money and will economise by laying only three roads. A road connects two towns and no town must be isolated. A town may, of course, be situated at the end of more than one road.

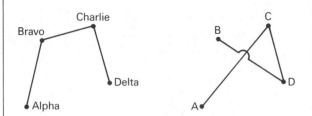

The two examples shown above were submitted to the Department. The second one shows that bridges may be necessary in some cases. How many different drawings can you submit?

(b) Five other towns, India, Juliet, Kilo, Lima and Mike, present a similar problem. No three of them are in a straight line and each town must be at the end of at least one road. Four roads are needed altogether.

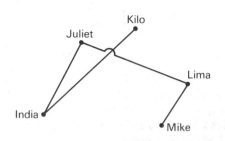

The above drawing which was submitted as an example shows again that bridges will be constructed wherever roads cross each other. Submit as many solutions as possible.

(c) A local authority called Triangle has a different problem. They already have three roads on the outside of the district, which accounts for its name.

The local architect says 'Please notice that we have two roads meeting at each point. We could have three roads meeting at each point if they were laid as I have shown in my second drawing.'

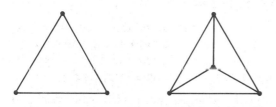

'However,' he continued, 'we would like to have four roads meeting at each point. You must retain the triangular shape of the district with the three outside roads, but you may do as you wish on the inside. You can have as many points as you like and the roads need not be straight.' Can you supply such a network?

The architect did not ask for five roads meeting at each point, but this is possible. Can you draw such a network in case he does ask for it?

Investigation 4

Many years ago mathematicians invented some simple games for two players based on the picking up of coins or counters from various piles or arrangements. Three of these games are described in the following notes.

First game:
Eleven counters are laid out in a line:

In turn, players may pick up 1, 2 or 3 counters at one time. The one who is left to pick up the last counter loses the game.

Second game:
This is the original game of **Nim**. Counters are stacked in heaps: there can be any number of heaps and a heap may contain any number of counters.

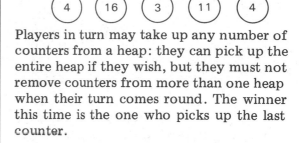

Players in turn may take up any number of counters from a heap: they can pick up the entire heap if they wish, but they must not remove counters from more than one heap when their turn comes round. The winner this time is the one who picks up the last counter.

Third game:
This is a variation of Nim, called **Kayles**. Twelve counters are needed and they are arranged with eleven touching each other and one standing alone as shown below:

In turn, the players may remove at one time any one counter. They are allowed to remove two provided these counters are touching each other. Counters must not be moved except when picked up. The player who picks up the last counter is the winner.

Investigate each game. It may help to work with another player.

In any of these games, does it make a difference whether a player is first or second to pick up counters?

Can you work out any winning moves in the first game?

In the second game, you may like to investigate these arrangements:

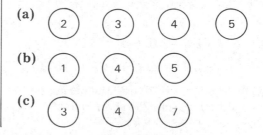

If you cannot form any conclusions about the third game, look at the position in which the first player removes the fourth counter in the line of eleven to leave 1, 3, 7.

You might also examine the situation when 2, 3, 4 are left.

Investigation 5

This investigation will involve the perimeter of a triangle. It may therefore be more convenient — though not essential — to begin with a triangle whose sides are whole numbers.

The sides in this example are $AC = 5$ cm, $AB = 8$ cm and $BC = 7$ cm. The perimeter is therefore 20 cm.

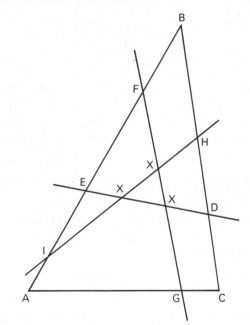

Point D is 2 cm from C and E is 3 cm from point A.

$$DB + BE = 5 + 5 = 10 \text{ cm}$$

$$DC + CA + AE = 2 + 5 + 3 = 10 \text{ cm}$$

Line ED was drawn to divide the perimeter into two equal parts. We shall call it a **bisector**.

Point F was chosen so that it is 2 cm from B. Point G was positioned 1 cm from C.

$$FA + AG = 6 + 4 = 10 \text{ cm}$$

$$FB + BC + CG = 2 + 7 + 1 = 10 \text{ cm}$$

Line FG is therefore also a bisector.

Similarly, the line HI is drawn as a bisector.

Where two bisectors intersect, we find a point marked X.

Investigate the points X by constructing several more bisectors passing through the given triangle and form an opinion regarding the region containing all points X.

Repeat the experiment with other triangles.

Is it possible for three bisectors to pass through any single point X?

Can you construct a triangle in which some of the points X exist externally to the triangle?

Are there any special triangles whose properties result in the points X occupying significant positions?

Investigation 6

An old game of patience called Mousetrap was played by one player using a set of cards bearing consecutive numbers. The player decides how many cards to use each time the game is played and the cards are then laid out in any order, face upwards in a circle formation.

Working clockwise from the top, the player touches the cards one at a time, counting 1, 2, 3, etc. If the spoken number is the same as the number on the card that is touched, the card counts as a **hit** and is picked up.

Touching cards continues at the next card but counting restarts at 1. If only one card remains, it is picked up immediately. The player aims to pick up all the cards or as many as possible.

This game starts with nine cards numbered 1 to 9 but laid out in random order from the 7 at the top.

Start at the top:

Say 1, touch 7; say 2, touch 5;
say 3, touch 8; say 4, touch 4: Hit.
Pick up 4.
Say 1, touch 6; say 2, touch 2: Hit.
Pick up 2.
Say 1, touch 3; say 2, touch 9;
say 3, touch 1; say 4, touch 7;
say 5, touch 5: Hit.
Pick up 5.

The sequence continues. Decide whether this particular arrangement allows you to pick up all the cards.

Now you can investigate the game of Mousetrap.

Begin with a set of four cards. If you test the arrangement shown below you should find that it provides a winning sequence.

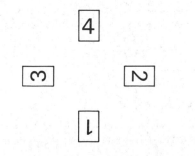

How many arrangements are possible?

How many would provide 4 hits?

In how many ways could they be arranged so that no cards can be picked up at all?

Investigate the pattern of four cards further.

Examine other arrangements. Can you discover a method of laying down any number of cards in a winning sequence?

Investigation 7

Sixteen bus stops are served by four different services. The single fare between two stops is shown in pence. The district has factories, offices, shops and housing estates.

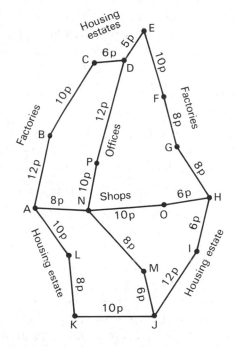

Bus routes:

Northern Circle ABCDEFGHONA both ways
Southern Circle ANOHIJKLA both ways
Centre Line EDPNMJ and return
Mini Line ANOH and return

Mr Carter lives at K and works in a factory at F, five days a week.

Mrs Arnold lives at J and does not work. Her mother, Mrs Jones, a wheelchair-bound invalid of 70 years, lives at E.

Mr Brent, an office worker, lives at I. His son, Peter, who lives with him, works in a factory at C: they both work five days a week.

Mrs Downton is a shop worker at O, six days a week and lives at D. Her colleague at work, Miss Savage, lives at M.

Special rates are available from the bus company as follows:

> Daily Wander card, £1.50: unlimited travel on any route

> Weekly Saver card, £4.50: one route; 7 return journeys

> Monthly Rover card, £25: unlimited travel on any route

> Annual Season card, £250: unlimited travel on any route

> Senior Citizens and Disabled Book, £5: 40 return journeys

Investigate the economy offered by the special rates.

Examine the possibilities for each person and give reasons why a person may or may not decide to use the discount offers.

Comment on the value of each special offer.

Suggest alterations to the bus routes.

Investigation 8

Before trying this exercise you need to be familiar with the 'Four Colour Theorem' which states that not more than four colours are ever needed to colour a network of regions, such that no two regions of the same colour share a common boundary. The outside of the network beyond all other boundaries is also classified as a region and must be coloured, following the same boundary rules.

This network requires the maximum of four colours which are shown in the diagram as A, B, C and D:

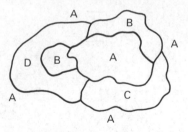

Map-makers have known about the Four Colour Theorem for a long time, but it was not recorded mathematically until A.F. Möbius described it in 1840.

You might like to try to discover the minimum number of colours required in the network shown in the next diagram.

Remember, regions sharing a common boundary must have a different colour.

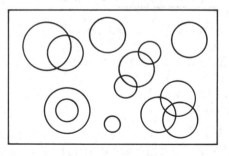

The theorem can now be related to faces of solids. Faces that share a common edge must have different colours.

Tetrahedron

Octahedron

Cube

Dodecahedron

You will find it necessary to use the nets or even construct the solids.

What is the minimum number of colours in each case for colouring the faces so that no two touching faces have the same colour?

Does the theorem still apply to solids?

If each face of a tetrahedron is divided into four equilateral triangles, can the sixteen regions be coloured with four colours or fewer, so that no two touching regions have the same colour?

If each of the faces of a cube is bisected by a diagonal, is the colouring of the cube affected in any way?

Investigation 9

This investigation concerns the behaviour of fractions and mixed numbers.

(a) If the numbers $1\frac{3}{5}$ and $2\frac{2}{3}$ are added, the answer is $4\frac{4}{15}$.

If the same numbers are multiplied, the answer is again $4\frac{4}{15}$.

Were those numbers specially chosen? Can you find other pairs which behave in the same way? Is there a general rule?

(b) The numbers $3\frac{1}{5}$ and $\frac{4}{5}$ not only give a total of 4, but the first divided by the second also gives an answer of 4.

Is there a pattern that would pinpoint other pairs of numbers displaying the same property?

(c) The difference between $\frac{1}{5}$ and $\frac{1}{4}$ is $\frac{1}{20}$. When $\frac{1}{5}$ and $\frac{1}{4}$ are multiplied, the answer is again $\frac{1}{20}$. Investigate this property.

(d) The numbers 6 and $\frac{6}{7}$ behave in the same way as the two fractions in the last statement. Their difference and their product are the same. Do they belong to the same pattern?

Investigation 10

Thirteen stations on a rail network are served by three circular routes each of which is a single track only. Trains travel clockwise only and are unable to overtake or arrive at a station together.

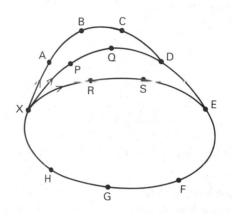

Outer route: XABCDEFGHX
Centre route: XPQDEFGHX
Inner route: XRSEFGHX

All trains travel at the same constant speed and the travelling time between any two consecutive stations is 10 minutes. The first train to set off each day on each route starts at X.

Investigate possible timetables for the three routes, ignoring time spent in a station. Apart from stopping at a station, no train must be brought to a standstill.

What happens if only one train is used on each route and they start together at X?

What is the regularity of service for each station?

Can the service be maintained non-stop from 6 a.m. to 10 p.m. each day?

Consider two or three trains on each route.

Further Studies

Further Study 1

We all have the responsibility of under-standing the safety factor built into electri-cal circuits and we need to appreciate the running costs of various pieces of equip-ment.

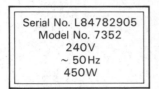

```
Serial No. L84782905
  Model No. 7352
       240V
      ~ 50Hz
       450W
```

The two important figures here are 240 V and 450 W.

It is an easy matter for us to decide on the fuse required to protect any particular item of equipment and the electrical circuitry.

$$\text{Amps} = \frac{\text{Watts}}{\text{Volts}}$$

In other words, the power of the appliance measured in watts is divided by the mains voltage.

Example 1

A lighting circuit has four 100 W bulbs and eight 60 W bulbs.

$$\text{Power} = (4 \times 100) + (8 \times 60)$$

$$= 880 \text{ watts}$$

$$\text{Amps} = \frac{880}{240}$$

$$= 3.67 \text{ amps}$$

It is necessary to use a fuse larger than 3.67 amps in this circuit. A 5 amp fuse is included in lighting circuits.

Fuses of various types — 2 amp, 3 amp, 5 amp and 13 amp — can be obtained at electrical stores.

Investigate these appliances:

electric iron   cooker ring
video recorder   freezer
television set   electric typewriter
immersion heater   microwave oven
electric drill   two-bar electric fire
music centre   cooker oven
home computer   electric kettle
lawn mower

If the power rating is not printed on the equipment, make enquiries at your local electrical dealers. Power in kilowatts must be multiplied by 1000.

You could also calculate the running cost of each appliance.

$$\text{Time} = \frac{1000}{\text{Watts}}$$

The time in hours for which a piece of equipment can be operated on one unit of electricity is found by dividing 1000 by the rating in watts. Assume the cost of 1 unit to be 6 pence.

Example 2

A 100 watt bulb:

$$\text{Time} = \frac{1000}{100}$$

$$= 10 \text{ hours}$$

A 100 watt bulb will run for 10 hours at a cost of 6 pence.

Example 3

An electric fire rated at 2500 watts:

$$\text{Time} = \frac{1000}{2500}$$

$$= 0.4 \text{ hours}$$

$$= 24 \text{ minutes}$$

In 24 minutes, the fire would already have a running cost of 6 pence.

Carry out similar calculations for other appliances.

Calculate the running costs per hour. Comment on your results.

Further Study 2

If you have constructed the five Platonic solids you may like to attempt the two models outlined in the following notes.

The name of the first solid is a cuboctahedron. Its net is formed of six squares and eight equilateral triangles:

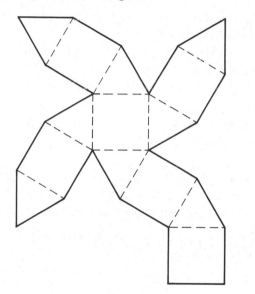

A convenient length for each line on the net would be 3 cm. Tabs will be required if liquid glue is used in the assembly of this solid.

The second solid is called a rhombicuboctahedron.

There are 18 squares and 8 equilateral triangles here. The three axes of symmetry that pass through A, B and C indicate how the squares will be located in the net. The triangles can then be attached in pairs to four of the squares.

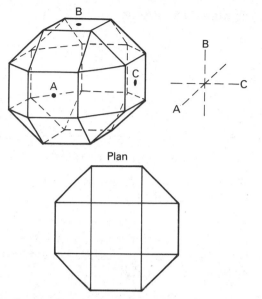

Plan

Work out a net for constructing this solid.

Further Study 3

It is possible to obtain reasonably accurate values for the heights of buildings and other structures even without instruments. The following notes outline some methods which may be employed.

(a) The top and bottom of the object — a wall, a fence, etc. — can be sighted using marks on a rod held vertically at arm's length. This is demonstrated in the following diagram:

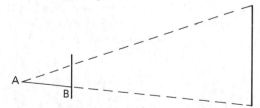

What is point A. What is length AB? How many separate lengths do we need to record?

Use the method to find the height of an outside object.

Study what happens when the arm is bent.

(b) Shadows are very useful:

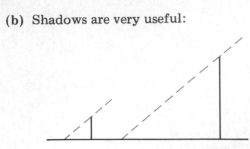

What do the various lines represent?

How could this be used to calculate the height of something?

(c) If access to a building is easy but the distance from the observer to the building cannot be measured, a vertical sighting rod can still be used:

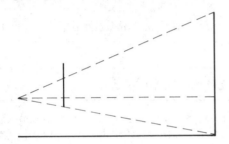

Explain the diagram and use the method to measure a height near your centre.

Further Study 4

More accurate surveys can be carried out with simple home-made instruments which can be constructed from wood, metal or even stiff cardboard.

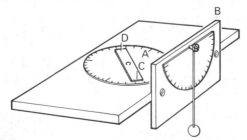

This is a rough sketch of two instruments which will allow angles to be measured on a horizontal and a vertical plane. Sights must be designed for AB and also for CD.

Explain the kind of survey for which the completed model could be used.

Further Study 5

The name of Pythagoras is probably mentioned more often than any other in mathematics. Who was Pythagoras? When and where did he live? What did he do? Who were the Pythagoreans and how did Pythagoras sub-divide the study of number? What is a Pythagorean triple? Seek out the answers to these and other questions in your local reference library.

The theorem attached to the name of Pythagoras reads:

> In any right-angled triangle, the square on the hypotenuse is equal to the sum of the squares on the other two sides.

If in that statement we replace the word 'square' with 'equilateral triangle', is the statement still true?

Could we also substitute the words 'semi-circle', 'similar triangle', or 'similar parallelogram' without contradicting the theorem? Provide practical proof of your conclusions.

The next diagram shows an elastic band stretched on a pin-board.

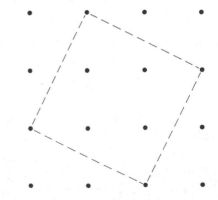

How is this diagram connected with Pythagoras? Produce a drawing to accompany your answer.

Pythagoras and other mathematicians applied themselves to the problem of calculating the area of a circle.

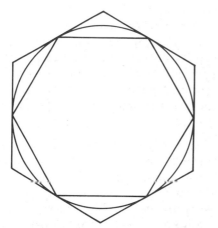

Explain with calculations how this diagram relates to those early experiments dealing with the area of a circle.

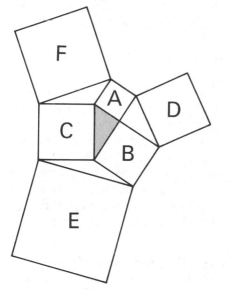

Explain the last diagram and comment on the total area of D, E and F.

Further Study 6

These experiments are better carried out by two people. A large piece of paper is required, together with ten sticks or straight lengths of wire. Lengths of drinking straw may also be used.

Across the width of the paper draw parallel lines 4 cm apart. Cut each of the ten pieces of stick to a length of 3 cm. Hold all ten sticks about 30 cm (the length of a ruler) above the centre of the paper and drop them so that they all fall on the paper.

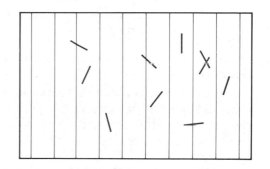

Note the number of sticks that make contact with a line or each other. Carry out the experiment ten times. This is equivalent to having dropped 100 sticks. Calculate the total of sticks which made contact with a line on the paper or touched each other.

Let x represent the total number of sticks dropped (100 in this case). Let y represent the number which contacted lines or other sticks.

In the following formula, A represents the length of a stick and B the distance between the lines on the paper:

$$P = \frac{2Ax}{By}$$

Calculate the value of P correct to two decimal places.

Repeat the experiment with different figures for A and B such that B is always greater than A and A is greater than a half of B.

In some experiments, try using a different value for x.

Comment on your results.

Further Study 7

A **palindrome** is a word that reads the same backwards as forwards: for example, HANNAH, LEVEL, etc. The number 36763 is also palindromic. Read through these notes about palindromic dates.

(a) We may consider 1st January in the year 1 to be the first palindromic date, i.e. 1.1.1. If dates are written in this way, 7.6.7 was also palindromic.

(b) It is usual, however, to write 01 when abbreviating a year such as 1901. In that case, the first palindromic year would have been 10th January in the year 1, which we might have written 10.1.01. Another of similar kind was 30.5.03.

(c) If we introduce a zero into the years, we probably should do the same with the day and month: in this system, all dates will have six figures. The first palindromic date would then have been 10.11.01.

(d) If we are not restricted to 6-figure dates, we may include 15.9.51 as a palindrome.

(e) 29th February is interesting when searching for palindromic dates.

(f) If the whole year is used, we can find other palindromes, such as 15.6.1651 (i.e. 1561651), 13.01.1031, etc.

Investigate the incidence of all the different kinds of palindromic dates and supply lists of the various types.

Further Study 8

Take a look into the future by examining some present-day statistics. Figures will be required from various reference books.

(a) Relate population figures to space available on the Earth. Look up the different population statistics and ground areas for a selection of the world's nations and compare the density of population of the UK with other countries such as Brazil, the Netherlands and Australia. Compare individual cities in different parts of the world. Calculate the space available per person, or present the figures in graphical form.

(b) In the year 1960, the world population was estimated to be three thousand and three million, or 3 003 000 000. You will find that modern reference books generally use 'billion' to represent 'a thousand million': this statistic might then be quoted as 'just over 3 billion'. In 1987, it was announced that the '5 billionth' human being had just been born and 11 July 1987 was designated 'Baby Five Billion Day' by the United Nations.

(c) In 1960, it was deduced that 4 babies were being born every second in the world and only 2 people were dying every second. Consider the implications of this estimate. Does this rate of birth and death tally with the population figure of 5 000 000 000 in 1987? Has the rate of population growth accelerated or slowed down in recent years? From your results, work out the likely world population figure for the year 2000.

(d) List the quantities and costs of the daily food intake of an average person (yourself?) in the UK. Work out some annual figures and relate these to the situation in other countries — especially in the Third World.

(e) From figures gathered at your own home, estimate the volume of water used daily and weekly in the UK.

(f) Project some figures from the weight of the weekly waste accumulated by a typical UK household.

Further Study 9

A group of 25 people recorded their heights and weights. The statistics for each person were recorded by crosses on a scatter diagram with axes scaled in centimetres and kilograms. When all 25 crosses were correctly placed, some of the group proposed drawing a straight line through the crosses while others suggested that a curve should be drawn. These two proposals are represented by the broken lines in the scatter diagram:

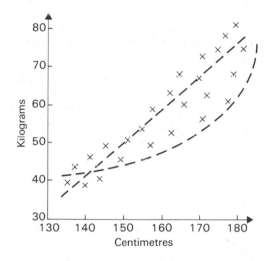

(a) Why did they think that any lines were necessary on the graph? Which proposal do you think is more useful — a straight line or a curve? Construct a similar graph using statistics gathered from a group of at least 25 people. Explain the reason for any line you decide to draw and comment on the need for such a line.

(b) Obtain a set of statistics concerning the amount of television watched by a group of people whose ages range from 10 years to retirement age. Record the age in years and the weekly viewing figures in hours for each member of the group. Produce a scatter diagram of your results. Would a line or curve of any kind be relevant to this situation? If so, how will you decide on its position and what significance can you attach to it?

Further Study 10

The line PQ with a length of 5.5 cm is divided into two parts at R, such at PR is 3.4 cm and RQ is 2.1 cm.

P ————— 3.4 ————— R —— 2.1 —— Q

Ratio PR : RQ = 3.4 : 2.1

= 1 : 1.62 (2 decimal places)

Ratio PQ : PR = 5.5 : 3.4

= 1 : 1.62 (2 decimal places)

This is called a **Golden Section,** and the ratio is the **Golden Ratio.**

The Golden Ratio is identified by the Greek letter phi (ϕ). If expressed to an accuracy of eight decimal places, the Ratio is 1.618 033 99. There are many properties to be discovered about the Golden Ratio.

A **Golden Rectangle** is one whose long and short sides are in the Golden Ratio. If a square is cut from one end, the remaining shape is also a Golden Rectangle:

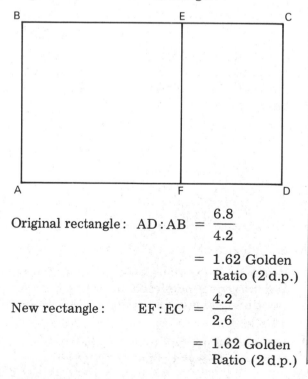

Original rectangle: $AD : AB = \dfrac{6.8}{4.2}$

= 1.62 Golden Ratio (2 d.p.)

New rectangle: $EF : EC = \dfrac{4.2}{2.6}$

= 1.62 Golden Ratio (2 d.p.)

Use a rectangle measuring 11 cm by 6.8 cm to demonstrate the property of the Golden Rectangle and continue the process of slicing off a square from each new rectangle as far as this practical demonstration will allow (your rectangles will eventually become too small to draw accurately).

A **Golden Triangle** is an isosceles triangle in which the Golden Ratio is formed by one of the equal sides and the base.

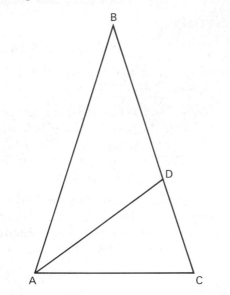

This Golden Triangle has a base of 4.2 cm and equal sides of 6.8 cm. Note the size of the angles in a Golden Triangle. Angle A is bisected by AD.

Original triangle: AB:AC = 6.8:4.2

= 1:1.62

Triangle ADC: AD:DC = 4.2:2.6

= 1:1.62

Thus triangle ADC is also a Golden Triangle.

Examine the triangle ABD in a similar way.

Use an isosceles triangle with a base of 5.5 cm and two equal sides each measuring 8.9 cm. Continue the process to obtain other Golden Triangles.

Now research the Golden Ratio further along the following lines:

(a) Find the ratio of successive pairs of numbers in the Fibonacci series.

(b) Calculate the positive root of the equation $x^2 - x - 1 = 0$.

(c) Work out the following expression on your calculator:

$$\frac{\sqrt{5} + 1}{2}$$

You now have a very accurate value of the Golden Ratio on the calculator display. Find the reciprocal of phi $\left(\frac{1}{\phi}\right)$.

Comment on the result. No other number displays the property you have discovered.

(d) Construct a regular pentagon with sides of length 3 cm. Extend each of the sides as shown in the diagram to produce a pentagram.

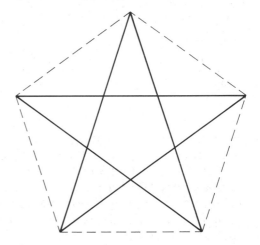

Join the five vertices of the pentagram with straight lines to produce a larger outer pentagon. How many Golden Triangles can you find altogether inside the large pentagon? Calculate the ratio of a side of the outer pentagon to a side of the inner pentagon. What do you notice about this ratio?

(e) Repeat the operation in part (c) to obtain an accurate value for ϕ, which is now required. Work out the square of the Golden Ratio (ϕ^2). What do you notice? If you have already completed part (d), comment on the result in the light of your discovery about the two pentagons. Can you show why this result comes about?

(f) Construct a regular decagon (ten-sided polygon) and draw its circumscribed circle. Calculate the ratio of the circle's radius to the side of the decagon.

(g) Ask several people to draw some free-hand rectangles. Tell them to continue sketching until they find a rectangle whose dimensions appear most pleasing to the eye. Calculate the ratio of length to breadth for the most pleasing rectangles produced.

(h) Find out how the Golden Ratio is linked to various famous paintings and other artistic creations, in particular Salvador Dali's painting *The Sacrament of the Last Supper*.

(i) Can you find any connections between the Golden Ratio and the world of plants and animals?

Problems

Problem 1

(a) Last year our district had only three football teams. Each team played each of the others once only, so there were three matches. The winning team in any match received 3 points, and for a draw each team was awarded 1 point. This is a copy of the league table at the end of last season:

	Played	Goals for	Goals against	Points
Attackers	2	4	3	4
Bounders	2	1	1	3
Crackers	2	3	4	1

The score sheets have unfortunately been lost. Can you supply the scores in the three matches?

(b) This year another team joined the league and once again each team played every other team only once, making a total of 6 matches. The final league table for this season looked like this:

	Played	Goals for	Goals against	Points
Attackers	3	4	1	9
Bounders	3	2	2	4
Crackers	3	1	2	3
Dribblers	3	0	2	1

The Secretary has been very careless and mislaid the scores again! Please supply the scores for the six matches.

Problem 2

(a) The magic wheel shown in the first diagram uses the numbers 1 to 11 with number 1 at the centre. Any line of three numbers produces a total of 14.

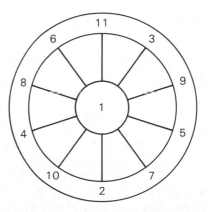

There are two other numbers that could be placed at the centre to produce magic wheels, but the total across the spokes will not be 14. Find these two other number patterns.

(b) The second magic wheel uses the numbers 1 to 21. Number 16 is at the centre and the magic total this time is 59.

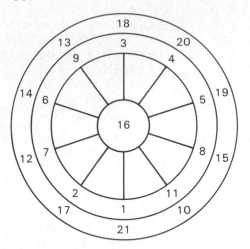

Four numbers other than 16 can be placed at the centre to produce a magic result across the spokes. Here is a clue to help you find the other patterns:

If you add all the numbers from 1 to 21 and subtract 16, the result is divisible by 5.

Find the other two magic wheels.

Explain how the numbers from 1 to 31 might be used to construct another larger magic wheel.

Problem 3

(a) In a knock-out competition, contestants are paired up and the losers are eliminated. The winners are paired up again and the process continues until two people are left to contest the final. What kind of number sequence will be best to ensure that numbers of competitors can be easily matched up when such a contest is held?

If the number is not in that sequence, a player at any time in the competition who has no opponent receives a bye and automatically moves on to the next round.

Is there a general rule to decide the total number of matches that will be played in any knock-out competition?

(b) In one knock-out competition, there are 16 players, numbered 1 to 16. The best player is Number 16, the next best player is Number 15, and so on down to Number 1, who is the worst player. A player always beats another who has a lower number.

The pairing in the first round is done by drawing numbers from a hat.

Would Number 16 always win the contest?

Would Number 15 always reach the final?

How many players have a chance of being in the final?

Problem 4

The 28 dominoes in a set are arranged in a rectangle and a drawing is made. The artist puts in the numbers but forgets to show how the dominoes were placed. However, he does remember where the dominoes 6-6, 5-5, 6-4 and 5-2 had been positioned.

2	3	1	5	3	3	0	4
2	4	4	6	6	4	5	2
6	5	0	0	2	4	3	3
0	5	0	1	0	4	1	5
3	3	6	1	4	6	1	4
1	5	5	2	0	0	1	6
2	1	6	6	3	2	2	5

In the top left-hand corner, for instance, is it the 2-3 domino or the 2-2? Which domino was placed in the top right-hand corner: the 0-4 or the 4-2?

Find the positions of all the rest of the dominoes on the diagram.

Problem 5

We have found a copy of Jake the Peg's treasure map and the instructions needed to locate his treasure. Unfortunately, the map is badly torn with vital landmarks missing. You will have to reconstruct the map for yourself using Jake's scribbled notes.

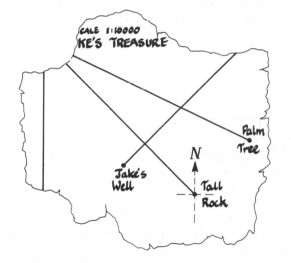

This is what Jake wrote:

(a) The scale be on my map.

(b) The Palm Tree be 300 paces due North-East of Tall Rock. [Jake forgot to say his pace was exactly one metre.]

(c) The Stockade [missing], be located due North-West of Tall Rock and 900 paces from the Palm Tree.

(d) Ye will find Parrot Point [missing] due South of my Stockade and 850 paces from the Palm Tree.

(e) Eagle's Nest [missing] be 400 paces due North of the Palm Tree.

(f) My well be sited at equal distances from the Palm Tree and the Stockade. It be 700 paces from Eagle's Nest.

(g) I buried the treasure this-a-way:
The centre of the path from Palm Tree to Stockade be P.
The centre of the path from Eagle's Nest to my well be Q.
The centre from Parrot Point to Stockade be R.
The centre from Tall Rock to Stockade be S.
Where QR do cross PS be where I buried my treasure.

Now, me hearties, when ye have drawn yon map and found my treasure, ye must answer this question:

When I buried it, how far were I from Tall Rock?

Problem 6

Our local blacksmith was a shrewd character. He put new shoes on one of our horses and each shoe was attached by five nails. He said 'You may have the shoes free of charge — just pay me for the nails. Give me one penny for the first nail, 2p for the second nail, 4p for the third nail and so on, doubling up for each successive nail.' We said 'That's not enough. Please accept £1000.'

Supply some figures to show why the blacksmith refused the £1000.

Problem 7

The same blacksmith was given an interesting task by the landscape gardener. He wanted five regular shapes constructed from lengths of wire: an equilateral triangle, a square, a regular pentagon, a regular hexagon and a regular octagon. Each shape was to have the same perimeter.

The blacksmith's tape measure could only measure to an accuracy of half a centimetre. He made each shape as accurately as possible with the minimum amount of wire. What length was the side in each shape and how much wire did he use altogether?

Problem 8

This arrangement of the figures 1 to 9 is interesting:

$$9 \ 8 \ 1$$
$$6 \ 5 \ 4$$
$$3 \ 2 \ 7$$

The middle row is double the bottom row and the top row is three times the bottom row.

There are three other ways of arranging those figures so that the same conditions apply. How many can you discover?

Problem 9

Six cards are marked with two 1s, two 2s and two 3s:

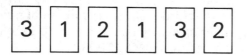

This arrangement forms a very special pattern:

There is   ONE card between the two 1s,

there are TWO cards between the two 2s,

and   THREE cards between the two 3s.

This is your problem:

(a) To those cards are added two 4s. Arrange them in a similar way to include in the pattern FOUR cards between the two 4s.

(b) Add two 5s and also two 6s. (Do not waste time attempting the puzzle with these twelve cards because it is impossible!) Add two 7s. Now the puzzle can be solved using all fourteen cards. There are many solutions, so you should be able to find at least one.

Problem 10

This sum shows eight odd numbers adding to make a total of 20:

$$5 + 5 + 3 + 3 + 1 + 1 + 1 + 1 = 20$$

There are ten other ways of using eight odd numbers to give a total of 20. See if you can find all ten.

Answers

Answers to Chapter 1

Exercise 1.1

1.

```
-10 -8 -6     -2  0   3 4 5  7     10
```

2. (a) ⌐3   (b) −57   (c) +72
3. 7, +3, 2, 0, 5
4. −9, −5, −4, −3, −1
5. +25, 18, 12, 3, −2, −6, −8, −15
6. +5, 7, +11
7. −5 and −9

Exercise 1.2

1. 40	2. 60	3. 90
4. 90	5. 80	6. 700
7. 700	8. 900	9. 400
10. 900	11. 3000	12. 5000
13. 8000	14. 9000	15. 8000

Exercise 1.3

1. 16	2. 18	3. 17
4. 30	5. 12	6. 16
7. 6	8. 36	9. 84
10. 48	11. 140	12. 240

Exercise 1.4

1. 13	2. −9	3. −32
4. −16	5. −4	6. −8
7. −14	8. 4	9. 3
10. −3		

Exercise 1.5

1. 1	2. 1	3. 8
4. 1	5. −6	6. 2
7. 4	8. −5	

Exercise 1.6

1 −42	2. −42	3. 6
4. −15	5. −60	6. 24
7. −40	8. 60	9. −4
10. −3	11. 4	12. −120
13. −3	14. 2	15. −2

Exercise 1.7

1. 19	2. 9	3. 7
4. 27	5. 33	6. 5
7. 23	8. 25	9. 7
10. 2	11. −5	12. −12
13. 56	14. −15	15. −20
16. −2	17. −12	18. −3
19. 25	20. 2	

Exercise 1.8

1. 768	2. 256	3. 0
4. 0	5. 18	6. 0
7. 13	8. −6	9. −32
10. 0	11. −42	12. 33

Exercise 1.9

1. 250, 1250	2. 17, 21	3. 27, 33
4. 22, 11	5. 15, 21	6. −11, −16
7. 21, 34	8. 78, 158	9. 165, 489
10. −162, 486	11. −6, −8	12. 3, 5
13. 6, 8	14. 21, 89	15. 82, 244
16. 245, 2189	17. 9, 1	

Miscellaneous Exercise 1

Section A

1. (a) 9, 5, 1   (b) 16, 8, 0
2. 900 000   3. 45   4. 36
5. 9, −9
6. (a) 40   (b) −13   (c) −4
    (d) −300
7. −5, −3, 0, 3   8. 7866
9. (5 + 2) × (7 − 4), 10
10. 3, 6, 96
11. (a) 44   (b) −26
12. (a) −   (b) −

Section B

1. (a) 21   (b) 20
2. 3
3. (a) 2, −2   (b) 1, 3
4. (a) 2000   (b) 900
5. (a) 1561   (b) 65   (c) 126
6. 0   7. 5   8. −119, −362

Multi-Choice Questions 1

1. B	2. B	3. B
4. D	5. D	6. B
7. A	8. B	9. A

Mental Test 1

1. 23	2. 7	3. 12
4. 10	5. −9	6. 30
7. 6	8. 9	9. −21
10. −13		

Answers to Chapter 2

Exercise 2.1

1. Odd	2. Even	3. Even
4. Odd	5. Odd	6. Even
7. Even	8. Odd	

9. (a) 17, 59, 121, 259   (b) 36, 98, 136
10. 36   11. 125   12. 49
13. −125
14. (a) 64   (b) 8   (c) 81
    (d) 32
15. (a) +4   (b) +12   (c) +13
16. (a) 2   (b) 3   (c) 6
17. 18   18. 64   19. 90
20. 30   21. 100   22. 288
23. 7   24. 2   25. 76

Exercise 2.2

1. 12, 42, 66
2. (a) No   (b) No   (c) Yes
   (d) No   (e) Yes   (f) No
   (g) Yes   (h) Yes
3. (a) No   (b) Yes   (c) Yes
   (d) No   (e) Yes
4. 2, 3, 6, 7   5. 2, 3, 4, 6, 8, 9, 12

Exercise 2.3

1. (a) 1, 3, 5, 15   (b) 1, 2, 4, 8, 16, 32, 64
   (c) 1, 2, 3, 4, 6, 8, 12, 16, 24, 48
   (d) ±1, ±2, ±3, ±4, ±6, ±8, ±12, ±24
   (e) ±1, ±2, ±7, ±14, ±49, ±98
2. 23, 29
3. (a) 24   (b) 27, 33, 45, 49 or 61
   (c) 61   (d) 49   (e) 27 or 45
4. (a) 21   (b) 7
5. (a) $2^3 \times 3$ (b) $2^3 \times 3^2$ (c) 5×3^2
6. (a) 24   (b) 60   (c) 12
   (d) 24   (e) 40   (f) 100
   (g) 160   (h) 120   (i) 420
   (j) 5040
7. 1, 2, 4, 5, 10, 20;
   1, 2, 3, 4, 6, 8, 12, 24;
   HCF = 4

8. 6, 12, 18, 24, 30, 36, 42, 48, 54, 60;
   10, 20, 30, 40, 50, 60;
   common multiples are 30 and 60;
   LCM = 30
9. (a) 4   (b) 12   (c) 5
   (d) 13   (e) 6   (f) 14
10. (a) $2^2 \times 3^2 \times 5$ (b) $2^3 \times 3^3 \times 5 \times 7$
    (c) $3^2 \times 5^2 \times 7$
11. ±1, ±2, ±3, ±4, ±6, ±7, ±12, ±14, ±21, ±28, ±42, ±84
12. 1, 2, 3, 4, 5, 6, 8, 10, 12, 15, 20, 24, 30, 40, 60, 120

Miscellaneous Exercise 2

Section A

1. (a) 55, 60   (b) 60, 63, 81   (c) 60, 122
2. $2^3 \times 3^2 \times 5$
3. (a) 30   (b) 60
4. (a) 1, 2, 3, 6, 7, 14, 21, 42
   (b) 15, 20, 25
5. (a) 61 Yes   (b) 5291 No
6. (a) 57   (b) 72
7. (a) $2^3 \times 5^2$
   (b) 6, 12, 18, 24, 30, 36, 42, 48, 54
   (c) 29, 31, 37

Section B

1. (a) 2, 3, 4, 6, 12   (b) 4, 12, 24
2. (a) $2^3 \times 7$ (b) $2^2 \times 3 \times 11$
3. 13 500
4. (a) 17   (b) 89   (c) 8000
   (d) 16
5. (a) ±1, ±2, ±7, ±14, ±49, ±98
   (b) ±1, ±2, ±3, ±4, ±6, ±8, ±12, ±16, ±24, ±48
   (c) ±1, ±2, ±3, ±4, ±6, ±7, ±9, ±12, ±18, ±21, ±28, ±36, ±42, ±63, ±84, ±126, ±252
6. 1, 2, 3, 4, 6, 9, 12, 18, 27, 36, 54, 108
7. 6, 12, 18, 24, 30, 36, 42, 48, 56
8. 5

Multi-Choice Questions 2

1. D	2. D	3. A
4. B	5. B	6. B
7. D	8. A	9. D
10. A		

Mental Test 2

1. 81	2. −125	3. 16
4. 4	5. 36	6. 1
7. 3^5		

8. ±1, ±2, ±3, ±4, ±6, ±8, ±12, ±16, ±24, ±48
9. 7, 14, 21, 28, 35
10. 2×3^2
11. 24   12. 3

Answers to Chapter 3

Exercise 3.1

1. $4\frac{1}{2}$ 2. $3\frac{2}{3}$ 3. $4\frac{3}{5}$
4. $2\frac{5}{7}$ 5. $4\frac{7}{8}$ 6. $\frac{13}{4}$
7. $\frac{13}{5}$ 8. $\frac{47}{8}$ 9. $\frac{67}{20}$
10. $\frac{41}{8}$ 11. $\frac{2}{5}$ 12. $\frac{3}{4}$
13. $\frac{1}{3}$ 14. $\frac{3}{5}$ 15. $\frac{5}{12}$

Exercise 3.2

1. $\frac{1}{3}, \frac{2}{5}, \frac{3}{7}$ 2. $\frac{2}{3}, \frac{7}{10}, \frac{3}{4}$ 3. $\frac{21}{32}, \frac{11}{16}, \frac{3}{4}, \frac{7}{8}$
4. $\frac{13}{20}, \frac{7}{10}, \frac{3}{4}, \frac{4}{5}$ 5. $\frac{13}{15}$ 6. $\frac{5}{8}$
7. $\frac{53}{72}$ 8. $1\frac{11}{12}$ 9. $1\frac{13}{120}$
10. $5\frac{7}{8}$ 11. $8\frac{4}{15}$ 12. $8\frac{23}{56}$
13. $13\frac{1}{30}$ 14. $13\frac{11}{40}$ 15. $\frac{5}{12}$
16. $\frac{7}{24}$ 17. $\frac{7}{20}$ 18. $1\frac{3}{8}$
19. $2\frac{5}{48}$ 20. $1\frac{13}{16}$ 21. $2\frac{7}{12}$
22. $\frac{31}{40}$

Exercise 3.3

1. $\frac{5}{14}$ 2. $\frac{3}{20}$ 3. $1\frac{13}{27}$
4. $1\frac{3}{32}$ 5. $\frac{3}{11}$ 6. $\frac{10}{21}$
7. $15\frac{2}{5}$ 8. 108   9. 18
10. 72   11. 12   12. $\frac{1}{5}$
13. $\frac{9}{32}$ 14. $1\frac{1}{4}$ 15. $1\frac{1}{6}$
16. $1\frac{1}{2}$ 17. 8   18. $\frac{1}{6}$
19. $\frac{2}{3}$ 20. $\frac{3}{4}$

Exercise 3.4

1. $\frac{5}{9}$ 2. $2\frac{1}{2}$ 3. $\frac{5}{6}$
4. $\frac{2}{3}$ 5. $2\frac{1}{2}$ 6. $\frac{2}{3}$
7. $1\frac{1}{3}$ 8. $1\frac{1}{2}$

Exercise 3.5

1. £150   2. 4
3. (a) $\frac{13}{20}$ (b) $\frac{7}{20}$
4. $262\frac{1}{2}$ ℓ   5. 330
6. 160 p   7. 30 minutes
8. £15 950

Exercise 3.6

1. $\frac{1}{16}$ 2. $\frac{9}{16}$ 3. $\frac{9}{49}$
4. $\frac{8}{27}$ 5. $1\frac{7}{9}$ 6. $5\frac{19}{25}$
7. $-\frac{27}{64}$ 8. $\frac{25}{36}$ 9. $1\frac{7}{25}$
10. $\frac{2}{9}$ 11. $\frac{2}{3}$ 12. $\frac{9}{10}$
13. $\frac{5}{7}$ 14. $\frac{5}{8}$ 15. $\frac{1}{6}$
16. 13   17. 12   18. 3
19. 12   20. 2

Miscellaneous Exercise 3

Section A

1. (a) $\frac{3}{4}$ (b) $\frac{3}{20}$ (c) $\frac{9}{20}$
   (d) $\frac{12}{13}$
2. (a) $\frac{2}{7}$ (b) $\frac{5}{22}$ (c) $\frac{5}{24}$
   (d) $4\frac{1}{12}$
3. $\frac{1}{2}$ 4. $3\frac{3}{8}, 3\frac{3}{4}, 4\frac{1}{8}$ 5. 20
6. $\frac{3}{16}$ 7. $\frac{17}{32}$ 8. $1\frac{1}{2}$
9. £2500   10. $5\frac{1}{2}$

Section B

1. $3\frac{1}{2}$ 2. £750   3. $106\frac{2}{3}$ ft
4. A receives £200, B receives £200, C receives £800
5. $2\frac{3}{10}$ 6. £160   7. $\frac{1}{16}$
8. £1800   9. £5175   10. $31\frac{1}{4}$

Multi-Choice Questions 3

1. A   2. A   3. C
4. D   5. B   6. C
7. A   8. A

Mental Test 3

1. $\frac{5}{8}$ 2. $\frac{11}{20}$ 3. $\frac{3}{10}$
4. $\frac{7}{8}$ 5. $\frac{5}{8}$ 6. $\frac{1}{10}$
7. $\frac{2}{7}$ 8. $-1\frac{3}{4}$ 9. $\frac{3}{8}$
10. $-\frac{1}{2}$ 11. $\frac{1}{8}$ 12. 8
13. 12   14. $\frac{5}{8}$ 15. $\frac{19}{5}$
16. $\frac{3}{8}$ 17. $\frac{8}{15}$ 18. $\frac{11}{16}$

Answers to Chapter 4

Exercise 4.1

1. (a) 0.7   (b) 0.38   (c) 0.087
   (d) 0.017   (e) 0.09   (f) 8.06
   (g) 826.003   (h) 24.029

2. (a) $\frac{7}{1000}$ (b) $\frac{7}{100}$ (c) $\frac{7}{10}$

3. (a) $\frac{2}{10}$ (b) $4\frac{6}{10}$ (c) $3\frac{58}{100}$

4. (a) 0.203, 2.03, 20.3, 203
    (b) 0.002 63, 0.263, 26.3, 263, 2630
    (c) 0.0534, 0.534, 5.34, 53.4

Exercise 4.2

1. 29.856   2. 2.146   3. 22.28
4. 683.097   5. 9163.8   6. 26.7858
7. 3.822 594   8. 13.6   9. 1.38
10. 112.4   11. 911.515   12. 51.409
13. 101.86   14. 0.3877   15. 11.309 22
16. 1.063 14   17. 0.001 36
18. (a) (i) 2.5   (ii) 25   (iii) 250
     (b) (i) 59.2   (ii) 592   (iii) 5920
     (c) (i) 381.73   (ii) 3817.3   (iii) 38.173
     (d) (i) 4812   (ii) 48 120   (iii) 481 200
19. (a) (i) 27.8   (ii) 2.78   (iii) 0.278
     (b) (i) 2.923   (ii) 0.2923
        (iii) 0.029 23
     (c) (i) 63.842   (ii) 6.3842
        (iii) 0.638 42
     (d) (i) 0.0057   (ii) 0.000 57
        (iii) 0.000 057
20. (a) 12.9   (b) 1.83   (c) 0.078
     (d) 2.25   (e) 4.9   (f) 0.75
     (g) 0.006   (h) 9.2   (i) 1.86
     (j) 0.006   (k) 0.002
     (l) 0.000 006   (m) 3.1   (n) 0.02
     (o) 0.4

Exercise 4.3

1. (a) 19.37   (b) 19.4
2. (a) 0.007 52   (b) 0.008   (c) 0.01
3. (a) 4.970   (b) 4.97
4. (a) 153.262   (b) 153.26   (c) 153.3
5. 1.33   6. 0.016   7. 23.53
8. 0.0025
9. (a) 24.94   (b) 25
10. (a) 0.007 33   (b) 0.0073   (c) 0.007
11. (a) 35.60   (b) 35.6
12. (a) 35 680   (b) 35 700   (c) 36 000
13. (a) 13 360 000   (b) 13 400 000
     (c) 13 000 000
14. (a) 48.9   (b) 4   (c) .5
     (d) 600   (e) .0073   (f) 108.07
15. 308.9   16. 224   17. 33
18. 47   19. 13   20. 70

Exercise 4.4

1. $20 + 40 + 430 = 490$; 487
2. $80 - 60 - 10 = 10$; 9.96
3. $20 \times 0.5 = 10$; 13
4. $40 \times 0.25 = 10$; 11
5. $0.7 \times 0.1 \times 2 = 0.14$; 0.16

6. $90 \div 30 = 3$; 2.931
7. $0.09 \div 0.03 = 9 \div 3 = 3$; 2.6
8. $(30 \times 30) \div 0.03 = 30\,000$; 21 000
9. $(1.5 \times 0.01) \div 0.05 = 0.3$; 0.342
10. $(30 \times 30) \div (10 \times 3) = 30$; 29.2

Exercise 4.5

1. (a) 0.25   (b) 0.875
     (c) 2.593 75   (d) 3.234 35
2. (a) (i) 0.555 56   (ii) 0.555 556
     (b) (i) 0.177 78   (ii) 0.177 778
     (c) (i) 0.353 54   (ii) 0.353 535
     (d) (i) 0.212 12   (ii) 0.212 121
     (e) (i) 0.428 43   (ii) 0.428 428
     (f) (i) 0.563 64   (ii) 0.563 636
     (g) (i) 0.567 17   (ii) 0.567 167
     (h) (i) 0.032 32   (ii) 0.032 323
3. (a) $0.\dot{2}$ (b) $\dot{4}\dot{5}$ (c) $0.1\dot{3}$
     (d) $0.40\dot{9}$
4. (a) $\frac{3}{10}$ (b) $\frac{13}{20}$ (c) $\frac{3}{8}$
     (d) $\frac{7}{16}$ (e) $1\frac{3}{4}$ (f) $7\frac{9}{25}$
5. 0.0175   6. -1.748 7. 4.5945
8. 0.276 875

Exercise 4.6

1. 5, 198
2. $-9, -8, -4$
3. $1.57, \frac{1}{7}, -5.625, \sqrt{16}, -3\frac{1}{2}$
4. 0, 6, 10
5. $-4, -3\frac{3}{4}, -3.4, 0, \frac{5}{8}, 2.5, 3, 7$

Exercise 4.7

1. 0.1221   2. 0.1082   3. 0.011 21
4. 0.000 140 0   5. 6.506   6. 25.34
7. 641.0   8. 0.000 061 19
9. 0.004 283   10. 53.12   11. 0.2311
12. 7.458   13. 0.001 023   14. 0.2230
15. 2.384   16. 0.3796   17. 0.030 68
18. 15.21

Miscellaneous Exercise 4

Section A

1. (a) 3, 19   (b) $-100, 0, 3, 19$
     (c) $-100, -58.3, 0, \frac{1}{2}, 3, 19, 14.3, 97.2$
2. (a) 3.317   (b) 3.32
3. (a) $\frac{22}{25}$ (b) $\frac{13}{500}$
4. (a) $2\frac{9}{10}$ (b) 2.9
5. (a) 0.1778   (b) 0.215 15
6. (a) $\frac{5}{6}$ (b) 0.833
7. $-0.1404, 0.042 12$
8. 0.0325

Section B

1. (a) 13, 72, 105   (b) $-97, 0, 13, 72, 105$
 (c) $-97, -85.2, 0, \frac{3}{5}, 13, 14.8, 72, 105$
2. 293.1
3. (a) 0.070   (b) 0.0704
4. (a) $0.\dot{2}$ (b) $0.7\dot{2}$ (c) $3.\dot{1}42\,85\dot{7}$
5. (a) $\frac{4}{5}$ (b) $3\frac{13}{20}$ (c) $\frac{3}{8}$ (d) $\frac{11}{16}$
6. (a) $\frac{11}{18}$ (b) $0.6\dot{1}$
 (c) 0.611 11
7. $8, 3, 3\frac{15}{16}, -0.7, -\frac{5}{6}, -7.2$

Multi-Choice Questions 4

1. C   2. B   3. C
4. B   5. A   6. D
7. B   8. A

Mental Test 4

1. 3.84   2. 0.000 12   3. 0.36
4. 500   5. 0.09
6. (a) 0.08   (b) 0.076
7. 25   8. 0.15   9. 0.001
10. 2.2   11. $\frac{2}{3}$ 12. $\frac{8}{100}$
13. 0.02   14. 0.05   15. 4

Answers to Chapter 5

Exercise 5.1

1. (a) 48   (b) 89   (c) 1530
 (d) 8300   (e) 28 100   (f) 1940
2. (a) 800   (b) 890   (c) 32.5
 (d) 80   (e) 896.4   (f) 0.54
3. (a) 9   (b) 0.058   (c) 0.008
 (d) 0.047   (e) 0.000 52
4. (a) 3.25   (b) 6.2   (c) 27.6
 (d) 118.2
5. (a) 4.4   (b) 2.9   (c) 403
6. (a) 189.6   (b) 183.6   (c) 459 360
7. (a) 6.4   (b) 9.3   (c) 8.7
8. (a) 6.98   (b) 11 200   (c) 0.5679
9. (a) 750   (b) 1.83
 (c) 6 300 000
10. (a) 4900   (b) 9   (c) 350 000
11. (a) 8.57   (b) 0.045   (c) 0.197
12. (a) 72   (b) 4.16
13. (a) 3.25   (b) 582.4   (c) 16 352
14. (a) 6   (b) 9.2
15. (a) 3.9   (b) 248
16. (a) 430   (b) 80
17. (a) 300   (b) 34
18. (a) 5.5   (b) 0.09
19. (a) 1.25   (b) 50
20. (a) 26   (b) 9   (c) 48
21. (a) 0.65   (b) 2

Exercise 5.2

1. (a) 9.2048   (b) 246.888
2. 7   3. 20   4. 4.5
5. 40   6. 2.70   7. 4.921 875

Exercise 5.3

1. 6.315   2. 93 462.05   3. 2.072
4. 62.8   5. 102.7   6. 120.65
7. 17.85
8. (a) 47   (b) 90
9. 468 complete jars   10. 5000

Exercise 5.4

1. £15.97   2 £4.42   3. £100.64
4. £2.39   5. £9.11   6. £701.69
7. £1098.81   8. £9.80   9. £5.72
10. £2.49

Miscellaneous Exercise 5

Section A

1. (a) 2800   (b) 2056   (c) 6.72
 (d) 28.4
2. 68
3. (a) 3000   (b) 900   (c) 3.5
4. 13, 25 cm   5. $\frac{9}{20}$
6. 800 mg, 0.88 g, 80 g, 0.8 kg
7. £3.68   8. 17.3

Section B

1. 2 200 000 00   2. 0.08
3. (a) 50 000 000   (b) 0.000 04
4. 28   5. 14   6. £16.79
7. £4.27

Multi-Choice Questions 5

1. C   2. C   3. D
4. C   5. D   6. B
7. C   8. C

Mental Test 5

1. 800   2. 45 000   3. 0.45
4. 0.007   5. 8200   6. 19 700
7. 4   8. 4   9. 5280
10. 0.08   11. 730   12. 18.7
13. 0.265   14. 400   15. 560
16. 160   17. 6   18. 2
19. 20   20. 1800   21. 10 m
22. 30 mm   23. 100   24. £3.99
25. £3.13

Answers to Chapter 6

Exercise 6.1

1. $1:5$
2. $1:3$
3. $3:2$
4. $5:6$
5. $8:7$
6. $\frac{5}{4}$
7. $1:2:4$
8. $4:5:6:7$
9. $\frac{5}{7}$
10. $\frac{1}{2}$
11. $\frac{2}{3}$
12. $\frac{1}{20}$
13. $\frac{1}{50}$
14. $\frac{12}{1}$
15. $\frac{1}{400}$

Exercise 6.2

1. $9:4$
2. $4:3$
3. $4:3$
4. $15:16$
5. $11:6$
6. $16:39$
7. $96:235$
8. $47:5$
9. $93:5$
10. $139:5$

Exercise 6.3

1. £500, £300
2. 24 kg, 36 kg, 60 kg
3. 168 mm, 588 mm, 924 mm
4. £280
5. 15 kg, 22.5 kg, 37.5 kg
6. £88
7. £10.80
8. (a) 1200 ℓ   (b) 2880 ℓ

Exercise 6.4

1. 48 km
2. 7 hours
3. 16 kg
4. $13\frac{1}{2}$ days
5. 6
6. 10 hours
7. (a) £1   (b) £8
8. 6

Exercise 6.5

1. (a) 75 miles   (b) 6 gal
2. (a) 12 ℓ/min   (b) 84 ℓ   (c) 7.5 min
3. (a) 30 mile/gal   (b) 150 miles   (c) 4.5 gal
4. (a) 150 miles   (b) 3.5 hours
5. (a) 18.9 t   (b) 0.044 m^3
6. (a) 2 m/s   (b) 2 min 55 s
7. (a) 240 newtons   (b) 2.5 mm^2

Exercise 6.6

1. 57.75
2. 5500
3. 101.6
4. 177.17
5. 7658.38
6. 3.66
7. 66.80
8. £15.48

Miscellaneous Exercise 6

Section A

1. 30
2. £178
3. 9
4. £3.48
5. 24
6. 187.5
7. 4 h
8. 8
9. 46
10. 56.25
11. £34
12. 27

Section B

1. (a) $3:5:10$ (b) $4:450$ (c) $89:5$
2. $15:34$
3. (a) £2.67, £8.00, £13.33
   (b) £2.88, £4.80, £7.68, £8.64
4. (a) 7.5 cm   (b) 12.25 cm
5. £248
6. (a) 91   (b) 50
7. (a) 4200   (b) 3896   (c) £29.51
8. (a) £96, £124.80, £19.20
   (b) £1440, £3120, £211.20

Multi-Choice Questions 6

1. D
2. B
3. B
4. A
5. A
6. C
7. C
8. B
9. A
10. C

Mental Test 6

1. 25 kg
2. $2:5$
3. $\frac{2}{3}$
4. $\frac{20}{1}$
5. £280, £120
6. 27 p
7. 4 days
8. 8
9. 80
10. $73:10$

Answers to Chapter 7

Exercise 7.1

	Fraction	Decimal	Percentage
1.	$\frac{1}{4}$	0.25	25%
2.	$\frac{11}{20}$	0.55	55%
3.	$\frac{7}{8}$	0.875	87.5%
4.	$\frac{2}{3}$	$0.\dot{6}$	$66.\dot{6}\% = 66\frac{2}{3}\%$
5.	$\frac{2}{25}$	0.08	8%
6.	$\frac{24}{125}$	0.192	19.2%
7.	$\frac{3}{20}$	0.15	15%
8.	$\frac{27}{100}$	0.27	27%
9.	$\frac{5}{8}$	0.625	62.5%
10.	$\frac{33}{400}$	0.0825	8.25%
11.	$\frac{79}{800}$	0.098 75	$9.875\% = 9\frac{7}{8}\%$
12.	$\frac{1}{15}$	0.06	$6.\dot{6}\% = 6\frac{2}{3}\%$

Exercise 7.2

1. (a) 10   (b) 24   (c) 6
   (d) 2.4   (e) 21.315   (f) 2.516
   (g) 4   (h) 3

2. (a) 12.5%   (b) 20%   (c) 16%
   (d) 16.292%   (e) 45.455%
3. (a) 60%   (b) 27%
4. 115 cm   5. 88.67 cm
6. (a) £7.20   (b) £13.20   (c) £187.50
7. 584 kg   8. 39 643

Exercise 7.3

1. £800   2. £20   3. £1000
4. 800   5. £1600   6. 2000 kg
7. £8000

Miscellaneous Exercise 7

Section A

1. £13   2. 90%   3. £65.25
4. £144   5. 1100   6. 28%
7. (a) £120   (b) £110.40
8. (a) 55%   (b) 54

Section B

1. 18.75%, £3380
2. (a) £400   (b) £480
3. £180   4. £2700   5. 46.7 ℓ
6. (a) 19.6%   (b) 11 803
7. 72%

Multi-Choice Questions 7

1. C   2. C   3. D
4. C   5. B   6. C
7. A   8. A

Mental Test 7

1. 60%   2. $\frac{2}{5}$ 3. 0.35
4. 98   5. 16%   6. 30
7. 40   8. £5   9. 67%
10. 120   11. £120   12. 210 kg

Answers to Chapter 8

Exercise 8.1

1. £4   2. £228   3. £177.12
4. (a) £6.25   (b) £7.50   (c) £10
5. (a) £2.52   (b) £3.15   (c) £37.80
   (d) £138.60
6. £147   7. £23.50   8. £39
9. £240   10. £294
11. (a) £443.33   (b) £500   (c) £800
    (d) £508

Exercise 8.2

1. £765   2. £450   3. £64.97
4. £928   5. £800
6. (a) £9500   (b) £2375
7. (a) £2690   (b) £1893
8. £6951
9. (a) $8800   (b) $14 800   (c) $4520
10. (a) £7260   (b) £24 200

Miscellaneous Exercise 8

Section A

1. (a) £8000   (b) £2400
2. (a) £6   (b) £9   (c) £303
3. £1500   4. £19
5. (a) £3020   (b) £4980   (c) £415
6. £7960
7. (a) £140   (b) £5.39   (c) £116.31
8. £225

Section B

1. (a) £4444   (b) £7400   (c) £11 050
2. 1.5%   3. 100   4. 5 hours
5. (a) £6800   (b) £30 200   (c) £11 810

Multi-Choice Questions 8

1. D   2. D   3. A
4. B   5. B   6. A
7. C   8. B

Mental Test 8

1. £5   2. £200   3. £4.50
4. £20   5. £30   6. £150
7. £100   8. £500

Answers to Chapter 9

Exercise 9.1

1. £48   2. £990   3. £1960
4. 4 years   5. 3 years   6. 4 years
7. 7%   8. 12%   9. 10%
10. £4000   11. £200   12. £1200
13. £850   14. 4 years

Exercise 9.2

1. £367.33   2. £7312.16
3. £3766.11   4. £12 795.67
5. £2503.65   6. £16 115.44
7. £23 167.81   8. £6856
9. £45 060   10. £647.50
11. £175.35   12. £3425

Exercise 9.3

1. £3796.88   2. £1853.70   3. £3348.07
4. £9288.08   5. £3543.74

Miscellaneous Exercise 9

Section A

1. £330   2. £12 884.08
3. £5000   4. £5570.02
5. 16.8%; £3461.12

Section B

1. £31 000   2. 4 years
3. £11 108.12   4. £3276.80
5. Building society £412.50; Bond £385

Multi-Choice Questions 9

1. B   2. C   3. B
4. C   5. B

Answers to Chapter 10

Exercise 10.1

1. 25%   2. 33.3%   3. 20%
4. 18.6%   5. £100   6. £60
7. £45   8. £104.50   9. £4.55
10. £1023   11. £276   12. £138
13. £52.17   14. £50

Exercise 10.2

1. £190   2. £240.50   3. £1.25
4. (a) £85 000   (b) £1.40 in the £1
5. £3 860 000   6. £313 600   7. £4 644 000
8. 2.9 p

Exercise 10.3

1. £45   2. £65   3. £128
4. £5.25   5. £117   6. £14.32
7. (a) £6000   (b) £5241
8. £600
9. (a) £1388.58   (b) 11.6 p
10. 16.9 p

Miscellaneous Exercise 10

Section A

1. £260.40   2. £72   3. £13.50
4. £4.05   5. £17.50   6. £2700
7. (a) £835 000   (b) £7 181 000
8. £120   9. £64.84   10. £1.88

Section B

1. 25%   2. 40%   3. £19.20
4. £60   5. £1800
6. (a) £24 600   (b) (i) £144   (ii) £1.44
7. £71.40
8. (a) 643 gal   (b) £1183.12   (c) 17.4 p
9. (a) £240   (b) 60%
10. (a) £80.50   (b) £12.60   (c) £66.01

Multi-Choice Questions 10

1. B   2. D   3. C
4. B   5. B   6. B
7. D   8. B

Mental Test 10

1. £2   2. £500   3. 20%
4. £10   5. £30   6. £160
7. £5000   8. £20   9. £160
10. £230

Answers to Chapter 11

Exercise 11.1

1. £34.06
2. (a) £375.84   (b) £32.73
3. £43.24   4. £292.80   5. £239.20
6. (a) £28 000   (b) £439.04
7. £600   8. £289.17

Exercise 11.2

1. £120
2. (a) £120   (b) £12   (c) £132
   (d) £13.20
3. (a) £128   (b) £512   (c) £61.44
   (d) £573.44   (e) £143.36
4. (a) £260   (b) £60
5. (a) £24   (b) £264   (c) £66
   (d) £150   (e) 16%
6. £60; $26\frac{2}{3}$%
7. £240
8. (a) £1200   (b) £266.67

Exercise 11.3

1. £35.30   2. 95 therms
3. (a) 323   (b) 323 therms   (c) £134.72

4. (a)

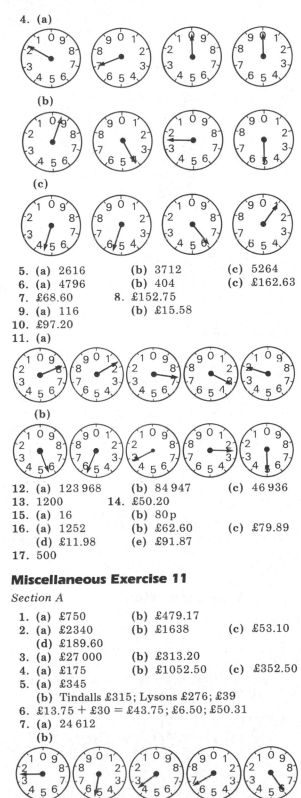

(b)

(c)

5. (a) 2616   (b) 3712   (c) 5264
6. (a) 4796   (b) 404   (c) £162.63
7. £68.60   8. £152.75
9. (a) 116   (b) £15.58
10. £97.20
11. (a)

(b)

12. (a) 123 968   (b) 84 947   (c) 46 936
13. 1200   14. £50.20
15. (a) 16   (b) 80p
16. (a) 1252   (b) £62.60   (c) £79.89
     (d) £11.98   (e) £91.87
17. 500

Miscellaneous Exercise 11

Section A

1. (a) £750   (b) £479.17
2. (a) £2340   (b) £1638   (c) £53.10
     (d) £189.60
3. (a) £27 000   (b) £313.20
4. (a) £175   (b) £1052.50   (c) £352.50
5. (a) £345
     (b) Tindalls £315; Lysons £276; £39
6. £13.75 + £30 = £43.75; £6.50; £50.31
7. (a) 24 612
     (b)

Section B

1. £118
2. (a) £12.80   (b) 12.8p
     (c) £41.60, 5.94p   (d) 6.46p, 0.52p
3. 38.4%   4. $44.28
5. Tariff 2; £25.26   6. 300 therms
7. £60

Multi-Choice Questions 11

1. C   2. B   3. B
4. C   5. D   6. D
7. D

Mental Test 11

1. £50   2. £400   3. £18 000
4. £1100   5. £100   6. £36.20
7. £60   8. 24

Answers to Chapter 12

Exercise 12.1

1. (a) 240   (b) 210
2. (a) 540   (b) 135   (c) 10 800
3. 5 h 50 min   4. 18 h 15 min
5. 17 h 15 min   6. 12 h 39 min
7. 11 h 47 min   8. 12 h 6 min
9. 12 h 39 min   10. 12 h 47 min

Exercise 12.2

1. 0830 bus; 2 min   2. 0830 hours; 29 min
3. (a) 25 min   (b) 15 min
4. (a) 1742 hours   (b) 1805 hours   (c) 23 min
5. 1255 hours; 45 min   6. 426
7. 438 and 426   8. 426 and 421; Thrupp

Exercise 12.3

1. 75 km/h   2. 4 h   3. 350 km
4. 26 km/h   5. 9 km/h   6. 75 km/h
7. 35.14 mile/h   8. 80 km/h

Miscellaneous Exercise 12

Section A

1. (a) 2125 hours   (b) 55 min
     (c) 32.7 mile/h
2. (a) 1 h 20 min   (b) 64 km/h
3. (a) 3 h 6 min   (b) 50 mile/h
4. (a) 36 min   (b) 2 h 24 min   (c) 3 h
     (d) 39 km/h
5. (a) 16 m/s   (b) 57.6 km/h

Section B

1. (a) 6.25 s   (b) 13 m
2. 4 h 12 min   3. 60 km/h   4. 24 s
5. (a) 64 km   (b) 12.40 p.m.   (c) 16 km
    (d) 82 km/h

Multi-Choice Questions 12

1. B   2. B   3. C
4. C   5. D   6. D
7. A   8. D

Mental Test 12

1. 8 min   2. 420 s
3. 4 h   4. 180 min
5. 15 days   6. 3 h 15 min
7. 4 h 10 min   8. 3 h 50 min
9. 4 h   10. 40 mile/h
11. 200 km   12. 1500 miles
13. 1 h 20 min   14. 1048 hours
15. 37; X49   16. No
17. 60, X49. Change at Monmouth

Answers to Chapter 13

Exercise 13.1

1. $7x$ 2. $4x - 3$ 3. $5x + y$
4. $\dfrac{x+y}{z}$ 5. $\dfrac{x}{2}$ 6. $8xyz$
7. $\dfrac{xy}{z}$ 8. $3x - 4y$

Exercise 13.2

1. 9   2. 3   3. 3
4. 18   5. 45   6. 6
7. 45   8. 30   9. 23
10. 38   11. 33   12. $\frac{33}{}$
13. 28   14. 1   15. $\frac{3}{4}$
16. 5   17. 5   18. 7.7

Exercise 13.3

1. 4   2. 81   3. 54
4. 32   5. 1152   6. 74
7. 20   8. 3024   9. 48
10. 18.96

Exercise 13.4

1. $18x$ 2. $2x$ 3. $-3x$
4. $-6x$ 5. $-5x$ 6. $5x$
7. $-5a$ 8. $12m$ 9. $5b^2$
10. ab 11. $14xy$ 12. $-3x$
13. $-6x^2$ 14. $7x - 3y + 6z$

15. $9a^2b - 3ab^3 + 4a^2b^2 + 11b^4$
16. $1.2x^3 + 0.3x^2 + 6.2x - 2.8$
17. $9pq - 0.1qr$
18. $0.4a^2b^2 - 1.2a^3 - 5.5b^3$
19. $10xy$ 20. $12ab$ 21. $12m$
22. $4pq$ 23. $-xy$ 24. $6ab$
25. $-24mn$ 26. $-12ab$ 27. $24pqr$
28. $60abcd$ 29. $2x$ 30. $-\dfrac{4a}{7b}$
31. $-\dfrac{5a}{8b}$ 32. $\dfrac{a}{b}$ 33. $\dfrac{2a}{b}$
34. $2b$ 35. $3xy$ 36. $-2ab$
37. $2ab$ 38. $\dfrac{7ab}{3}$ 39. a^2
40. $-b^2$ 41. $-m^2$ 42. p^2
43. $6a^2$ 44. $5X^2$ 45. $-15q^2$
46. $-9m^2$ 47. $9pq^2$ 48. $-24m^3n^4$
49. $-21a^3b$ 50. $10q^4r^6$ 51. $30mnp$
52. $-75a^3b^2$ 53. $-5m^5n^4$

Exercise 13.5

1. $3x + 12$ 2. $2a + 2b$ 3. $9x + 6y$
4. $\dfrac{x}{2} - \dfrac{1}{2}$ 5. $10p - 15q$ 6. $7a - 21m$
7. $-a - b$ 8. $-a + 2b$ 9. $-3p + 3q$
10. $-7m + 6$ 11. $-4x - 12$ 12. $-4x + 10$
13. $-20 + 15x$ 14. $2k^2 - 10k$ 15. $-9xy - 12y$
16. $ap - aq - ar$ 17. $4abxy - 4acxy + 4dxy$
18. $3x^4 - 6x^3y + 3x^2y^2$ 19. $-14P^3 + 7P^2 - 7P$
20. $2m - 6m^2 + 4mn$ 21. $5x + 11$
22. $14 - 2a$ 23. $x + 7$ 24. $16 - 17x$
25. $7x - 11y$ 26. $\dfrac{7y}{6} - \dfrac{3}{2}$
27. $-8a - 11b + 11c$ 28. $7x - 2x^2$
29 $3a - 9b$ 30 $-x^3 + 18x^2 - 9x - 15$

Miscellaneous Exercise 13

Section A

1. 42   2. $\frac{1}{4}$
3. (a) $11a$ (b) $5a$ (c) $6a^2$
    (d) 4
4. (a) $22x - 6y$ (b) $6y - x$ (c) $-2y$
5. (a) $30a^3$ (b) $-2x$ (c) $3a^2$

Section B

1. $-\frac{9}{4}$ 2. $12x - 5y - 14z$
3. (a) 16   (b) 0   (c) 6
4. 53
5. (a) -1 (b) 21   (c) 8
6. 73   7. $-12\frac{5}{18}$ 8. 80

Multi-Choice Questions 13

1. D   2. B   3. D
4. D   5. A   6. D
7. B   8. C

Mental Test 13

1. $11a$ 2. $2x$ 3. $5y + 8$
4. $5p$ 5. $12ab$ 6. $30a^3$
7. $2a$ 8. $6a^3$ 9. $6x + 9y$
10. $-3x + 4y$ 11. $-3x + 6y$ 12. $-2x$

Answers to Chapter 14

Exercise 14.1

1. $5(x + y)$ 2. $5(p - q)$
3. $6(x + 3y)$ 4. $a(x - y)$
5. $4(p - 3q)$ 6. $2x(2 - 3y)$
7. $4x(2x - 1)$ 8. $x(ax - b)$
9. $(a - b)(x + 2)$ 10. $(p + q)(m - n)$
11. $5(a - 2b + 3c)$

Exercise 14.2

1. $x^2 + 4x + 3$ 2. $2x^2 + 11x + 15$
3. $10x^2 + 16x - 8$ 4. $a^2 - 3a - 18$
5. $6x^2 - 13x - 5$ 6. $x^2 - 5x + 6$
7. $8x^2 - 14x + 3$ 8. $x^2 - 2x - 3$
9. $6x^2 - 5x - 6$ 10. $2x^2 - 3x - 20$

Exercise 14.3

1. $(x + y)(a + b)$ 2. $(b + c)(a - d)$
3. $(r - 2s)(2p + q)$ 4. $(2x + 3y)(2a - 2b)$
5. $(3m + 2n)(x - y)$ 6. $(p + q)(ab - cd)$
7. $(3x - 1)(mn - pq)$ 8. $(l - 1)(k^2l - mn)$

Exercise 14.4

1. $(x + 2)(x + 3)$ 2. $(x - 3)(x - 5)$
3. $(x + 1)(x - 6)$ 4. $(x - 2)(x - 3)$
5. $(x + 2)(x - 9)$ 6. $(x + 5)(x - 3)$
7. $(2x + 5)(x + 1)$ 8. $(2x + 3)(x + 5)$
9. $(3x - 2)(x + 1)$ 10. $(3x - 14)(x + 2)$
11. $(2x + 1)(x - 3)$ 12. $(5x - 3)(2x + 5)$
13. $(3x - 7)(2x + 5)$ 14. $(5x - 1)(x - 2)$
15. $(2x - 1)(x - 5)$ 16. $(3x - 5)(2x + 1)$

Exercise 14.5

1. $x^2 + 8x + 16$ 2. $x^2 - 14x + 49$
3. $4x^2 + 20x + 25$ 4. $9x^2 - 24x + 16$
5. $x^2 - 9$ 6. $1 - x^2$
7. $4x^2 - 25$ 8. $9x^2 - 4y^2$
9. $(x + 3)^2$ 10. $(x + 4)^2$
11. $(3x - 2)^2$ 12. $(5x - 1)^2$
13. $(x + 1)(x - 1)$ 14. $(x + 4)(x - 4)$
15. $(2x + 3)(2x - 3)$ 16. $(5x + 6)(5x - 6)$

Exercise 14.6

1. $(a - b)(a - b - 2x)$ 2. $(x + y)(3 + x + y)$
3. $(x + y)(x - y - 1)$ 4. $(a + b)(a - b + 3)$
5. $a(R + r)(R - r)$ 6. $y(2y + 3a)(2y - 3a)$
7. $\frac{1}{3}\pi r^2(2r + h)$
8. $(x - 1 + 2y)(x - 1 - 2y)$

Miscellaneous Exercise 14

Section A

1. (a) $2ab(c - 3d)$ (b) $(a - b)(4 - c)$
2. $(x - 4)(x - 1)$
3. $4(x + 3)(x - 3)$
4. $(x - 2)^2$
5. (a) $5(2 - 3x^2)$ (b) $4x(1 - 2y)$

Section B

1. (a) $(3x + 4)(3x - 4)$ (b) $x(9x - 16)$
   (c) $(x + 2)(x - 11)$
2. (a) $(2x + 3)^2$ (b) $(3x + 4)(2x - 5)$
   (c) $(3p + 2q)(3p - 2q)$
3. $(a - 4c)(3a + 2b)$
4. $7(1 + 3a)(1 - 3a)$
5. $25x^2 - 70x + 49 = (5x - 7)^2$
6. $2x(x + 1)(x - 2)$
7. $(c - 2b)(a + 3b)$
8. (a) $3(2x + y)(2x - y)$ (b) $(a - 2c)(b + d)$

Multi-Choice Questions 14

1. B   2. D   3. C
4. D   5. D   6. D

Answers to Chapter 15

Exercise 15.1

1. $\dfrac{2}{ab}$ 2. $\dfrac{3y}{2x}$ 3. $\dfrac{4qs}{r}$

4. $\dfrac{6a^2d^3}{bc}$ 5. $\dfrac{b}{c}$ 6. $\dfrac{9s^2}{2t}$

7. $\dfrac{8acz}{3y^3}$ 8. $\dfrac{b^2c}{a}$ 9. $\dfrac{21b^2}{10ac}$

10. $\dfrac{9qs}{pr}$

Exercise 15.2

1. $\dfrac{47x}{60}$ 2. $\dfrac{a}{36}$ 3. $\dfrac{1}{2q}$

4. $\dfrac{32}{15y}$ 5. $\dfrac{9q - 10p}{15pq}$ 6. $\dfrac{9x^2 - 5y^2}{6xy}$

7. $\dfrac{15xz - 4y}{5z}$ 8. $\dfrac{40 - 11x}{40}$ 9. $\dfrac{19m - n}{7}$

10. $\dfrac{a + 11b}{4}$ **11.** $\dfrac{8n - 3m}{3}$ **12.** $\dfrac{5x - 2}{20}$

13. $\dfrac{x - 14}{12}$ **14.** $\dfrac{13x - 21}{30}$

Miscellaneous Exercise 15

Section A

1. $\dfrac{3}{2b}$ **2.** $\dfrac{27}{5q}$ **3.** $\dfrac{7x}{12}$

4. $\dfrac{5x - 6}{6}$ **5.** $\dfrac{x - 4}{2x - 5}$

Section B

1. $\dfrac{x + 1}{6}$ **2.** $\dfrac{2}{3}$ **3.** $-\dfrac{1}{3p}$

4. $\dfrac{3ab^2c^2d}{2}$ **5.** $\dfrac{21q^2}{10pr}$ **6.** $\dfrac{7}{20}$

7. $\dfrac{8x + 29}{4}$

Multi-Choice Questions 15

1. B   **2.** A   **3.** D
4. B   **5.** C

Answers to Chapter 16

Exercise 16.1

1. 5   **2.** 10   **3.** 4
4. 15   **5.** 8   **6.** 3
7. 1   **8.** $\frac{2}{5}$ **9.** 3
10. 5   **11.** $\frac{7}{4}$ **12.** $\frac{7}{2}$
13. $\frac{27}{5}$ **14.** $\frac{23}{14}$ **15.** -24
16. $\frac{17}{36}$ **17.** 12   **18.** $\frac{39}{56}$
19. $\frac{20}{11}$ **20.** 2   **21.** -15
22. $-\frac{11}{2}$ **23.** $\frac{135}{106}$ **24.** 3
25. $\frac{25}{18}$ **26.** -5 **27.** $\frac{13}{6}$

Exercise 16.2

1. $(m - 3)$ years   **2.** $\dfrac{2a + 15b}{25}$

3. £$\dfrac{nx}{m}$ **4.** $6p + 4q + r$

5. $5x - 3y$
6. (a) $4(x + y)$ **(b)** $3xy$
7. $(5a + 10b + 50c)$ pence

8. $\dfrac{7x}{3y} + z$

9. (a) $\dfrac{100b}{a}$ **(b)** £$\dfrac{bc}{a}$

10. $\dfrac{100y}{x}$ kg

Exercise 16.3

1. 20 at 8 p and 5 at 7 p   **2.** 17
3. 5 m and 7 m   **4.** 3
5. 17, 18 and 19   **6.** 7 at 5 p and 20 at 10 p
7. 11   **8.** 9, 4 and 12 cm

Miscellaneous Exercise 16

Section A

1. (a) 4   **(b)** -5
2. 10 at 10 p and 15 at 5 p
3. 19, 20 and 21
4. 4
5. $6 + 16(w + z)$

Section B

1. 8 at £7.50 and 10 at £6
2. 5   **3.** $\frac{15}{28}$
4. (a) $-11\frac{1}{5}$ **(b)** $25\frac{1}{2}$
5. 5

6. (a) $\dfrac{py}{1000}$ **(b)** $\dfrac{py}{1000n}$ **(c)** $\dfrac{a}{y}$

7. (a) 10   **(b)** 32

Multi-Choice Questions 16

1. C   **2.** B   **3.** D
4. B   **5.** D   **6.** D
7. C

Mental Test 16

1. 4   **2.** 3   **3.** -6
4. 8   **5.** 3   **6.** -6
7. 3   **8.** 3   **9.** $2k$
10. $\dfrac{a + b}{c}$ **11.** $\dfrac{8k}{x}$
12. (a) $36x$ **(b)** $3x$

Answers to Chapter 17

Exercise 17.1

1. 18   **2.** 1150   **3.** 11
4. 11.71   **5.** 0.2

Exercise 17.2

1. $d = \dfrac{C}{a}$

2. $P = \dfrac{c}{V}$

3. $y = \dfrac{b}{x}$

4. $E = IR$

5. $a = \dfrac{Sp}{t}$

6. $b = a - 8$

7. $x = \dfrac{7}{y} - 4$

8. $k = \dfrac{5}{3 - x}$

9. $m = \dfrac{2E}{v^2}$

10. $x = \dfrac{y - c}{m}$

11. $t = \dfrac{v - u}{a}$

12. $h = \dfrac{3V}{ab}$

13. $y = \dfrac{M}{5} - x$

14. $n = N - 2pC$

15. $h = \dfrac{S}{ar} - r$

16. $n = \dfrac{t - a}{d} + 1$

17. $y = x - \dfrac{A}{3}$

18. $k = d - \dfrac{v^2}{200}$

19. $y = 4 - 3x$

20. $x = \dfrac{2(1 - y)}{3y + 5}$

21. $n = \dfrac{2 - 3k}{k - 3}$

22. $R = \dfrac{gT^2}{4\pi^2} + H$

23. $b = \dfrac{ca^2}{1 - a^2}$

24. $v = \sqrt{\dfrac{2Kg}{m}}$

25. $A = 4\pi r^2$

26. $p = \dfrac{m^2}{q^2}$

27. $a = x - \dfrac{x}{x + b}$

28. $x = \dfrac{5y - a}{1 + by}$

29. $y = \dfrac{5 - 2x}{3x + 4}$

30. $k = \sqrt{\dfrac{T^2gh}{4} - h^2}$

Miscellaneous Exercise 17

Section A

1. (a) $h = \dfrac{V}{A}$ (b) $h = 5$

2. $n = \dfrac{S + 360}{180}$; $n = 6$

3. $y = \dfrac{b + d}{m}$; $y = 3\frac{1}{3}$

4. $A = 180 - B - C$

5. $a = \dfrac{V - 16t^2}{t}$; $a = 11$

Section B

1. $n = \dfrac{Ir}{E - IR}$

2. $K = \dfrac{1 + p^2}{1 - p^2}$

3. $x = \dfrac{k^2}{p^2}$

4. $x = \dfrac{y - p}{1 + py}$

5. (a) $T = 352$ (b) $a = 8$

Multi-Choice Questions 17

1. D
2. D
3. B
4. A
5. B

Answers to Chapter 18

Exercise 18.1

1. $x = 2, y = 4$
2. $x = 4, y = 1$
3. $x = 2, y = 5$
4. $x = 3, y = 2$
5. $x = 6, y = 10$
6. $x = 12, y = 3$
7. $x = 8, y = 5$
8. $x = 5, y = 3$

Exercise 18.2

1. 27 and 15
2. 27
3. 350
4. (a) $x + y = 42$ (b) $x = 6y$
   (c) $x = 36, y = 6$
5. $x = 8$ and $y = 5$

Miscellaneous Exercise 18

Section B

1. $x = 2, y = -1$
2. $q = -2$
3. $x = 2, y = 14$
4. 8
5. 81 and 27
6. (a) 5 (b) -7
   (c) $x = -1, y = 6$
7. £5 per hour

Multi-Choice Questions 18

1. B
2. C
3. D
4. C
5. B

Answers to Chapter 19

Exercise 19.1

1. ± 2
2. ± 4
3. ± 3
4. ± 5
5. $7, -3$
6. $0, -5$
7. $\frac{5}{3}, -\frac{9}{2}$
8. $2, 3$
9. $2, -5$
10. $\frac{7}{2}, -\frac{5}{3}$
11. $\frac{3}{4}, -\frac{1}{2}$
12. $0, 4$
13. -3
14. $-\frac{5}{3}$
15. $\frac{3}{2}$

Exercise 19.2

1. ± 3.74
2. ± 2.41
3. ± 1.73
4. $(x + 2)^2$
5. $(3x + 2)^2$
6. $(2x - 3)^2$
7. $(5x - 1)^2$
8. $(x + 2)^2 - 1$
9. $(x + 3)^2 - 11$
10. $(x - 4)^2 - 13$
11. $(x - 4)^2 - 18$
12. $4(x + 1)^2 - 3$

13. $3(x + \frac{2}{3})^2 - \frac{19}{3}$
14. $10(x - 1)^2 - 1$
15. $2 - (x - 2)^2$
16. $9 - 2(x - 1)^2$
17. $\frac{13}{4} - 4(x - \frac{1}{4})^2$
18. 1.18 or -0.43
19. 1.62 or -0.62
20. 0.57 or -2.91
21. 0.21 or -1.35
22. 1 or -0.20
23. 3.89 or -0.39
24. -3.77 or 0.44
25. -9.18 or 2.18
26. -11 or 6
27. 3.30 or -0.30

Exercise 19.3

1. 3 or $\frac{1}{2}$
2. 2 or $\frac{1}{4}$
3. 2 or -5
4. 6 or -1
5. 6.16 or -0.16
6. -0.72 or -2.78
7. ± 2

Exercise 19.4

1. $x = 1, y = 2$ or $x = 2, y = 1$
2. $x = -17, y = -20$ or $x = 9, y = 6$
3. $x = 7, y = 2$ or $x = 7.5, y = 1.5$
4. $x = 9, y = 3$ or $x = 4, y = 8$
5. $x = 3.2, y = -0.4$ or $x = 4, y = -2$
6. $x = 2.2, y = 5.4$ or $x = 3, y = 5$
7. $x = 4, y = \frac{1}{2}$ or $x = \frac{1}{3}, y = 6$
8. $x = 1.5, y = -10.5$ or $x = -5, y = -30$

Exercise 19.5

1. 6 or -7
2. 9 m, 8 m
3. 8.02 cm
4. 15 cm
5. 4 m × 1 m
6. 13.35 m
7. 5, 6 and 7
8. 10 cm
9. $\frac{2}{5}$
10. 8.83 m or 3.17 m
11. $d(d - 2) + (d - 4)^2 = 1.48$; $d = 11$ cm

Miscellaneous Exercise 19

Section B

1. (a) $x = 0$ or -4 (b) $y = \pm 2$
2. $x = \pm \frac{2}{3}$
3. 1.69
4. (a) $x^2 + 2xy + y^2$ (b) $xy = 6$
5. $x = 5, y = 2$ or $x = -3, y = -2$
6. $x = 24, y = 1.75$ or $x = -7, y = -6$
7. $x = 3$ or $-\frac{1}{2}$

Multi-Choice Questions 19

1. D
2. D
3. C
4. C
5. B
6. B
7. A
8. D

Answers to Chapter 20

Exercise 20.1

1. c^7
2. b^{12}
3. m^4
4. a^2
5. a^6
6. $25a^6b^2c^8$
7. $\frac{b^6}{c^{12}}$
8. $\frac{q^2}{p^3}$
9. 256
10. $\frac{1}{25}$
11. 125
12. $\frac{1}{32}$
13. 2
14. 2
15. 4
16. 1
17. 2
18. 1.024
19. 8
20. 4
21. 1
22. 8
23. 3
24. 2
25. 3

Exercise 20.2

1. 3.59×10^2
2. 7.28×10^3
3. 1.94×10^4
4. 8.036×10^6
5. 6×10^{-2}
6. 5.6×10^{-3}
7. 9×10^{-6}
8. 7.25×10^{-4}
9. 320
10. 9450
11. 1 870 000
12. 0.002
13. 0.567
14. 0.0326
15. 0.000 56
16. 5.3×10^3
17. 7.5×10^3
18. 6×10^7
19. 3×10^2
20. 5.03×10^{-2}
21. 4.8×10^{-2}
22. 8×10^{-7}
23. 4×10^2
24. 2.66×10^9
25. 9.21×10^8
26. 2.88×10^{-4}
27. 1.37
28. 1.46×10^2
29. $2.46 \times 10^{-4} m$
30. 3.24×10^5
31. 9.227×10^3
32. 6.259×10^5
33. $6.760 325 \times 10^2$
34. 3.269×10^{-2}
35. 6.154×10^5
36. 7.654×10^3
37. 4.0871×10^{-3}
38. 9.1062×10^{-4}

Miscellaneous Exercise 20

Section A

1. 1.2×10^6
2. 2.41×10^4
3. 0.005 01
4. (a) 600 000   (b) 3200   (c) 2800
   (d) 15
5. (a) 9   (b) 1   (c) 0.1
6. 1.2×10^3 by 327
7. 8.93×10^{-4}
8. (a) $n = 3$ (b) $n = 3$ (c) $n = 4$
   (d) $n = -3$

Section B

1. $1, \frac{1}{5}, 100$
2. 30
3. 12
4. $x = 1$
5. $16\frac{1}{32}$
6. $p = 3$
7. 5.7×10^{-2}
8. (a) 16   (b) 4   (c) 1

Multi-Choice Questions 20

1. B   2. B   3. C
4. A   5. C   6. C
7. C

Multi-Choice Questions 21

1. C   2. D   3. B
4. B   5. D   6. A
7. B

Answers to Chapter 21

Exercise 21.1

1. 84 in^2; 38 in   2. 3 m
3. 168 m^2 4. 6 m
5. 40 yd^2 6. 10.5 cm^2
7. 47.9 cm^2 8. 10 ft
9. 2480 mm^2 10. 15 ft
11. 215 m^2 12. 115 cm^2
13. (a) 22 cm   (b) 88 cm   (c) 132 cm
14. (a) 380 cm^2 (b) 1020 cm^2
     (c) 1660 cm^2
15. 854 mm^2 16. 4.43 in
17. (a) 4.19 cm   (b) 23.2 cm
18. (a) 108° (b) 71.7°
19. 109 in^2
20. 67.5 cm, 506 cm^2

Exercise 21.2

1. 45 cm^3
2. (a) 270 cm^2 (b) 14 cm^2 (c) 298 cm^2
     (d) 210 cm^3
3. 4928 in^3; 1940 in^2
4. 440 cm^3
5. 268 in^3; 201 in^2
6. 51.3 cm^3 7. 720 cm^3
8. 7285 cm^3 9. 15 500 ℓ
10. 33 900 ℓ   11. 600 mℓ; 120
12. 5.30 ℓ   13. 25.5 m side
14. 8.24 in   15. 24 cm

Exercise 21.3

1. 154 cm^2 2. 45 in^2 3. 5 cm
4. 452 cm^2, 904 cm^3; 1017 cm^2, 3052 cm^3
5. 8 cm   6. 9.35 in
7. 58.88 cm^2, 32.71 cm^3; 529.9 cm^2; 883.2 cm^3
8. (a) 3 cm   (b) 16 cm, 12 cm

Miscellaneous Exercise 21

Section A

1. (a) 32 in   (b) 52 in^2
2. (a) 9 cm   (b) 28 cm^2
3. 476.28 cm^2 4. 24 000 ℓ   5. 47

Section B

1. (a) 42 860 cm^2 (b) 796 ℓ
2. 35, 321.6 m^2 3. $5\frac{1}{4}$ in; $16\frac{1}{2}$ in
4. 12 cm   5. 750 m^3

Answers to Chapter 22

Exercise 22.1

1. 800 m   2. 1:500
3. 24 cm   4. 15 cm
5. 15.2 km   6. 120 000 m^2
7. 300 ha   8. 1040 ha
9. 24 000 cm^2
10. (a) 7.2 m^2 (b) 720 cm^2 (c) 43.2 m^3
     (d) 43 200 cm^3
11. (a) 1 920 000 cm^3 (b) 240 000 m^3
     (c) 6000 m^2
12. 20 m^3

Exercise 22.2

1. 1:25   2. 1:40   3. 100 m
4. 30 m   5. 1:8   6. 1:20
7. 15 m

Miscellaneous Exercise 22

1. 500 m
2. 1.75 cm by 2.25 cm by 1.25 cm
3. (a) 10 m   (b) 14 cm   (c) 20 m^2
4. (a) 80 m   (b) 5 m^2
     (c) 288 000 m^3
5. (a) 0.5 m   (b) 648 m^2 (c) 19.44 m^3
6. (a) 10 000 cm^2 (b) 100 cm^3
7. 10 ft, 1167 ft^2, 4398 ft^3
8. (a) 120 m   (b) 120.4 m   (c) 8.3%

Answers to Chapter 23

Exercise 23.1

1. (a) £4.07   (b) £70.56   (c) £3.85
     (d) £81.15   (e) £24.40   (f) £43.44
2. (a) 28.3°C   (b) 45.9°C   (c) 131°F
     (d) 155.7°F
3. (a) 488 miles   (b) 130 miles
     (c) 63 miles
4. (a) 14.2%   (b) 1:14
5. (a) 24.13 km   (b) 30 mile/h   (c) 93.3 km
     (d) 914.12 km/h
6. (a) 38°C   (b) 176°F   (c) 18°C
     (d) 118°F
7. (a) 1.5 kg   (b) 258 kg   (c) 4.0 lb
8. (a) 1.54 kg/cm^2 (b) 30 lb/in^2
     (c) 43 lb/in^2

Exercise 23.2

1. $x = 6$, $y = 7$
2. Private motoring 15.7%, rail 39.3%, other 45%
3. Unskilled workers 52.9%, craftsmen 29.4%, draughtsmen 5.9%, clerical staff 11.8%
7. Africa 80°, Asia 73°, Europe 13°, N. America 63°, S. America 49°, Oceania 23°, U.S.S.R. 59°
8. Clothing 81°, furniture 108°, stationery 27°, sports equipment 54°, household goods 90°

Exercise 23.3

1. **(a)** 40   **(b)** 11   **(c)** 35
6. **(a)** 12 000
7. 800 items   **8.** 16 000 articles

Miscellaneous Exercise 23

Section A

1. Bus 57%, private motoring 31%, other 12%
2. Food and drink 137°, housing 61°, transport 42°, clothing 48°, other 73°
5. **(a)** 126°
   (b) White 2000, Green 1250, Brown 1750
6. Vegetables 139 tons, grain 111 tons, potatoes 69 tons, fruit 56 tons, roots 42 tons, other 83 tons
7. £28 000, £20 000, £8000
8. **(a)** 528 km   **(b)** 352 km
   (c) Barcelona and Cacares
   (d) 314 miles   **(e)** 53.4 mile/h

Multi-Choice Questions 23

1. A   2. B   3. B
4. B   5. D   6. B
7. A   8. B

Mental Test 23

1. **(a)** 110 miles   **(b)** 240 miles
   (c) 110 miles   **(d)** 72 miles
2. **(a)** $-40°F$ **(b)** $68°F$ **(c)** $38°C$
3. **(a)** 2   **(b)** 17
4. **(a)** 25%   **(b)** $\frac{1}{3}$
5. **(a)** 36°   **(b)** 90°
6. **(a)** 90 000   **(b)** 120 000
   (c) 150 000
7. **(a)** 1981   **(b)** 50 000
   (c) 35 000   **(d)** 200 000

Answers to Chapter 24

Exercise 24.2

1. **(a)** 18, 30, 46   **(b)** 3, 5, 8
2. **(a)** 12, 27, 75 metres
   (b) 1.2, 3.1, 4.2 seconds
3. $Q = 12.5$ **4.** £19   **5.** $y = 2.3$

Exercise 24.3

1. **(a)**   **(b)**

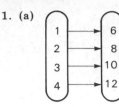 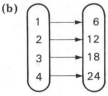

2.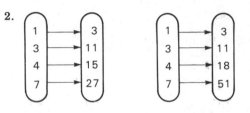

3. **(a)** -30 **(b)** -9 **(c)** 26

4.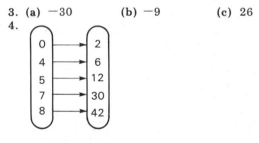

5. **(a)** 14   **(b)** -8 **(c)** -6

Exercise 24.4

2. $y = (3x + 5)^2$ **4.** $y = 14$
5. $F = \frac{9}{5}C + 32$ **(a)** $C = \frac{5}{9}(F - 32)$
   (b)(i) $-\frac{10}{9}°C$ **(ii)** $68°F$

Exercise 24.5

1. $f: x \quad \dfrac{x}{3}$ **2.** $f: x \quad \dfrac{x + 3}{2}$

3. $f: x \quad 2x + 3$ **4.** $f: x \quad \dfrac{x + 15}{6}$

5. $f: x \quad \dfrac{2x + 5}{2 - x}$

6. **(a)** 0   **(b)** $\frac{5}{6}$

7. **(a)** $\frac{10}{3}$ **(b)** 2

8. **(a)** $\frac{5}{2}$ **(b)** $\frac{5}{4}$

9. (a) 9   (b) -2 (c) -27
    (d) -131 (e) -8

10. $\dfrac{x-9}{15}$; $\dfrac{x+1}{15}$ 11. $x = 13$

12. $x = \frac{16}{17}$

Miscellaneous Exercise 24

Section A

1. (a) £23   (b) 240
    (c) £5 and 5p per unit
2. (b) $y = 4$
3. (a) 9 units   (b) 18 square units
4. (a) A(0, 6), B(-12, 0)
    (b) 36 square units
5.

Section B

1.

2. $f^{-1}(x) = \dfrac{x+4}{3-x}$; $f^{-1}(2) = 6$

3. $x = -7$
4. (a) $gf(2) = -2$ (b) $fg(2) = 9$
5. $f(-2) = 26$; $f(0) = -8$; $f(3) = 16$

6. $f^{-1}(x) = \dfrac{x+15}{10}$; $f^{-1}(-2) = \frac{13}{10}$

7. $(gf)^{-1}$: $x \to \dfrac{x+1}{15}$

8. (a) $gf(x) = 2x - 1$ (b) $fg(x) = 2x - 3$

Multi-Choice Questions 24

1. D   2. C   3. B
4. B   5. A   6. C
7. B

Answers to Chapter 25

Exercise 25.1

1. Right-angled triangle, 9 square units
2. Rectangle, 30 square units
3. Parallelogram, 20 square units
4. Trapezium, 8 square units
5. Right-angled triangle, $7\frac{1}{2}$ square units

Exercise 25.2

1. 4.24   2. 3.61   3. 6.40
4. 5.83   5. 4.47
6. (1.5, 5.5)   7. (-2, 6.5)   8. (2.5, 6)

Exercise 25.4

1. 1, 3   2. -3, 4
3. -5, -2 4. 4, -3
5. $m = 4, c = 13$ 6. $m = 2, c = -2$
7. $m = 2, c = 1$ 8. $m = 3, c = -2$
9. $m = -2, c = -3$ 10. $m = -3, c = 4$
11. $m = 5, c = 7$ 12. $a = 3, b = 4$
13. $a = 5, b = 1$ (approx.)
14. $a = 0.5, b = 3$ (approx.)
15. $m = 1.3, c = 20$ 16. $E = 4I$

Exercise 25.5

1. (a) $a = 4.38, b = 28.49$
    (b) 5.8 s   (c) 39.43 m
2. $a = 1.285, b = 3.714$
    (a) 13.4   (b) 180%
3. (a) 0.22, $b = 1045$
    (b) 45.5 km/h   (c) 35%
4. (a) $a = 7.19, b = 2.22$
    (b) 0.56 kg   (c) 25%

Exercise 25.6

1. a   2. b   3. c
4. a   5. c   6. b
7. c

Miscellaneous Exercise 25

Section A

1. (b) 2   (c) $y = 2x - 3$
2. (a) $y = 5x + 4$ (b) $y = -11$
3. $c = 15$

4. $y = \dfrac{x}{5} + 4$

5. $y = 2x + 8$

Section B

1. (a) $a = 5$ (b) $y = 3x - 2$
2. (0, -3)
3. $y = 2x - 1$
4. (a) $\frac{5}{3}$ (b) $3y - 8x = 0$
    (c) $(1\frac{1}{2}, 5\frac{1}{2})$
5. $P = 2.8M + 4.2$
6. (a) $a = 18.7, b = 27$
    (b) 27.5 hours

Multi-Choice Questions 25

1. C	2. A	3. A
4. C	5. C	6. D
7. A	8. D	9. C
10. A	11. B	12. C

Answers to Chapter 26

Exercise 26.1

6. $x = 3$ or 4 7. $x = 4$ 8. $x = \pm 3$

9. $x = -5.4$ or 3.7

10. (a) $x = -6.54$ or -0.46
    (b) $x = -7.27$ or 0.27 (c) $x = -6$ or -1

11. (a) $x = -1$ or $\frac{1}{3}$
    (b) $x = -1.39$ or 0.72 (c) $x = -1.22$ or 0.55

12. (a) $x = \pm 3$ (b) $x = \pm 2.24$
    (c) no solution

Exercise 26.2

1. (a) ± 1.15 (b) -0.72 or 1.39
   (c) 0 or 2.33

2. -6.16 or 0.16

3. $x = 4, y = 1$

4. $x = 7, y = 3$

5. $x = 3, y = 2$

6. -0.54 or 1.40

7. $0, 3, 11.25$
   (a) 0 and 2 (b) $0.84; 2x^3 - 8x = 0$

8. $4, 4.75, 5, 3.85$ or 0.65

9. $x^2 - x - 11 = 0; x = 3.85$

10. $1, -1.75, -2; \dfrac{x^2}{4} + \dfrac{24}{x} - 12 = \dfrac{x}{3} - 2;$

    $x = 2.6$ or 5.4

Exercise 26.3

4. (b) 1.89 (c) 1.50×1.089^x
   (d) 3.52 millions   (e) 1.94 millions

5. (a)

T	1	2	3	4	5
N	2	4	8	16	32

   (b) 6

6. $3\,600\,000$

7. (b) 10% (c) £10 700
   (d) 11.5 years

Miscellaneous Exercise 26

Section A

1. $x = -1\frac{1}{2}$ or 2

2. (d) $x = -3$ or 2

3. (i) $x = 6, y = 4$ (ii) $y = x - 4$

4. (b) $x = 0.65$ or -4.65 (c) $x = 0$ or -6

5. Year

Year	1960	1970	1980	1990	2000
$N \times 10^6$	2	3.26	5.31	8.64	14.08

   6.77×10^6

Section B

1. (c) (i) $x = 2$ or 4 (ii) $x = 0.55$ or 5.45

2. (a) $0, -0.5, 0, 1.6$ (c) 4.30 or 0.70

3. (a) $x = \frac{1}{3}$ or -1 (b) $x = 0.72$ or -1.39
   (c) $x = -0.55$ or -1.22

4. (a) $x = \pm 2$ (b) $x = 1.39$ or -0.72
   (c) $x = 0$ or 2.33

5. (a)

x	0	1	2	3	4	5	6
y	1	2	4	8	16	32	64

   (b) 4.64 (c) 7.5

Answers to Chapter 27

Exercise 27.1

1. $-5, 19$ 2. $-4, 8$ 3. 3
4. 5.5 or -6.5 5. $-4, 6, 1.25$ 6. 8

Exercise 27.2

1. (a) 0.71 thousand per min
   (b) 1.4 thousand per min

2. -0.078

3. 0.078 g/s

4. (a) 0.09 millions per year
   (b) 0.15 millions per year

5. (a) 0.75 mmHg/h   (b) -0.2 mmHg/h

6. C

Exercise 27.3

1. $-3\frac{1}{3}$ 2. $13\frac{1}{4}$ 3. $9, -23$

4. (a) -2.25 (b) 3.3 or -0.3
   (c) 2.4 or -0.4

5. $2\frac{1}{4}$ 6. $\frac{7}{2}, -\frac{1}{2}, +\frac{1}{2}; -1\frac{13}{24}, \frac{5}{6}$

7. 262.6 cm^3 8. $7.37 \times 7.37 \times 3.68$ m

9. $x = \frac{2}{3}, k = \frac{1}{3}$

10. 625 m; 38.23×11.77 m

Exercise 27.4

1. $x = -1; -4$ 2. $x = 2; 11$ 3. $x = 1; 3$
4. $x = 1; -5$ 5. $x = 1\frac{1}{2}; 4\frac{3}{4}$ 6. $x = -2; 15$

Exercise 27.5

1. 136 2. 13 3. $5\frac{1}{6}$
4. $18\frac{2}{3}$ 5. $3\frac{3}{4}$

Exercise 27.6

1. 60 km/h
2. 5 hours
3. 300 km
4. 50 km/h
5. 36.4 km/h
6. 10.6 km/h
7. 22.1 km/h
8. 150 m/s
9. 24 m/s
10. $\frac{2}{27}$ m/s
11. 70 m/s

Exercise 27.7

1. (a) 60 m   (b) 6 m   (c) 80 m
   (d) 120 m   (e) 300 m
2. (a) acceleration 5 m/s^2
   (b) acceleration $\frac{1}{2}$ m/s^2
   (c) retardation 1.5 m/s^2
   (d) retardation 2.5 m/s^2
3. 150 m
4. (a) 0.2 m/s^2 (b) 4 m/s   (c) 210 m
5. (a) 1 m/s^2 (b) 3 m/s^2 (c) 5 m/s
   (d) 250 m
6. (a) 1 m/s^2 (b) 2 m/s^2 (c) 0.5 m/s^2
   (d) 30 m/s   (e) 875 m
7. (b) 2 m/s^2 (c) 1.6 m/s^2 (d) 384 m
8. (b) 239 m
9. (b) (i) 2.5 m/s^2 (ii) 10 m/s^2
10. (a) 1.3 m/s^2 (b) 12.4 s   (c) 135 m

Exercise 27.8

1. (a) $y = kx^2$ (b) $U = kV$ (c) $S = \dfrac{k}{T}$
   (d) $h = \dfrac{k}{m^2}$
2. (a) $\frac{81}{8}$ (b) $\frac{32}{81}$ (c) $\frac{8}{9}$
3. 1.06
4. 1.33
5. $\sqrt{2}:1$

Miscellaneous Exercise 27

Section A

1. (a) 1 m/s^2 (b) $1\frac{1}{3}$ m/s^2 (c) 5 m/s
   (d) 287.5 m
2. (b) (i) 8 m/s^2 (ii) 16 m/s^2 (c) 144 m
3. (b) 16.8 m/s   (c) $k = 2.8$
4. (a) 150 m   (b) 750 m
5. (a) 28 km   (b) 50 km
   (c) 12.5 km/h

Section B

1.
x	0	3	6	9	11	13	15	18
A	0	142	243	304	322	322	304	243
; 324 cm^2

2.
x	-2	-1	0	1	2	3	4
y	-18	-2	2	0	-2	2	18
;
   (a) gradient = 8
   (b) $(2, -2)$ min, $(0, 2)$ max
3. 36 square units
4. $y = 20$ 5. $x = 3$ 6. (b)

Answers to Chapter 28

Exercise 28.1

1. $x > 2$ 2. $x < 6$ 3. $x \leqslant 2$
4. $x \geqslant 4$ 5. $x > 1$ 6. $x > 7$
7. $x \leqslant 2$ 8. $x \leqslant -5$ 9. $\{3, 4\}$
10. $\{3, 4, 5, 6, 7, 8, 9\}$ 11. $\{1, 2, 3, 4, 5, 6\}$
12. $\{1, 2, 3, 4, 5\}$ 13. $\{-3, -2, -1, 0, 1\}$
14. $\{-4, -3, -2\}$ 15. $\{3, 4, 5, 6, 7, 8, 9\}$
16. $\{-2, -1, 0, 1, 2, 3, 4, 5, 6, 7, 8\}$
17. $\{3, 4, 5\}$ 18. $\{2, 3\}$
19. $\{-2, -1, 0, 1, 2, 3\}$ 20. $\{-1, 0\}$

Miscellaneous Exercise 28

Section A

1. 11
2. (a) $=$ (b) $>$ (c) $<$
   (d) $>$ (e) $>$
3. (a) $>$ (b) $<$ (c) $=$
   (d) $>$
4. (a) $\{3\}$ (b) $\{5, 6, 7, 8\}$
5. (a) $\{-2, -1, 0, 1, 2\}$
   (b) $\{-3, -2, -1, 0, 1, 2, 3, 4\}$

Section B

1. $x = 17$
2. 10
3. $30 + 3n$
5. (a) (i) $y - x = 2$ (ii) $y + x = 4$
   (b) $(1, 3)$
   (c) $x \geqslant 0, y \geqslant 0, x + 2y \leqslant 4$
6. 5, 6, 7, 8, 9
7. $(-2.3, 5.3)$ or $(1.3, 1.7)$

Multi-Choice Questions 28

1. C   2. C   3. C
4. A   5. C   6. D
7. C

Answers to Chapter 29

Exercise 29.1

1. $\{3, 5, 7, 9, 11\}$
2. $\{2, 4, 6, 8\}$
3. $\{2, 4, 6, 8, 10, 12, 14, 16\}$
4. (a) Infinite   (b) Finite   (c) Finite
   (d) Null   (e) Null
5. 5
6. 8
7. (a) True   (b) False   (c) False
8. (a) $\{3, 5, 9, 11, 13, 15\}$ (b) $\{2, 6, 8, 14\}$
   (c) $\{2, 3, 5, 11, 13\}$ (d) $\{3, 6, 9, 15\}$

9. $\emptyset, \{a\}, \{b\}, \{c\}, \{d\}, \{a, b\}, \{a, c\}, \{a, d\}, \{b, c\},$
   $\{b, d\}, \{c, d\}, \{a, b, c\}, \{a, c, d\}, \{a, b, d\},$
   $\{b, c, d\}, \{a, b, c, d\}$
10. 64, 63
11. $g \subset a, h \subset b, e \subset c, f \subset d$
12. 256
13. (a) $\{3, 5, 7, 9, 11, 13, 15, 17\}$
    (b) $\{2, 3, 5, 7, 11, 13, 17\}$
    (c) $\{3, 6, 9, 12, 15\}$
15. $B \subset A$ 16. $\{1, 2, 9, 12\}$
17. (a) $A = C$ (b) A and C, B and D

Exercise 29.2

1. (a) $\{1, 2, 3, 4, 5, 6, 7, 8, 9, 10, 11\}$
   (b) $\{3, 4, 6, 7, 8, 9\}$
   (c) $\{2, 5, 6, 8, 9\}$
   (d) $\{1, 2, 5, 10, 11\}$
   (e) $\{1, 3, 4, 7, 10, 11\}$
   (f) $\{6, 8, 9\}$
   (g) $\{2, 3, 4, 5, 6, 7, 8, 9\}$
   (h) $\{1, 10, 11\}$
2. $X \cap Y = \{4, 7\}$

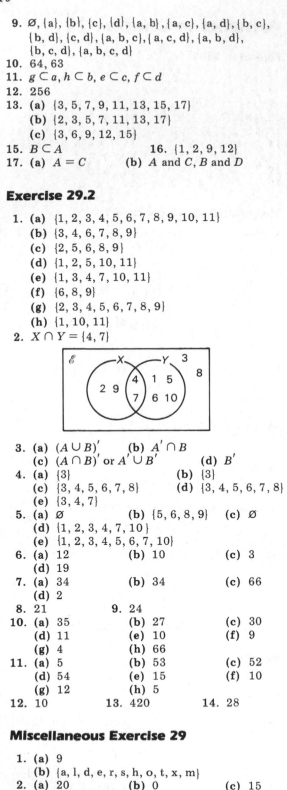

3. (a) $(A \cup B)'$ (b) $A' \cap B$
   (c) $(A \cap B)'$ or $A' \cup B'$ (d) B'
4. (a) $\{3\}$ (b) $\{3\}$
   (c) $\{3, 4, 5, 6, 7, 8\}$ (d) $\{3, 4, 5, 6, 7, 8\}$
   (e) $\{3, 4, 7\}$
5. (a) $\emptyset$ (b) $\{5, 6, 8, 9\}$ (c) $\emptyset$
   (d) $\{1, 2, 3, 4, 7, 10\}$
   (e) $\{1, 2, 3, 4, 5, 6, 7, 10\}$
6. (a) 12   (b) 10   (c) 3
   (d) 19
7. (a) 34   (b) 34   (c) 66
   (d) 2
8. 21   9. 24
10. (a) 35   (b) 27   (c) 30
    (d) 11   (e) 10   (f) 9
    (g) 4   (h) 66
11. (a) 5   (b) 53   (c) 52
    (d) 54   (e) 15   (f) 10
    (g) 12   (h) 5
12. 10   13. 420   14. 28

Miscellaneous Exercise 29

1. (a) 9
   (b) $\{a, l, d, e, r, s, h, o, t, x, m\}$
2. (a) 20   (b) 0   (c) 15
3. (a) 17   (b) 5   (c) 8

4. (a) $\{25\}$ (b) $\emptyset$
   (c) $\{25, 27, 36, 37, 47, 49, 57, 64, 67, 77\}$
5. (a)

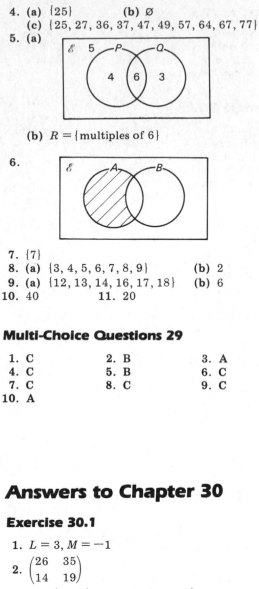

   (b) $R = \{$multiples of 6$\}$

6.

7. $\{7\}$
8. (a) $\{3, 4, 5, 6, 7, 8, 9\}$ (b) 2
9. (a) $\{12, 13, 14, 16, 17, 18\}$ (b) 6
10. 40   11. 20

Multi-Choice Questions 29

1. C   2. B   3. A
4. C   5. B   6. C
7. C   8. C   9. C
10. A

Answers to Chapter 30

Exercise 30.1

1. $L = 3, M = -1$
2. $\begin{pmatrix} 26 & 35 \\ 14 & 19 \end{pmatrix}$
3. (a) $\begin{pmatrix} 7 & 0 \\ 4 & 3 \end{pmatrix}$ (b) $\begin{pmatrix} 1 & 5 \\ 2 & -6 \end{pmatrix}$
   (c) $\begin{pmatrix} 14 & 0 \\ 8 & -2 \end{pmatrix}$ (d) $\begin{pmatrix} 10 & 4 \\ 12 & 2 \end{pmatrix}$
4. (a) $PQ = \begin{pmatrix} 2 & 1 \\ 3 & 1 \end{pmatrix}$
   $RS = \begin{pmatrix} -6 & 3 \\ 1 & -2 \end{pmatrix}$
   $PQRS = \begin{pmatrix} -11 & 4 \\ -17 & 7 \end{pmatrix}$
   $P^2 - Q^2 = \begin{pmatrix} 6 & 3 \\ 9 & 3 \end{pmatrix}$
   (b) $a = -2, b = 5$

5. $A^2 = \begin{pmatrix} 16 & -9 \\ -12 & 13 \end{pmatrix}$

$B = \begin{pmatrix} -18 & 12 \\ 16 & -14 \end{pmatrix}$

6. $\begin{pmatrix} 9 \\ -1 \end{pmatrix}$

7. $p = 5, q = -1$

8. (a) $\begin{pmatrix} 3 & 1 \\ 2 & 2 \end{pmatrix}$ (b) $\begin{pmatrix} 1 & 3 \\ 2 & 3 \end{pmatrix}$

(c) $\begin{pmatrix} -1 & 2 \\ 1 & -1 \end{pmatrix}$

9. (a) $\begin{pmatrix} 7 & 3 \\ 3 & 4 \end{pmatrix}$ (b) $\begin{pmatrix} 1 & -1 \\ 4 & 12 \end{pmatrix}$ (c) $\begin{pmatrix} 13 & 6 \\ 12 & 4 \end{pmatrix}$

10. $AD = \begin{pmatrix} -7 & 11 \\ 2 & -6 \end{pmatrix}$

$BA = \begin{pmatrix} -9 & -1 \\ -16 & -4 \end{pmatrix}$

11. $x = 4, y = -1$

12. (a) $x = 1, y = 2$ (b) $x = 3, y = 4$
(c) $x = 2, y = 1$

Exercise 30.2

1. $\begin{pmatrix} 4 & 4 & 4 \\ 3 & 2 & 7 \\ 6 & 3 & 3 \\ 5 & 2 & 5 \\ 2 & 4 & 6 \end{pmatrix} \begin{pmatrix} 3 \\ 1 \\ 0 \end{pmatrix} = \begin{pmatrix} 16 \\ 11 \\ 21 \\ 17 \\ 10 \end{pmatrix}$

2. (a) $\begin{pmatrix} 6 & 4 & 2 \\ 3 & 6 & 1 \end{pmatrix}$

(b) $\begin{pmatrix} 150 \\ 120 \\ 375 \end{pmatrix}$ (c) $\begin{pmatrix} 2130 \\ 1545 \end{pmatrix}$

(d) The number of passengers that each airline can carry

3. $\begin{pmatrix} 2 & 3 & 8 \\ 1 & 2 & 5 \\ 0 & 2 & 9 \end{pmatrix} \begin{pmatrix} 6 \\ 4 \\ 1 \end{pmatrix} = \begin{pmatrix} 32 \\ 19 \\ 17 \end{pmatrix}$

4. (a) $P = \begin{pmatrix} 6 & 4 & 3 \\ 8 & 5 & 2 \end{pmatrix}$ (b) $C = \begin{pmatrix} 4 \\ 5 \\ 12 \end{pmatrix}$

(c) $PC = \begin{pmatrix} 80 \\ 81 \end{pmatrix}$

(d) $161 = $ total amount spent

5. $\begin{pmatrix} 5 & 7 & 2 \\ 6 & 5 & 3 \end{pmatrix} \begin{pmatrix} 1 \\ 2 \\ 3 \end{pmatrix} = \begin{pmatrix} 25 \\ 25 \end{pmatrix}$

Miscellaneous Exercise 30

1. $\begin{pmatrix} 0.4 & 0.1 \\ -0.2 & 0.2 \end{pmatrix}$ 2. $\begin{pmatrix} 0 \\ -4 \end{pmatrix}$

3. $\begin{pmatrix} -1 & 6 \\ -1 & 1 \end{pmatrix}$ 4. $x = 2, y = 5$

5. (a) $\begin{pmatrix} 5 & 0 \\ 0 & 5 \end{pmatrix}; \begin{pmatrix} 5 & 0 \\ 0 & 5 \end{pmatrix}$ (b) $\frac{1}{5}\begin{pmatrix} 7 & 4 \\ -8 & -1 \end{pmatrix}$

7. $a = 1, b = 2, c = 4$

8. (a) $\begin{pmatrix} 2 \\ 3 \end{pmatrix}$ (b) $\begin{pmatrix} -16 \\ 6 \end{pmatrix}$

9. $x = 25$

10. $\begin{pmatrix} 18 & -8 \\ 19 & 1 \end{pmatrix}$

11. $(2 \quad 3 \quad 1)\begin{pmatrix} 5 \\ 4 \\ 6 \end{pmatrix} = (28)$; total cost $= £28$

Multi-Choice Questions 30

1. A 2. A 3. A
4. A 5. D 6. B
7. C

Answers to Chapter 31

Exercise 31.1

1. (a) $(-1, 7)$ (b) $(0, 5)$ (c) $(-6, 1)$
(d) $(-4, 9)$ (e) $(5, 4)$
2. $A'(-1, -6), B'(2, -6), C'(2, -3)$
3. $A'(4, 9), B'(8, 9), C'(8, 7), D'(4, 7)$
4. $A(13, 9), B(21, 9), C(21, 13), D(13, 13)$
5. $X'(6, 2), Y'(10, 2), Z'(6, 4)$
6. $A(4, 1), B(6, 0), C(5, -1)$
7. $A'(0, -3), B'(1, -5), C'(1, -2)$
8. (a) $\begin{pmatrix} -5 \\ 3 \end{pmatrix}$ (b) $\begin{pmatrix} -7 \\ 0 \end{pmatrix}$ (c) $\begin{pmatrix} 3 \\ -5 \end{pmatrix}$
(d) $\begin{pmatrix} 6 \\ 2 \end{pmatrix}$ (e) $\begin{pmatrix} 1 \\ -3 \end{pmatrix}$

Exercise 31.2

1. (a)
A'	B'	C'	D'
0	2	2	0
-1	-1	-3	-3

(b)
A'	B'	C'	D'
0	-2	2	0
1	1	3	3

(c)

| A′ | B′ | C′ | D′ |

$$\begin{pmatrix} -1 & -1 & -3 & -3 \\ 0 & -2 & -2 & 0 \end{pmatrix}$$

(d)

| A′ | B′ | C′ | D′ |

$$\begin{pmatrix} 1 & 1 & 3 & 3 \\ 0 & 2 & 2 & 0 \end{pmatrix}$$

2. (a)

| A′ | B′ | C′ | D′ |

$$\begin{pmatrix} 2 & 7 & 6 & 3 \\ -1 & -1 & -3 & -3 \end{pmatrix}$$

(b)

| A′ | B′ | C′ | D′ |

$$\begin{pmatrix} -2 & -7 & -6 & -3 \\ 1 & 1 & 3 & 3 \end{pmatrix}$$

(c)

| A′ | B′ | C′ | D′ |

$$\begin{pmatrix} 1 & 1 & 3 & 3 \\ 2 & 7 & 6 & 3 \end{pmatrix}$$

3. (a) $(2, 5)$ **(b)**

| W′ | X′ | Y′ | Z′ |

$$\begin{pmatrix} 3 & 3 & 5 & 5 \\ 2 & 4 & 4 & 2 \end{pmatrix}$$

4.

| A | B | C |

$$\begin{pmatrix} 2 & 4 & 3 \\ 3 & 4 & 5 \end{pmatrix}$$

5. (a) $y = x$
(b) $A'(-1, 1)$, $B'(-4, 4)$, $y = -x$

Exercise 31.3

1.

| A′ | B′ | C′ | D′ |

$$\begin{pmatrix} -2 & -3 & -5 & 0 \\ 1 & 3 & 6 & 4 \end{pmatrix}$$

2.

| A′ | B′ | C′ |

$$\begin{pmatrix} -3 & -4 & -5 \\ 2 & 4 & 3 \end{pmatrix}$$

3. $(6, 1), 270°$
4. $(0, 2)$
5.

| A | B |

$$\begin{pmatrix} 2 & 4 \\ 1 & 1 \end{pmatrix}$$

6.

| X′ | Y′ | Z′ |

$$\begin{pmatrix} 3 & 3 & 5 \\ -3 & -6 & -6 \end{pmatrix}$$

7.

| A′ | B′ | C′ | D′ |

$$\begin{pmatrix} -1 & -1 & -3 & -3 \\ 2 & 4 & 4 & 2 \end{pmatrix}$$

Exercise 31.4

1.

| A′ | B′ | C′ |

$$\begin{pmatrix} -1 & -3 & -2 \\ 6 & 8 & 10 \end{pmatrix}$$

2. (a) $\begin{pmatrix} 9 & 4 \\ 18 & 8 \end{pmatrix}$ **(b)** $(-31, -62)$

3. (a) $\begin{pmatrix} 1 & 4 \\ 0 & -1 \end{pmatrix}$ **(b)**

| A′ | B′ | C′ |

$$\begin{pmatrix} 50 & 80 & 95 \\ -10 & -15 & -20 \end{pmatrix}$$

4. $(-6.83, 0.66)$
5. $(-2, -3), (-2, -1), (-4, -1), (-4, -3)$
6. (a) $(6, 2)$ **(b)** $(6, 2)$ **(c)** $(4, -1)$
(d) $(-1, -1)$ **(e)** $(1, 2)$ **(f)** $(9, 4)$
(g) $(7, 1)$
7. $A(3, 2)$, $B(5, 2)$, $C(4, 3)$

Exercise 31.5

1. (a) $k = 3$ **(b)** $(-2, 1)$
2. (a) $k = 3$ **(b)** $(0, 0)$
3. $A'(4, 0)$, $B'(4, 4)$, $C'(8, 6)$, $D'(8, 0)$; 4
4. (a) 2   **(b)** $(-3, -4)$
5. (a) $A'(8, 12)$, $B'(16, 4)$, $C'(16, 16)$
(b) $A'(1, 1\frac{1}{2})$, $B'(2, \frac{2}{3})$, $C'(2, 2)$
(c) $A'(-4, -6)$, $B'(-8, -2)$, $C'(-8, -8)$
6.

| A′ | B′ | C′ | D′ |

$$\begin{pmatrix} -2\frac{1}{2} & -2\frac{1}{2} & \frac{1}{2} & -1 \\ 2\frac{1}{2} & 4 & 5\frac{1}{2} & 2\frac{1}{2} \end{pmatrix}$$

7. $k = 3$; $AB = 3$ cm, $AC = 5$ cm; 30.7 cm^2

Miscellaneous Exercise 31

Section A

1. $A'(-2, 4)$
2. (a) $A(4, -3)$ **(b)** $B(-4, 3)$
3. $(6, 3)$
4. $P'(5, 2)$
5. $(2, 7)$
6. $A'(0, 0)$, $B'(4, 0)$, $C'(4, 2)$

Section B

1. $(-4, -3)$
2. $(-4, 6)$
3. $A'(8, 2)$, $B'(6, 2)$, $C'(8, 1)$;
$A''(8, 2)$, $B''(6, -2)$, $C''(8, -1)$
4. (a) $(1, 2), (2, 1), (0, 1), (-1, 2)$
(b) parallelogram
(c) $(2, -4), (4, -2), (0, -2), (-2, -4)$
(e) parallelogram
(f) $\frac{1}{4}$
5. (a)

| A | B | C | D |

$$\begin{pmatrix} 0 & 2 & 1 & 0 \\ 0 & 0 & 1 & 1 \end{pmatrix}$$

(b)

| A′ | B′ | C′ | D′ |

$$\begin{pmatrix} 0 & 0 & -1 & -1 \\ 0 & -2 & -1 & 0 \end{pmatrix}$$

(c)   A″   B″   C″   D″
$$\begin{pmatrix} 4 & 4 & 3 & 3 \\ 4 & 2 & 3 & 4 \end{pmatrix}$$

6. (a) $T = \begin{pmatrix} 1 & 3 & 5 \\ 5 & 7 & 4 \end{pmatrix}$

   (b) $Q = \begin{pmatrix} -1 & -3 & -5 \\ 2 & 4 & 1 \end{pmatrix}$

   (c) $P = \begin{pmatrix} 2 & 4 & 1 \\ -1 & -3 & -5 \end{pmatrix}$

   (d) $R = \begin{pmatrix} -2 & -4 & -1 \\ -1 & -3 & -5 \end{pmatrix}$

7. (a)   A′   B′   C′
$$\begin{pmatrix} 4 & 4 & 8 \\ 4 & 6 & 4 \end{pmatrix}$$

   (b)   A′   B′   C′   D′
$$\begin{pmatrix} 4 & 8 & 8 & 4 \\ 0 & 0 & 4 & 4 \end{pmatrix}$$

Multi-Choice Questions 31

1. B   2. B   3. A
4. C   5. B   6. D
7. B

Answers to Chapter 32

Exercise 32.1

4. 2
5. (a) isosceles triangle (b) equilateral triangle
6. 5
7. A1 B1 D1 F1 G0 H2 J0
8. (a) 0   (b) 2   (c) 0
   (d) 0   (e) 1   (f) 0
   (g) 4   (h) 4   (i) ⊥

Exercise 32.2

1. (a) 2   (b) 2   (c) Yes
2. (a) 5   (b) 5   (c) No
3. (a) 2   (b) 2   (c) Yes
4. (a) 8   (b) 8   (c) Yes
5. (a) 5   (b) 5   (c) No
6. (a) 2   (b) 2   (c) Yes
7. (a) 1   (b) 1   (c) No
8. (a) 3   (b) 3   (c) No
9. (a) infinite number   (b) infinite number
   (c) 9
10. 5

Exercise 32.3

3. (a) (i) 12 red, 4 white   (ii) 16 red, 6 white
   (b) 44   (c) 23
5. 1, 6, 15, 20, 15, 6, 1; 1, 7, 21, 35, 35, 21, 7, 1

Pattern number	1	2	3	4	5	6	7	8
Total number of tiles	8	16	32	64	128	256	512	1024

6. (a) (i) 12   (ii) 16   (iii) 36
   (b) $W = 2(S + 3)$ (c) 166   (d) 150
7. 15, 21, 28

Rows	1 & 2	1, 2 & 3	1, 2, 3 & 4
Total number of tins	3	6	10

Rows	1, 2, 3, 4 & 5
Total number of tins	15

12. (a) 15   (b) 36

Multi-Choice Questions 32

1. C   2. D   3. B
4. B   5. D   6. C
7. D   8. B

Answers to Chapter 33

Exercise 33.1

1. 135°   2. 54°   3. 60°
4. 63°   5. 18°   6. 135°
7. 288°   8. 288°   9. 90°
10. 136.8°
11. A acute   B reflex   C obtuse   D acute
    E reflex   F acute   G obtuse   H reflex
12. (a) 27°   (b) 67°   (c) 143°
    (d) 72°

Exercise 33.2

1. 20°   2. 100°   3. 35°
4. 70°, 110°, 110°, 70°
5. 65°   6. 80°   7. 54°
8. 130°   9. 65°   10. 230°, 32°
16. $y = -\frac{1}{2}x + 10$ 17. $y = -\frac{1}{3}x + 10\frac{2}{3}$

Exercise 33.3

1. (a) $039°$ (b) $306°$ (c) $213°$
   (d) $108°$
2. (a) $225°$ (b) $135°$ (c) $315°$
   (d) $270°$ (e) $180°$
3. $S48°W$ $(228°)$
4. $235°$ 5. $N65°W$ $(295°)$
6. $126°$ 7. $248°$ 8. $300°, 75°$
9. $140°$
10. $78\,km, 190°, 10°$
11. (a) $90\,km$ (b) $150\,km$
12. $43\,km, 25\,km$
13. $326°$
14. $12\,km, 11\,km$
15. $44\,m, S64°E$ $(116°)$

Miscellaneous Exercise 33

Section A

1. (a) $110°$ (b) $350°$
2. $x = 145°, y = 85°$
3. (a) $400\,m, 742\,m$ (b) $828\,m$
   (c) $154°$
4. (a) $2.9\,km$ (b) $2\,h\,48\,min$ (c) $18\,km$
5. $15\,km, 67°, 247°$

Section B

1. $a = 105°, b = 49°, c = 131°, d = 131°$
2. (b) $097°$ (b) $249°, 16\,km/h$
3. (a) (i) $045°$ (ii) $345°$ (b) $26\,km$
4. $p = 82°, q = 108°, r = 72°, s = 54°, t = 54°,$
   $u = 126°$
5. $9.4\,km, 312°$

Multi-Choice Questions 33

1. B   2. C   3. C
4. D   5. D   6. B
7. B   8. C   9. D

Answers to Chapter 34

Exercise 34.1

1. $x = 49°, y = 131°$ 2. $x = 77°, y = 81°$
3. $x = 63°, y = 98°$ 4. $x = 37°, y = 127°$
5. $x = 140°, y = 60°$ 6. $x = 80°, y = 70°$

Exercise 34.2

1. $a = 10\,cm$
2. $b = 22.4\,cm$
3. $c = 2.65\,cm$
4. (a) $3.87\,cm$ (b) $4.24\,cm$ (c) $5.29\,cm$
5. (a) $7.42\,cm$ (b) $3.71\,cm$ (c) $6.54\,cm$
6. (a) $60°$ (b) $40°$ (c) $40°$

7. (a) $x = 70°, y = 40°, z = 35°$
   (b) $x = 110°, y = 70°, z = 70°$
8. $13.86\,cm, 110.9\,cm^2$
9. $9.53\,cm$
10. $a = 80°, b = 40°, c = 40°, d = 60°, e = 40°$

Exercise 34.4

1. (a), (b), (d), (f), (h)
2. $CD = 8\,cm, EC = 5\,cm$
3. $ADF \equiv DFE \equiv FEB \equiv CED;$
   $DGJ \equiv GHJ \equiv HJE \equiv GHF$
4. $TP = 8\,cm, TQ = 6\,cm$
5. $8\,cm$
6. (a) $8\,cm$ (b) $11.3\,cm$ (c) $11.3\,cm$
   (d) $40°$ (e) $20°$
7. $5\,cm$
8. $AB = 4.0\,cm, CD = 4.0\,cm$
9. (a) $80°$ (b) $40°$ (c) $6\,cm$
10. $PO = 8.51\,cm, \angle TPO = 20°$

Exercise 34.5

1. $AFI; CK; BDM; HJ$ 2. $15.7\,cm$
3. (a) $6\,cm$ (b) $\frac{2}{3}$
4. (a) $\angle BGD = \angle BHC; \angle BDG = \angle BCH;$
   $\angle DEG = \angle HBC$
   (b) $\angle AFG = \angle CFH; \angle GAE = \angle HCE;$
   $\angle AGE = \angle CHE$
   (c) $1:2$
5. $14\,cm$
6. $6\,cm$
7. (a) PBC (b) PDC
8. $QR = 5\frac{1}{3}\,cm; ST = 6\frac{3}{4}\,cm$

Exercise 34.6

1. $72\,cm^2$
2. $36\,cm^2$
3. (a) $192\,cm^2$ (b) $\frac{1}{9}$ (c) $24\,cm$
4. (a) $548\,cm^2$ (b) $27.4\,cm$ (c) $54.8\,cm$
5. (a) $12\,cm$ (b) $28.1\,cm^2$
6. (a) $15\,cm^2$ (b) $2.4\,cm^2$ (c) $7.81\,cm$
   (d) $3.12\,cm$
7. (a) $6\frac{2}{3}\,cm$ (b) $9:16$

Miscellaneous Exercise 34

Section A

1. $13.4\,cm$
2. $53.2\,cm$
3. (a) $42°$ (b) $42°$ (c) $138°$
   (d) isosceles
4. (a) $4, 4$ and $4\,cm$
   (b) $5\,cm, 5\,cm$ and $2\,cm$ etc.
   (c) $3\,cm, 4\,cm$ and $5\,cm$
5. $x = 103°, y = 52°, z = 51°$
6. $p = 70°, q = 20°, r = 60°, s = 120°$
7. $11.3\,cm$

Section B

1. (a) $180° - 2x°$ (b) $x°$
2. (a) $k = 3$ (b) DF $= 10.5$ cm
   (c) $7\,\text{cm}^2$
3. (a) $60°$ (b) $30°$
4. (a) 6 cm (b) 10 cm
5. $105°$
6. BC $= 7$ cm, $8.89\,\text{cm}^2$
7. $30°$
8. (a) 6 cm (b) 2.5 cm
   (c) $31.5\,\text{cm}^2$
10. $x = 20°, y = 62°, z = 98°$

Multi-Choice Questions 34

1. C	2. B	3. B
4. A	5. C	6. A
7. A	8. B	9. D
10. D		

Mental Test 33

1. $80°$ 2. $70°$ 3. $120°, 60°$
4. (a) $60°$ (b) $120°$ (c) $60°$
   (d) $70°$

5.

Triangle	$\angle$A	$\angle$B	$\angle$C	x
(a)	$70°$	$70°$	$40°$	$110°$
(b)	$60°$	$60°$	$60°$	$120°$
(c)	$30°$	$30°$	$120°$	$150°$
(d)	$80°$	$80°$	$20°$	$100°$
(e)	$50°$	$50°$	$80°$	$130°$
(f)	$60°$	$60°$	$60°$	$120°$
(g)	$45°$	$45°$	$90°$	$135°$

6. $60°$
7. 13 cm

Answers to Chapter 35

Exercise 35.1

1. $x = 40°, y = 100°, z = 71°$
2. $143°$
3. $93°$
4. $x = 39°, y = 105°$
5. $65°$
6. $100°$
7. Yes, Yes, Yes
8. $32°$
9. $100°, 20°$
11. 7.5 cm
14. (a) $x = 42°$ (b) $y = 42°$
    (c) $\angle$ACB $= 96°$ (d) isosceles
15. AO $= 12$ cm, B $= 5$ cm, AB $= 13$ cm

Exercise 35.2

1. (a) $540°$ (b) $1080°$ (c) $1440°$
   (d) $1800°$
2. (a) $108°$ (b) $135°$ (c) $144°$
   (d) $150°$
3. $131°$
4. 12
5. $6°$
6. 24
7. 7
8. $132°, 75°$
9. $72°, 108°$
10. $36°, 10$
11. $n = 30$
12. 12

Exercise 35.3

1. (b) 52 m
2. (b) 132 m
3. (a) 349 m, 213 m (b) 116 m
4. (a) BD $= 95$ m, AD $= 70$ m (b) $3792\,\text{m}^2$
6. (b) $20.8\,\text{cm}^2$
7. $128.6°$

Miscellaneous Exercise 35

Section A

1. $x = 109°, y = 127°$
2. $a = 116°, b = 64°$
3. $\angle$DAC $= 65°, \angle$BCD $= 130°, \angle$ABC $= 50°,$
   $\angle$DOC $= 90°$
4. 11.7 cm
5. 11.3 cm
6. 10.2 cm
7. 10.5 m
8. (a) AXD $= 90°$ (b) AX $= 3$ cm
   (c) trapezium

Section B

1. parallelogram 2. $105°$
3. $60°$ 4. 4.3 cm, 9.0 cm
5. $157.5°$ 6. $61°$
7. 124 m

Multi-Choice Questions 35

1. A	2. C	3. B
4. B	5. C	6. B
7. C	8. B	9. C
10. C		

Mental Test 35

2. 60°
3. $x = 60°$, $y = 90°$
4. 10 cm
5. 7
6. $\angle ACD = 30°$, $\angle ACB = 75°$
7. (a) 9   (b) 9   (c) 140°
8. 8
9. 144°
10. 8

Answers to Chapter 36

Exercise 36.1

1. 38°
2. 120°
3. 61°
4. 50°
5. $C = 110°$, $D = 75°$
6. $x = 27°$, $y = 58°$
7. $x = 83°$, $y = 70°$
8. 9.75 cm
9. $a = 32°$, $b = 42°$
10. $x = 110°$, $y = 20°$
11. 5.83 cm

Exercise 36.2

1. 3.32 cm
2. 11.49 cm
3. 9.24 cm
4. 8.33 cm
5. 1.35 cm
6. 1.24 in
7. 40 cm
8. 1.53 cm

Exercise 36.3

1. 17.9 cm
2. $a = 78°$, $b = 12°$
3. 22.9 cm
4. $\angle AOB = 66°$, $\angle OBD = 33°$, $\angle CBD = 57°$
5. 11.3 cm
6. 59.9 cm
7. 24 cm
8. 57.0 cm

Miscellaneous Exercise 36

Section A

1. (a) 35°   (b) 125°
2. (a) 63°   (b) 54°
3. $\angle AOB = 124°$, $\angle OAB = 28°$
4. 5.4 cm
5. $\angle BAE = 102°$, $\angle DAE = 90°$
6. $\angle PQR = 65°$, $\angle QRS = 90°$, $\angle RSP = 115°$, $\angle SPQ = 90°$
7. $\angle ABC = 48°$, $\angle AOC = 96°$

Section B

1. (a) 45°   (c) 70°
2. (a) 6.6 cm   (b) 21.0 cm^2
    (c) 37.2 cm^2

3. 5 cm
4. (a) 66°
    (b) △BFC and △AFD; △BFA and △CFD
5. 22.22 cm
6. 46.5 mm
7. 1.5 cm

Multi-Choice Questions 36

1. C   2. C   3. A
4. D   5. B   6. B
7. C   8. B

Mental Test 36

1. (a) 100°   (b) 90°   (c) $\angle ABC$
2. 50°
3. 40°
4. 40°
5. (a) 90°   (b) 30°   (c) 30°
    (d) 40°
6. 20°
7. (a) 80°   (b) 10°
8. 60°
9. 5 cm
10. 4 cm

Answers to Chapter 38

Exercise 38.1

1. (a) 3.381 cm   (b) 10.13 cm   (c) 25.94 in
2. (a) 41.8°   (b) 40.8°   (c) 22.4°
3. 28.3 cm
4. 0.795 m
5. 21.6 cm
6. 7.47 cm
7. 44.7°, 44.7°, 90.6°
8. $b = 3.76$ cm, $c = 5.18$ cm
9. AD $= 3.08$ m, AB $= 3.94$ m, BC $= 3.91$ m, CD $= 6.35$ m
10. AC $= 87.2$ m, $\angle ACB = 27.3°$

Exercise 38.2

1. (a) 9.33 cm   (b) 2.64 cm   (c) 5.29 ft
2. (a) 60.7°   (b) 69.3°   (c) 53.3°
3. 66.1°, 66.1°, 47.8°, 3.84 cm
4. 2.88 cm
5. 1.97 cm
6. $\angle BAC = 92°$, BC $= 8.74$ cm
7. BD $= 4.53$ m, AD $= 2.11$ m, AC $= 2.39$ m, BC $= 5.65$ m
8. AB $= 11.92$ cm, BC $= 5.60$ cm, AD $= 6.84$ cm

Exercise 38.3

1. (a) 4.35 cm   (b) 9.29 in   (c) 4.43 m
2. (a) 59°   (b) 15.9°   (c) 22.7°
3. 7.70 cm
4. 2.78 cm
5. 33.3 cm
6. 2.86 m
7. 2.09 cm
8. AB = 3.73 m, BD = 5.47 m, BC = 5.87 m

Exercise 38.4

4. 0.7431
5. 0.4540
6. (a) $\frac{3}{4}$ (b) $\frac{1}{3}$ (c) $\frac{3}{4}$
7. (a) 0.5696   (b) 0.2061   (c) 2.5611
8. 41.81°
9. 48.59°

Exercise 38.5

1. 11.55 m   2. 9.60 m   3. 51°
4. 110 m   5. 10.63 m   6. 24.8 m
7. 1287 m   8. 38.67°   9. 189 m
10. 74.88 m, 20.7°
11. 2.97 m   12. 11.78 m
13. (a) 70.0 m   (b) 30 m   (c) 32.6 m
14. 53.26 m
15. (a) 4848 m   (b) 8746 m   (c) 4746 m
    (d) 45.6°   (e) 6784 m

Exercise 38.6

1. (a) 92.65 km   (b) 151.6 km
2. 25 km, 43.3 km
3. 326.3°
4. 11.55 km, 10.77 km
5. 61.3 m, 091.33°
6. 24.72 km, 338.85°
7. 24.41 km, 14.00 km
8. 127 km, 049.6°
9. 11.5 km, N61.8°E
10. 15.23 km, 066.8°

Miscellaneous Exercise 38

Section A

1. (a) AD = 6.90 cm   (b) BC = 9.14 cm
   (c) 82.8 cm^2
2. (a) AN = 3.716 cm   (b) 16.73 cm^2
   (c) 33.46 cm^2
3. (a) WX = 3.830 cm   (b) ∠YZV = 70.53°
4. (a) x = 6.587   (b) tan y = 0.8
   (c) p = 12.81 cm
5. 6.67 cm
6. 6.22 km
7. (a) ∠CAD = 77°   (b) DE = 24 in
8. (a) sin ∠BDC = 0.7   (b) AB = 10 cm

Section B

1. (a) 1 cm   (b) 4 : 1
2. (a) 8.7 cm   (b) 6.2 cm
3. 15 cm
4. (a) 11.92 cm   (b) 3.420 cm
5. 14.4 cm
6. 64.15°
7. 31°
8. (a) BC = 90.4 km   (b) AC = 150 km

Multi-Choice Questions 38

1. C   2. D   3. B
4. C   5. C   6. B
7. D   8. B

Answers to Chapter 39

Exercise 39.1

1. (b) 13 cm   (d) 15.26 cm
2. (a) 4 cm   (c) 71.57°
   (d) 12.65 cm   (e) 50.6 cm^2
   (f) 266.4 cm^2
3. 12.73 cm
4. 4.528 cm
5. 14, q = 25.38°
6. (a) 34.92°   (b) 33°   (c) 31.31°
7. (a) 17.32 cm   (b) 60°   (c) 61.52°
   (d) 19.71 cm
8. (a) 61.93°   (b) 12 000
9. (a) 55.15°   (b) 58.95°
10. (a) 1524 m   (b) 43.96°

Answers to Chapter 40

Exercise 40.1

θ	sin θ	cos θ
1. 108°	0.9511	−0.3090
2. 163°	0.2924	−0.9563
3. 95°	0.9962	−0.0872
4. 115°	0.9063	−0.4226
5. 134°	0.7193	−0.6947
6. 168°	0.2079	−0.9781
7. 146°	0.3592	−0.8290
8. 175°	0.0872	−0.9962
9. 31.08°	148.92°	
10. 12.52°	167.48°	
11. —	—	
12. 28.15°		
13.	139.58°	
14.	94.66°	

Exercise 40.2

1. $C = 71°, b = 5.90, c = 9.985$
2. $A = 48°, a = 71.52, c = 84.18$
3. $B = 56°, a = 3.741, b = 9.527$
4. $B = 46°, b = 13.61, c = 5.845$
5. $C = 67°, a = 1.508, c = 2.362$
6. $C = 63.53°, a = 9.485, b = 11.56$
7. $B = 135.63°, a = 9.394, c = 14.44$
8. $B = 81.90°, b = 9.948, c = 3.609$
9. $A = 53.65°, b = 2390, a = 2125$
10. $A = 13.87°, B = 144.02°, b = 17.17$
11. $C = 49.95°, A = 44.77°, a = 10.69$
12. $C = 36.87°, B = 93.82°, b = 30.26$
13. $B = 48.53°, C = 26.40°, c = 4.244$
14. 21.74
15. 117.5
16. 11.25
17. 9.046
18. 21.78

Exercise 40.3

1. $c = 10.15, A = 50.17°, B = 69.83°$
2. $a = 11.81, B = 44.60°, C = 79.40°$
3. $b = 4.987, A = 82.55°, C = 60.15°$
4. $A = 38.22°, C = 60°, B = 81.78°$
5. $A = 24.70°, B = 44.92°, C = 110.38°$
6. $A = 34.55°, B = 18.03°, C = 127.42°$
7. $c = 18.07, A = 35.90°, B = 26.10°$
8. $a = 18.34, B = 18.92°, C = 29.08°$

Exercise 40.4

1. $540.4 \, \text{cm}^2$
2. $737.5 \, \text{cm}^2$
3. (a) $7.547 \, \text{cm}^2$ (b) $8.071 \, \text{m}^2$
4. (a) $143 \, \text{cm}^2$ (b) $53.7 \, \text{m}^2$
   (c) $43,61 \, \text{cm}^2$
5. (a) $16.90 \, \text{cm}^2$ (b) $31.49 \, \text{cm}^2$
   (c) $15.72 \, \text{cm}^2$
6. $89.15 \, \text{cm}^2$
7. $66.81 \, \text{cm}^2$
8. $19.31 \, \text{cm}^2$

Miscellaneous Exercise 40

1. $30°$
2. $6.93 \, \text{cm}^2$
3. $38.30 \, \text{cm}$
4. $64.16°$
5. $66.26°$
6. (a) $5.736 \, \text{cm}$ (b) $34.41 \, \text{cm}^2$
7. (a) $400 \, \text{m}, 742 \, \text{m}$
   (b) $828 \, \text{m}$ (c) $153.65°$ (S26.35°E)

Answers to Chapter 41

Exercise 41.1

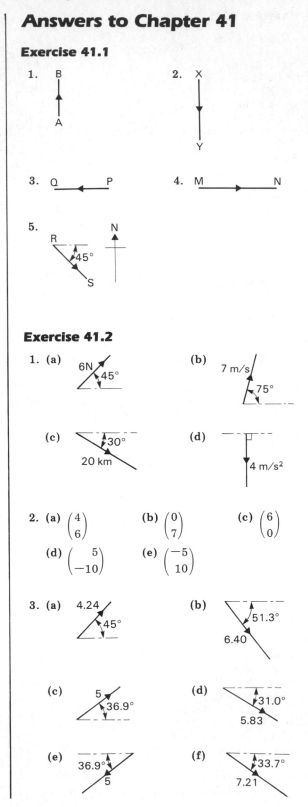

Exercise 41.2

1. (a) 6N, 45°   (b) 7 m/s, 75°   (c) 20 km, 30°   (d) 4 m/s²

2. (a) $\begin{pmatrix} 4 \\ 6 \end{pmatrix}$ (b) $\begin{pmatrix} 0 \\ 7 \end{pmatrix}$ (c) $\begin{pmatrix} 6 \\ 0 \end{pmatrix}$
   (d) $\begin{pmatrix} 5 \\ -10 \end{pmatrix}$ (e) $\begin{pmatrix} -5 \\ 10 \end{pmatrix}$

3. (a) 4.24, 45°   (b) 51.3°, 6.40   (c) 5, 36.9°   (d) 31.0°, 5.83   (e) 36.9°, 5   (f) 33.7°, 7.21

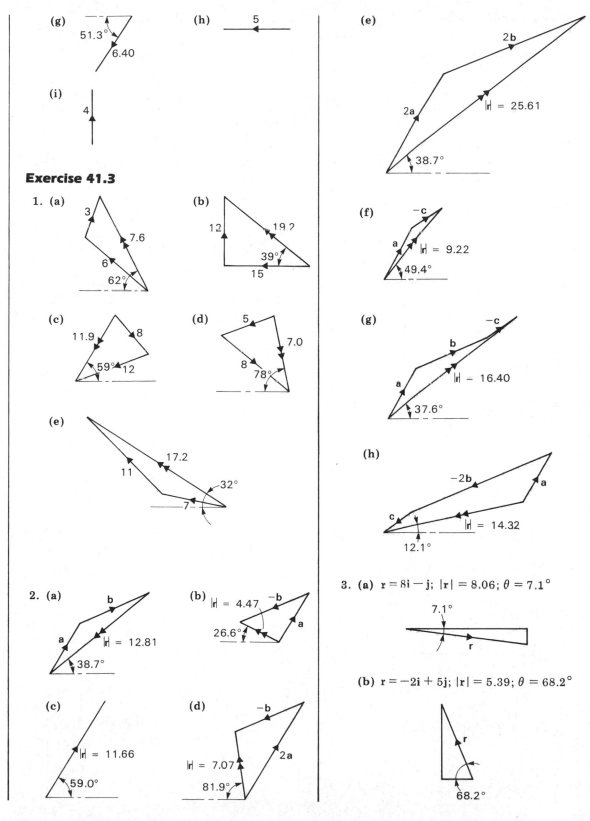

(g) 51.3° 6.40

(h) 5

(i) 4

Exercise 41.3

1. (a) 3 7.6 6 62°

(b) 12 19.2 39° 15

(c) 11.9 8 59° 12

(d) 5 7.0 8 78°

(e) 17.2 11 32° 7

(e) 2b 2a $|r| = 25.61$ 38.7°

(f) −c a $|r| = 9.22$ 49.4°

(g) −c b a $|r| = 16.40$ 37.6°

(h) −2b a c $|r| = 14.32$ 12.1°

2. (a) b a $|r| = 12.81$ 38.7°

(b) $|r| = 4.47$ −b a 26.6°

(c) $|r| = 11.66$ 59.0°

(d) −b 2a $|r| = 7.07$ 81.9°

3. (a) $r = 8i − j$; $|r| = 8.06$; $\theta = 7.1°$

7.1° r

(b) $r = −2i + 5j$; $|r| = 5.39$; $\theta = 68.2°$

r 68.2°

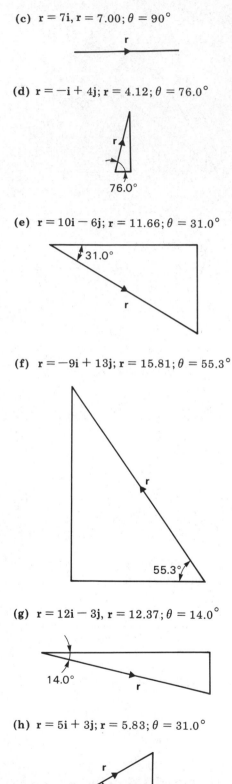

(c) $r = 7i, r = 7.00; \theta = 90°$

(d) $r = -i + 4j; r = 4.12; \theta = 76.0°$

76.0°

(e) $r = 10i - 6j; r = 11.66; \theta = 31.0°$

31.0°

(f) $r = -9i + 13j; r = 15.81; \theta = 55.3°$

55.3°

(g) $r = 12i - 3j, r = 12.37; \theta = 14.0°$

14.0°

(h) $r = 5i + 3j; r = 5.83; \theta = 31.0°$

31.0°

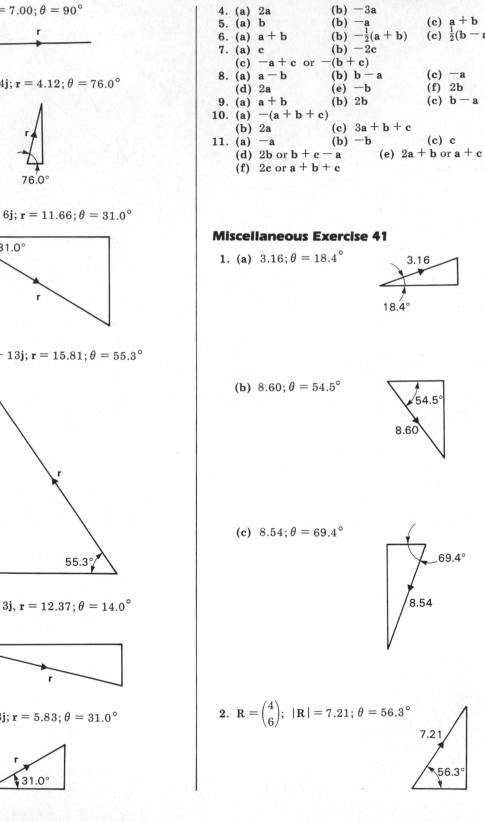

4. **(a)** 2a　　　　**(b)** −3a
5. **(a)** b　　　　**(b)** −a　　　　**(c)** a + b
6. **(a)** a + b　　**(b)** $-\frac{1}{2}(a + b)$　**(c)** $\frac{1}{2}(b - a)$
7. **(a)** c　　　　**(b)** −2c
　　(c) −a + c or −(b + c)
8. **(a)** a − b　　**(b)** b − a　　**(c)** −a
　　(d) 2a　　　**(e)** −b　　　**(f)** 2b
9. **(a)** a + b　　**(b)** 2b　　　**(c)** b − a
10. **(a)** −(a + b + c)
　　(b) 2a　　　**(c)** 3a + b + c
11. **(a)** −a　　　**(b)** −b　　　**(c)** c
　　(d) 2b or b + c − a　　**(e)** 2a + b or a + c
　　(f) 2c or a + b + c

Miscellaneous Exercise 41

1. **(a)** $3.16; \theta = 18.4°$

3.16

18.4°

(b) $8.60; \theta = 54.5°$

54.5°

8.60

(c) $8.54; \theta = 69.4°$

69.4°

8.54

2. $R = \begin{pmatrix} 4 \\ 6 \end{pmatrix}$; $|R| = 7.21; \theta = 56.3°$

7.21

56.3°

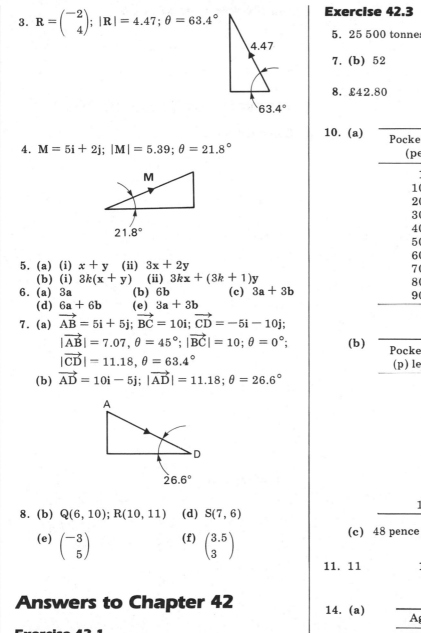

3. $R = \begin{pmatrix} -2 \\ 4 \end{pmatrix}$; $|R| = 4.47$; $\theta = 63.4°$

4.47

63.4°

4. $M = 5i + 2j$; $|M| = 5.39$; $\theta = 21.8°$

M

21.8°

5. (a) (i) $x + y$ **(ii)** $3x + 2y$
   (b) (i) $3k(x + y)$ **(ii)** $3kx + (3k + 1)y$
6. (a) $3a$ **(b)** $6b$ **(c)** $3a + 3b$
   (d) $6a + 6b$ **(e)** $3a + 3b$
7. (a) $\overrightarrow{AB} = 5i + 5j$; $\overrightarrow{BC} = 10i$; $\overrightarrow{CD} = -5i - 10j$;
       $|\overrightarrow{AB}| = 7.07$, $\theta = 45°$; $|\overrightarrow{BC}| = 10$; $\theta = 0°$;
       $|\overrightarrow{CD}| = 11.18$, $\theta = 63.4°$
   (b) $\overrightarrow{AD} = 10i - 5j$; $|\overrightarrow{AD}| = 11.18$; $\theta = 26.6°$

A

D

26.6°

8. (b) Q(6, 10); R(10, 11) **(d)** S(7, 6)

(e) $\begin{pmatrix} -3 \\ 5 \end{pmatrix}$ **(f)** $\begin{pmatrix} 3.5 \\ 3 \end{pmatrix}$

Answers to Chapter 42

Exercise 42.1

5. (a) 15.015 and 14.985 mm **(b)** 0.03 mm
   (c) 14.96 − 14.98 mm
6. Discrete **(b)**, **(d)** and **(g)**

Exercise 42.2

1. £28   **2.** 175 cm   **3.** £82.40
4. 199.92 mm **5.** 163.1 cm   **6.** 5
7. 4.5   **8.** 57   **9.** 5
10. no mode   **11.** 3, 5, 8   **12.** 4
13. 200 mm   **14.** 4

Exercise 42.3

5. 25 500 tonnes   **6.** 20 kN

7. (b) 52   **(c)** 46   **(d)** 57

8. £42.80   **9.** 29.76 mm

10. (a)

Pocket money (pence)	Number of children
1–9	0
10–19	5
20–29	7
30–39	17
40–49	26
50–59	16
60–69	10
70–79	10
80–89	5
90–99	4

(b)

Pocket money (p) less than	Number of children
10	0
20	5
30	12
40	29
50	55
60	71
70	81
80	91
90	96
100	100

(c) 48 pence

11. 11   **12.** 6 mm   **13.** £375

14. (a)

Age group	Frequency
Under 10	9
Under 20	17
Under 30	24
Under 40	31
Under 50	38
Under 60	45
Under 70	50
Under 80	53
Under 100	54

(c) 16 years, 33 years, 53 years
(d) 18.5 years

Miscellaneous Exercise 42

Section A

1.

Age	Frequency
13	2
14	10
15	6
16	6
17	4
18	2

2.

Score	Frequency
0–3	4
4–7	8
8–11	19
12–15	10
16–19	5
20–23	2
24–27	1
28–31	1

3.

Height	Frequency
150	2
151	3
152	4
153	5
154	6
155	7
156	8
157	6
158	5
159	4
160	3
161	1

4. 1

5. $19\,^{\circ}$C

6. (a)

Max. temp. $^{\circ}$C	2	3	4	5	6	7
No. of days	3	5	3	5	3	6

   (b) $7\,^{\circ}$C   (c) $5\,^{\circ}$C

7. 6

Section B

1. 16; 26; 5 grams
2. (a) 200   (b) 23   (c) 22.6
3. 80
4. (a) 168–172 cm (b) 168 cm
5. (a) 7.54   (b) 7.3
6. (a) 13   (b) 16   (c) 11
   (d) 16   (e) $2\frac{1}{2}$
7. (a) 32–33 cm   (b) 319.5–339.5 mm

Multi-Choice Questions 42

1. B	2. B	3. D
4. B	5. C	6. A
7. C	8. C	9. A
10. C		

Answers to Chapter 43

Exercise 43.1

1. (a) $\frac{1}{6}$ (b) $\frac{1}{3}$ (c) $\frac{1}{2}$

2. (a) $\frac{1}{52}$ (b) $\frac{1}{13}$ (c) $\frac{4}{13}$

   (d) $\frac{1}{26}$

3. (a) $\frac{1}{4}$ (b) $\frac{3}{8}$ (c) $\frac{5}{8}$

4. (a) $\frac{1}{5}$ (b) $\frac{1}{2}$ (c) $\frac{7}{10}$

5. (a) $\frac{1}{9}$ (b) $\frac{1}{6}$ (c) $\frac{13}{18}$

6. (a) $\frac{1}{15}$ (b) $\frac{1}{15}$ (c) $\frac{1}{10}$

7. (a) $\frac{7}{50}$ (b) $\frac{9}{25}$ (c) $\frac{1}{10}$

8. (a) $\frac{7}{50}$ (b) $\frac{6}{50}$ (c) $\frac{7}{10}$

Exercise 43.2

1. $\frac{5}{52}$

2. (a) $\frac{1}{12}$ (b) $\frac{1}{4}$

3. (a) $\frac{4}{25}$ (b) $\frac{9}{25}$ (c) $\frac{12}{25}$

4. $\frac{3}{190}$

5. (a) 0.32   (b) (i) 0.0273   (ii) 0.0546

6. (a) $\frac{1}{13}$ (b) $\frac{1}{17}$

7. (a) $\frac{4}{27}$ (b) $\frac{5}{27}$

8. (a) $\frac{1}{3}$ (b) $\frac{5}{12}$

   (c) (i) $\frac{1}{24}$ (ii) $\frac{1}{12}$

9. (a) $\frac{1}{6}$ (b) (i) $\frac{1}{36}$ (ii) $\frac{1}{18}$

   (c) $\frac{1}{6}$

10. (a) $\frac{1}{6}$ (b) $\frac{1}{2}$

    (d) (i) $\frac{1}{9}$ (ii) $\frac{5}{18}$

11. (a) $\frac{1}{10}$ (b) $\frac{3}{10}$ (c) $\frac{7}{10}$

    (d) $\frac{1}{15}$ (e) $\frac{8}{15}$

12. (a) $A - \frac{5}{12}$; $B - \frac{5}{11}$; $C - \frac{7}{11}$; $D - \frac{4}{11}$;

    (b) $\frac{7}{22}$

13. (a) $\frac{4}{35}$ (b) $\frac{4}{35}$ (c) $\frac{6}{35}$

    (d) $\frac{12}{35}$ (e) $\frac{4}{35}$

14. (b) $0; \frac{1}{25}; \frac{2}{25}; \frac{1}{5}; \frac{1}{25}$

    (c) (i) 4   (ii) 9   (iii) 16

15. (a) $q^5 + 5q^4 p + 10q^3 p^2 + 10q^2 p^3 + 5qp^4 + p^5$
    (b) $q^7 + 7q^6 p + 21q^5 p^2 + 35q^4 p^3 + 35q^3 p^4$
       $+ 21q^2 p^5 + 7qp^6 + p^7$
    (c) $q^9 + 9q^8 p + 36q^7 p^2 + 84q^6 p^3 + 126q^5 p^4$
       $+ 126q^4 p^5 + 84q^3 p^6 + 36q^2 p^7 + 9qp^8 + p^9$

16. (a) $\frac{1}{16}$ (b) $\frac{4}{16}$ (c) $\frac{6}{16}$

    (d) $\frac{4}{16}$ (e) $\frac{1}{16}$

17. (a) $\frac{1}{169}$ (b) $\frac{9}{169}$

18. (a) 0.8145   (b) 0.0135   (c) 0.0140

Multi-Choice Questions 43

1. B   2. C   3. B
4. B   5. B   6. D
7. C   8. A   9. A
10. B

Answers to Chapter 44

Investigations

1. (a) 8 moves. The centre square cannot be used.
   (b) One square vacant in 4 × 4; all squares
      used in 4 × 3 and 5 × 4.

2.

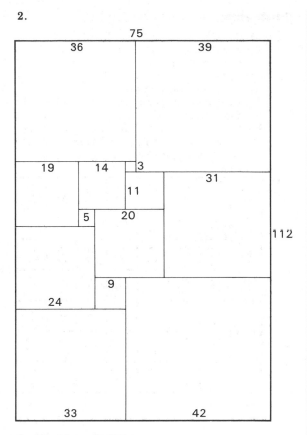

3. (a) 16 possibilities
   (b) 125 possibilities
   (c) Diagrams are equivalent to the edges of
      the octahedron and icosahedron.

4. First game: middle counter is significant.
   Second game: avoid leaving a single heap of
   counters.
   Third game: keeping the odd counter is
   significant.

5. Region contained within three parabolas.
   Three bisectors are possible.

6. Five give 4 hits, three give 2 hits, seven give
   1 hit, nine give 0 hits.

9. (a) $\frac{x+y}{x}$; $\frac{x+y}{y}$ (b) $\frac{x^2 + 2x + 1}{x + 2}$; $\frac{x + 1}{x + 2}$

   (c) $\frac{1}{x}$; $\frac{1}{x - 1}$ (d) x; $\frac{x}{x + 1}$

10. Routes begin to coincide before 10 p.m. For
    example, after 520 min, Inner and Centre are
    at E; after 590 min, Inner and Outer are at E;
    after 630 min, Inner and Outer are at X.
    Intervals other than multiples of 10 are
    required.

Problems

1. **(a)** A v B, 1-0; A v C, 3-3; B v C, 1-0
   (b) A v B, 2-1; A v C, 1-0; A v D, 1-0
   B v C, 1-0; B v D, 0-0; C v D, 1-0
2. **(a)** 6, 11   **(b)** 1, 6, 11, 21
3. **(a)** Sequence: powers of 2.
   Number of matches is always one less than
   the number of players.
   (b) Number 16 always wins; Number 15 may
   be beaten by Number 16 before the final;
   any player from 8 to 16 may be in the
   final.
4.

2	3	1	5	3	3	0	4
2	4	4	6	6	4	5	2
6	5	0	0	2	4	3	3
0	5	0	1	0	4	1	5
3	3	6	1	4	6	1	4
1	5	5	2	0	0	1	6
2	1	6	6	3	2	2	5

5. 430 paces
6. A total of £10 485.75 for 20 nails
7. Sides: 20 cm, 15 cm, 12 cm, 10 cm and $7\frac{1}{2}$ cm
   Each perimeter: 60 cm; Total: 300 cm
8. 576, 384, 192; 657, 438, 219; 819, 546, 273
9. **(a)** 41312432   **(b)** 34673245261715
10. 13 and seven 1s; 11, 3 and six 1s; 9, 5 and six 1s;
    9, 3, 3 and five 1s; 7, 7 and six 1s;
    7, 5, 3 and five 1s; 7, three 3s and four 1s;
    three 5s and five 1s; 5, four 3s and three 1s;
    six 3s and two 1s

Index